AF577129

MATHEMATICS IN INDUSTRY 11

THE EUROPEAN CONSORTIUM
FOR MATHEMATICS IN INDUSTRY

SUBSERIES

G. Ciuprina
D. Ioan
Editors

Scientific Computing in Electrical Engineering

With 231 Figures, 112 in Color, and 33 Tables

Editors

Gabriela Ciuprina
Daniel Ioan

Politehnica University of Bucharest
Electrical Engineering Department
Spl. Independentei 313
060042, Bucharest, Romania
Email: gabriela@lmn.pub.ro
daniel@lmn.pub.ro

Library of Congress Control Number: 2007926783

Mathematics Subject Classification (2000):
65-06, 65Lxx, 65Mxx, 65Nxx, 65Yxx, 78-06

ISBN 978-3-540-71979-3 Springer Berlin Heidelberg New York

Springer is a part of Springer Science+Business Media

springer.com

Typeset by the editors and SPi using a Springer LaTeX macro-package
Cover design: *design & production* GmbH, Heidelberg
Printed on acid-free paper SPIN: 12049363 46/3142/YL - 5 4 3 2 1 0

Preface

The sixth international conference on Scientific Computing in Electrical Engineering (SCEE) was another event in the SCEE series, aiming to bring together scientists from universities and industry with the goal of intensive discussions about modeling and simulation of electronic circuits and electromagnetic fields. It was held in Sinaia, Romania, from 17th to 22nd September 2006 and it was endorsed by Philips Research Laboratories, Eindhoven (*http://www.philips.nl*), Infineon Technologies from Munich (*http://www.infineon.com*), ST Microelectronics (*http://www.st.com*), Computer Simulation Technology (*http://www.cst.com*), IEEE Romania Section (*http://www.ieee.ro*), Romanian Ministry of Education and Research by the CEEX program (*http://www.mct-excelenta.ro*).

The history of SCEE begun in 1997, as a national German meeting held in Darmstadt and then in Berlin (1998), both under the auspices of the DMV (Deutscher Mathematiker Verein). In 2000, the first truly international workshop was organized in Warnemünde by the University of Rostock, Germany (*http://www.scee-2000.uni-rostock.de/*). In 2002, the 4th SCEE conference was jointly organized by the Eindhoven University of Technology (TU/e) and Philips Research Laboratories Eindhoven, The Netherlands (*http://www.win.tue.nl/scee2002/*). In 2004, the 5th SCEE conference took place in Capo D'Orlando, Italy, organized by Universita di Catania and Consorzio Catania Ricerche (*http://www.dmi.unict.it/scee2004/*). A SCEE Summer School on Computational methods for microelectronics was organized in 2005 as a follow up of the SCEE04 conference (*http://unict.it/sceeschools*). The 6th conference was organized by "Universitatea Politehnica din Bucuresti (UPB), Centrul de Inginerie Electrica Asistata de Calculator (CIEAC) - Laboratorul de Metode Numerice (LMN)" in Sinaia, Romania (*http://www.scee06.org/*).

As on all previous occasions, the conference was supported both from the industrial sector and academia, thus being guaranteed the relevance of work to practical situations and challenging open problems.

One of the main aims of the SCEE events is to strengthen the interaction between electrical or electronic engineers and the mathematics community. This aim is also illustrated by the SCEE logo which has some lines that might be interpreted as field lines or wave fronts and part of a bracket which stands for mathematical bracket but also symbolizes the idea of connecting together several communities mathematicians and engineers, university and industry. This logo was designed by Ramona Weyde-Ferch for SCEE 2000.

The conference provided an excellent opportunity to the European Community for project meetings (*www.chameleon-rf.org, www.comson.org*) or to discuss new research projects in the EU seventh research program FP7.

The conference topics were: **Computational Electromagnetics** (Modelling and parameter extraction, Discretization and Solution Methods, Applications :Antennas, Microwave, Interconnects and on-chip passive structures), **Circuit Simulation** and Design (Reduced Order Modeling, Numerical Integration Techniques, TCAD/EDA tools and techniques, Applications: Radio Frequency, Power Electronics,Optical Networks), **Coupled Problems** (Field-circuit coupled problems, Multi-physics (coupling, Coupling with electrical, thermal and mechanical problems, Application: Co-Simulation, Electromagnetic Compatibility, Bio-engineering), **Mathematical and Computational Methods** (Inverse Problems, Optimization, Multi-Scale Schemes, Solutions methods for large linear systems, Differential-Algebraic Equations, Grid Computing, Grid Computing).

The Program Committee consisted of:

- Prof. A. M. Anile - Universita di Catania, Italy
- Dr. A. Bossavit - Ecole Superiore delectricite Gif sur Yvette, France
- Assoc. Prof. Dr. G. Ciuprina - Univ. Politehnica din Bucuresti, Romania
- Dr. U. Feldmann - Infineon Technologies AG, Germany
- Prof. Dr. M. Günther - Bergische Universitat Wuppertal, Germany
- Prof. Dr. D. Ioan - Univ. Politehnica din Bucuresti, Romania
- Prof. Dr. U. Langer - Johannes Kepler Univ., Austria
- Dr. E. J. W. ter Maten - Philips Research, The Netherlands
- Prof. Dr. U. van Rienen - Univ. Rostock, Germany
- Prof. Dr. W. H. A. Schilders - Philips Research, Eindhoven Univ. of Technology, The Netherlands
- Prof. Dr. T. Weiland - Technische Univ. Darmstadt, Germany

The Program Committee selected invited speakers from industry and academia for each of the four topics. Thus, SCEE 2006 was honoured by the presence of the following **invited speakers**:

- Prof. Athanasios C. Antoulas, (Rice University - Electrical and Computer Engineering Dpt. ECE, Houston, Texas - USA): "Approximation of large-scale dynamical systems: An overview and some new results";
- Dr. Janne Roos, (Helsinki University of Technology, Circuit Theory Lab -APLAC - Finland): "Overview of Circuit-Simulation Activities at TKK CTL";
- Prof. Luis Miguel Silveira, (Technical University of Lisbon (IST), School of Engineering, Department of Electrical and Computer Engineering, INESC-ID, Lisbon - Portugal): "Outstanding Challenges in Model Order Reduction";
- Dr. Francois Henrotte, (RWTH - Aachen University - Institut fur Elektrische Maschinen, Germany): "The energy viewpoint in computational electromagnetics";
- Dr. Irina Munteanu, (CST - Germany): "RF & Microwave Simulation with the Finite Integration Technique - From component to system design";
- Dr. Herbert De Gersem, (Technical University Darmstadt, Computational Electromagnetics Lab. - TEMF - Germany): "Transient field-circuit coupled models with switching elements for the simulation of electric energy transducers";
- Dr. Andrea Marmiroli, (STMicroelectronics, - Italy): "Technology and Device modelling in micro and nanoelectronics: current and future challenges";
- Prof. Barbara Wohlmuth, (Stuttgart University - Institut fur Angewandte Analysis und Numerische Simulation IANS - Germany): "Advances in Mathematical and Computational Methods Applied in Electrical Engineering";
- Prof. Piet Hemker, (Centre for Mathematics and Computer Science - CWI, Dpt. Modelling, Analysis and Simulation, Amsterdam, Univ. of Amsterdam, Dpt. of Mathematics, - The Netherlands): "Space mapping and defect correction for efficient optimization:.

Overall, there were about 100 contributions (40 oral presentations and 60 posters) including the talks of the Invited Speakers. As in previous editions, there were sessions dedicated to short oral introduction of poster, where each contributor was given two minutes to advertise his/her work.

It has always been the policy of these conferences to encourage participants from all countries, and this conference has been remarkably succesfull, there were about 90 participants from 14 countries. This confirmed that SCEE 2006 was a truly international event.

The papers appearing in this book represent a selection of papers presented at the conference. Each paper was carefully referreed by two or three referees chosen by the Program Committee. The Program Commitee supervised the reviewing iterative process, aiming to improve the published form of the articles.

The selected papers have been organized according to the scientific area. Therefore, there are four parts, respectively devoted to Coupled Problems, Circuit Simulation, Electromagnetism and General Mathematical Computational Methods.

We would like to thank the referees of the papers who have spent a lot of time in order to ensure a high quality scientific level of the papers in this book and also to their effort to help us in completing the reviewing process according to the time schedule.

The local organizing committee is greatly indebted to the financial support received from the sponsors and to all the people whose enthusiasm and hard work ensured the success of the conference. Special thanks go to Prof. Mihai Iordache, the Dean of the Electrical Engineering Faculty of the Politehnica University of Bucharest for his constant and precious support. Finally, we would like to thank Ph.D. students Diana Mihalache and Alexandra Stefanescu for the care they have shown in assembling all the information into this book.

Bucharest,
March, 2007

Gabriela Ciuprina
Daniel Ioan

Contents

Part II Circuit Simulation and Design

Part III Computational Electromagnetics

List of Contributors

V. Anghel
Politehnica University of Bucharest
Spl. Independentei 313
060042, Bucharest, Romania.

Paula Anghelita
Research and Development Institute
for Electrical Industry
apel2@icpe.ro.

Athanasios C. Antoulas
Rice University
Department of Electrical and
Computer Engineering
Houston, TX, USA
aca@rice.edu.

Simone Bächle
Technical University of Berlin
Institute of Mathematics
MA 4-5 Straße des 17. Juni 136
10623 Berlin, Germany
baechle@math.tu-berlin.de.

M. Bebendorf
University of Leipzig
Mathematical Institute
D-04109 Leipzig, Germany
bebendorf@math.uni-leipzig.de.

T. Bechtold
Philips Semiconductors - NXP
Eindhoven
tamara.bechtold@nxp.com.

T.G.J. Beelen
DMS, NXP Semiconductors B.V.
High Tech Campus 48
5656 AE Eindhoven, The Netherlands
bratislav.tasic@philips.com.

Livio Belegante
National Institute of R&D
for Optoelectronic
camelia@inoe.inoe.ro.

Galina Benderskaya
Technische Universität Darmstadt
Schloßgartenstraße 8,
D-64289 Darmstadt, Germany
DeGersem@temf.tu-darmstadt.de.

Roberto Beneduci
University of Calabria and INFN-Gruppo
c.Cosenza
Italy
rbeneduci@unical.it.

A. Blaszczyk
ABB Corporate Research
5405 Baden-Daettwil, Switzerland
Andreas.Blaszczyk@ch.abb.com.

Diana Bogusevschi
Dublin City University
diana@eeng.dcu.ie.

Conor Brennan
Dublin City University
brennanc@eeng.dcu.ie.

Gianpietro Carnevale
STMicroelectronics
20041 Agrate Brianza, Italy.

Emil Carstea
National Institute of R&D for Optoelectronic.

Mircea Ciobanu
National Institute of R&D for Optoelectronic.

Jeni Ciuciu
National Institute of R&D for Optoelectronic.

Gabriela Ciuprina
Politehnica University of Bucharest
Electrical Engineering Department
Spl. Independentei 313
060042 Bucharest, Romania
lmn@lmn.pub.ro.

M. Clemens
Helmut-Schmidt-University
Department of Electrical Engineering
D-22043 Hamburg, Germany.

N. Constantin
Politehnica University of Bucharest
Spl. Independentei 313
060042 Bucharest, Romania.

C. P. Cristescu
Politehnica University of Bucharest
Spl. Independentei 313
060042 Bucharest, Romania
cpcris@physics.pub.ro.

H. Dağ
Isk University
Information Technologies Department
Istanbul, Turkey
dag@isikun.edu.tr.

Carlo de Falco
Bergische Universität Wuppertal
and Qimonda AG, München
defalco@math.uni-wuppertal.de.

Herbert De Gersem
Technische Universität Darmstadt
Schloßgartenstraße 8
D-64289 Darmstadt, Germany
DeGersem@temf.tu-darmstadt.de.

Georg Denk
Qimonda AG, München.

Stefan Diaconescu
Politehnica University of Bucharest
Spl. Independentei 313
060042, Romania.

Gabriel Dimitriu
Politehnica University of Bucharest
Spl. Independentei 313
060042, Romania.

C. R. Drago
Università di Catania
Dipartimento di Matematica e Informatica
Viale A. Doria 6
I-95125 - Catania
drago@dmi.unict.it.

Gheorghe Dumitrescu
Research and Development Institute for Electrical Industry
apel2@icpe.ro.

Falk Ebert
Technical University of Berlin
Institute of Mathematics
MA 4-5 Straße des 17. Juni 136
10623 Berlin, Germany
ebert@math.tu-berlin.de.

David Echeverría
Centrum voor Wiskunde en Informatica
Kruislaan 413, NL 1098 SJ Amsterdam
The Netherlands
D.Echeverria@cwi.nl.

A. El Guennouni
Magma Design Automation
Eindhoven, The Netherlands.

Vincenzo Enea
STMicroelectronics
Stradale Primosole 50
I-95121 Catania, Italy.

Jorge Fernández Villena
INESC ID / Instituto Superior Técnico
Technical University of Lisbon
Rua Alves Redol, 9
1000-029 Lisboa, Portugal
jorge@algos.inesc-id.pt.

Lelia Festila
Technical University of Cluj-Napoca
Str Ctin Dacovicuiu,nr 15
400020 Cluj-Napoca.

Paulo Flores
INESC ID / Instituto Superior Técnico
Technical University of Lisbon
Rua Alves Redol, 9
1000-029 Lisboa, Portugal.

Mariana Funieru
Technical Universität Darmstadt
Institut für Theorie Elektromagnetische Felder
Schloßgartenstraße 8
D-64289 Darmstadt, Germany
funieru@temf.tu-
darmstadt.de.

N. Gödel
Helmut-Schmidt-University
Department of Electrical Engineering
D-22043 Hamburg, Germany.

Michael Günther
Bergische Universität Wuppertal
Departement of Mathematics
D-42097 Wuppertal, Germany
guenther@math.uni-wuppertal.de.

M. Gavan
Politehnica University of Bucharest
Spl. Independentei 313
060042, Bucharest, Romania.

Alexandru Gherega
University Politehnica Bucharest
Spl. Independentei 313
060042, Bucharest, Romania.

Andrea Ghetti
STMicroelectronics
20041 Agrate Brianza, Italy.

C. Gramsch
hagenuk KMT GmbH
Rderaue 41
01471 Radeburg, Germany
Gramsch.C@sebakmt.com.

J. Greb
Bergische Universität Wuppertal
Fachbereich Mathematik und Naturwissenschaften
Gaußstr. 20
D-42119 Wuppertal, Germany.

Raimond Grimberg
National Institute of R&D for Technical Physics
47 D. Mangeron Blv., Iasi
700050, Romania
grimberg@phys-iasi.ro.

S. Grossmann
Technical University Dresden
Institute of Electrical Power Systems and
High Voltage Engineering
01062 Dresden, Germany
Grossmann@ieeh.et.tu-dresden.de.

Kay Hameyer
Institute of Electrical Machines
RWTH Aachen University
Schinkelstrae 4
D-52056 Aachen, Germany .

Pieter W. Hemker
Centrum voor Wiskunde en Informatica
Kruislaan 413, NL 1098 SJ Amsterdam
The Netherlands
P.W.Hemker@cwi.nl.

Francois Henrotte
RWTH Aachen University
Institute of Electrical Machines
Schinkelstrae 4
D-52056 Aachen, Germany
fh@iem.rwth-aachen.de.

I. Ibragimov
University of Saarland
PF 15 11 50, 66041 Saarbrücken, Germany
ilgis@num.uni-sb.de.

Z. Ilievski
Technische Universiteit Eindhoven
Z.Ilievski@tue.nl.

Daniel Ioan
Politehnica University of Bucharest
Electrical Engineering Department
Spl. Independentei 313
060042 Bucharest, Romania
lmn@lmn.pub.ro.

Felicia Ionescu
Politehnica University of Bucharest
Spl. Independentei 313
060042 Bucharest, Romania
fionescu@tech.pub.ro.

Tudor C. Ionescu
Rijksuniversiteit Groningen
t.c.ionescu@rug.nl.

Valentin Ionita
Politehnica University of Bucharest
Electrical Eng. Dept.
Spl. Independentei 313
060042 Bucharest, Romania
vali@mag.pub.ro.

Roxana Ionutiu
Rice University
Department of Electrical and Computer Engineering
Houston, TX, USA
rlonutiu@rice.edu.

H.H.J.M. Janssen
NXP Semiconductors Research
High Tech Campus 5
5656 AE, Eindhoven, The Netherlands
rick.janssen@nxp.com.

Daniele Kroell
STMicroelectronics
Stradale Primosole 50
I-95121 Catania, Italy .

Tuomo Kujanpää
Helsinki University of Technology
Circuit Theory Laboratory
P.O.Box 3000
FI-02015 TKK,Finland
tuomo.kujanpaa@tkk.fi.

S. Kurz
Helmut-Schmidt-University
Department of Electrical Engineering
D-22043 Hamburg, Germany .

Ulrich Langer
Johannes Kepler University
Institute of Computational Mathematics
Altenberger Str. 69
4040 Linz, Austria
ulanger@numa.uni-linz.ac.at.

Sanda Lefteriu
Rice University
Department of Electrical and Computer Engineering
Houston, TX, USA
slefteri@rice.edu.

Sorin Leitoiu
National Institute of R&D for Technical Physics
47 D. Mangeron Blv., Iasi
700050, Romania.

H. Löbl
Technical University Dresden
Institute of Electrical Power Systems and High Voltage Engineering
01062 Dresden, Germany
Loebl@ieeh.et.tu-dresden.de.

Vasile Manoliu
Politehnica University of Bucharest
Electrical Engineering Faculty
Spl. Independentei 313
060042, Bucharest, Romania
vasilem@amotion.pub.ro.

Andrea Marmiroli
STMicroelectronics
20041 Agrate Brianza, Italy
andrea.marmiroli@st.com.

Giovanni Mascali
University of Calabria and INFN-Gruppo c.Cosenza
Italy
g.mascali@unical.it.

Gheorghe Mates
Technical University of Cluj-Napoca
Department of Electrotechnics
C. Daicoviciu 15
400020 Cluj-Napoca, Romania.

R.M.M. Mattheij
Technische Universiteit Eindhoven
Den Dolech 2, 5600 MB
The Netherlands.

Michele Messina
STMicroelectronics
Stradale Primosole 50
I-95121 Catania, Italy
michele.messina@st.com.

Diana Mihalache
Politehnica University of Bucharest
Electrical Engineering Department
Spl. Independentei 313
060042 Bucharest, Romania
lmn@lmn.pub.ro.

Calin Munteanu
Technical University of Cluj-Napoca
Department of Electrotechnics
C. Daicoviciu 15
400020 Cluj-Napoca, Romania
Calin.Munteanu@et.utcluj.ro.

I. Munteanu
Computer Simulation Technology, Bad Nauheimer Straße 19
D-64289 Darmstadt, Germany
munteanu@cst.com.

Neag Marius
Technical University of Cluj-Napoca
Str Ctin Dacovicuiu,nr 15
400020 Cluj-Napoca, Romania
Marius.Neag@bel.utcluj.ro.

Liviu Nedelea
Technical University of Cluj-Napoca
Str Ctin Dacovicuiu,nr 15
400020 Cluj-Napoca, Romania.

Anca Nemuc
National Institute of R&D for Optoelectronic.

Doina Nicolae
National Institute of R&D for Optoelectronic .

J. Niehof
NXP Semiconductors Research
High Tech Campus 5
5656 AE, Eindhoven
The Netherlands
jan.niehof@nxp.com.

Constantin Nitu
Politehnica University of Bucharest
Slp. Independentei 313,
060042 Bucharest, Romania .

Smaranda Nitu
Politehnica University of Bucharest
Slp. Independentei 313,
060042 Bucharest, Romania
snitu@apel.apar.pub.ro.

Clemens Pechstein
Johannes Kepler University
Special Research Program SFB F013
Altenberger Str. 69
4040 Linz, Austria
clemens.pechstein@numa.uni-linz.ac.at.

Lucian Petrescu
Politehnica University of Bucharest
Electrical Eng. Dept.
Slp. Independentei 313,
060042 Bucharest, Romania .

Roland Pulch
Bergische Universität Wuppertal
Fachbereich Mathematik und Naturwissenschaften
Gaußstr. 20
D-42119 Wuppertal, Germany
pulch@math.uni-wuppertal.de.

Enrique S. Quintana-Ortí
Universidad Jaume I
Depto. de Ingeniería y Ciencia de Computadores
12.071–Castellón, Spain
quintana@icc.uji.es.

Gregorio Quintana-Ortí
Universidad Jaume I
Depto. de Ingeniería y Ciencia de Computadores
12.071–Castellón, Spain
gquintan@icc.uji.es.

Adina Racasan
Technical University of Cluj-Napoca
Department of Electrotechnics
C. Daicoviciu 15
400020 Cluj-Napoca, Romania
Adina.Racasan@et.utcluj.ro.

Claudia Racasan
Technical University of Cluj-Napoca
Department of Electrotechnics
C. Daicoviciu 15
400020 Cluj-Napoca, Romania .

Alfredo Remón
Universidad Jaume I
Depto. de Ingeniería y Ciencia de Computadores
12.071–Castellón, Spain
remon@icc.uji.es.

V. Rischmüller
Robert Bosch GmbH
PF 10 60 50
70049 Stuttgart, Germany
volker.rischmueller@de.bosch.com.

S. Rjasanow
University of Saarland
PF 15 11 50
66041 Saarbrücken, Germany
rjasanow@num.uni-sb.de.

Vittorio Romano
University of Catania
romano@dmi.unict.it.

Cesare Ronsisvalle
STMicroelectronics
Stradale Primosole 50
I-95121 Catania, Italy .

Janne Roos
Helsinki University of Technology
Circuit Theory Laboratory
P.O.Box 3000
FI-02015 TKK, Finland
janne@ct.tkk.fi.

Ioana - Gabriela Sîrbu
University of Craiova
Electrical Engineering Faculty
Decebal Blv. No. 107
200440-Craiova, Romania
osirbu@elth.ucv.ro.

Adriana Savin
National Institute of R&D for Technical Physics
47 D.Mangeron Blvd
700050 Iasi, Romania .

Jacquelien M. A. Scherpen
Rijksuniversiteit Groningen
j.m.a.scherpen@rug.nl.

W. H. A. Schilders
NXP Semiconductors Research
High Tech Campus 5
5656 AE, Eindhoven
The Netherlands
wil.schilders@nxp.com.

Reinhart Schultz
Qimonda AG, München .

Thorsten Sickenberger
Humboldt-Universität zu Berlin
Institut für Mathematik
10099 Berlin
sickenbergermath.hu-
berlin.de.

João M. S. Silva
INESC ID / Instituto Superior Técnico
Technical University of Lisbon
Rua Alves Redol, 9
1000-029 Lisboa, Portugal
jmss@algos.inesc-id.pt.

L. Miguel Silveira
INESC ID / Instituto Superior Técnico
Technical University of Lisbon
Rua Alves Redol, 9
1000-029 Lisboa, Portugal
lms@algos.inesc-id.pt.

Stefan Sorohan
Politehnica University of Bucharest
Spl. Independentei 313
060042, Bucharest, Romania
sorohan@form.resist.pub.ro.

Salvatore Spinella
Consorzio Catania Ricerche
Via A. Sangiuliano 262
I95124 Catania, Italy
spins@unical.it.

T. Steinmetz
Helmut-Schmidt-University
Department of Electrical Engineering
D-22043 Hamburg, Germany
t.steinmetz@hsu-hh.de.

K. Straube
Robert Bosch GmbH
PF 10 60 50
70049 Stuttgart, Germany
katharina.straube@de.bosch.com.

Michael Striebel
Infineon Technologies Austria AG
Siemensstr. 2
A-9500 Villach, Austria
michael.striebel2@infineon.com.

Camelia Tailanu
National Institute of R&D for Optoelectronic
camelia@inoe.inoe.ro.

B. Tasić
DMS, NXP Semiconductors B.V
High Tech Campus 48
5656 AE Eindhoven
The Netherlands
bratislav.tasic@philips.co.

E.J.W. ter Maten
Philips Semiconductors
High Tech Campus 48
5656 AE Eindhoven
The Netherlands
jan.ter.maten@philips.com.

Marina Topa
Technical University of Cluj-Napoca
Str Ctin Dacovicuiu,nr 15
400020 Cluj-Napoca.

Vasile Topa
Technical University of Cluj-Napoca
Str Ctin Dacovicuiu,nr 15
400020 Cluj-Napoca.

K.J. van der Kolk
Delft University of Technology
EEMCS, Circuits and Systems Group
Mekelweg 4
NL-2628 CD Delft
keesjan@cas.et.tudelft.nl.

N.P. van der Meijs
Delft University of Technology
EEMCS, Circuits and Systems Group
Mekelweg 4
NL-2628 CD Delft
nick@cas.et.tudelft.nl.

A. Verhoeven
Technische Universiteit Eindhoven
Den Dolech 2
5600 MB, The Netherlands
averhoev@win.tue.nl.

T. Voß
University of Groningen
Faculty of Mathematics and Natural Sciences
Nijenborgh 4
9747 AG Groningen, The Netherlands
t.voss@rug.nl.

A.J. Vollebregt
Bergische Universität Wuppertal.

Thomas Weiland
Technical Universität Darmstadt
Institut für Theorie Elektromagnetische Felder
Schloßgartenstraße 8
D-64289 Darmstadt, Germany
thomas.weiland@temf.tu-darmstadt.de.

G. Wimmer
Helmut-Schmidt-University
Department of Electrical Engineering
D-22043 Hamburg, Germany.

Renate Winkler
Humboldt-Universität zu Berlin
Institut fürMathematik
10099 Berlin
winkler@math.hu-berlin.de.

H. Xu
Technische Universiteit Eindhoven.

E. F. Yetkin
Istanbul Technical University
Informatics Institute
Istanbul, Turkey
fatih@be.itu.edu.tr.

Part I

Coupled Problems

Comparison of Model Reduction Methods with Applications to Circuit Simulation*

Roxana Ionutiu, Sanda Lefteriu, and Athanasios C. Antoulas

Department of Electrical and Computer Engineering, Rice University, Houston, TX, USA
rlonutiu@rice.edu, slefteri@rice.edu, aca@rice.edu

Summary. We compare different model reduction methods applied to the dynamical system of a coupled transmission line: balanced truncation (BT), truncation by balancing one gramian (or PMTBR - poor man's truncated balanced reduction), positive real balanced truncation (PRBT) and its Hamiltonian implementation (PRBT-Ham), PRIMA, spectral zero method (SZM) and its Hamiltonian implementation (SZM-Ham), and finally, optimal $\mathcal{H}_2$. Their performance is analyzed in terms of several criteria such as: preservation of controllability, observability, stability and passivity, relative $\mathcal{H}_2$ and $\mathcal{H}_\infty$ norms, and the computational cost involved.

1 Introduction

This paper presents different reduction methods together with results obtained by applying each method on a dynamical system given by a coupled transmission line. In Sect. 2, a modified nodal analysis (MNA)-similar representation of the system is derived. The model reduction methods are grouped in two main categories, *gramian based* and *Krylov based*, discussed in sections 3 and 4 respectively. Sect. 3 outlines the theory behind gramian based reduction methods: BT, PMTBR and PRBT. Krylov based reduction methods PRIMA, SZM and optimal $\mathcal{H}_2$ are described in Sect. 4. In Sect. 5 we compare all methods in terms of: preservation of some important properties like controllability, observability, stability and passivity, the relative $\mathcal{H}_2$ and $\mathcal{H}_\infty$ norms and in terms of the computational cost. In Sect. 6, error systems resulting from different methods are compared. This allows us to identify frequency ranges where one particular method approximates the original system more accurately. Sect. 7 presents additional results obtained with the optimal $\mathcal{H}_2$ method. Finally, Sect. 8 summarizes our analysis and motivates further research.

2 State-space representation

The model reduction problem of transmission lines has been studied extensively, see for instance [8]. Our system consists of two transmission lines with inductive

* This work was supported in part by the NSF through Grants CCR-0306503, ACI-0325081, and CCF-0634902. Invited Paper at SCEE-2006

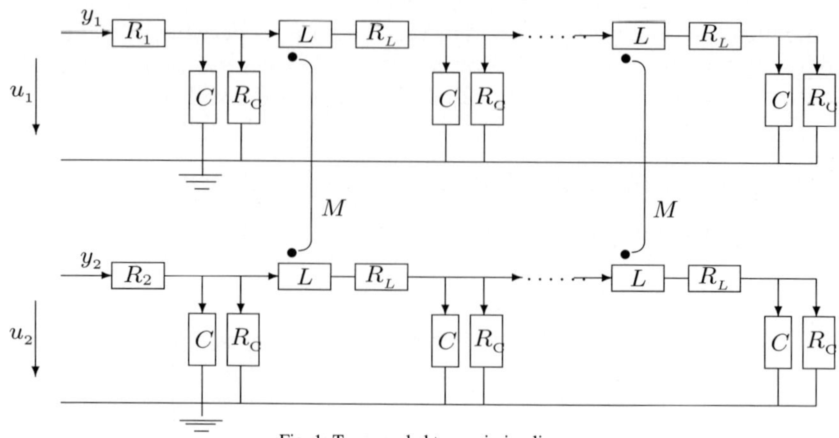

Fig. 1: Two coupled transmission lines

coupling as shown in Fig. 1. Each section consists of an inductor and its associated resistor, in series with a capacitor and its associated resistor. The first section has no inductor. All capacitor values C_i are equal. The same holds for the inductors L_i, the coupling inductors M_{ij}, the resistors associated with the capacitors R_{C_i}, the resistors associated with the inductors R_{L_i} and the input resistors, R_1 and R_2.

To simulate this circuit, the *state-space representation* of the system needs to be derived. Choosing the state variables as the currents through the inductors and the voltages across the capacitors, we obtain a system of order $n = 4N - 2$, where N is the number of sections of the circuit. The state-space representation in *modified nodal analysis (MNA)*-similar form is the following:

$$\left.\begin{aligned} \mathbb{C}\dot{\mathbf{x}}(t) &= \mathbf{G}\mathbf{x}(t) + \mathbf{B}\mathbf{u}(t) \\ \mathbf{y}(t) &= \mathbf{L}\mathbf{x}(t) + \mathbf{D}\mathbf{u}(t) \end{aligned}\right\} \tag{1}$$

where $\mathbb{C} \in \mathbb{R}^{n\times n}$, $\mathbf{G} \in \mathbb{R}^{n\times n}$, $\mathbf{B} \in \mathbb{R}^{n\times 2}$, $\mathbf{L} \in \mathbb{R}^{2\times n}$, $\mathbf{D} \in \mathbb{R}^{2\times 2}$ and $\mathbf{x}(t) \in \mathbb{R}^n$, $\mathbf{u}(t) \in \mathbb{R}^2$, $\mathbf{y}(t) \in \mathbb{R}^2$.

The problem will be studied under the following simplifying assumptions:

(1) the equations are in an MNA-similar form so that the resulting $\mathbb{C}$ matrix in (1) is nonsingular and positive definite (this means that all variables are state variables and none is redundant). In general, $\mathbb{C}$ resulting from circuit simulation is singular, due to additionally generated variables at the nodes between L_i and R_{L_i}.

(2) The transmission line has one input and one output, that is $u_2 = 0$ and only y_1 is observed, so that $\mathbf{u} = u_1$ and $\mathbf{y} = y_1$.

These assumptions are made to ease certain technical issues and allow a comparison of all reduction methods enumerated above; for instance, the optimal $\mathcal{H}_2$ method is currently available for single-input-single-output (SISO) systems only. None of these assumptions is essential for the validity of the results presented. Similar results for a system with MNA equations (where $\mathbb{C}$ is singular), using in part results from [5], will be reported in a future analysis.

For simplicity we will show the form of the equations by deriving them for $N = 3$ sections, namely for a circuit with 6 capacitors and 4 inductors, resulting in 10 states. In particular, the elements of the first line, from left to right will be

$$R_1,\ C_1,\ R_{C_1};\quad L_1,\ R_{L_1},\ C_2,\ R_{C_2};\quad L_2,\ R_{L_2},\ C_3,\ R_{C_3},$$

and those of the second line from left to right

$$R_2,\ C_4,\ R_{C_4};\quad L_3,\ R_{L_3},\ C_5,\ R_{C_5};\quad L_4,\ R_{L_4},\ C_6,\ R_{C_6}.$$

The state variables are:

$$\mathbf{x}_{C_1},\ \mathbf{x}_{L_1},\ \mathbf{x}_{C_2},\ \mathbf{x}_{L_2},\ \mathbf{x}_{C_3}, \mathbf{x}_{C_4},\ \mathbf{x}_{L_3},\ \mathbf{x}_{C_5},\ \mathbf{x}_{L_4},\ \mathbf{x}_{C_6},$$

and the state is chosen as:

$$\mathbf{x} = \begin{pmatrix} \mathbf{x}_C \\ \mathbf{x}_L \end{pmatrix},\quad \mathbf{x}_C = \begin{pmatrix} \mathbf{x}_{C_1} \\ \mathbf{x}_{C_2} \\ \mathbf{x}_{C_3} \\ \mathbf{x}_{C_4} \\ \mathbf{x}_{C_5} \\ \mathbf{x}_{C_6} \end{pmatrix},\quad \mathbf{x}_L = \begin{pmatrix} \mathbf{x}_{L_1} \\ \mathbf{x}_{L_3} \\ \mathbf{x}_{L_2} \\ \mathbf{x}_{L_4} \end{pmatrix}.$$

The associated system matrices are[2]:

$$\mathbb{C} = \begin{pmatrix} \tilde{\mathbb{C}} & 0 \\ 0 & \tilde{\mathbb{L}} \end{pmatrix},\ \mathbf{G} = \begin{pmatrix} -\mathbf{R}_C & \tilde{\mathbb{E}} \\ -\tilde{\mathbb{E}}^* & -\mathbf{R}_L \end{pmatrix},\ \mathbf{B} = \left(\tfrac{1}{R_1}\ 0\ 0\ 0\ 0\ 0\ 0\ 0\ 0\ 0 \right)^*,$$

$\mathbf{L} = -\mathbf{B}^*$ and $\mathbf{D} = \frac{1}{R_1}$, where:

$$\tilde{\mathbb{C}} = \mathrm{diag}(C_1, C_2, C_3, C_4, C_5, C_6),\ \tilde{\mathbb{L}} = \begin{pmatrix} L_1 & M_{13} & & \\ M_{13} & L_3 & & \\ & & L_2 & M_{24} \\ & & M_{24} & L_4 \end{pmatrix} \text{ and}$$

$$\tilde{\mathbb{E}} = \begin{pmatrix} -1 & 0 & 0 & 0 \\ 1 & 0 & -1 & 0 \\ 0 & 0 & 1 & 0 \\ 0 & -1 & 0 & 0 \\ 0 & 1 & 0 & -1 \\ 0 & 0 & 0 & 1 \end{pmatrix},\quad \begin{aligned} &\mathbf{R}_C = \mathrm{diag}\left(\tfrac{1}{R_1} + \tfrac{1}{R_{C_1}}, \tfrac{1}{R_{C_2}}, \tfrac{1}{R_{C_3}}, \tfrac{1}{R_2} + \tfrac{1}{R_{C_4}}, \tfrac{1}{R_{C_5}}, \tfrac{1}{R_{C_6}}\right) \\ &\mathbf{R}_L = \mathrm{diag}(R_{L_1}, R_{L_3}, R_{L_2}, R_{L_4}). \end{aligned}$$

The values of the elements used in the simulation are as follows: the input resistors are $R_1 = R_2 = 10\Omega$, the capacitors are $C_i = 5.4 \cdot 10^{-12} F$ and the associated resistors $R_{C_i} = 10^3 \Omega$, $(i = 1, \ldots, 6)$, the inductors are $L_i = 0.25 \cdot 10^{-9} H$, $(i = 1, \ldots, 4)$, the mutual inductors are $M_{ij} = 0.2 L_i$ $(i = 1, 2,\ j = 3, 4)$ of that value. The associated resistors are zero $R_{L_i} = 0$, $(i = 1, \ldots, 4)$.

3 Gramian based methods

Gramian based methods involve diagonalization of gramians by congruence. These can either be the positive definite solutions to the Lyapunov equations (called *controllability* and *observability gramians*) or the positive definite solutions to algebraic Ricccati equations (called *positive real controllability* and *observability gramians*). The methods that we discuss are balanced truncation (BT) in Sect. 3.1 which

[2] For a matrix $\mathbf{M}$, $\mathbf{M}^*$ denotes transposition followed by complex conjugation if the matrix is complex.

performs simultancous diagonalization of the controllability and the observability gramians, an equivalent of poor man's truncated balanced reduction (PMTBR) in Sect. 3.2 in which only one of the gramians is diagonalized and positive real balanced truncation (PRBT) in Sect. 3.3 in which positive definite solutions to the algebraic Ricatti equations are simultaneously diagonalized.

3.1 Balanced truncation (BT)

The idea behind balanced truncation is to simultaneously diagonalize the two infinite gramians, $\mathcal{P}$ and $\mathcal{Q}$ [1]. These are the solutions to the controllability and observability *Lyapunov equations* respectively, which are associated with the state space formuation (1). The mathematical model of the system may come in two representations: standard state-space and MNA-similar representation (or invertible descriptor form), respectively. We describe the application of model reduction methods for both cases of models.

Standard state-space representation

The standard state-space representation $(\mathbf{A}_{ss}, \mathbf{B}_{ss}, \mathbf{C}_{ss}, \mathbf{D}_{ss})$ is obtained from (1) by inverting the $\mathbb{C}$ matrix.

$$\mathbf{A}_{ss} = \mathbb{C}^{-1}\mathbf{G}, \mathbf{B}_{ss} = \mathbb{C}^{-1}\mathbf{B}, \mathbf{C}_{ss} = -\mathbf{B}^*, \mathbf{D}_{ss} = \mathbf{D}$$

The controllability and observability gramians are given by the symmetric positive definite solutions to the controllability and observability Lyapunov equations:

$$\mathbf{A}_{ss}\mathcal{P} + \mathcal{P}\mathbf{A}_{ss}^* + \mathbf{B}_{ss}\mathbf{B}_{ss}^* = 0 \tag{2}$$

$$\mathbf{A}_{ss}^*\mathcal{Q} + \mathcal{Q}\mathbf{A}_{ss} + \mathbf{C}_{ss}^*\mathbf{C}_{ss} = 0 \tag{3}$$

BT is performed in two steps. First, the balancing projection is computed (both gramians become equal and diagonal, with the Hankel singular values (HSVs) on the diagonal). Second, the states which are equally difficult to reach and to observe are truncated. This amounts to eliminating the states corresponding to the HSVs which are below a certain tolerance. Setting a tolerance for the reduced system a priori defines the number of states to be kept. The procedure is the following.

1. Compute the Cholesky factors of $\mathcal{P} = \mathbf{U}\mathbf{U}^*$ and $\mathcal{Q} = \mathbf{L}\mathbf{L}^*$
2. Compute the singular value decomposition of the product $\mathbf{U}^*\mathbf{L}$
$$\mathbf{U}^*\mathbf{L} = \mathbf{Z}\mathbf{S}\mathbf{Y}^* \tag{4}$$
The diagonal elements: $\mathbf{S} = \mathrm{diag}(\sigma_1, \ldots \sigma_n)$, $\sigma_1 \geq \sigma_2 \geq \ldots \geq \sigma_n$, where $\sigma_i = \sqrt{\lambda_i(\mathcal{P}\mathcal{Q})}$ are the *Hankel singular values* of the system. Choosing only the first k singular values and the first k columns of $\mathbf{Z}$ and $\mathbf{Y}$ gives the reduced system of order k after applying the projection $\mathbf{\Pi}$
3. $\mathbf{\Pi} = \mathbf{V}\mathbf{W}^*$ where $\mathbf{V} = \mathbf{U}\mathbf{Z}_k\mathbf{S}_k^{-\frac{1}{2}}$, $\mathbf{V} \in \mathbb{R}^{n \times k}$, $\mathbf{W} = \mathbf{L}\mathbf{Y}_k\mathbf{S}_k^{-\frac{1}{2}}$, $\mathbf{W} \in \mathbb{R}^{n \times k}$
4. Compute the representation of the reduced system:
$$\tilde{\mathbf{A}}_{ss} = \mathbf{W}^*\mathbf{A}_{ss}\mathbf{V}, \tilde{\mathbf{B}}_{ss} = \mathbf{W}^*\mathbf{B}_{ss}, \tilde{\mathbf{C}}_{ss} = \mathbf{C}_{ss}\mathbf{V}, \tilde{\mathbf{D}}_{ss} = \mathbf{D}_{ss}$$

The corresponding diagonalized controllability and observability gramians are given by $\tilde{\mathcal{P}} = \mathbf{W}^*\mathcal{P}\mathbf{W} = \mathbf{S}_k$, $\tilde{\mathcal{Q}} = \mathbf{V}^*\mathcal{Q}\mathbf{V} = \mathbf{S}_k$ where $\mathbf{S}_k$ is the matrix containing the largest k HSV's on the diagonal.

Descriptor form representation

The MNA-similar representation is precisely (1). For simplicity, we rename the matrices in (1) to match the standard descriptor system representation:[3]

$$\mathbf{E}_{ds} = \mathbb{C}, \mathbf{A}_{ds} = \mathbf{G}, \mathbf{B}_{ds} = \mathbf{B}, \mathbf{C}_{ds} = \mathbf{L}, \mathbf{D}_{ds} = \mathbf{D}$$

The gramians are now the solutions to the following Lyapunov equations:

$$\mathbf{A}_{ds}\mathcal{P}\mathbf{E}_{ds}^* + \mathbf{E}_{ds}\mathcal{P}\mathbf{A}_{ds}^* + \mathbf{B}_{ds}\mathbf{B}_{ds}^* = 0 \tag{5}$$

$$\mathbf{A}_{ds}^*\hat{\mathcal{Q}}\mathbf{E}_{ds} + \mathbf{E}_{ds}^*\hat{\mathcal{Q}}\mathbf{A}_{ds} + \mathbf{C}_{ds}^*\mathbf{C}_{ds} = 0, \tag{6}$$

where $\mathcal{P}$ in (5) is precisely the solution of (2), while the original observability gramian corresponding to the solution of (3) is obtained by means of the congruence transformation

$$\mathcal{Q} = \mathbf{E}_{ds}^*\hat{\mathcal{Q}}\mathbf{E}_{ds}$$

The balancing and truncation procedures follow as described in Sect. 3.1, where (4) is replaced by:

$$\mathbf{U}^*\mathbf{E}_{ds}\mathbf{L} = \mathbf{Z}\mathbf{S}\mathbf{Y}^*$$

The system representation in the new basis now becomes:

$$\tilde{\mathbf{E}}_{ds} = \mathbf{W}^*\mathbf{E}_{ds}\mathbf{V} = \mathbf{I}_k, \tilde{\mathbf{A}}_{ds} = \mathbf{W}^*\mathbf{A}_{ds}\mathbf{V},$$

$$\tilde{\mathbf{B}}_{ds} = \mathbf{W}^*\mathbf{B}_{ds}, \tilde{\mathbf{C}}_{ds} = \mathbf{C}_{ds}\mathbf{V}, \tilde{\mathbf{D}}_{ds} = \mathbf{D}_{ds}.$$

Gramians $\mathcal{P}$ and $\mathcal{Q}$ are simultaneously diagonalized as mentioned in Sect. 3.1.

Solving the Lyapunov equation

There are many methods for solving the Lyapunov equation $\mathbf{A}\mathcal{P} + \mathcal{P}\mathbf{A}^* = \mathbf{Q}$ [1]. We will use the so-called *square-root method*, which directly computes $\mathbf{U}$ such that $\mathcal{P} = \mathbf{U}\mathbf{U}^*$. In MATLAB, this is implemented by `lyapchol`. Another important tool is the *sign function method*, which is discussed next.

The Lyapunov equation is a particular form of the Sylvester equation $\mathbf{AX} + \mathbf{XB} = \mathbf{C}$. To treat this generalized case, consider a matrix of the type

$$\mathbf{Z} = \begin{pmatrix} \mathbf{A} & -\mathbf{C} \\ \mathbf{0} & -\mathbf{B} \end{pmatrix},$$

where $\mathbf{A} \in \mathbb{R}^{n\times n}$, $\Re(\lambda_i(\mathbf{A})) < 0$, $\mathbf{B} \in \mathbb{R}^{k\times k}$, $\Re(\lambda_i(\mathbf{B})) < 0$, and $\mathbf{C} \in \mathbb{R}^{n\times k}$. The sign function iteration $\mathbf{Z}_{n+1} = (\mathbf{Z}_n + \mathbf{Z}_n^{-1})/2$, $\mathbf{Z}_0 = \mathbf{Z}$ converges to

$$\lim_{j\to\infty} \mathbf{Z}_j = \begin{pmatrix} -\mathbf{I}_n & 2\mathbf{X} \\ \mathbf{0} & \mathbf{I}_k \end{pmatrix}$$

where $\mathbf{X}$ is the solution to the equation $\mathbf{AX} + \mathbf{XB} = \mathbf{C}$.

For the Lyapunov equation $\mathbf{A}\mathcal{P} + \mathcal{P}\mathbf{A}^* = \mathbf{Q}$, the starting matrix is

[3] As mentioned earlier, our analysis of the system in descriptor form is restricted to the case in which matrix $\mathbf{E}_{ds} = \mathbb{C}$ is invertible.

$$\mathbf{Z} = \begin{pmatrix} \mathbf{A} & -\mathbf{Q} \\ \mathbf{0} & -\mathbf{A}^* \end{pmatrix}, \ \mathbf{A} \in \mathbb{R}^{n\times n}, \ \Re(\lambda_i(\mathbf{A})) < 0 \Rightarrow \mathbf{Z}_j = \begin{pmatrix} \mathbf{A}_j & -\mathbf{Q}_j \\ \mathbf{0} & -\mathbf{A}_j^* \end{pmatrix}$$

where the iterations can be written as follows

$$\mathbf{A}_{j+1} = \tfrac{1}{2}\left(\mathbf{A}_j + \mathbf{A}_j^{-1}\right), \ \mathbf{A}_0 = \mathbf{A}; \quad \mathbf{Q}_{j+1} = \tfrac{1}{2}\left(\mathbf{Q}_j + \mathbf{A}_j^{-1}\mathbf{Q}_j\mathbf{A}_j^{-*}\right), \ \mathbf{Q}_0 = \mathbf{Q}.$$

The limits of these iterations are $\mathbf{A}_\infty = -\mathbf{I}_n$ and $\mathbf{Q}_\infty = 2\mathcal{P}$ where $\mathcal{P}$ is the solution of the Lyapunov equation $\mathbf{A}\mathcal{P} + \mathcal{P}\mathbf{A}^* = \mathbf{Q}$.

Often, the constant term in the Lyapunov equation above is provided in factored form $\mathbf{Q} = \mathbf{R}\mathbf{R}^*$. As a consequence, it is possible to obtain the solution in factored form. In particular, the $(j+1)^{st}$ iterate in factored form is

$$\mathbf{Q}_{j+1} = \mathbf{R}_{j+1}\mathbf{R}_{j+1}^* \text{ where } \mathbf{R}_{j+1} = \tfrac{1}{\sqrt{2}}\left[\mathbf{R}_j, \mathbf{A}_j^{-1}\mathbf{R}_j\right] \Rightarrow \mathbf{Q}_\infty = \mathbf{R}_\infty\mathbf{R}_\infty^* = 2\mathcal{P}$$

$\mathbf{R}_\infty$ has infinitely many columns, although its rank cannot exceed n. This can be avoided by performing at each step a rank revealing RQ factorization $\mathbf{R}_j\mathbf{P}_j = \mathbf{T}_j\mathbf{U}_j$ with $\mathbf{P}_j$ the permutation matrix and $\mathbf{T}_j\mathbf{P}_j = \left[\Delta_j^*, \mathbf{0}\right]^*$. Δ_j is upper triangular and $\mathbf{U}_j\mathbf{U}_j^* = \mathbf{I}_j$. Thus, at the jth step, $\mathbf{R}_j$ is replaced by Δ_j which has as many columns as the rank of $\mathbf{R}_j$. For accelerating convergence, the eigenvalues of $\mathbf{A}$ can be scaled [3]: at each step, $\mathbf{A}_j$ is replaced by $\frac{1}{\gamma_j}\mathbf{A}_j$ where the factors γ_j can be chosen as $\gamma_j = |det(\mathbf{A}_j)|^{\frac{1}{n}}$ in order to minimize the distance of the geometric mean of the eigenvalues of $\mathbf{A}_j$ from 1.

Convergence of the iteration which uses scaling is quadratic. The time required to compute the Cholesky factor by MATLAB's `lyapchol` function versus the iterative implementation of the sign function method in [3] is as follows: on a Pentium M at 1.3Ghz with 768MB RAM, `lyapchol` runs in 0.751s for a matrix $\mathbf{A}$ of dimension 242, while the implementation in [3] requires 5.423s and converges in $16 \approx \sqrt{242}$ steps. Even if, in theory, no scaling should also give quadratic convergence, in practice, due to numerical issues, convergence occurs after 20 steps.

3.2 Truncation by diagonalization of one gramian or poor man's truncated balanced reduction (PMTBR)

For the standard state-space representation, the procedure is the following [1].

1. Compute the gramian to be diagonalized (controllability gramian $\mathcal{P}$ in our case)
2. Compute the eigenvalue decomposition of $\mathcal{P} = \mathbf{V}\Sigma\mathbf{V}^*$
3. Choose the eigenvectors corresponding to the largest k eigenvalues to obtain the transformation $\mathbf{T} = \mathbf{V}_k^*$
4. The reduced system is

$$\tilde{\mathbf{A}}_{ss} = \mathbf{T}\mathbf{A}_{ss}\mathbf{T}^*, \ \tilde{\mathbf{B}}_{ss} = \mathbf{T}\mathbf{B}_{ss}, \ \tilde{\mathbf{C}}_{ss} = \mathbf{C}_{ss}\mathbf{T}^*, \ \tilde{\mathbf{D}}_{ss} = \mathbf{D}_{ss}$$

PMTBR is presented in [10] and uses numerical quadrature to approximate the gramian $\mathcal{P}$, without solving the Lyapunov equation. The algorithm used in our analysis, however, diagonalizes the exact solution $\mathcal{P}$ of the Lyapunov equation. As mentioned in Sect. 3.1, the solution to the Lyapunov equation can be computed either by using the *sign function method* or by using MATLAB's `lyapchol` function.

3.3 Positive real balanced truncation (PRBT)

Coupled transmission lines such as the one in Fig. 1 are passive systems, with *positive real* transfer functions (further information on passivity and positive realness is provided in [1]). We are therefore interested in reduced order models that are passive. In general, BT is not a passivity preserving method, since the resulting reduced system may have a non-positive real transfer function. PRBT, however, is a passivity preserving method. It yields reduced order models with positive real transfer functions by simultaneously diagonalizing the positive definite solutions $\mathcal{P}$ and $\mathcal{Q}$ of the controllability and observability algebraic *Riccati equations* respectively. This desirable result cannot be guaranteed with BT, where the solutions to the Lyapunov equations are diagonalized, rather than the solutions the Riccati equations. Riccati equations have a different form depending on whether the system is in standard state-space form or in descriptor form.

Historical note: this method was first introduced by Ober [6] and rediscovered by Phillips, Daniel and Silveira [9]. For an overview see also [1].

Standard state-space representation

The controllability and observability positive real Riccati equations are:

$$\mathbf{A}_{ss}\mathcal{P} + \mathcal{P}\mathbf{A}_{ss}^* + (\mathcal{P}\mathbf{C}_{ss}^* - \mathbf{B}_{ss})\Delta(\mathcal{P}\mathbf{C}_{ss}^* - \mathbf{B}_{ss})^* = 0 \qquad (7)$$

$$\mathbf{A}_{ss}^*\mathcal{Q} + \mathcal{Q}\mathbf{A}_{ss} + (\mathcal{Q}\mathbf{B}_{ss} - \mathbf{C}_{ss}^*)\Delta(\mathcal{Q}\mathbf{B}_{ss} - \mathbf{C}_{ss}^*)^* = 0 \qquad (8)$$

where $\Delta = (\mathbf{D}_{ss} + \mathbf{D}_{ss}^*)^{-1}$.
The procedure is the same as for BT (see Sect. 3.1), except that now balancing is performed on the minimal solutions of the Riccati equations. The diagonal elements of $\mathbf{S}$ in (4) are the *positive real singular values* of the system, which we denote by π_i: $\mathbf{S} = \mathrm{diag}(\pi_1, \ldots \pi_n)$, where $\pi_1 \geq \pi_2 \geq \ldots \geq \pi_n$.

Descriptor form representation

The corresponding algebraic Riccati equations in descriptor form are

$$\mathbf{A}_{ds}\mathcal{P}\mathbf{E}_{ds}^* + \mathbf{E}_{ds}\mathcal{P}\mathbf{A}_{ds}^* + (\mathbf{E}_{ds}\mathcal{P}\mathbf{C}_{ds}^* - \mathbf{B}_{ds})\Delta(\mathbf{E}_{ds}\mathcal{P}\mathbf{C}_{ds}^* - \mathbf{B}_{ds})^* = 0 \quad (9)$$

$$\mathbf{A}_{ds}^*\hat{\mathcal{Q}}\mathbf{E}_{ds} + \mathbf{E}_{ds}^*\hat{\mathcal{Q}}\mathbf{A}_{ds} + (\mathbf{E}_{ds}^*\hat{\mathcal{Q}}\mathbf{B}_{ds} - \mathbf{C}_{ds}^*)\Delta(\mathbf{E}_{ds}^*\hat{\mathcal{Q}}\mathbf{B}_{ds} - \mathbf{C}_{ds}^*)^* = 0 \quad (10)$$

where $\Delta = (\mathbf{D}_{ds} + \mathbf{D}_{ds}^*)^{-1}$. The observability gramian given by the solution of (8) is obtained via the congruence transformation $\mathcal{Q} = \mathbf{E}_{ds}^*\hat{\mathcal{Q}}\mathbf{E}_{ds}$.

Balancing and truncation are now performed on the solutions to (9) and (10) and the procedure follows as in 3.1.

Hamiltonian Riccati Balanced Truncation (PRBT-Ham)

Solutions to Riccati equations ((7),(8)) (or ((9),(10)) for MNA-similar form) can be obtained using the MATLAB function `care`. This can be applied to a system in usual state space form or in descriptor form. An alternative is to solve for $\mathcal{P}$ and $\hat{\mathcal{Q}}$ by means of the Hamiltonian eigenvalue problem [11]:

$$\begin{bmatrix} \mathbf{A}_{ds} - \mathbf{B}_{ds}\Delta\mathbf{C}_{ds} & -\mathbf{B}_{ds}\Delta\mathbf{B}_{ds}^* \\ \mathbf{C}_{ds}^*\Delta\mathbf{C}_{ds} & -\mathbf{A}_{ds}^* + \mathbf{C}_{ds}^*\Delta\mathbf{B}_{ds}^* \end{bmatrix} \begin{bmatrix} \mathbf{X} \\ \mathbf{Y} \end{bmatrix} = \begin{bmatrix} \mathbf{E}_{ds} & \\ & \mathbf{E}_{ds}^* \end{bmatrix} \begin{bmatrix} \mathbf{X} \\ \mathbf{Y} \end{bmatrix} \begin{bmatrix} \Lambda_- & \\ & \Lambda_+ \end{bmatrix} \qquad (11)$$

where $\Delta = (\mathbf{D}_{ds} + \mathbf{D}_{ds}^*)^{-1}$, and Λ_-, Λ_+ are the Hamiltonian eigenvalues, with negative and positive real parts respectively (i.e. the *stable* and *antistable spectral zeros* of the system). We can partition $\mathbf{X}$ and $\mathbf{Y}$ according to the stable and antistable eigenvalues of the Hamiltonian into

$$\begin{bmatrix} \mathbf{X} \\ \mathbf{Y} \end{bmatrix} = \begin{bmatrix} \mathbf{X}_- & \mathbf{X}_+ \\ \mathbf{Y}_- & \mathbf{Y}_+ \end{bmatrix}$$

The minimal solutions to (9) and (10) are given by:

$$\mathcal{P} = -\mathbf{X}_+(\mathbf{Y}_+)^{-1}\mathbf{E}_{ds}^{-*} \tag{12}$$

$$\hat{\mathcal{Q}} = -\mathbf{Y}_-(\mathbf{X}_-)^{-1}\mathbf{E}_{ds}^{-1} \tag{13}$$

and are the same as the ones resulting from the MATLAB `care` routine. The stabilizing solution (corresponding to the stable spectral zeros) is $\hat{\mathcal{Q}}$ while $\mathcal{P}$ is the antistabilizing solution (corresponding to the antistable spectral zeros). Both $\hat{\mathcal{Q}}$ and $\mathcal{P}$ are obtained form the same Hamiltonian eigenvalue computation (11). The original positive real observability gramian as solution to (8) is $\mathcal{Q} = \mathbf{E}_{ds}^*\hat{\mathcal{Q}}\mathbf{E}_{ds}$, so the positive real Hankel singular values are $\pi_i = \sqrt{\lambda_i(\mathcal{P}\mathcal{Q})}$, i.e. the diagonal elements of $\mathbf{X}_+(\mathbf{Y}_+)^{-1}\mathbf{Y}_-(\mathbf{X}_-)^{-1}$. We see that the positive real Hankel singular values can be computed without any inversion of $\mathbf{E}_{ds}$. The reduction procedure follows as in Sect. 3.1 using the computed (12) and (13).

If the system is the in usual state space form rather than in descriptor form, $\mathbf{E}_{ds}$ in (11) is simply replaced by $\mathbf{I}$. The resulting solutions $\mathcal{P}$ and $\hat{\mathcal{Q}}$ computed as (12) and (13) respectively, are precisely the positive real gramians solving (7) and (8). They are also the same as the solutions obtained with the MATLAB `care` routine in the usual state space form. The reduction procedure follows as in Sect. 3.3.

NOTE: The gramians used in balanced truncation, i.e. the solutions to the Lyapunov equations ((2), (3)) (and correspondingly ((5), (6)) for descriptor form) can be obtained using (11) with $\Delta = \mathbf{I}$, $\mathbf{C} = \mathbf{0}$ (for controllability) and $\mathbf{B} = \mathbf{0}$ (for observability).

4 Krylov based methods

Krylov based reduction methods exploit the use of Krylov subspace iterations to achieve system approximation by *moment matching* [1]. Three such methods are: PRIMA, the spectral zero method (SZM) and optimal $\mathcal{H}_2$. As outlined next, PRIMA matches k moments at zero by means of an *orthogonal projection*. SZM matches $2k$ moments of the original system, at k stable spectral zeros and their mirror images (the corresponding k antistable spectral zeros), by means of an *oblique projection*. Finally, using an oblique projection, the optimal $\mathcal{H}_2$ method matches $2k$ moments of the original system at the mirror images of the k poles of the reduced system (2 moments are matched at each pole). Hence an iteration is required.

4.1 PRIMA

For PRIMA, the moments of the transfer function $\mathbf{H}(s) = \mathbf{L}(s\mathbb{C} - \mathbf{G})^{-1}\mathbf{B} + \mathbf{D}$ are defined as the coefficients of the Taylor expansion of $\mathbf{H}(s)$ around $s_0 = 0$: $\mathbf{H}(s) = \mathbf{M}_0 + \mathbf{M}_1 s + \mathbf{M}_2 s^2 + \ldots$, where

$$\mathbf{M}_0 = \mathbf{D} - \mathbf{L}\mathbf{G}^{-1}\mathbf{B} \text{ and } \mathbf{M}_k = (-1)^{(k+1)}\mathbf{L}(\mathbb{C}^{-1}\mathbf{G})^{-(k+1)}\mathbb{C}^{-1}\mathbf{B}, \text{ for } k > 0.$$

PRIMA computes a kth order reduced system by matching k moments of the original system. This is achieved by computing the orthogonal projection $\mathbf{\Pi} = \mathbf{X}_k\mathbf{X}_k^*$ such that $\mathbf{X}_k^*\mathbb{C}^{-1}\mathbf{G}\mathbf{X}_k = \mathbf{H}_k$ with $\mathbf{H}_k$ upper Hessenberg; the *column span* of $\mathbf{X}_k$ is the same as the *column span* of:

$$[\mathbb{C}^{-1}\mathbf{B},\ (\mathbb{C}^{-1}\mathbf{G})^{-1}\mathbb{C}^{-1}\mathbf{B},\ (\mathbb{C}^{-1}\mathbf{G})^{-2}\mathbb{C}^{-1}\mathbf{B},\ \ldots,\ (\mathbb{C}^{-1}\mathbf{G})^{-(k-1)}\mathbb{C}^{-1}\mathbf{B}].$$

The procedure is as follows [7].

1. Solve $\mathbf{G}\mathbf{R} = \mathbf{B}$ for $\mathbf{R}$.
2. $(\mathbf{X}_0, \mathbf{T})$ =QR($\mathbf{R}$); QR Factorization of $\mathbf{R}$
3. For $i = 1, 2, \ldots, k$
   Set $\mathbf{V} = \mathbb{C}\mathbf{X}_{i-1}$
   Solve $\mathbf{G}\mathbf{X}_i^{(0)} = \mathbf{V}$ for $\mathbf{X}_i^{(0)}$
   For $j = 1, 2, \ldots, i$
   $\mathbf{H} = \mathbf{X}_{i-j}^*\mathbf{X}_i^{(j-1)}$
   $\mathbf{X}_i^{(j)} = \mathbf{X}_i^{(j-1)} - \mathbf{X}_{i-j}\mathbf{H}$
   $(\mathbf{X}_i, \mathbf{T})$ = QR($\mathbf{X}_i^{(i)}$); QR Factorization of $\mathbf{X}_i^{(i)}$
4. Set $\mathbf{X} = [\mathbf{X}_0\ \mathbf{X}_1, \ldots, \mathbf{X}_{i-1}]$ and truncate $\mathbf{X}$ so that it has k columns only
5. Compute $\hat{\mathbb{C}} = \mathbf{X}^*\mathbb{C}\mathbf{X}$, $\hat{\mathbf{G}} = \mathbf{X}^*\mathbf{G}\mathbf{X}$, $\hat{\mathbf{B}} = \mathbf{X}^*\mathbf{B}$ and $\hat{\mathbf{L}} = \mathbf{L}\mathbf{X}$

4.2 Spectral zero method (SZM)

With PRIMA, system approximation was achieved by matching k moments of the transfer function at zero. In the general case, using the *rational Krylov* approach [1], reduced systems are obtained which match moments at preassigned *interpolation points* in the complex plane. SZM is a rational Krylov reduction method, in which the interpolation points are chosen as a subset of the spectral zeros of the original system [2], [11]. This selection guarantees the stability and passivity of the reduced system [2], [11]. The spectral zeros are given by Λ in (11). The real spectral zeros s_i come in pairs $(s_i, -s_i)$ while the complex spectral zeros come in quadruples of the form:

$$\begin{aligned} s_i &= \Re(s_i) + j \cdot \Im(s_i), & & \\ s_{i+1} &= \Re(s_i) - j \cdot \Im(s_i) &=& \ s_i^*, \\ s_{i+2} &= -\Re(s_i) + j \cdot \Im(s_i) &=& -s_i^*, \\ s_{i+3} &= -\Re(s_i) - j \cdot \Im(s_i) &=& -s_i, \end{aligned}$$

where without loss of generality, we assume $\Re(s_i) < 0$.

The usual procedure

The usual procedure for obtaining a kth order reduced system with SZM is as follows.

1. Construct matrices $\mathbf{V}$ and $\mathbf{W}$ using $2k$ interpolation points:

$$\mathbf{V} = \left[(s_1\mathbf{E}_{ds} - \mathbf{A}_{ds})^{-1}\mathbf{B}_{ds},\ (s_2\mathbf{E}_{ds} - \mathbf{A}_{ds})^{-1}\mathbf{B}_{ds},\ \cdots,\ (s_k\mathbf{E}_{ds} - \mathbf{A}_{ds})^{-1}\mathbf{B}_{ds} \right]$$

$$\mathbf{W} = \begin{bmatrix} \mathbf{C}_{ds}(-s_1\mathbf{E}_{ds} - \mathbf{A}_{ds})^{-1} \\ \mathbf{C}_{ds}(-s_2\mathbf{E}_{ds} - \mathbf{A}_{ds})^{-1} \\ \vdots \\ \mathbf{C}_{ds}(-s_k\mathbf{E}_{ds} - \mathbf{A}_{ds})^{-1} \end{bmatrix}$$

where s_i, $i = 1, 2, \ldots, k$ are k spectral zeros that we select a priori. In our case, we selected the spectral zeros which are closest to the real axis.

2. The reduced system: $\tilde{\mathbf{E}}_{ds} = \mathbf{W}\mathbf{E}_{ds}\mathbf{V}$, $\tilde{\mathbf{A}}_{ds} = \mathbf{W}\mathbf{A}_{ds}\mathbf{V}$, $\tilde{\mathbf{B}}_{ds} = \mathbf{W}\mathbf{B}_{ds}$, $\tilde{\mathbf{C}}_{ds} = \mathbf{C}_{ds}\mathbf{V}$, $\tilde{\mathbf{D}}_{ds} = \mathbf{D}_{ds}$, matches the chosen $2k$ spectral zeros of the original system, s_i and $-s_i$, $i = 1, \ldots, k$.

Note: Since no inversion of $\mathbf{E}_{ds}$ is involved, the SZM method is also applicable to systems with singular $\mathbf{E}_{ds}$. Also, the oblique projection that reduces the system is $\mathbf{\Pi} = \bar{\mathbf{V}}\bar{\mathbf{W}}$, where: $\mathbf{W}\mathbf{E}_{ds}\mathbf{V} = \mathbf{L}\mathbf{U}$, $\bar{\mathbf{V}} = \mathbf{V}\mathbf{U}^{-1}$ and $\bar{\mathbf{W}} = \mathbf{L}^{-1}\mathbf{W}$.

Hamiltonian spectral zero method (SZM-Ham)

An alternative way to build $\mathbf{V}$ and $\mathbf{W}$ is without inverting the matrix $s_i\mathbf{E}_{ds} - \mathbf{A}_{ds}$ for each spectral zero s_i we choose. This is achieved as presented in [11], by solving the Hamiltonian eigenvalue problem (11).

Once the eigenvectors and eigenvalues of the Hamiltonian are obtained, the spectral zeros which are closest to the real axis are chosen. Matrices $\mathbf{V}$ and $\mathbf{W}$ are now computed from the eigenvectors corresponding to the k chosen spectral zeros:

$$\mathbf{W}^* = \mathbf{Y}_k^*,\ \mathbf{V} = \mathbf{X}_k$$

The reduced system is $\tilde{\mathbf{E}}_{ds} = \mathbf{W}^*\mathbf{E}_{ds}\mathbf{V}$, $\tilde{\mathbf{A}}_{ds} = \mathbf{W}^*\mathbf{A}_{ds}\mathbf{V}$, $\tilde{\mathbf{B}}_{ds} = \mathbf{W}^*\mathbf{B}_{ds}$, $\tilde{\mathbf{C}}_{ds} = \mathbf{C}_{ds}\mathbf{V}$, $\tilde{\mathbf{D}}_{ds} = \mathbf{D}_{ds}$.

Note: As for SZM, $\mathbf{E}_{ds}$ may be singular, so the SZM-Ham method also applies to the general case of descriptor systems since it involves no inversion of $\mathbf{E}_{ds}$. We emphasize that SZM-Ham gives the same reduced model as the usual procedure in Sect. 4.2, the difference is only in how $\mathbf{W}$ and $\mathbf{V}$ are computed.

4.3 Optimal $\mathcal{H}_2$

The optimal $\mathcal{H}_2$ method, as the name suggests, produces reduced order models which minimize the $\mathcal{H}_2$ norm of the error system. The problem formulation follows [4].

Given an n-dimensional single-input, single-output dynamical system in the MNA-similar form (1) (where $\mathbb{C}$ may be singular), with transfer function $\mathbf{H}(s) = \mathbf{L}(s\mathbb{C} - \mathbf{G})^{-1}\mathbf{B}$, find a stable reduced system of order $k < n$ such that its transfer function $\mathbf{H}_k(s) = \mathbf{L}_k(s\mathbb{C}_k - \mathbf{G}_k)^{-1}\mathbf{B}_k$ minimizes the $\mathcal{H}_2$ error, i.e.:

$$\mathbf{H}_k(s) = \arg \min_{\deg(\hat{\mathbf{H}})=k} \| \mathbf{H}(s) - \hat{\mathbf{H}}(s) \|_{\mathcal{H}_2}, \ \| \mathbf{H}(s) \|^2_{\mathcal{H}_2} := \frac{1}{2\pi}\int_{-\infty}^{\infty} |\mathbf{H}(j\omega)|^2 d\omega$$

The reduced order model that achieves this is constructed using the *iterative rational Krylov algorithm* (**IRKA**) [4]:

1. Make an initial shift selection $\sigma_i \in \mathbb{C}$, $i = 1, \ldots, k$

2. Construct $\mathbf{W} = [\,(\sigma_1\mathbb{C}^* - \mathbf{G}^*)^{-1}\mathbf{L}^*, \cdots, (\sigma_k\mathbb{C}^* - \mathbf{G}^*)^{-1}\mathbf{L}^*\,]$ and $\mathbf{V} = [\,(\sigma_1\mathbb{C} - \mathbf{G})^{-1}\mathbf{B}, \cdots, (\sigma_k\mathbb{C} - \mathbf{G})^{-1}\mathbf{B}\,]$

3. Repeat:

   a) $\mathbb{C}_k = \mathbf{W}^*\mathbb{C}\mathbf{V}, \;\; \mathbf{G}_k = \mathbf{W}^*\mathbf{G}\mathbf{V}$

   b) $\sigma_i \longleftarrow \; -\lambda_i(\mathbf{G}_k, \mathbb{C}_k)$ for $i = 1, \cdots, k$

   c) $\mathbf{W} = [\,(\sigma_1\mathbb{C}^* - \mathbf{G}^*)^{-1}\mathbf{L}^*, \cdots, (\sigma_k\mathbb{C}^* - \mathbf{G}^*)^{-1}\mathbf{L}^*\,]$

   d) $\mathbf{V} = [\,(\sigma_1\mathbb{C} - \mathbf{G})^{-1}\mathbf{B}, \cdots, (\sigma_k\mathbb{C} - \mathbf{G})^{-1}\mathbf{B}\,]$

   until $\sum_{i=1}^{k} |\sigma_i - \overline{\sigma}_i| < \epsilon$, where σ_i and $\overline{\sigma}_i$, $i = 1, \ldots, k$, are the shifts at iterations j and $j+1$ respectively and ϵ is the desired convergence tolerance.

4. Project the system matrices

$$\mathbb{C}_k = \mathbf{W}^*\mathbb{C}\mathbf{V}, \; \mathbf{G}_k = \mathbf{W}^*\mathbf{G}\mathbf{V}, \; \mathbf{B}_k = \mathbf{W}^*\mathbf{B}, \; \mathbf{L}_k = \mathbf{L}\mathbf{V}$$

Upon convergence, the reduced order model satisfies the necessary $\mathcal{H}_2$ optimality conditions:

$$\mathbf{H}(-\hat{\lambda}_i) = \mathbf{H}_k(-\hat{\lambda}_i), \;\; \frac{d}{ds}\mathbf{H}(s)\,|_{s=-\hat{\lambda}_i} = \frac{d}{ds}\mathbf{H}_k(s)\,|_{s=-\hat{\lambda}_i} \quad i = 1, \ldots, k,$$

where $\hat{\lambda}_i$ are the eigenvalues of $(\mathbf{G}_k, \; \mathbb{C}_k)$ (*Ritz values*). The reduced system therefore matches $2k$ moments of the original at the mirror images of the reduced order poles. Initial shifts σ_i can be arbitrarily chosen and influence the convergence rate. Since this algorithm produces a *locally optimal* reduced model, some initial shifts may not lead to convergence. Future work will investigate the optimal choice of initial shifts, how they influence the convergence rate of the Ritz values and the approximation error of the resulting reduced model.

Note: If the initial shift selection in step 1. is a subset of the spectral zeros of the original system, reducing the system directly after step. 2. makes optimal $\mathcal{H}_2$ equivalent to SZM. The resulting reduced system will, however, not be optimal in the $\mathcal{H}_2$ norm. $\mathcal{H}_2$ norm optimality is guaranteed only through the iterative procedure; this however cannot guarantee passivity for the reduced model like SZM does.

Table 1: Classification of all methods

Reduction Method	Type	Iterative	Moments Matched	Projection
Balanced truncation (BT)	Gramian based	No	-	Oblique
One Gramian Method (PMTBR)	Gramian based	No	-	Orthogonal
Riccati Balanced Truncation (PRBT)	Gramian based	No	-	Oblique
Hamiltonian Riccati Balanced Truncation (PRBT-Ham)	Gramian based	No	-	Oblique
PRIMA	Krylov based	No	k moments at 0	Orthogonal
Spectral Zero Method (SZM)	Krylov based	No	$2k$ spectral zeros	Oblique
Hamiltonian Spectral Zero Method (SZM-Ham)	Krylov based	No	$2k$ spectral zeros	Oblique
Optimal $\mathcal{H}_2$	Krylov based	Yes	$2k$ moments at mirror images of reduced order poles	Oblique

A classification of all methods used in our analysis is presented in Tab. 1.

5 Comparison of all methods: performance

A first indication of how easily these systems can be approximated is given by the Hankel singular values and the positive real Hankel singular values. Figure 2 shows a logarithmic plot of the normalized Hankel singular values and the eigenvalues of the gramians for the system associated with the circuit in Fig. 1 with $N = 61$ sections, resulting in $n = 242$ states. The eigenvalues of $\mathcal{P}$ and $\mathcal{Q}$ decay at about the same rate. The Hankel singular values and the positive real Hankel singular values (see Sect. 3.1 and 3.3) also decay at about the same rate, but not as fast as the controllability or observability gramian eigenvalues.

Since the gramian eigenvalues decay much faster than the Hankel singular values, a reduction method which balances the system by diagonalizing only one gramian seems justifiable. However, as will be shown in Sect. 6, the decay rate of the gramian eigenvalues do not provide sufficient information for the efficiency of the reduction algorithm. For example, a method that balances the system by simultaneously diagonalizing both gramians is more efficient, even though it exploits the slower decay rate of the Hankel singular values. The relative $\mathcal{H}_\infty$ and $\mathcal{H}_2$ norms of the associated error systems in Sect. 6 support the above statement.

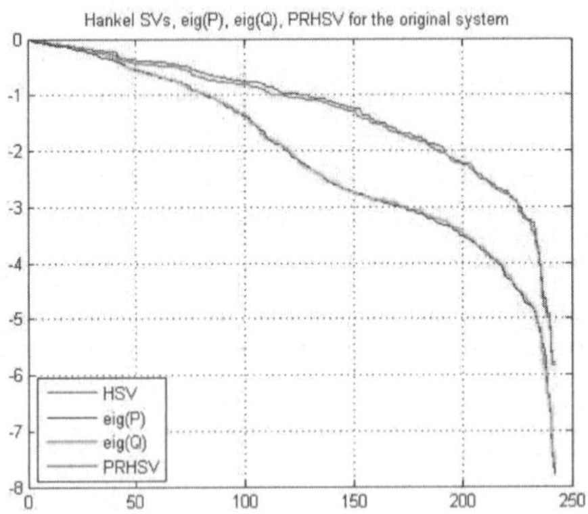

Fig. 2: Hankel singular values, positive real Hankel singular values, eigenvalues of $\mathcal{P}$, eigenvalues of $\mathcal{Q}$.

Fig. 2 also shows the trade-off between accuracy and complexity [1]. Choosing a larger order k of the reduced system by truncating the last $n-k$ states gives a smaller approximation error. In particular, for an error of 10%, one needs to keep about half the states, namely 120. We conclude that our circuit is difficult to approximate as the decay of both Lyapunov and Riccati Hankel singular values is slow.

We further investigate whether the properties of the original system are preserved. We would like to check which of the methods used produce a reduced system which is controllable, observable, stable and passive.

Controllability and *observability* are equivalent to the controllability and observability gramians $\mathcal{P}$ and $\mathcal{Q}$ having full rank. *Stability* is equivalent to all poles lying in the left half plane. *Passivity* is ensured by the nonexistence of spectral zeros on the $j\omega$ axis.

Starting from an initial system with $N = 61$ sections ($n = 242$ states), both reduced systems ($k = 11$ and $k = 21$) preserve the same characteristics.

Table 2: Preservation of Controllability, Observability, Stability and Passivity for each Reduction Method

Reduction Method	Controllable	Observable	Stable	Passive
Balanced truncation (BT)	Yes	Yes	Yes	Yes
One Gramian Method (PMTBR)	Yes	Yes	Yes	Yes
Riccati Balanced Truncation (PRBT)	Yes	Yes	Yes	Yes
Hamiltonian Riccati Balanced Truncation (PRBT-Ham)	Yes	Yes	Yes	Yes
PRIMA	**No**	**No**	Yes	Yes
Spectral Zero Method (SZM)	Yes	Yes	Yes	Yes
Hamiltonian Spectral Zero Method (SZM-Ham)	Yes	Yes	Yes	Yes
Optimal $\mathcal{H}_2$	Yes	Yes	Yes	Yes

Tab. 2 shows that all methods produce controllable and observable reduced systems, except for PRIMA. However, all reduced systems are stable. Even though only PMTBR, PRBT, PRIMA and SZM are *passivity preserving* methods for MNA representations, all the methods preserve passivity for our system.

Table 3: Relative norms of the error systems, $k = 21$, $k = 11$

Reduction Method $N = 61, n = 242$	$\mathcal{H}_\infty$ $k = 21$	$\mathcal{H}_2$ $k = 21$	$\mathcal{H}_\infty$ $k = 11$	$\mathcal{H}_2$ $k = 11$
Balanced Truncation (BT)	0.4746	0.4230	0.5409	0.4599
One Gramian (PMTBR)	0.7204	0.5284	0.5867	0.5322
Riccati Balanced Truncation (PRBT)	0.5247	0.5318	0.6486	0.7068
Hamiltonian Riccati Balanced Truncation (PRBT-Ham)	0.5247	0.5318	0.6486	0.7068
PRIMA	0.8519	0.6762	0.8147	0.8134
Spectral Zero Method (SZM)	0.6498	0.7259	0.6946	0.8392
Hamiltonian Spectral Zero Method (SZM-Ham)	0.6498	0.7259	0.6946	0.8392
Optimal $\mathcal{H}_2$	**0.3554**	**0.2676**	**0.3561**	**0.2909**

To assess the performance of these methods, Tab. 3 collects the relative norms of the error systems, that is $\frac{\|\Sigma_{orig} - \Sigma_k\|}{\|\Sigma_{orig}\|}$. The surprising result in Tab. 3 is that optimal $\mathcal{H}_2$ yields the smallest relative error both in the $\mathcal{H}_\infty$ and $\mathcal{H}_2$ norms; it is superior for instance to BT, which is usually considered as the overall best approximation method. From Tab. 3, it is also evident that relative errors for PRBT and PRBT-Ham are identical. The same holds for SZM and SZM-Ham. This shows that SZM and PRBT are equivalent to their Hamiltonian counterparts respectively. Reduction by diagonalizing one gramian method yields the best approximant out of all the passivity preserving methods. PRIMA gives a reduced system which approximates the original one very well for low frequencies (much better than any other method). This is a consequence of the fact that the expansion point in the Arnoldi algorithm is the origin.

The $\mathcal{H}_\infty$ norm of a stable system is the maximum singular value of the transfer function or the value of the highest peak in the frequency response plot. Notice that, in case of one gramian and PRIMA, a larger dimension of the reduced system does not yield a smaller $\mathcal{H}_\infty$ relative norm of the error system: the value for the reduced system of size $k = 21$ is larger than the value obtained for the reduced system of size $k = 11$. This could be explained by the fact that a higher dimension of the reduced system yields peaks of higher amplitude. The $\mathcal{H}_2$ norm of a stable SISO system with the $\mathbf{D}$ term equal to 0 is a measure of the area frequency response over the entire frequency range. Notice that there is no norm which captures all aspects of the reduced system. The ones used here are the most popular.

Table 4: Elapsed Times

Reduction Method $N = 61, n = 242$	Elapsed Time (s) $k = 21$	Elapsed Time(s) $k = 11$
Balanced Truncation (BT)	1.915538	1.956464
One Gramian (PMTBR)	0.972908	0.974607
Riccati Balanced Truncation (PRBT)	65.47625	66.20935
Hamiltonian Riccati Balanced Truncation (PRBT-Ham)	*39.64582*	*40.97727*
PRIMA	**0.355047**	**0.189727**
Spectral Zero Method (SZM)	12.208282	8.749104
Hamiltonian Spectral Zero Method (SZM-Ham)	*5.017118*	*4.746759*
Optimal $\mathcal{H}_2$	136.5	111.35

Elapsed times are useful for comparing the computational cost of each method versus the quality of the resulting reduced system. The computational times in Tab. 4 were obtained on a Pentium M at 1.3Ghz with 768MB RAM. The most expensive method is optimal $\mathcal{H}_2$; it requires a certain number of iteration steps to converge, depending on the initial shift selection. PRBT is also expensive, when implemented using MATLAB's `care` function for obtaining the positive definite solutions to the algebraic Riccati equations. On the other hand, PRIMA is the most computationally efficient, having the complexity of an iterative Arnoldi algorithm. This is more computationally efficient than performing eigenvalue decompositions, singular value decompositions or solving the Lyapunov or algebraic Riccati equations, which are needed in the other reduction methods.

Another aspect worth noticing in Tab. 4 is that, indeed, for the spectral zero method, computing the projectors from the eigenvectors of the Hamiltonian matrix (SZM-Ham) requires about half the time needed to compute the projectors by inverting the matrix $s_i E_{ds} - A_{ds}$ for each spectral zero s_i we choose (SZM). A similar performance improvement is achieved when the Riccati solutions in PRBT were computed with the Hamiltonian eigenvalue problem (PRBT-Ham) rather than with the MATLAB `care` routine (PRBT). Considering that relative error norms for PRBT and PRBT-Ham are identical (the same holds for SZM and SZM-Ham), we conclude that performing PRBT (and SZM) via the Hamiltonian approach is more efficient.

Also, from Tab. 4 we notice the small difference between elapsed times for the two reduced dimensions ($k = 21$ and $k = 11$), since most of the computational effort is used in computing the projectors, not in obtaining the reduced systems themselves.

6 Comparison of all methods: plots

We first provide pairwise comparisons of error systems resulting from applying each reduction procedure on the circuit in Fig. 1. Next, the preservation of stability and passivity is shown in the distributions of poles and spectral zeros for the original and each reduced system (Figs. 13-18). Figures for methods PRBT-Ham and SZM-Ham are omitted, because they are identical to figures for PRBT and SZM respectively. The original system has $n = 242$ states (resulting from $N = 61$ sections) and we reduce it to dimension $k = 21$.

6.1 Error systems

Fig. 3 shows the frequency response of the original system together with all reduced systems. Fig. 4 shows the frequency response of the systems obtained by taking the difference between the original and each of the reduced systems.

Comparing the errors for BT and PRBT in Fig. 5 shows that the first one is a better approximant of the original system since the error plot is almost always below the error plot for PRBT. However, we notice that the shapes of the plots are almost the same, with the second one shifted up by a few decibels.

Comparing the error systems for PRBT and PRIMA in Fig. 6 shows that, even though PRIMA gives small error for low frequencies, PRBT performs better in the middle range, where the response is harder to capture because of the large number of oscillations.

Comparing BT with PMTBR essentially means comparing diagonalization of only one gramian versus simultaneous diagonalization of both controllability and observability gramians. Fig. 7 shows that BT gives a smaller approximation error. Diagonalizing both gramians therefore leads to a better approximation than diagonalizing only one gramian. This is because after simultaneous diagonalization, truncation is performed on states that are equally difficult to reach and to observe.

From Fig. 8, it is clear that the spectral zero method performs comparably to balanced truncation. The advantage of the spectral zero method over balanced truncation is that it guarantees the passivity of the reduced system, irrespective of whether the system is in MNA-similar form. As shown in figures 9 and 10, the spectral zero method also performs similarly to the other two passivity preserving methods, one gramian and PRIMA.

Inspecting Fig. 22, we emphasize that with randomly chosen initial shifts, optimal $\mathcal{H}_2$ yields an approximation error smaller than BT. Fig. 11 shows that optimal $\mathcal{H}_2$ gives a smaller approximation error than PRIMA, except for low frequencies, as expected. Optimal $\mathcal{H}_2$ also provides a better approximant than SZM, as seen from Fig. 12.

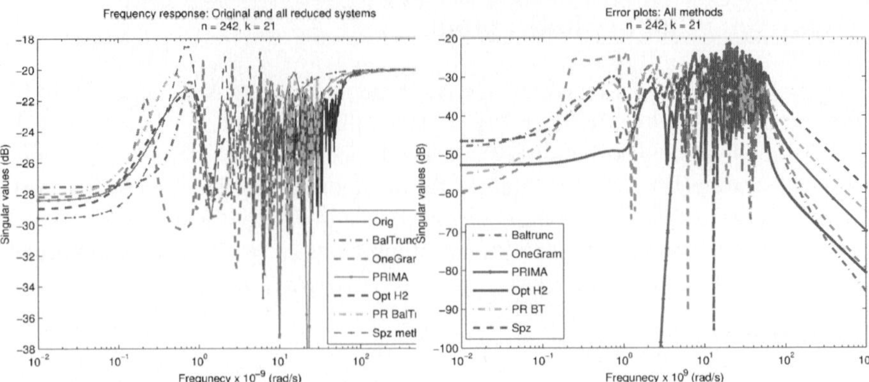

Fig. 3: Frequency response of original and all reduced systems

Fig. 4: Error for all reduced systems

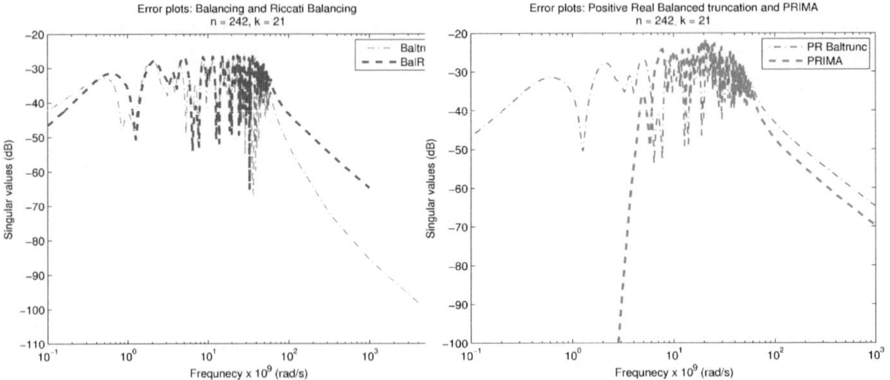

Fig. 5: Error systems: Balanced truncation and Positive Real Balanced Truncation

Fig. 6: Error systems: Positive Real Balanced Truncation and PRIMA

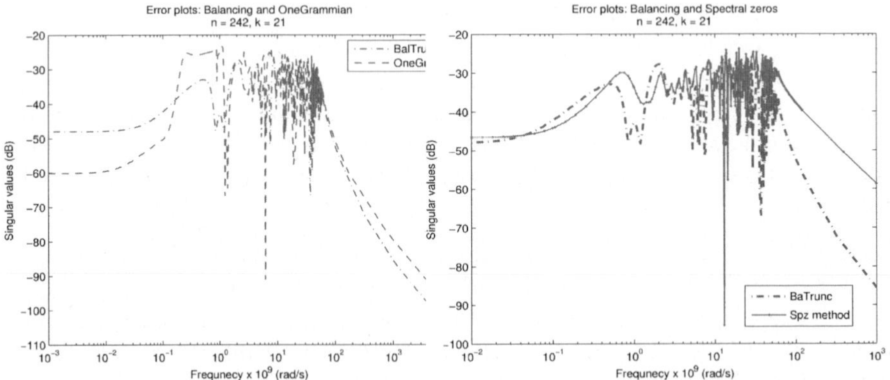

Fig. 7: Error systems: Balanced truncation and One Gramian

Fig. 8: Error systems: Balanced truncation and Projection using spectral zeros

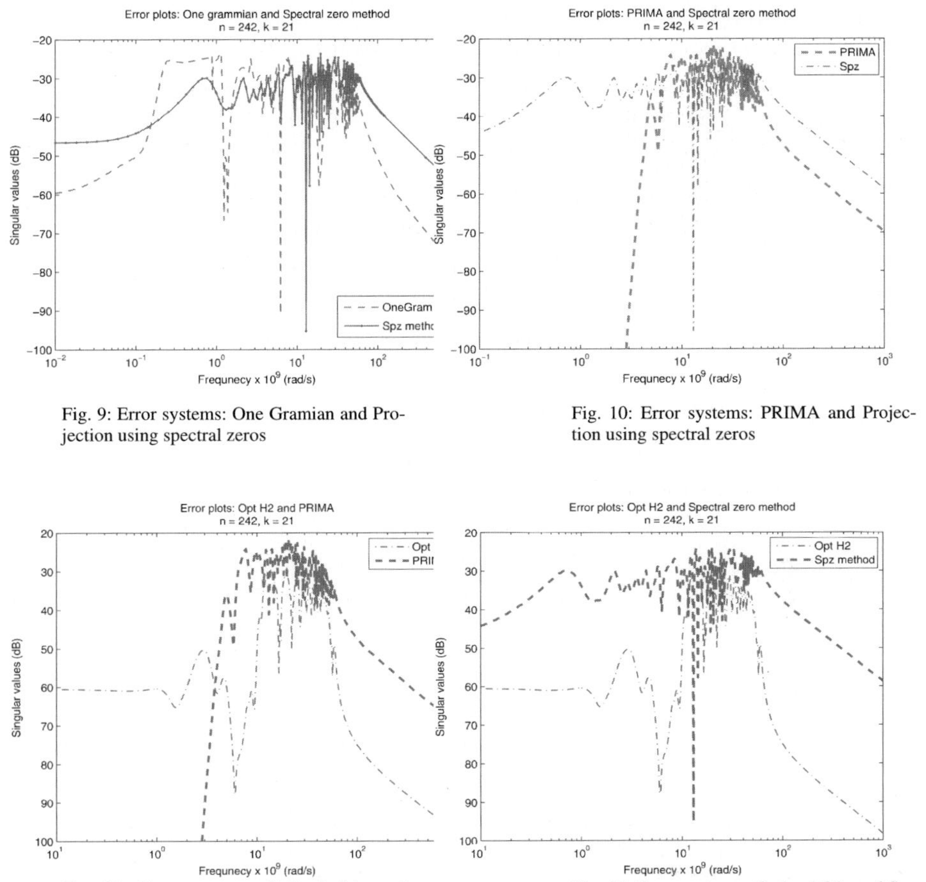

Fig. 9: Error systems: One Gramian and Projection using spectral zeros

Fig. 10: Error systems: PRIMA and Projection using spectral zeros

Fig. 11: Error systems: Optimal $\mathcal{H}_2$ and PRIMA

Fig. 12: Error systems: Optimal $\mathcal{H}_2$ and Spz method

6.2 Poles and spectral zeros of reduced systems

The following figures show the location of poles and spectral zeros of the original, stable and passive system together with the poles and spectral zeros of each reduced system. As already indicated in Tab. 2, all reduction methods yielded stable and passive reduced order systems. This is shown in Figs. 13-18, where the poles of all reduced system lie in the left half plane and the spectral zeros of all reduced systems are located away from the $j\omega$ axis.

No spectral zero matching or similarity in pole distribution occurs for reduced models obtained with BT, one gramian or PRBT, as seen from Figs. 13, 14 and 15. In particular, Fig. 13 shows that poles and spectral zeros resulting from BT are scattered, while in Fig. 14, the poles and spectral zeros from one gramian are clustered close to the $j\omega$ axis. Spectral zeros resulting from PRBT are aligned along some of the spectral zeros of the original system, as seen in Fig. 15.

However, comparing Figs. 16, 17, and 18, we identify a similarity between the distribution of spectral zeros and poles resulting from PRIMA, SZM, and optimal $\mathcal{H}_2$ respectively. The poles of these reduced systems follow a pattern, being located close to the real axis. Furthermore, the spectral zeros resulting from optimal $\mathcal{H}_2$, match some of the spectral zeros of the original system, similarly to the spectral zero method and PRIMA.

Fig. 16 shows that PRIMA preserves some of the poles as well as some of the spectral zeros of the original system. This is the only method for which some poles of the reduced system are close to the poles of the original system.

In Fig. 18, as expected, the spectral zeros of the reduced system match the chosen spectral zeros of the original system with smaller imaginary parts.

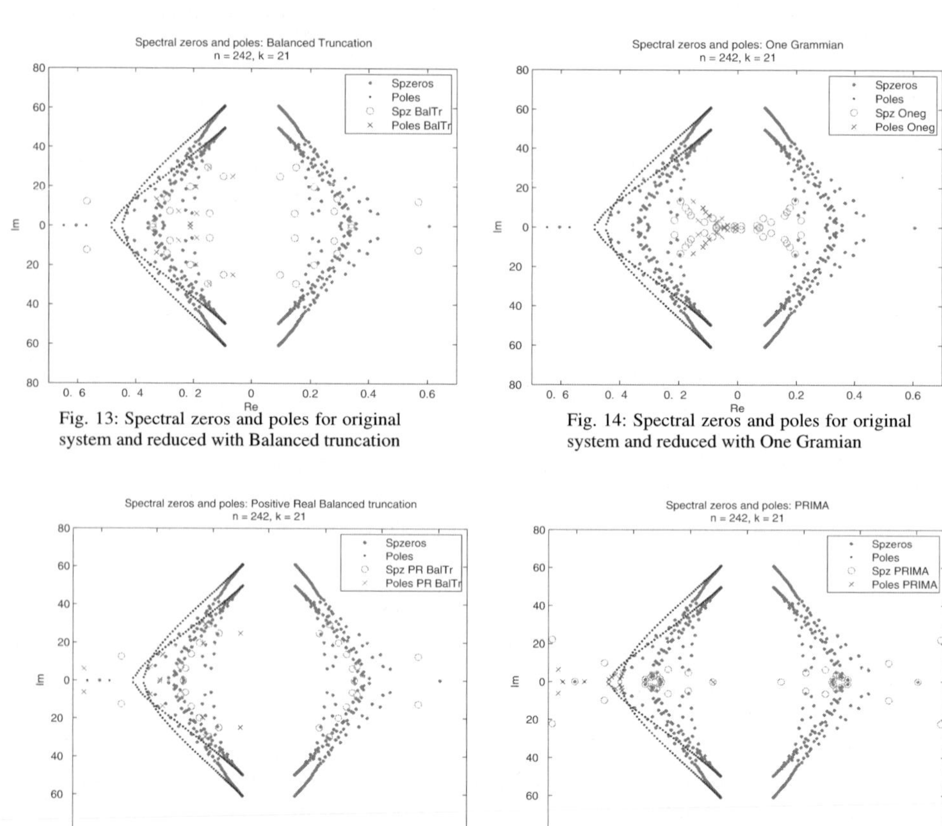

Fig. 13: Spectral zeros and poles for original system and reduced with Balanced truncation

Fig. 14: Spectral zeros and poles for original system and reduced with One Gramian

Fig. 15: Spectral zeros and poles for original system and reduced with Riccati Balanced truncation

Fig. 16: Spectral zeros and poles for original system and reduced with PRIMA

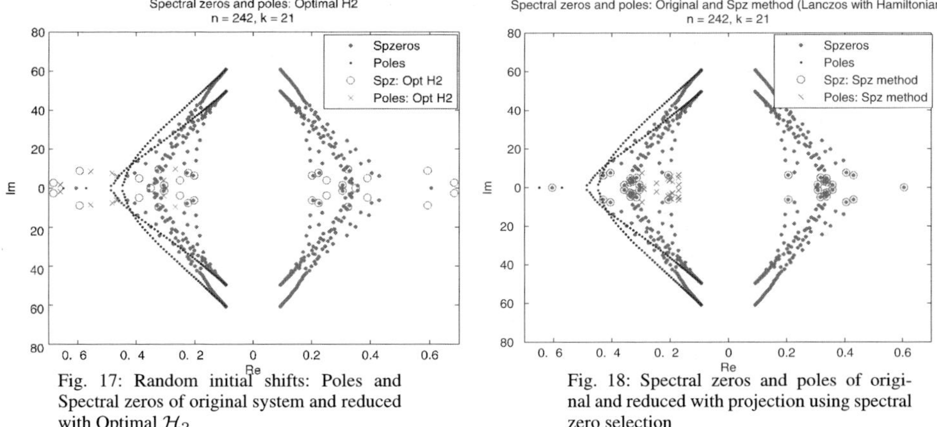

Fig. 17: Random initial shifts: Poles and Spectral zeros of original system and reduced with Optimal $\mathcal{H}_2$

Fig. 18: Spectral zeros and poles of original and reduced with projection using spectral zero selection

7 Optimal $\mathcal{H}_2$ results: errors, convergence and initial shifts

We compare reduced order models obtained with balanced truncation and optimal $\mathcal{H}_2$, since these methods yield the smallest relative $\mathcal{H}_2$ error norms. We approximate the initial order $n = 242$ system with reduced models of orders $k = 21$ and $k = 11$. The corresponding relative error norms are found in Tab. 3, Sect. 5. From these results, it is clear that optimal $\mathcal{H}_2$ is the overall best method with respect to both relative error norms: $\mathcal{H}_\infty$ and $\mathcal{H}_2$.

The selection of initial shifts can be determined to influence the convergence rate, the approximation error and the distribution of poles and spectral zeros for the reduced system. Results were obtained for two different sets of initial shifts: the poles resulting from BT and randomly generated complex shifts. The table below summarizes the number of iterations needed for IRKA to converge, with a threshold difference of 10^{-4} between successively generated shifts. Choosing random complex numbers as initial shifts yielded convergence which was almost twice as fast.

Intial Shift Choice	Red. order: $k = 21$	Red. order: $k = 11$
Poles from BT	49 steps	85 steps
Random complex	**28 steps**	**47 steps**

1. $n = 242,\ k = 21$

   Figs. 19-22 show that the reduced system obtained with optimal $\mathcal{H}_2$ using randomly generated shifts approximates the original system more accurately than when the initial shifts are the poles of the reduced system obtained from balanced truncation. The error systems in Fig. 22 show that when initial shifts are randomly chosen, optimal $\mathcal{H}_2$ yields a better approximant than BT. This is not the case when initial shifts are the poles from BT, as seen from the error systems in Fig. 21. Additionally, when the initial shifts are randomly chosen, some of the spectral zeros of the reduced system closely match spectral zeros of the original system, as shown in figure 24. This appealing behavior is not present in Figure 23, where the initial shifts are the poles resulting from BT: no spectral zeros are matched and the corresponding approximation error is larger.

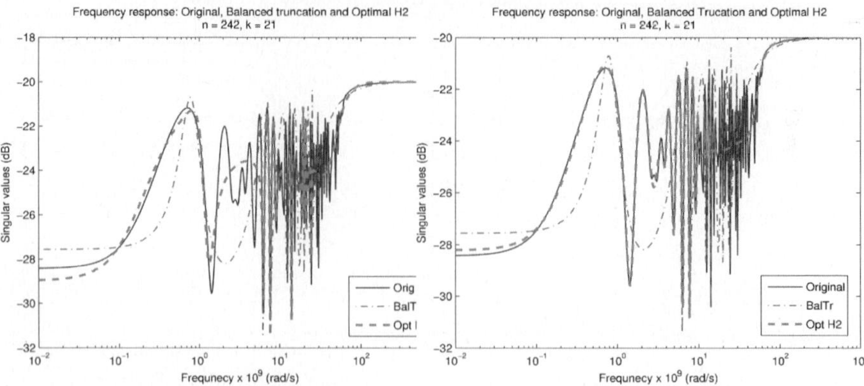

Fig. 19: BT poles as initial shifts: Frequency response of original and reduced systems

Fig. 20: Random intial shifts: Frequency response of original and reduced systems

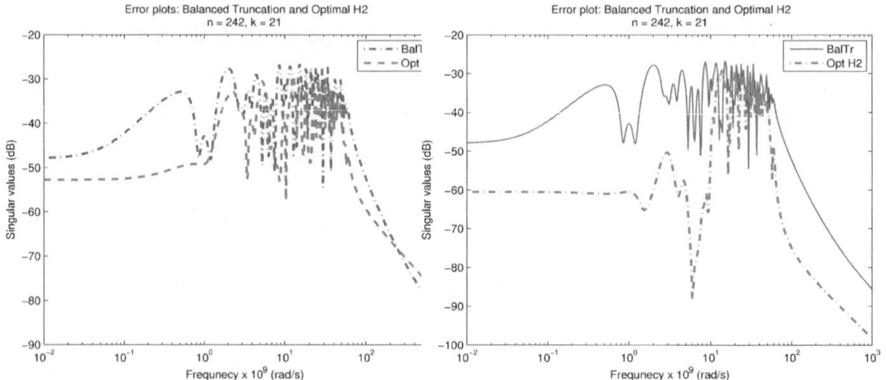

Fig. 21: BT poles as initial shifts: Error systems

Fig. 22: Random initial shifts: Error systems

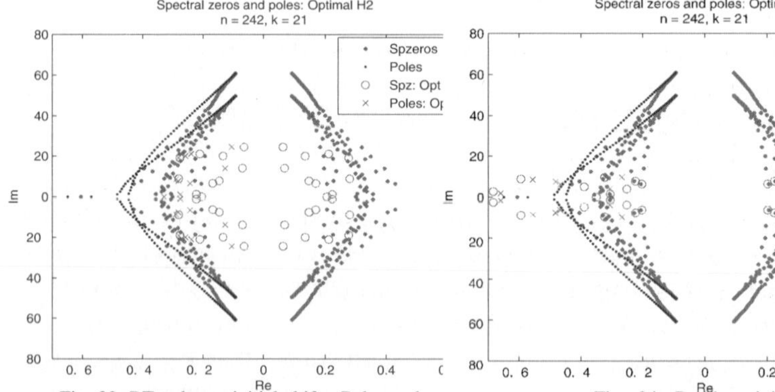

Fig. 23: BT poles as initial shifts: Poles and Spectral zeros of original system and reduced with Optimal $\mathcal{H}_2$

Fig. 24: Random initial shifts: Poles and Spectral zeros of original system and reduced with Optimal $\mathcal{H}_2$

2. $n = 1002, \ k = 71$

   Promising results for optimal $\mathcal{H}_2$ were also obtained on a system of dimension $n = 1002$ ($N = 251$). As shown in Figs. 25 and 26, the reduced model of dimension $k = 71$ clearly approximates the original more accurately than the reduced model obtained with BT. Again, optimal $\mathcal{H}_2$ is superior to BT with respect to both relative error norms, $\mathcal{H}_2$ and $\mathcal{H}_\infty$.

Reduction Method: $n = 1002, k = 71$	$\mathcal{H}_\infty$	$\mathcal{H}_2$
Balanced Truncation	0.1488	0.1124
Optimal $\mathcal{H}_2$	**0.09467**	**0.06408**

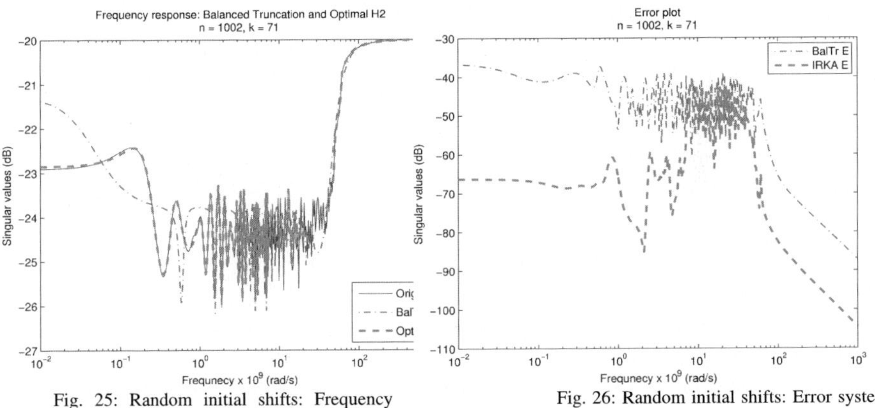

Fig. 25: Random initial shifts: Frequency response: original, reduced with balanced truncation and Optimal $\mathcal{H}_2$

Fig. 26: Random initial shifts: Error systems for balanced truncation and Optimal $\mathcal{H}_2$

8 Conclusion and further research

This paper compares several model reduction methods used in circuit simulation, in particular for systems in invertible descriptor form. The methods are grouped in two categories, gramian and Krylov based respectively. Theoretical considerations for all methods are outlined, and their performance is evaluated by reducing the dynamical system of a coupled transmission line. Approximation error and computational cost analysis for each method shows that while some methods yield better reduced systems, others are computationally cheaper. Furthermore, not all methods that yield small relative approximation errors preserve important properties of the original system, such as controllability, observability or passivity.

Optimal $\mathcal{H}_2$ is the overall best in terms of both relative $\mathcal{H}_\infty$ and $\mathcal{H}_2$ norms, but requires the highest computational complexity and cannot guarantee passivity for the reduced system. It can also be applied to the general class of descriptor systems, where $\mathbb{C}$ in (1) may be singular. Further research is needed for determining how the choice of initial shifts in optimal $\mathcal{H}_2$ influences the distribution of poles and spectral zeros of the reduced system and the convergence rate.

Among passivity preserving methods, SZM provides the best trade-off between approximation error, computational cost, and preservation of stability and passivity. Furthermore, SZM can be applied to the general class of descriptor systems. However, determining the optimal selection of spectral zeros in SZM is an open problem.

SZM overcomes the limitations of PRIMA: controllability and observability loss for the reduced system (due to possible pole-zero cancelations) and larger approximation error for high frequencies. PRIMA however provides the best fit for low frequencies from all methods considered. SZM is also computationally cheaper than PRBT. PRBT on the other hand yields an approximation error comparable to BT and has the benefit of preserving passivity.

Since our analysis is conducted on a SISO circuit with invertible descriptor form ($\mathbb{C}$ in (1) was invertible), a further step would be to reproduce these results for a system in general descriptor form, where $\mathbb{C}$ is singular, partly using the work in [5]. Applying these reduction methods on a MIMO network is currently under investigation.

References

[1] Antoulas, A.C.: *Approximation of large-scale dynamical systems*, SIAM, Philadelphia (2005)

[2] Antoulas, A.C.: *A new result on passivity preserving model reduction*, Systems and Control Letters, vol. **54**, 361-374 (2005)

[3] Benner, P., Quintana-Orti, E.S.: *Solving stable generalized Lyapunov equations with the matrix sign function*, Numerical Algorithms, vol. **20**, 75-100, (1999)

[4] Gugercin, S., Antoulas, A.C.: Beattie, C.A.: *A rational Krylov iteration for optimal H_2 model reduction*, Proceedings of the 17th Int. Symposium on Mathematical Theory of Networks and Systems, MTNS06, Kyoto, 1665-1667 (July, 2006)

[5] Mehrmann, V., Stykel, T.: *Balanced truncation model reduction for large-scale systems in descriptor form*, Dimension Reduction of Large-Scale Systems, P. Benner, V. Mehrmann and D. Sorensen, Edtrs., LNCSE vol. **45**, Springer Verlag, Heidelberg (2005)

[6] Ober, R.J.: *Balanced parametrization of classes of linear systems*, SIAM Journal of Control and Optimization, vol. **29**, 1251-1287 (1991)

[7] Odabasioglu, A., Celik M.: *PRIMA. Passive reduced-order interconnect macromodelling algorithm*, IEEE Trans. on Computer-Aided Design of Integrated Circuits and Systems vol. **17**, 8, 645-654 (1998)

[8] Palenius, T.: *Efficient Time-Domain Simulation of Interconnects Characterized by Large RLC Circuits or Tabulated S Parameters*, Licentiate Thesis (2004)

[9] Phillips, J.R.; Daniel, L.; Silveira, L.M.: *Guaranteed passive balancing transformations for model order reduction*, IEEE Trans. on Computer-Aided Design of Integrated Circuits and Systems vol. **22**, 8, 1027 - 1041 (2003)

[10] Phillips, J.R., Silveira, L. M.: *Poor man's TBR: a simple model reduction scheme*, IEEE Trans. on Computer-Aided Design of Integrated Circuits and Systems vol. **24**, 1, 43-55 (2005)

[11] Sorensen, D.C.: *Passivity preserving model reduction via interpolation of spectral zeros*, Systems and Control Letters, vol. **54**, 347-360 (2005)

Transient Field-Circuit Coupled Models with Switching Elements for the Simulation of Electric Energy Transducers*

Herbert De Gersem, Galina Benderskaya, and Thomas Weiland

Technische Universität Darmstadt, Institut für Theorie Elektromagnetischer Felder, Schloßgartenstraße 8, D-64289 Darmstadt, Germany
DeGersem@temf.tu-darmstadt.de

This paper deals with the transient simulation of large, nonlinear magnetoquasistatic field models which are monolithically coupled to electric circuits. Solid- and stranded-conductor models embedded in the field model are connected to the external circuit. In order to guarantee the numerical efficiency of the field-circuit coupled formulation, conductor models coupling the circuit to the field at a reference cross-section, have to be preferred over conductor models that couple the whole conductor volume to the circuit. The circuit is formulated in terms of both voltage drops and currents in order to avoid fill-in in the field matrix parts. For time stepping, an error-controlled, adaptive singly diagonally Runge-Kutta method is applied. A dense output solution is used to detect and localise switching events in the circuit. The actual time step is restricted to the time instant of switching at which consistent initial conditions are determined before restarting the time integration. The transient field-circuit coupling is applied to the models of a capacitor motor and a three-phase transformer.

1 Introduction

Contemporary designs of electrical-energy convertors force the machine to operate at higher flux densities and higher frequencies, leading to higher levels of ferromagnetic saturation and eddy currents, respectively. Two- and three-dimensional field simulation are indispensable to resolve these kinds of local effects. However, the power-electronic components connected to the device and, in the two-dimensional case, the interconnections between the different conductors, are commonly excluded from the field model. To uniquely define the field model, the voltage drops along the massive conductors and the currents through the coils have to be known a-priori, which is impractical. The interaction of the device with external excitation and load circuits can be very complicated such that engineers are obliged to iterate between the field and the circuit model, an approach which is called *simulator coupling*. Simulator coupling especially performs well when the time constants considered by both simulators are different by orders of magnitude. For situations where this is not the

* Invited Paper at SCEE-2006

case, a *monolithic coupling*, i.e., combining both the field and the circuit model into a single system of equations, is recommended. Monolithic coupling is especially valuable for coupling electromagnetic field and circuit simulation because the coupling itself is linear and can therefore be adequately represented at the algebraic level. Monolithic coupling requires, however, all stages of the coupling process to be designed carefully. A bad coupling approach and implementation leads to systems of equations that cause difficulties at the algebraic level, which may cause the performance of the monolithic coupling to be degenerated to the one of a simulator coupling.

The paper exemplarily describes the coupling of an electric circuit to a magnetoquasistatic field model. Couplings in other physical disciplines can be developed similarly. The field model is discretised by the finite-element method or the finite-integration technique. The coupling is designed from the field point-of-view, adding a few circuit equation to a large system of field equations, without too much influencing typical field simulation techniques to loose their performance. Hence, the approach is complementary to coupling procedures where field-simulation actions are embedded in an established circuit simulator.

2 Discrete Magnetoquasistatic Formulation

The magnetic flux density $\mathbf{B}$ is forced to be divergence-free by stating $\mathbf{B} = \nabla \times \mathbf{A}$ with $\mathbf{A}$ the magnetic vector potential. The integration of the Faraday-Lenz law yields the electric field strength $\mathbf{E} = -\frac{\partial}{\partial t}\mathbf{A} - \nabla\phi$ with the gradient of the electric scalar potential ϕ as an integration constant. The material properties are expressed in their easiest form. $\mathbf{B}$ is related to the magnetic field strength $\mathbf{H}$ by the reluctivity ν, i.e., $\mathbf{H} = \nu\mathbf{B}$ where ν may depend on $\mathbf{B}$. The current density $\mathbf{J}$ is related to $\mathbf{E}$ by the conductivity σ, i.e., $\mathbf{J} = \sigma\mathbf{E}$. The combination of the material laws and the potentials within Ampères law directly leads to the magnetoquasistatic formulation

$$\nabla \times (\nu\nabla \times \mathbf{A}) + \sigma\frac{\partial \mathbf{A}}{\partial t} = -\sigma\nabla\phi\,. \tag{1}$$

The righthandside is called the source current density $\mathbf{J}_{\mathrm{s}} = -\sigma\nabla\phi$.

In the case of the finite-integration technique (FIT), (1) is transferred to a staggered grid pair $(G, \tilde{G})$ [35, 36]. Here, only the special case of a structured, orthogonal grid pair is considered. The degrees of freedom are the magnetic vector potentials integrated along the edges L_i of the primary grid G, collected into the algebraic vector $\widehat{\mathbf{a}}$, i.e.,

$$\widehat{\mathbf{a}}_i = \int_{L_i} \mathbf{A} \cdot \mathrm{ds}\,. \tag{2}$$

The application of the primary curl operator $\mathbf{C}$ gives $\widehat{\widehat{\mathbf{b}}} = \mathbf{C}\widehat{\mathbf{a}}$ where the components of $\widehat{\widehat{\mathbf{b}}}$ are the magnetic fluxes through the primary facets S_p. The magnetic material law is expressed at the crossing points between primary facets and dual edges. The magnetic voltage $\widehat{\mathbf{h}}_p$ along a dual edge $\tilde{L}_p$ reads

$$\widehat{\mathbf{h}}_p = \int_{\tilde{L}_p} \mathbf{H} \cdot \mathrm{ds} \approx \mathbf{M}_{\nu,p,p}\widehat{\widehat{\mathbf{b}}}_p = \frac{\nu|\tilde{L}_p|}{|S_p|} \tag{3}$$

where the entries $\mathbf{M}_{\nu,p,p}$ are gathered in the diagonal *reluctivity matrix* $\mathbf{M}_\nu$. Similarly, the current $\widehat{\widehat{\mathbf{j}}}_q$ through a dual facet $\tilde{S}_q$ is related to the electric voltage $\widehat{\mathbf{e}}_q$ allocated at the associated primary edge by

$$\widehat{\widehat{\mathbf{j}}}_q = \int_{\tilde{S}_q} \mathbf{J} \cdot \mathrm{d}\mathbf{S} \approx \mathbf{M}_{\sigma,q,q} \widehat{\mathbf{e}}_q = \frac{\sigma |\tilde{S}_q|}{|L_q|} \tag{4}$$

with the *conductivity matrix* $\mathbf{M}_\sigma$. The discrete equivalent of Ampères law reads $\widehat{\widehat{\mathbf{j}}} = \widetilde{\mathbf{C}}\widehat{\mathbf{h}}$ where $\widetilde{\mathbf{C}}$ is the discrete curl operator at the dual grid. The operators $\mathbf{C}$ and $\widetilde{\mathbf{C}}$ do not incorporate any discretisation . The discretisation error is solely attributed to the material matrices $\mathbf{M}_\nu$ and $\mathbf{M}_\sigma$. The discrete counterpart of (1) reads

$$\widetilde{\mathbf{C}}\mathbf{M}_\nu \mathbf{C}\widehat{\mathbf{a}} + \mathbf{M}_\sigma \frac{d\widehat{\mathbf{a}}}{dt} = \widehat{\widehat{\mathbf{j}}}_{\mathrm{s}} \tag{5}$$

with $\widehat{\widehat{\mathbf{j}}}_{\mathrm{s}}$ the vector of the discrete source currents [7].
In the case of the finite-element (FE) method, the magnetic vector potential is expressed as a linear combination of n_{fe} edge elements $\mathbf{w}_j$. The FE formulation follows from weighting (1) by the test functions $\mathbf{w}_i$ and integrating by parts. The introduction of the discrete curl operators $\mathbf{C}$ and $\widetilde{\mathbf{C}}$ to the FE grid leads to the same formulation as (5) but with slightly different material matrices and source currents, here indicated by a superscript $\cdot^{(\mathrm{fe})}$:

$$\mathbf{M}^{(\mathrm{fe})}_{\nu,p,q} = \int_\Omega \nu \mathbf{z}_p \cdot \mathbf{z}_q \,\mathrm{d}\Omega \; ; \tag{6}$$

$$\mathbf{M}^{(\mathrm{fe})}_{\sigma,i,j} = \int_\Omega \sigma \mathbf{w}_i \cdot \mathbf{w}_j \,\mathrm{d}\Omega \; ; \tag{7}$$

$$\widehat{\widehat{\mathbf{j}}}^{(\mathrm{fe})}_{\mathrm{s,i}} = \int_\Omega (-\sigma \nabla \phi) \cdot \mathbf{w}_i \,\mathrm{d}\Omega \tag{8}$$

where Ω denotes the computational domain and $\mathbf{z}_p$ is the facet element associated with the primary grid facet S_p. In the following, a distinction between FIT and FE formulations is only made when absolutely necessary.

3 Conductor Models

3.1 Solid-conductor model

A massive conductor which covers the volume $\Omega_{\mathrm{sol},q}$ is excited by a voltage drop $u_{\mathrm{sol},q}$ between two electrodes (Fig. 1a). From the application of the Faraday-Lenz law along a closed contour passing along the massive conductor and through the voltage source, one finds that

$$u_{\mathrm{sol},q} = -\int_{\ell_{\mathrm{sol},q}} \nabla \phi \cdot \mathrm{d}\mathbf{s} \tag{9}$$

with $\ell_{\mathrm{sol},q}$ an arbitrary path between both electrodes. The potentials $\mathbf{A}$ and ϕ are, however, not unique, i.e., when $(\mathbf{A}, \phi)$ solves (1), so does $\left(\mathbf{A} + \psi, \phi + \frac{\partial}{\partial t}\psi\right)$ where

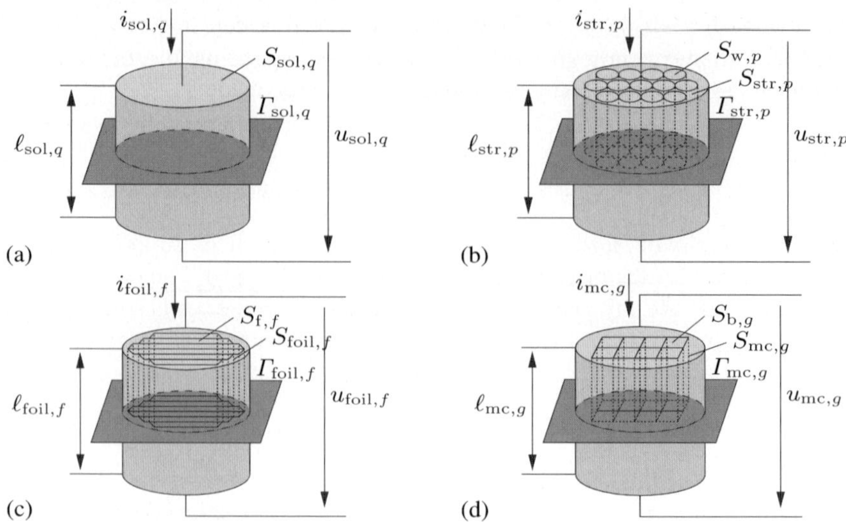

Fig. 1: (a) Solid-conductor model, (b) stranded-conductor model, (c) foil-conductor model and (d) multi-conductor model.

ψ is an arbitrary scalar field. As a consequence, the division of $\mathbf{J}$ into a *source-current density* $\mathbf{J}_{\mathrm{s}} = -\sigma\nabla\phi$ and an *eddy-current density* $\mathbf{J}_{\mathrm{e}} = -\sigma\frac{\partial}{\partial t}\mathbf{A}$ is not unique as well. Commonly, the formulation is closed by forcing the source-current density and the eddy-current density to be divergence-free. Then, the source-current density $\mathbf{J}_{\mathrm{s}} = -\sigma\nabla\phi_{\mathrm{s}}$ is obtained by solving the stationary-current problem $-\nabla\cdot(\sigma\nabla\phi_{\mathrm{s}}) = 0$ with the boundary conditions $\phi_{\mathrm{s}} = \phi_1$ and $\phi_{\mathrm{s}} = \phi_2$ at the electrodes such that $u_{\mathrm{sol},q} = \phi_2 - \phi_1$. The discrete equivalent reads

$$\widetilde{\mathbf{S}}\mathbf{M}_{\sigma}\widetilde{\mathbf{S}}^{T}\Phi_{\mathrm{s}} = 0 \tag{10}$$

where $\widetilde{\mathbf{S}}$ is the discrete divergence operator at the dual grid, $-\widetilde{\mathbf{S}}^{T}$ equals the discrete gradient operator at the primary grid and Φ_{s} is the vector of electrical scalar potentials allocated at the primary nodes. A particular field-circuit coupling scheme consists of solving (10) where a unit voltage drop between the electrodes is applied as a boundary condition. The resulting discrete source-current distribution is $\mathbf{M}_{\sigma}\mathbf{Q}_{\mathrm{sol},q} = \mathbf{M}_{\sigma}\widetilde{\mathbf{S}}^{T}\Phi_{\mathrm{s}}$ and defines a coupling operator $\mathbf{Q}_{\mathrm{sol},q}$ which allows to express the discrete source current generated by an arbitrary voltage drop $u_{\mathrm{sol},q}$ across the massive conductor q by $\widehat{\widehat{\mathbf{j}}}_{\mathrm{s}} = \mathbf{M}_{\sigma}\mathbf{Q}_{\mathrm{sol},q}u_{\mathrm{sol},q}$. The column vector $\mathbf{Q}_{\mathrm{sol},q}$ contain nonzero contributions for all primary edges in the massive-conductor volume. Hence, it represents a 3D-to-0D coupling between the field and the circuit. The number of nonzeros scales as $\mathcal{O}\left(n_{\mathrm{1D}}^{3}\right)$ where n_{1D} stands for the number of degrees of freedom in one spatial direction. The number of nonzeros in $\widetilde{\mathbf{C}}\mathbf{M}_{\nu}\mathbf{C}$ and $\mathbf{M}_{\sigma}$ scales by $\mathcal{O}\left(n_{\mathrm{1D}}^{3}\right)$ as well, such that the computation time for the application of $\mathbf{Q}_{\mathrm{sol},q}$ is expected to have the same complexity as the one for the application of the field model.

A second coupling strategy exploits the non-uniqueness of $\mathbf{A}$ and ϕ and even may consider potentials that are not continuous in parts of Ω [14]. The voltage drop is introduced as a step potential difference at an arbitrary reference cross-section $\Gamma_{\mathrm{sol},q}$

which cuts $\Omega_{\mathrm{sol},q}$ in two parts without touching the electrodes. In the discrete setting, the voltage drop is assigned to the primary edges corresponding to a set of dual facets covering $\Gamma_{\mathrm{sol},q}$. The discrete source-current vector then reads $\widehat{\widehat{\mathbf{j}}}_{\mathrm{s}} = \mathbf{M}_\sigma \tilde{\mathbf{Q}}_{\mathrm{sol},q} u_{\mathrm{sol},q}$ where $\tilde{\mathbf{Q}}_{\mathrm{sol},q}$ contains 0, 1 and -1 indicating the contribution and orientation of primary edges with respect to the reference cross-section $\Gamma_{\mathrm{sol},q}$. In contrast to $\mathbf{Q}_{\mathrm{sol},q}$, the coupling operator $\tilde{\mathbf{Q}}_{\mathrm{sol},q}$ represents a 2D-to-0D field-circuit coupling which has a complexity that only scales by $\mathcal{O}\left(n_{\mathrm{1D}}^2\right)$ and which consequently guarantees the efficiency of the field-circuit coupling scheme.
The total current $i_{\mathrm{sol},q}$ through the massive conductor q is integrated at $\Gamma_{\mathrm{sol},q}$ which, in the discrete setting, boils down to a summation of the currents through the dual facets covering $\Gamma_{\mathrm{sol},q}$:

$$i_{\mathrm{sol},q} = \tilde{G}_{\mathrm{sol},q} u_{\mathrm{sol},q} - \tilde{\mathbf{Q}}_{\mathrm{sol},q} \frac{d\widehat{\mathbf{a}}}{dt} \tag{11}$$

where $\tilde{G}_{\mathrm{sol},q} = \tilde{\mathbf{Q}}_{\mathrm{sol},q}^T \mathbf{M}_\sigma \tilde{\mathbf{Q}}_{\mathrm{sol},q}$ reflects the discrete conductance of the reference cross-section. A field-circuit coupling involving only the massive conductor q excited by a current source then reads

$$\begin{bmatrix} \widetilde{\mathbf{C}}\mathbf{M}_\nu \mathbf{C} & -\mathbf{M}_\sigma \tilde{\mathbf{Q}}_{\mathrm{sol},q} \\ \mathbf{0} & \tilde{G}_{\mathrm{sol},q} \end{bmatrix} \begin{bmatrix} \widehat{\mathbf{a}} \\ u_{\mathrm{sol},q} \end{bmatrix} + \begin{bmatrix} \mathbf{M}_\sigma & \mathbf{0} \\ -\tilde{\mathbf{Q}}_{\mathrm{sol},q}^T \mathbf{M}_\sigma & \mathbf{0} \end{bmatrix} \frac{d}{dt} \begin{bmatrix} \widehat{\mathbf{a}} \\ u_{\mathrm{sol},q} \end{bmatrix} = \begin{bmatrix} \mathbf{0} \\ i_{\mathrm{sol},q} \end{bmatrix} \tag{12}$$

which after time discretisation and appropriate scaling of the single circuit equation yields a symmetric, semi-positive-definite system of equations. A symmetric coupling can also be achieved using $\mathbf{Q}_{\mathrm{sol},q}$ instead of $\tilde{\mathbf{Q}}_{\mathrm{sol},q}$ and $G_{\mathrm{sol},q} = \mathbf{Q}_{\mathrm{sol},q}^T \mathbf{M}_\sigma \mathbf{Q}_{\mathrm{sol},q}$ instead of $\tilde{G}_{\mathrm{sol},q}$. In that case, $G_{\mathrm{sol},q}$ is the DC conductance of the massive conductor and the summation by $\mathbf{Q}_{\mathrm{sol},q}^T$ computes the current by averaging the currents evaluated at all possible discrete cross-sections of the massive conductor with the dual grid.
The 3D-to-0D coupling operator $\mathbf{Q}_{\mathrm{sol},q}$ is disadvantageous for reasons of numerical efficiency. In Table 1, the performance of the Conjugate-Orthogonal Conjugate-Gradient (COCG) solver [34], preconditioned by the Symmetric Successive Over-relaxation (SSOR) algorithm is compared for a coupling with $\mathbf{Q}_{\mathrm{sol},q}$ and a coupling with $\tilde{\mathbf{Q}}_{\mathrm{sol},q}$. For both test models, the number of iterations is smaller for the 3D-to-0D coupling than for the 2D-to-0D coupling, indicating the better condition of the system matrix resulting from the tighter 3D-to-0D coupling. This advantage causes the computation time for the small single-phase transformer model to be in favour of the 3D-to-0D coupling. For the larger three-phase transformer model, however, the matrix-vector multiplications by a denser system matrix adversely influences the overall computation time.
The coupling by $\mathbf{Q}_{\mathrm{sol},q}$ splits the current $\widehat{\widehat{\mathbf{j}}}$ in a source-current part $\widehat{\widehat{\mathbf{j}}}_{\mathrm{s}} = \mathbf{M}_\sigma \mathbf{Q}_{\mathrm{sol},q} u_{\mathrm{sol},q}$ and an eddy-current part $\widehat{\widehat{\mathbf{j}}}_{\mathrm{e}} = -\mathbf{M}_\sigma \frac{d}{dt}\widehat{\mathbf{a}}$ that are both free of divergence. The sparser coupling by $\tilde{\mathbf{Q}}_{\mathrm{sol},q}$ is related to a division of the divergence-free current $\widehat{\widehat{\mathbf{j}}}$ into two non-divergence-free parts $\widehat{\widehat{\mathbf{j}}}_{\mathrm{s}} = \mathbf{M}_\sigma \tilde{\mathbf{Q}}_{\mathrm{sol},q} u_{\mathrm{sol},q}$ and $\widehat{\widehat{\mathbf{j}}}_{\mathrm{e}} = -\mathbf{M}_\sigma \frac{d}{dt}\widehat{\mathbf{a}}$ for which a physical interpretation is cumbersome. Special care has to be taken for the algebraic solution of (12). The coupling operator $\tilde{\mathbf{Q}}_{\mathrm{sol},q}$ and hence also the discrete magnetic vector potential $\widehat{\mathbf{a}}$ do not mimic continuous fields. Hence, the system solver may experience increasingly worse condition numbers as the discretisation

Table 1: Iteration counts and solution times for SSOR-COCG applied to a field-circuit coupling with the coupling matrices $\mathbf{Q}_{\mathrm{sol},q}$ and $\mathbf{P}_{\mathrm{str},p}$ or with the coupling matrices $\tilde{\mathbf{Q}}_{\mathrm{sol},q}$ and $\tilde{\mathbf{P}}_{\mathrm{str},p}$.

	coupling matrices	number of iterations	solution time (s)
single-phase	$\tilde{\mathbf{Q}}_{\mathrm{sol},q}$ and $\tilde{\mathbf{P}}_{\mathrm{str},p}$	198	15
transformer	$\mathbf{Q}_{\mathrm{sol},q}$ and $\mathbf{P}_{\mathrm{str},p}$	127	12
three-phase	$\tilde{\mathbf{Q}}_{\mathrm{sol},q}$ and $\tilde{\mathbf{P}}_{\mathrm{str},p}$	756	145
transformer	$\mathbf{Q}_{\mathrm{sol},q}$ and $\mathbf{P}_{\mathrm{str},p}$	465	176

is refined. A stable system solver is mandatory to ensure that the solution for $\widehat{\mathrm{a}}$ alleviates the discontinuity of $\tilde{\mathbf{Q}}_{\mathrm{sol},q}$ such that the non-divergence-free source- and eddy-current densities combine to a physically sound, divergence-free discrete current distribution.

The field-circuit coupling approach can be understood as an *agglomeration* of *local* field quantities into *global* circuit quantities [14]. In the FIT and in other discretisation techniques closely related to differential geometry, this agglomeration is representable by a simple incidence relation [25, 14, 19].

3.2 Stranded-conductor model

When the wire diameter of a coil is significantly smaller than the expected skin depth, is not necessary to resolve each individual wire by the computational grid. Instead, the stranded-conductor model includes the assumption that the current is homogeneously distributed along the cross-section of the coil. The conventional treatment of coils in a 3D field model is to compute the discrete current distribution due to a unit current applied to coil p by a geometric algorithm, yielding the vector field $\mathbf{J}_{\mathrm{unit},p}$. In the FIT case, this continuous current is integrated over the dual facets, whereas in the FE case, edge elements are applied for weighting:

$$\mathbf{P}^{(\mathrm{fit})}_{\mathrm{str},p,i} = \int_{\tilde{S}_i} \mathbf{J}_{\mathrm{unit},p} \cdot \mathrm{d}S \; ; \tag{13}$$

$$\mathbf{P}^{(\mathrm{fe})}_{\mathrm{str},p,i} = \int_{\Omega_{\mathrm{str},p}} \mathbf{J}_{\mathrm{unit},p} \cdot \mathbf{w}_i \mathrm{d}\Omega \; . \tag{14}$$

In both cases, the applied current density reads $\widehat{\widehat{\mathbf{j}}} = \mathbf{P}_{\mathrm{str},p} i_{\mathrm{str},p}$. The coupling operator $\mathbf{P}_{\mathrm{str},p}$ connects all dual facets inside the coil volume $\Omega_{\mathrm{str},p}$ to the circuit and hence, also has the nature of a 3D-to-0D coupling, possibly causing a degeneration of the performance of the coupled simulation. Eddy currents are prohibited by omitting the eddy-current term in the field formulation. The voltage drop along the coil is

$$u_{\mathrm{str},p} = R_{\mathrm{str},p} i_{\mathrm{str},p} + \mathbf{P}^T_{\mathrm{str},p} \frac{d\widehat{\mathrm{a}}}{dt} \tag{15}$$

where $R_{\mathrm{str},p}$ is the DC resistance of the coil and $\mathbf{P}_{\mathrm{str},p}$ averages the voltage drop of all filamentary wires in the coil. The field-circuit coupling of a single coil p excited by a voltage source reads

$$\begin{bmatrix} \tilde{\mathbf{C}}\mathbf{M}_\nu \mathbf{C} & -\mathbf{P}_{\mathrm{str},p} \\ \mathbf{0} & \tilde{R}_{\mathrm{str},p} \end{bmatrix} \begin{bmatrix} \widehat{\mathrm{a}} \\ i_{\mathrm{str},p} \end{bmatrix} + \begin{bmatrix} \mathbf{0} & \mathbf{0} \\ \mathbf{P}^T_{\mathrm{str},p} & \mathbf{0} \end{bmatrix} \frac{d}{dt} \begin{bmatrix} \widehat{\mathrm{a}} \\ i_{\mathrm{str},p} \end{bmatrix} = \begin{bmatrix} \mathbf{0} \\ u_{\mathrm{str},p} \end{bmatrix} . \tag{16}$$

Also for coils, a 2D-to-0D coupling scheme can be developed [17, 14]. The homogeneous current distribution is only applied to the dual facets covering a reference cross-section $\Gamma_{\mathrm{str},p}$, i.e., $\widehat{\widehat{\mathbf{j}}} = \tilde{\mathbf{P}}_{\mathrm{str},p} i_{\mathrm{str},p}$ where the dimension-less coupling operator $\tilde{\mathbf{P}}_{\mathrm{str},p}$ contains the relative orientations of the participating dual facets with respect to $\Gamma_{\mathrm{str},p}$. The current distribution is forced to remain homogeneous throughout the entire coil by an anisotropic conductivity matrix added to the magnetoquasistatic field problem. In the FE case, the conductivity matrix reads

$$\mathbf{M}^{\mathrm{fe}}_{\sigma,\mathrm{coil},i,j} = \int_{\Omega_{\mathrm{str},p}} \sigma \left(\mathbf{w}_i \cdot \mathbf{t}_{\mathrm{str},p}\right) \left(\mathbf{w}_j \cdot \mathbf{t}_{\mathrm{str},p}\right) \mathrm{d}\Omega \tag{17}$$

where $\mathbf{t}_{\mathrm{str},p}$ denotes the direction of the wires of coil p. The summation by $\tilde{\mathbf{P}}_{\mathrm{str},p}$ corresponds to an integration of the electric field along a reference layer. $\tilde{R}_{\mathrm{str},p} = \tilde{\mathbf{P}}^T_{\mathrm{str},p} \mathbf{M}^{\dagger}_{\sigma,\mathrm{coil}} \tilde{\mathbf{P}}_{\mathrm{str},p}$ where $\dagger$ denotes a pseudo-inverse carried out for the nonzero parts of $\mathbf{M}_{\sigma,\mathrm{coil}}$ only, represent the resistance of the reference layer. The 2D-to-0D coupling of a stranded-conductor model is found by replacing $\mathbf{P}_{\mathrm{str},p}$ by $\tilde{\mathbf{P}}_{\mathrm{str},p}$ and $R_{\mathrm{str},p}$ by $\tilde{R}_{\mathrm{str},p}$ in (16). The same remarks concerning the algebraic solver apply as for the solid-conductor case.

3.3 Specialised conductor models

Massive conductors and wire coils are adequately modelled by solid- and stranded-conductor models respectively. In engineering practice, however, more complicated coils and winding schemes exist. Particular distribution transformers and inductors contain foil windings, which are constructed by rolling up sheets of conductive material. The current through the sheet cross-section remains constant. However, a significant redistribution of the current towards the tips of the sheet occurs. In particular devices, the eddy-current effects can also not be neglected in the individual wires of the windings. Especially when then number of turns becomes very large, it is not recommended to resolve the individual sheets or wires by the FE or FIT mesh, even if significant eddy-current effects are expected [9]. The discretisation for the magnetic vector potential should resolve the skin depth but should not necessarily adapt to the size of individual wires. The choice for a particular conductor model is motivated by the ratio of the conductor sizes d_x and d_y and the expected skin depths δ_x and δ_y (Fig. 2). The magnetic flux penetrates a stranded-conductor model because no eddy currents occur (Fig. 3a). For a solid-conductor model, the magnetic flux is expelled in both directions because of eddy-current effects (Fig. 3c), whereas in the foil-conductor case, the magnetic flux is only expelled in the direction towards the tips of the sheets (Fig. 3b).
For foil windings, dedicated foil-conductor models applicable within 2D and 3D FE models have been proposed in [9] and [16]. For windings with a rectangular wire cross-section, a multi-conductor model has been proposed in [10]. These methods avoid the explicit consideration of the separate turns by assuming a smooth variation of the turn voltages over the reference cross-section of the winding. This voltage drop is discretised at an additional mesh defined at the reference cross-section. A weak formulation is applied to force the currents through the turns to be the same at the control volumes corresponding to the turn-voltage discretisation. Such coil models

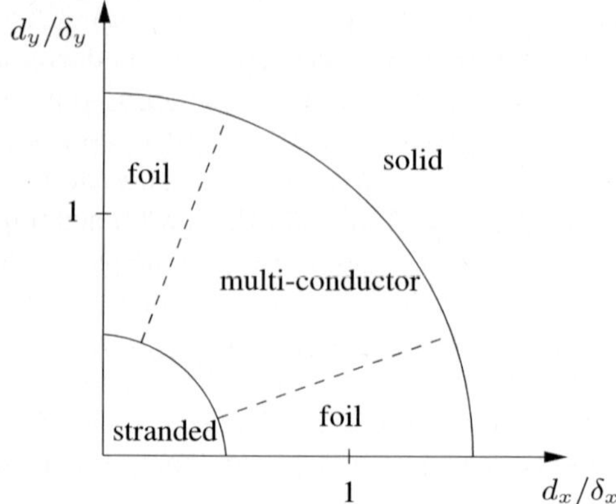

Fig. 2: Application range of the solid-, stranded-, foil- and multi-conductor models.

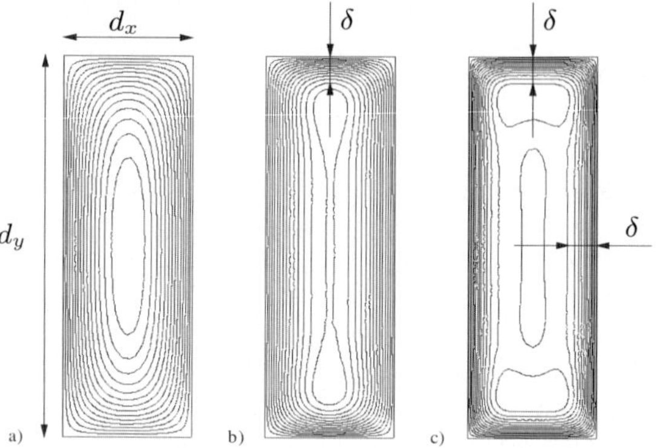

Fig. 3: Magnetic flux lines within (a) a stranded, (b) a foil and (c) a solid conductor of equal size and with the same number of Ampère-turns.

are especially efficient for windings where the spatial scale of the turn-voltage variation and the spatial scale of the eddy-current redistribution is substantially larger than the smallest dimension of the individual wires or sheets. More details about the mathematical formulation of the foil-conductor model can be found in [9] and [16]. The formulation of the wire-conductor case is developed in [10].

4 Field-Circuit Coupling

The relations between the currents and the voltage drops of solid- and stranded-conductor models connected within the circuit are expressed by (11) and (15) respectively. These relation can be interpreted as controlled current and voltage sources [13, 11]. The inversion of the expression is cumbersome because of the presence of the time derivative and especially because of the coupling to the field model. No additional circuit equations are needed if the voltage drops along the massive conductors and the currents through the coils are known on beforehand [23, 33]. If this is not the case, the voltage drops along the massive conductors and the currents through the stranded conductors should appear as degrees of freedom in the circuit system. A systematic description of the circuit problem accounting for this consists of a division of the circuit into a tree and a co-tree while forcing the solid conductors (together with the voltage sources and the capacitors) and the stranded conductors (together with the current sources and the inductors) to be part of the tree and the co-tree respectively. Here, we assume that such a decomposition is possible. When this is not the case, appropriate mitigation techniques are discussed in [13] and [11]. The fundamental cutset matrix $\mathbf{D}$ and the fundamental loop matrix $\mathbf{B}$ are partitioned with respect to the solid-conductor, capacitor and resistor tree branches by the subscript $\cdot_{\mathrm{two}}$, with respect to the stranded-conductor, inductor and resistor links by the subscript $\cdot_{\mathrm{lno}}$ and with respect to the independent voltage and current sources by the subscripts $\cdot_{\mathrm{twu}}$ and $\cdot_{\mathrm{lni}}$ respectively. The coupling operators are brought together into $\mathbf{Q}_{\mathrm{sol}}$ and $\mathbf{P}_{\mathrm{str}}$, possibly adding zero columns to account for non-coupled circuit branches. The conductances and resistances of the circuit resistors and the coupled solid- and stranded-conductor models are collected in the diagonal matrices $\mathbf{G}_{\mathrm{two}}$ and $\mathbf{R}_{\mathrm{lno}}$. Similarly, the capacitances and inductances of the circuit branches are gathered in $\mathbf{C}_{\mathrm{two}}$ and $\mathbf{L}_{\mathrm{lno}}$ respectively. The voltages and currents of the independent sources are denoted by u_{twu} and i_{lni} respectively. Then, the field-circuit coupling reads

$$\begin{bmatrix} \widetilde{\mathbf{C}}\mathbf{M}_\nu\mathbf{C} & -\mathbf{M}_\sigma\mathbf{Q}_{\mathrm{sol}} & -\mathbf{P}_{\mathrm{str}} \\ 0 & \mathbf{G}_{\mathrm{two}} & \mathbf{D}_{\mathrm{two,lno}} \\ 0 & \mathbf{B}_{\mathrm{lno,two}} & \mathbf{R}_{\mathrm{lno}} \end{bmatrix} \begin{bmatrix} \widehat{\mathbf{a}} \\ \mathbf{u}_{\mathrm{two}} \\ \mathbf{i}_{\mathrm{lno}} \end{bmatrix} + \begin{bmatrix} \mathbf{M}_\sigma & 0 & 0 \\ -\mathbf{Q}_{\mathrm{sol}}^T\mathbf{M}_\sigma & \mathbf{C}_{\mathrm{two}} & 0 \\ \mathbf{P}_{\mathrm{str}}^T & 0 & \mathbf{L}_{\mathrm{lno}} \end{bmatrix} \frac{d}{dt} \begin{bmatrix} \widehat{\mathbf{a}} \\ \mathbf{u}_{\mathrm{two}} \\ \mathbf{i}_{\mathrm{lno}} \end{bmatrix} = \begin{bmatrix} 0 \\ -\mathbf{D}_{\mathrm{two,lni}}\mathbf{i}_{\mathrm{lni}} \\ -\mathbf{B}_{\mathrm{lno,twu}}\mathbf{u}_{\mathrm{twu}} \end{bmatrix} . \tag{18}$$

Notice that the circuit description is organised such that no fill-in in the field system part appears.

5 Time Integration

5.1 Singly diagonally implicit Runge-Kutta method

The coupled system of equations, here abbreviated to $\mathbf{K}\mathbf{x}+\mathbf{M}\frac{d}{dt}\mathbf{x} = \mathbf{f}$, is integrated in time by an implicit Runge-Kutta method [24, 28, 6]. The stage vectors $\bar{\mathbf{x}}_i$ and the stage derivatives $\dot{\mathbf{x}}_i$ for n_{stage} stages $i = 1, \ldots, n_{\mathrm{stage}}$ of the algorithm relate the solution $\mathbf{x}_n$ at the old time instant t_n to two solutions $\mathbf{x}_{n+1}$ and $\tilde{\mathbf{x}}_{n+1}$ of different order of approximation at the new time instant $t_{n+1} = t_n + \tau_{n+1}$ by

$$\bar{\mathbf{x}}_i = \mathbf{x}_n + \tau_{n+1} \sum_{j=1}^{n_{\mathrm{stage}}} a_{ij}\dot{\mathbf{x}}_j \,, \quad i = 1, \ldots, n_{\mathrm{stage}} \tag{19}$$

$$\mathbf{x}_{n+1} = \mathbf{x}_n + \tau_{n+1} \sum_{j=1}^{n_{\mathrm{stage}}} b_j\dot{\mathbf{x}}_j \tag{20}$$

$$\tilde{\mathbf{x}}_{n+1} = \mathbf{x}_n + \tau_{n+1} \sum_{j=1}^{n_{\mathrm{stage}}} \tilde{b}_j\dot{\mathbf{x}}_j \,. \tag{21}$$

$1-\sqrt{2}/2$	$1-\sqrt{2}/2$	0	0	0
1	$\sqrt{2}/2$	$1-\sqrt{2}/2$	0	0
$\sqrt{2}/2$	$5-3\sqrt{2}$	$2\sqrt{2}-6$	$1-\sqrt{2}/2$	0
1	$\sqrt{2}/3+1/6$	$\sqrt{2}/6-1/3$	1/6	$1-\sqrt{2}/2$
order 3	$\sqrt{2}/3+1/6$	$\sqrt{2}/6-1/3$	1/6	$1-\sqrt{2}/2$
order 2	$\sqrt{2}/2$	$1-\sqrt{2}/2$	0	0
order 1	1/2	1/8	1/4	1/8

Fig. 4: Butcher table for the applied singly diagonally implicit Runge-Kutta method with four stages, a solution of 3rd order, an embedded solution of 2nd order and an embedded solution of 1st order.

The coefficients a_{ij}, b_j and $\tilde{b}_j$ are collected in a Butcher table [6]. Here, we consider a singly diagonally implicit Runge-Kutta method with four stages, achieving a solution of 3rd order and an embedded solution of 2nd order (SDIRK-3(2)) for which the coefficients are listed in Fig. 4. For each stage i, the system

$$\left(\mathbf{K}+\frac{1}{a_{ii}\tau_{n+1}}\mathbf{M}\right)\bar{\mathbf{x}}_i=\mathbf{f}\left(t_n+c_i\tau_{n+1}\right)+\frac{1}{a_{ii}\tau_{n+1}}\mathbf{M}\mathbf{x}_n+\mathbf{M}\sum_{j=1}^{i-1}\frac{a_{ij}}{a_{ii}}\dot{\mathbf{x}}_j \quad (22)$$

with c_i the coefficients of the left column in Fig. 4, has to be solved. The nonlinearity caused by the dependence of the reluctivity on the magnetic field is resolved by the successive-substitution approach or by the Newton method. The Kirchhoff voltage law (second row in (18)) is scaled by $a_{ii}\tau_{n+1}$ whereas the Kirchhoff current law (3rd row in (18)) is scaled by $-a_{ii}\tau_{n+1}$ in order to achieve a symmetric system of equations. The resulting system is indefinite and is solved by the Minimal Residual method [29] or the Quasi-Minimal Residual method [20] for symmetric, indefinite systems. The system is preconditioned by a block preconditioner using multigrid for the field part and an exact inverse for the circuit part [12]. The 2D examples given in the paper are preconditioned by a multigrid approach developed for field-circuit coupled systems [26].

5.2 Adaptive time-step selection

The difference of both solutions $\mathbf{y}=\mathbf{x}_{n+1}-\tilde{\mathbf{x}}_{n+1}$ is used to control the error of the time-integration process [8, 4]. The error, measured in the norm

$$\|\mathbf{y}\|_{\mathrm{err}}=\sqrt{\sum_j\left(\frac{\mathbf{y}_j}{|\mathbf{x}_{n+1,j}|+\delta_{\mathrm{abs}}}\right)^2} \quad (23)$$

where δ_{abs} is an absolute tolerance, is compared to a user-defined error tolerance ϵ_{tol} multiplied by an acceleration factor μ, typically set slightly larger than 1. If $\|\mathbf{y}\|_{\mathrm{err}}>\mu\epsilon_{\mathrm{tol}}$, the last time step is rejected, otherwise the time step is accepted. The last time step is repeated or a new time step is computed with the time-step length

$$\tau_{n+2}=\rho_{\mathrm{safety}}\left(\frac{\epsilon_{\mathrm{tol}}}{\|\mathbf{y}\|_{\mathrm{err}}}\right)^{1/(\tilde{p}+1)}\tau_{n+1} \quad (24)$$

where $\tilde{p}$ is the order of the embedded solution and ρ_{safety} is a safety factor, typically set to 0.9 [22].

5.3 Sinusoidal dynamics

Many electrotechnical devices are excited by sinusoidal voltages and currents. For that case, a bad performance of the above described error-controlled adaptive time-stepping scheme was observed [5]. This phenomenon is explained by the fact that every second term of the Taylor series expansion of harmonic functions vanishes at particular time instants. Then, the difference between the 3rd order accurate solution and the 2nd order accurate embedded solution is negligible which motivates the time integrator to put very large time steps. A possible alleviation of this problem consists of using an embedded solution that differs by two orders of approximation, e.g., an SDIRK-3(1) method (Fig. 4).

5.4 Time-integration over discontinuities

When field effects due to the switching of power electronic components are considered, the switching events have to be considered by the time integrator [1, 31, 15]. A next time step is computed under the assumption that no switching events occur [2]. Afterwards, a possible event is detected by a sign checking procedure in the case of a θ-type time integrator [30, 18] or by evaluating Sturm sequences in the case of a higher-order time integrator [32, 3]. The time step is reduced to the instant of switching. At this time instant, the field and circuit solutions are determined relying upon the dense output capabilities of the implicit Runge-Kutta method [6]. When due to the switching events, capacitors are short-circuited or inductive chains are opened, a direct redistribution of charge and flux, respectively, is carried out. It also makes sense to carry out direct redistributions in all capacitive loops and all inductive cutsets where the associate time constants are significantly smaller than the time constants of the field problem. Such strategy avoids irrelevant time steps to be carried out for the entire field-circuit coupled system. Another possibility would include the use of a multi-rate time stepping scheme [21], e.g., performing additional small time steps for the circuit, especially when a switching event has occurred. After the computation of consistent initial conditions, the time-integration procedure is restarted with a changed circuit [27]. In our implementation, we favour to change the topology of the circuit, and by that, also the structure and possibly also the size of the system matrix, instead of the approach where switches are modelled by highly nonlinear resistors, causing bad condition numbers of the systems of equations [18, 31].

6 Examples

The first example is a single-phase machine with a start/run capacitor (Fig. 5). Its 2D cross-section is discretised by a finite-element method, resolves local saturation and eddy-current effects by adaptive mesh refinement and models rotor motion by a sliding-surface technique. By transient simulation, the currents through the main and auxiliary windings at start-up are computed.

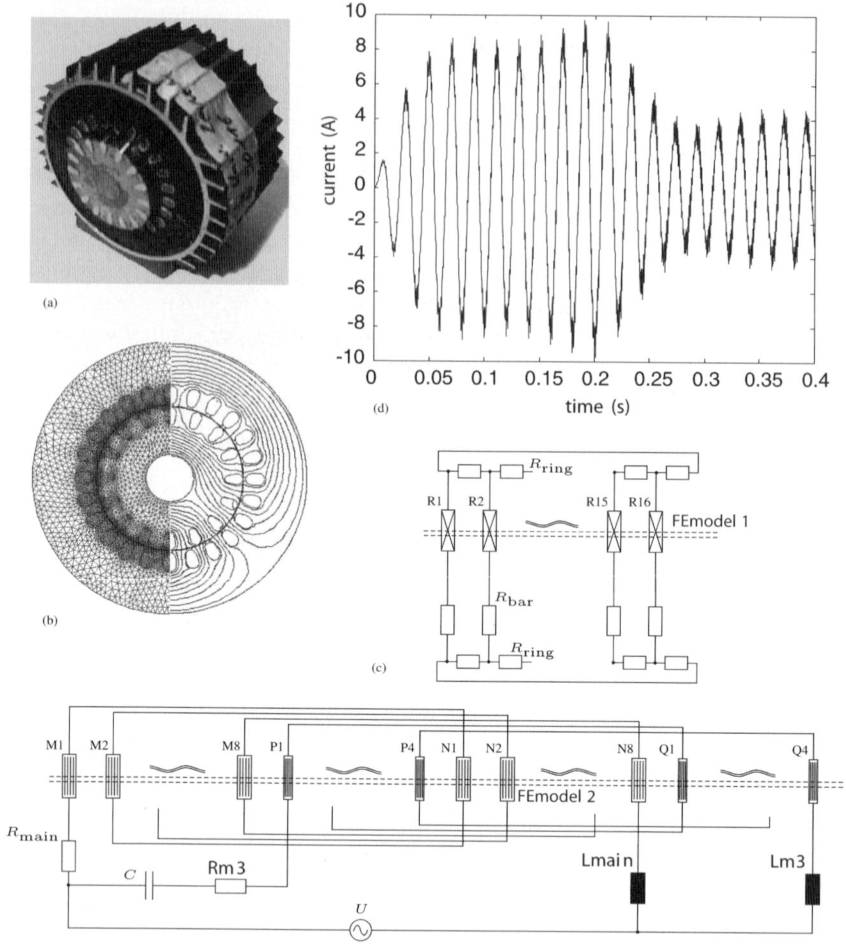

Fig. 5: Capacitor motor: (a) photograph; (b) finite-element mesh and magnetic flux lines at no-load operation; (c) external circuit with the applied sinusoidal voltage U, the capacitance C, the resistances R_{main} and R_{aux} and inductances L_{main} and L_{aux} modelling the end winding parts and the resistances R_{bar} and R_{ring} modelling the rotor ring and rotor-bar parts outside the finite-element model; (d) current through the main stator winding during start-up.

The second example is a three-phase transformer of which the primary side is connected to the grid and the second side is connected to a diode rectifier with an inductive load (Fig. 6). The detection and treatment of switching instants is carried out by a modified SDIRK-3(2) time integrator. Here, the capability of simulating immediate flux redistribution is exploited (Fig. 7).

7 Conclusions

Field-circuit coupling is extremely important to obtain reliable simulation results for electrical devices in an efficient way. The coupling between the degrees of freedom of the field formulation and the ones of the circuit formulation has to be designed such that no computational bottleneck arises, e.g., by too dense algebraic coupling

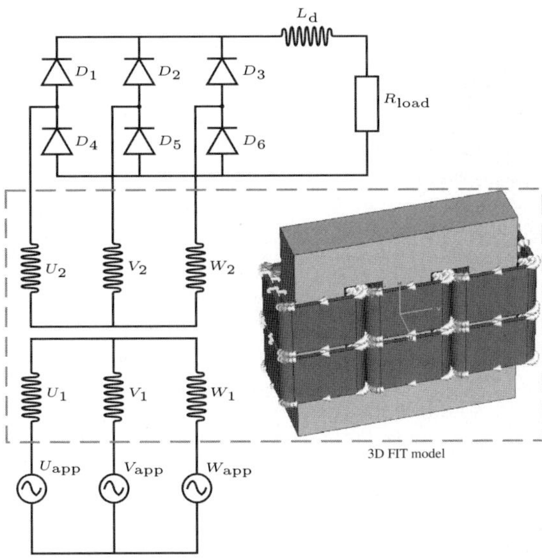

Fig. 6: 3D finite-integration model of a three-phase transformer connected to an external electric circuit for the power grid, diode rectifier and inductive load.

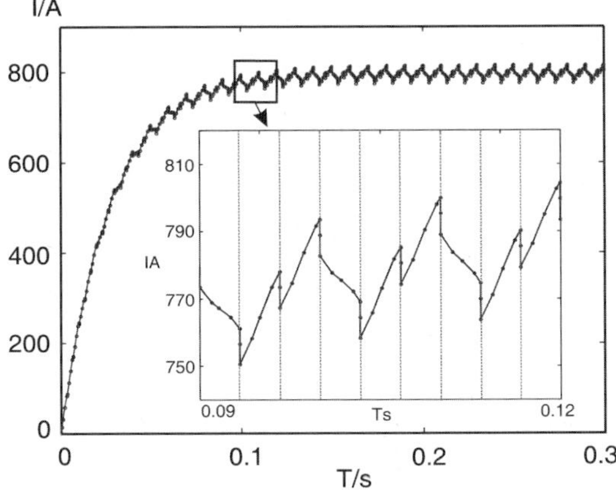

Fig. 7: Current through the first coil at the high-voltage side of the three-phase transformer.

matrices. Besides the traditional solid- and stranded-conductor models commonly applied for massive bars and wire coils within the field model, specialised conductor models exist and should be applied, e.g. for foil windings. An arbitrary connection of conductor models within an external circuit possibly incorporating switches is possible. The time integrator applied to the field-circuit coupled problem should detect and localise the switching events. The time step is restricted to this time instant. Fast dynamics due to (almost) short-circuited capacitors and opened inductive chains are resolved without superfluously evaluating the expensive field problem.

Acknowledgment

The work of H. De Gersem is supported by the Gesellschaft für Schwerionenforschung, Darmstadt, Germany. G. Benderskaya is supported by the Computational Engineering Research Center at the Technische Universität Darmstadt.

References

1. A. Arkkio. Finite element analysis of cage induction motors fed by static frequency converters. *IEEE Trans. Magn.*, 26(2):551–554, March 1990.
2. P. Barton and C. Pantelides. Modeling of combined discrete/continuous processes. *AIChE J.*, 40(6):966–979, 1994.
3. G. Benderskaya, M. Clemens, H. De Gersem, and T. Weiland. Embedded Runge-Kutta methods for field-circuit coupled problems with switching elements. *IEEE Trans. Magn.*, 41(5):1612–1615, May 2005.
4. G. Benderskaya, H. De Gersem, M. Clemens, and T. Weiland. Transient field-circuit coupled formulation based on the finite integration technique and a mixed circuit formulation. *COMPEL*, 23(4):968–976, 2004.
5. G. Benderskaya, H. De Gersem, and T. Weiland. Adaptive time integration for electromagnetic models with sinusoidal excitation. In *Proc. Int. Workshop Elec. Magn.*, pages 107–108, Aussois, France, June 2006.
6. F. Cameron. *Low-order Runge-Kutta methods for differential-algebraic equations*. PhD thesis, Tampere University of Technology, Tampere, Finland, 1999.
7. M. Clemens and T. Weiland. Transient eddy-current calculation with the FI-method. *IEEE Trans. Magn.*, 35(3):1163–1166, May 1999.
8. M. Clemens, M. Wilke, and T. Weiland. 3D transient eddy-current simulations using FI^2TD with variable time step size selection schemes. *IEEE Trans. Magn.*, 38(2): 605–608, March 2002.
9. H. De Gersem and K. Hameyer. A finite element model for foil winding simulation. *IEEE Trans. Magn.*, 37(5):3427–3432, September 2001.
10. H. De Gersem and K. Hameyer. A multi-conductor model for finite element eddy current simulation. *IEEE Trans. Magn.*, 38(2):533–536, March 2002.
11. H. De Gersem, K. Hameyer, and T. Weiland. Field-circuit coupled models in electromagnetic simulation. *J. Comput. Appl. Math.*, 168(1-2):125–133, July 2004.
12. H. De Gersem, R. Mertens, D. Lahaye, S. Vandewalle, and K. Hameyer. Solution strategies for transient, field-circuit coupled systems. *IEEE Trans. Magn.*, 36(4):1531–1534, July 2000.
13. H. De Gersem, R. Mertens, U. Pahner, R. Belmans, and K. Hameyer. A topological method used for field-circuit coupling. *IEEE Trans. Magn.*, 34(5):3190–3193, September 1998.
14. H. De Gersem and T. Weiland. Field-circuit coupling for time-harmonic models discretized by the finite integration technique. *IEEE Trans. Magn.*, 40(2):1334–1337, March 2004.
15. A. Demenko. Time-stepping FE analysis of electric motor drives with semiconductor convertors. *IEEE Trans. Magn.*, 30(5):3264–3267, September 1994.
16. P. Dular and C. Geuzaine. Spatially dependent global quantities associated with 2D and 3D magnetic vector potential formulations for foil winding modeling. *IEEE Trans. Magn.*, 38(2):633–636, March 2002.
17. P. Dular and J. Gyselinck. Modeling of 3-D stranded inductors with the magnetic vector potential formulation and spatially dependent turn voltages of reduced support. *IEEE Trans. Magn.*, 40(2):1298–1301, March 2004.
18. P. Dular and P. Kuo-Peng. An efficient time discretization procedure for finite element-electronic circuit equation coupling. *COMPEL*, 21(2):274285, 2002.

19. P. Dular, R. Specogna, and F. Trevisan. Coupling between circuits and A-χ discrete geometric approach. *IEEE Trans. Magn.*, 42(4):1043–1046, April 2006.
20. R. Freund and N.M. Nachtigal. A new Krylov-subspace method for symmetric indefinite linear systems. In W.F. Ames, editor, *Proceedings of the 14th IMACS World Congress on Computational and Applied Mathematics*, pages 1253–1256, 1994.
21. M. Günther and P. Rentrop. Multirate ROW methods and latency of electrical circuits. *Appl. Numer. Math*, 13:83–102, 1993.
22. K. Gustafsson. Control-theoretic techniques for stepsize selection in implicit Runge-Kutta methods. *ACM Trans. Math. Software*, 20(4):496–517, December 1994.
23. J. Gyselinck and J. Melkebeek. Numerical methods for time stepping coupled field-circuit systems. In *Proceedings of the International Conference on Modelling and Simulation of Electric Machines, Converters and Systems (ELECTRIMACS 96)*, volume 1, pages 227–234, Saint-Nazaire, France, September 1996.
24. E. Hairer and G. Wanner. *Solving ordinary differential equations, stiff and differential-algebraic problems*. Springer-Verlag, Berlin, 2 edition, 1996.
25. L. Kettunen. Fields and circuits in computational electromagnetism. *IEEE Trans. Magn.*, 37(5):3393–3396, September 2001.
26. D. Lahaye, K. Hameyer, and S. Vandewalle. An algebraic multilevel preconditioner for field-circuit coupled problems. *IEEE Trans. Magn.*, 38(2):413–416, March 2002.
27. G. Mao and L.R. Petzold. Efficient integration over discontinuities for differential-algebraic systems. *Comput. Math. Appl.*, 43:65–79, 2002.
28. A. Nicolet and F. Delincé. Implicit Runge-Kutta methods for transient magnetic field computations. *IEEE Trans. Magn.*, 32(3):1405–1408, May 1996.
29. C.C. Paige and M.A. Saunders. Solution of sparse indefinite systems of linear equations. *SIAM J. Numer. Anal.*, 12(4):617–629, 1975.
30. A.J. Preston and M. Berzins. Algorithms for the location of discontinuities in dynamic simulation problems. *Comput. Chem. Eng.*, 15(10):701–713, 1991.
31. N. Sadowski, B. Carly, Y. Lefevre, M. Lajoie-Mazenc, and S. Astier. Finite element simulation of electrical motors fed by current inverters. *IEEE Trans. Magn.*, 29(2):1683–1688, March 1993.
32. L.F. Shampine, I. Gladwell, and R.W. Brankin. Reliable solution of special event localisation problems for ODEs. *ACM Trans. Math. Soft.*, 17(1):11–25, March 1991.
33. I.A. Tsukerman, A. Konrad, G. Meunier, and J.C. Sabonnadière. Coupled field-circuit problems: trends and accomplishments. *IEEE Trans. Magn.*, 29(2):1701–1704, March 1993.
34. H.A. Van der Vorst and J.B.M. Melissen. A Petrov-Galerkin type method for solving $Ax = b$, where A is symmetric complex. *IEEE Trans. Magn.*, 26(2):706–708, March 1990.
35. T. Weiland. A discretisation method for the solution of Maxwell's equations for six-component fields. *Electr. Commun. AE*, 31:116–120, 1977.
36. T. Weiland. Time domain electromagnetic field computation with finite difference methods. *Int. J. Num. Mod.*, 9(4):295–319, July-August 1996.

Technology and Device Modeling in Micro and Nano-electronics: Current and Future Challenges*

Andrea Marmiroli, Gianpietro Carnevale and Andrea Ghetti

STMicroelectronics, 20041 Agrate Brianza, Italy andrea.marmiroli@st.com

Abstract
The number of physical effects that have to be taken into account to accurately model and design current and future micro- and nano-electronics devices is continuously increasing. At the same time, the importance of the coupling among them is increasing as well. An accurate simulation of such effects with strong interactions is often non-trivial and in many cases a satisfactory solution is not yet available. Two challenging problems are presented in more detail: the first one refers to the thermo-mechanical problem of silicon oxidation, the second is the electrical coupling which occurs in strained silicon substrate.

1 Introduction

The peculiar driving force of micro- and nano-electronic industry is the shrink of dimensions. This shrink allows to use less silicon and to pack more devices on the same wafer, reducing the production costs. At the same time it results in the increase of transistors' driven current and in the reduction of the total capacitance (as the coupling capacitance remains roughly constant and the capacitance between different layers reduces). Moreover, such shrinkage allows the reduction of the dimensions of the final equipments (cellular phones, portable computers, etc.), increasing the added value of the integrated circuits. Along this shrink path, the minimum features defined by today's technology are in few tens of nanometers range [ITR05].
A second driving force is the integration of different functions in the same integrated circuit or in the same package. This is driven by the reduction of dimensions of the final equipment and by the increase of performances of the same equipments, thanks to faster communications between different blocks.
A further driving force deals with innovation: to increase the added value of integrated circuit, new functions have to be included in the circuits.
Because of this continuous shrinkage, increased integration of functions and new features development, to design and to manufacture semiconductor devices, more and more physical mechanisms, which were previously negligible, have now to be taken into account. Among the most important there are:

* Invited Paper at SCEE-2006

- diffraction and interference effects in lithography
- modeling of coupled electro-thermal phenomena
- electrical behavior of strained silicon
- electro-magnetic coupling between conduction lines which are closer and closer
- resistance of the parasitic interconnect metal lines
- effects of power dissipation

In this paper, in the next section we will present different examples of coupled problems, with a particular emphasis on the modeling aspects. Next two examples are discussed in details to show our modeling approach: the coupling between thermal and mechanical effects which will be discussed in section 3 and the coupling between electrical and mechanical effects presented in section 4, with particular attention to the application of the simulation methodology and tool to a silicon nano-wire MOSFET case.

2 Coupled Problems: general case studies

The first case we address refers to the coupling between electrical and thermal effects which has to be taken into account to understand the phenomena involved in the principles of operation of Phase Change Memories (see [Pir05] and references therein). The principle is described in fig. 1:

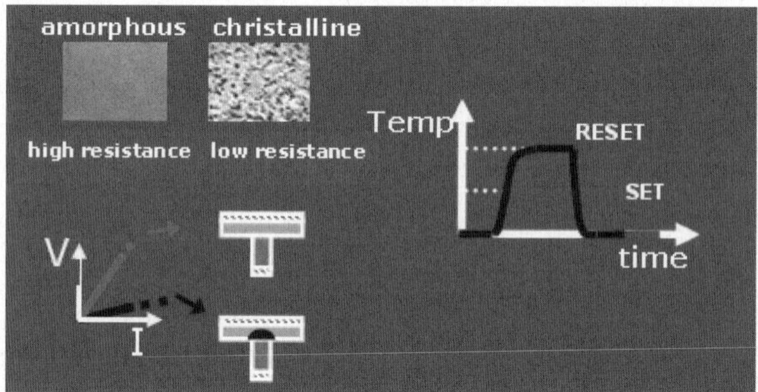

Fig. 1: Phase Change Memory (PCM) description: low and high resistivity of PCM material is associated to the crystalline and amorphous state, respectively. At the left side, bottom part of the figure the schematic pictures of a cross section of the bit architecture are sketched: the "T" shape at the lower side correspond to the low resistance state, with the material partially modified in the crystalline configuration. The upper "T" shape corresponds to the high resistance state, with the material in the amorphous phase.

a) Top left: the storing mechanism of the bit information is based on the different conductivity of the amorphous and of the poly-crystalline phases of the selected material (chalcogenide). The TEM photographs show the two different states.
b) Bottom left: the sensing mechanism exploits the different resistance of the two states. The two different current/voltage characteristics are reported.
c) Right: Finally the writing mechanism based on the joule effect due to the current flow in the chalcogenide film. The temperature profiles to reach the reset (amorphous) and set (crystalline) are reported.

The second case deals with the coupling between electrical and mechanical aspects. The conductivity in semiconductors depends on crystals strain. To accurately model the current flowing in the devices it is necessary to take such strain into account. Fig. 2 shows the impact of this effect (see [Fan05]). At the upper left corner a top view of the layout of an MOS transistor is reported. As shown at the lower right corner, the electrical behaviour depends, besides the obvious W, L values on the total active area dimension LOD (Length Of Device, is given by "2a+L"). The lower left corner shows the stress field in 2 dimensions, while the upper right corner shows the 1-Dimensional cutline for different LOD values.

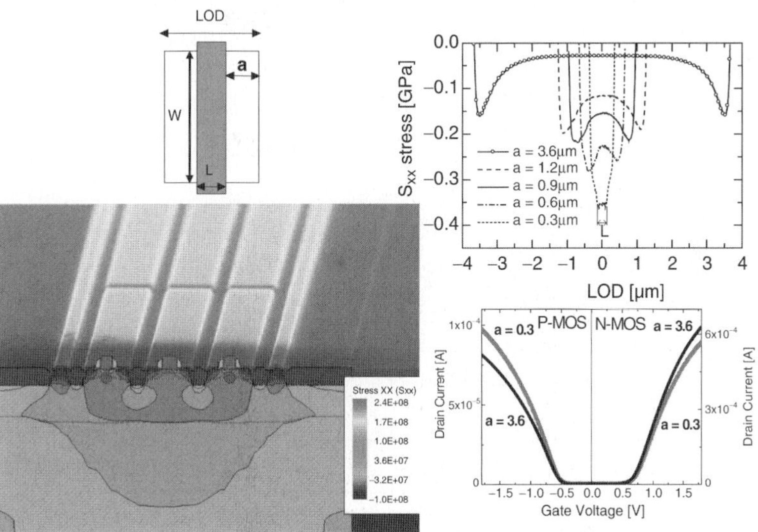

Fig. 2: Strain effect in mosfet transistors

3 Silicon Oxidation study

In this section we will discuss in detail the modeling of silicon oxidation. The modeling of such phenomenon has two main objectives: first to predict the exact shape of silicon and of silicon dioxide, then to evaluate the stress and strain in the two films as the electrical performances, some failure and degradation mechanisms are strongly dependent on the stress/strain level [Rim01], [Tho04]. An accurate modeling of silicon oxidation, has to deal with two linked problems: the diffusion of the oxidizing species in silicon oxide and the solution of the mechanical problem related to the formation of silicon dioxide (whose volume is twice as large as that of the original silicon). These two issues are strongly linked by means of a nonlinear dependence of the main physical quantities which are used in the diffusion-reaction problem: the diffusivity, the reaction rate and the oxide viscosity (respectively $\mathbf{D}^{diff}$, $\mathbf{K}^{react}$, ν^{oxi}) and the stress quantities which are calculated in the mechanical problem: pressure and maximum shear stress (P, σ), as shown in Fig.3.

This coupled algorithm has been published for the first time by Kao [Kao88] et al. who studied the oxidation of concave and convex silicon surfaces. During these studies they observed that the thickness of the oxide growth along cylinders with different radius was very different and that it was not related only to a geometrical effect. It was in particular noticed that the silicon oxide growth in a convex surface was even smaller with respect to the one growth on a flat silicon surface. This is in contrast with the expected consideration based on purely geometrical issues: a convex structure exposes more area than a flat surface, therefore the oxidant flux should be larger. It has been supposed that diffusivity of oxidant in the oxide layer and reaction rate and the silicon interface were both reduced by the stress field inside oxide and normal at the silicon surface. Furthermore, it was also supposed on the basis of previous studies on silica performed by Eyring [Eyr36] that silicon oxide viscosity was affected by the stress inside the material, reducing its value when larger amount of stress were accumulated in oxide. To our knowledge the non-linear relationship between stress and viscosity has not been further investigated.

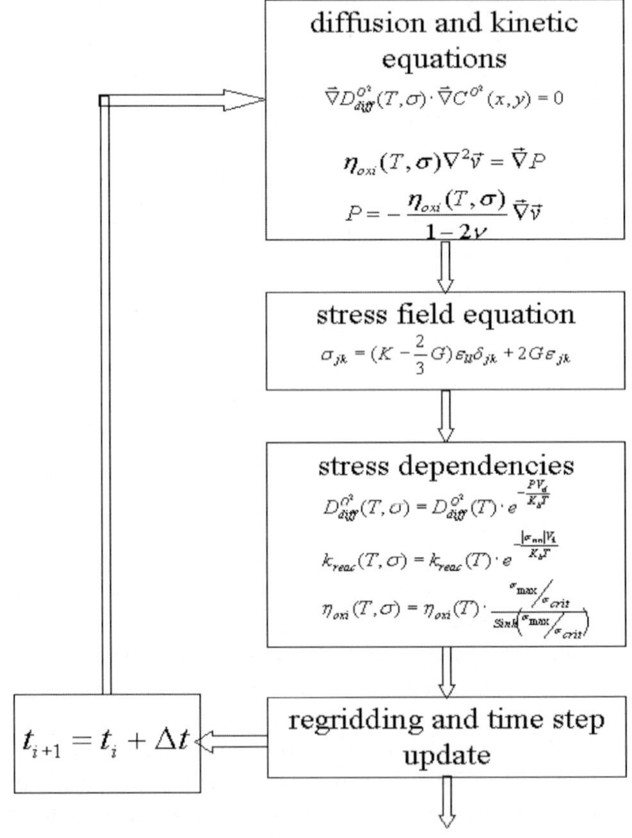

Fig. 3: Solution algorithm coupling oxidation rates and stress calculations

This model has been widely accepted by the scientific community and most of the simulation program adopted it for the correct calculation in two dimensions of the isolation oxide shape in flash memory arrays.
In a more recent past, it has become more and more important to evaluate the stress field distribution not only in the memory cell array, but also in the devices devoted to manage the internal voltage (circuitry or logic circuits). These structures are basically 3D (see fig.4), and are strongly affected by stress in silicon due to the impact on carrier mobility, as highlighted in the next section.

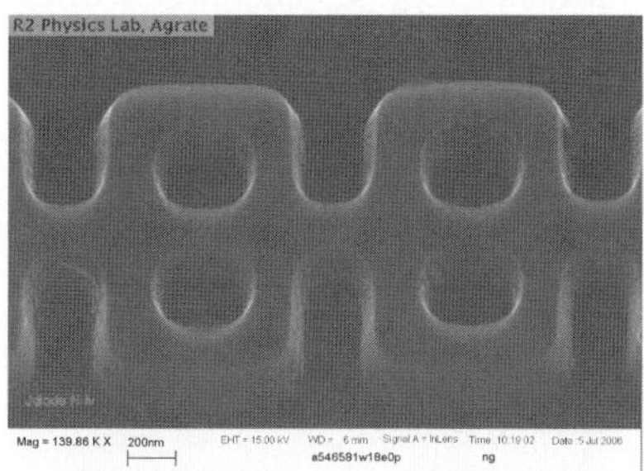

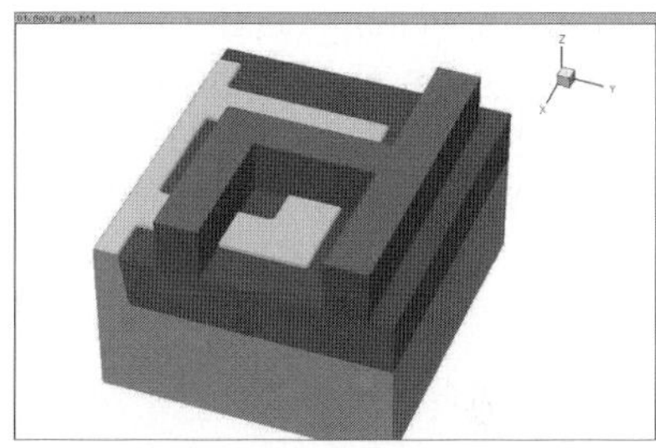

Fig. 4: A SEM picture of a silicon active area (oxide has been removed during stripping operation) and the corresponding 3D structure (pink region are silicon, brown is oxide and violet is poly-silicon material).

Unfortunately, the complexity of this task is largely increased by the computational/ mathematical problem of managing in three dimensions (3D) the moving boundaries describing the silicon-silicon oxide and the silicon-gas interfaces. At the present time, the commercial tool [Syn07] available for the industrial research activity only allows to manage the stress related effects due to thermal mismatch between different materials for a given fixed, but 3D, geometry. The simple continuous model suitable to calculate the strain induced by thermal mismatch between two materials is expressed by the following equation:

$$\varepsilon(x, y) = (\alpha_{sil} - \alpha_{ox}) \times (T_{fin} - T_{ini})$$

where α_{sil},α_{ox} are the linear coefficient thermal expansion for silicon and oxide, respectively, and T_{fin}, T_{ini} are the initial and final temperatures.
The strain, calculated at the interface between the materials, is a isotropic quantity acting in the plane parallel to the same interface.
The results in terms of accuracy and CPU time are largely affected by the solution given by the meshing strategy used in the program.
The state-of-the-art for what regards silicon oxidation in a 3D framework is represented by FEDOS [Tuv06] simulator program, a code which is developed at the University of Vienna and which is still under development. The models implemented in this code are based on a new approach, which calculates the growth of silicon oxide starting from a diffusion-reaction approach. In fact, one assumes the following mechanism: the oxidant species reaches the silicon interface after a diffusion step in silicon oxide and then the reaction with silicon is able to create a new product which is the silicon oxide molecule, consuming a single silicon atom and two oxygen atoms. The equation set is the following:

$$\frac{\partial \rho_{ox}}{\partial t} = D \frac{\partial^2 \rho_{ox}}{\partial x^2} - k \rho_{ox} \rho_{sil}$$

$$\frac{\partial \rho_{sil}}{\partial t} = -k \rho_{ox} \rho_{sil}$$

In the above equations D and k are respectively the diffusion of oxidant in silicon oxide and the reaction rate of oxidant at silicon / silicon dioxide interface, and ρ_{ox}, ρ_{sil} are the density of oxide and silicon, respectively.
For given process parameters (temperature, oxidant gas flow, time) this model incorporates oxidation when dealing in non-stationary regime. The largest difficulty is represented in the modeling of the silicon oxide interface: in fact, in order to solve the mechanical problem it is necessary to define a material interface for each time step after the solution of diffusion-reaction problem and to apply to the interface the correct velocity fields which are derived from the solution of the previous problem.

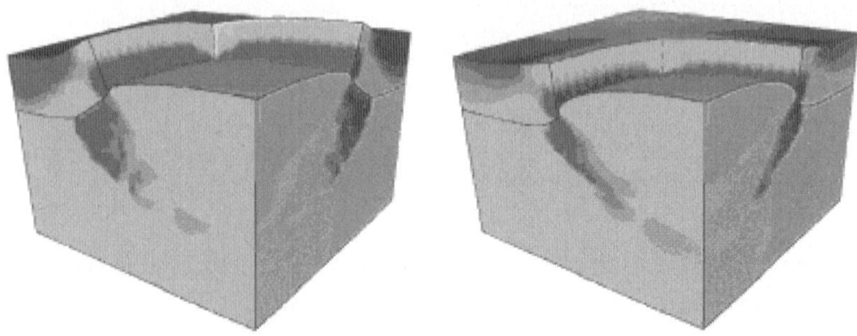

Fig. 5: At the left, the pressure distribution calculated with FEDOS without stress dependent oxidation. At the right stress dependent oxidation is included [Hol05].

4 Transport in NanoWire mosfets study

The second detailed case deals with device simulation of silicon NanoWire MOSFETs. NanoWire MOSFETs (NW MOSFETs) like the one reported in [Yan04] are gaining increasingly popularity due to their superior channel control. This is achieved by reducing the silicon channel to a thin wire surrounded as much as possible by the gate. This makes this kind of devices intrinsically 3D.
In addition, highly non-equilibrium transport still dominated by scattering is expected in this kind of devices [Gil05]. This complex non-stationary/ballistic transport can be accurately accounted for by semi-classical Monte Carlo (MC) simulation. However, for such small devices, quantum mechanical and strain-induced effects play a fundamental role that must be accounted for in conjunction with the real 3D geometry of the device. Therefore it is necessary to include quantum mechanical (QM) and strain effects in the framework of semi-classical 3D MC device simulation.
In this section we report on a new MC simulator (called MC++ [Ghe06]) that solves self-consistently in 1D, 2D or 3D, the Schrdinger Eq. for the QM correction of the potential, while mechanical strain effects are accounted for by an appropriate change

of the band structure. We will show that QM corrected 3D semi-classical Monte Carlo device simulation can accurately address all the above issues.

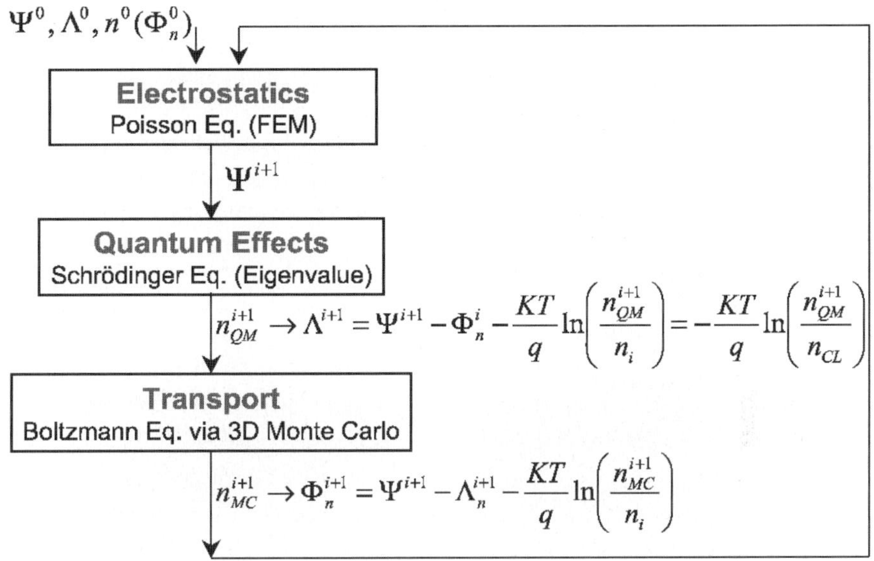

Fig. 6: Main blocks of the simulation program and their interactions. Simulation starts by reading an initial guess computed with conventional programs.

Fig. 6 graphically depicts the interaction among the main blocks of MC++. It solves the Schrdinger Eq. (SE) and the Poisson Eq. (PE) self-consistently with the semi-classical 3D Monte Carlo simulation of carrier transport through an iterative procedure. The linear PE is solved using standard box methods for the potential (Ψ) profile frequently enough (every 2fs) to assure time stability. The solution of the SE provides the QM correction term (Λ) of the potential accounting for charge quantization [Kat03]. Both Ψ and Λ act as driving force in the Boltzmann Transport Equation that is solved for via semi-classical 3D Monte Carlo simulation providing carrier/pseudo-potential profiles to be used in the solution of both PE and SE.
In case of 3D structures, the SE is solved using a “Quasi 3D” approach [Wan04]: the simulation domain is cut in several sections normal to the channel in which the 2D SE is solved for. Then, a continuous 3D description of the QM charge is recovered by interpolating the results of two adjacent sections. This approach is valid as long as the confinement region does not change shape, as in the case of NW-MOSFET [Wan04]. The 2D SE is solved as in [Abr00]. Assuming a rectangular domain with zero boundary conditions the solution can be expanded as $\Phi(x,y) = \sum_{ij}^{N} A_{ij} sin\left(k_x^i x\right) sin\left(k_y^j y\right)$. Hence, the 2D SE can be transformed into a standard eigenvalue problem (solved by highly optimized libraries [And99]) involving the Fourier transform of the potential that can be efficiently computed exploiting FFT algorithms [Fri05].
This methodology can be applied to more arbitrary geometries, as illustrated in Fig. 7 for the case of a circular well with radius R. First, the initial domain (Fig. 7.a) is

mapped onto a uniformly spaced tensor product grid (Fig. 7.b) needed by the FFT algorithm. Then, thc energy profile is interpolated on the new grid. Points outside the initial domain are assigned an arbitrary high value (Fig. 7.c, 7.d). This assures no wave penetration outside the original domain.

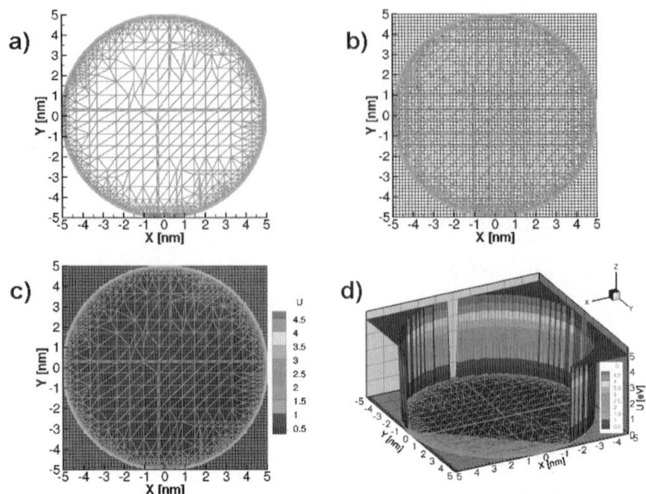

Fig. 7: Numerical solution of the 2D SE in the case of a circular well with R=5nm. a) initial finite element mesh; b) domain map to a uniform tensor product grid; c) contour plot of the energy profile; d) partial 3D view of the energy profile.

For the energy profile in Fig. 7, the 2D SE admits the following analytical solution:

$$\Psi_{mm} = A_{mm} J_m \left(z_{mm} r / R \right) \left\{ \begin{array}{l} cos(m\varphi) \\ sin(m\varphi) \end{array} \right. \tag{1}$$

$$E_{mm} = \frac{\hbar^2 z_{mm}^2}{2m^* R^2} \tag{2}$$

where J_m is the Bessel function of first kind of order m, while z_{mm} is the $n-th$ zero of J_m. Fig. 8 demonstrates the accuracy of this procedure by comparing quantitatively the numerical and analytical solution.

Both physical and phase spaces are discretized with a tetrahedral mesh. This allows for the greatest flexibility in describing device geometry and makes the free-flight equations linear [Bud94], i.e. easy and fast to be solved.

The silicon band structure is computed with the Empirical Pseudo-potential Method [Rid06] that accounts for strain-induced band structure distortion. The Density of State (DOS) is computed by directly calculating the area of the equi-energy surfaces that are also stored in memory to speed up the determination of the state after scattering [Bud94].

Scattering mechanisms are assumed to be isotropic and to depend on strain through the variation of the DOS. Scattering mechanisms include: elastic acoustic phonon scattering, inelastic optical phonon scattering, ionized impurity scattering (isotropic model of [Buf00], impact ionization. Scattering against an interface is treated empirically as a mixture of reflecting and randomizing scattering [Buf00].

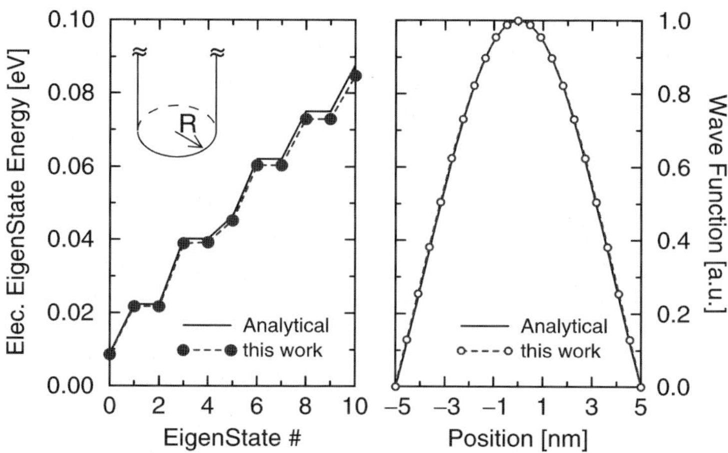

Fig. 8: Validation of the numerical solution of the 2D SE in the case of a circular well with R=5nm. Solid line: analytical solution; symbols: simulation. Left: eigenstate energy; right: first eigenstate wave function.

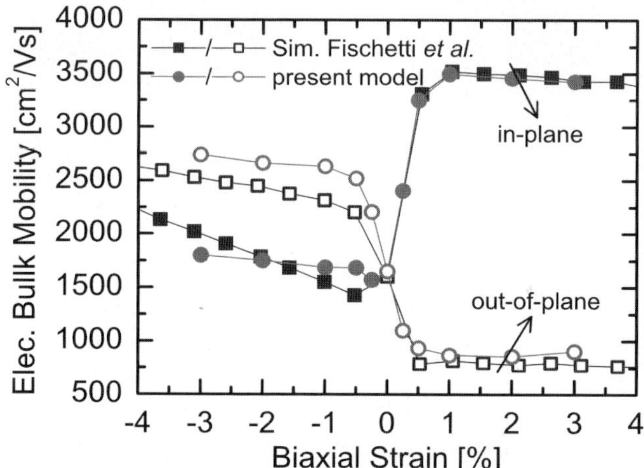

Fig. 9: Simulated electron bulk mobility (□/■) in comparison with calculation of [Fis96] (○/●) for un-doped silicon under biaxial strain. Closed/open symbols refer to in-plane/out-of-plane mobility.

Phonon scattering for electrons and holes has been extensively calibrated to reproduce a large variety of experiments including strain dependent mobility for electrons (Fig. 9) and holes (Fig. 10) [Fan05], [Ghe06], [Fer06].
As an application example we used MC++ to simulate the NW-MOSFET reported in [Yan04] and shown in Fig. 11. The actual iteration scheme is shown in Fig. 12.
The simulation starts by reading an initial bias profile computed with conventional QM, i.e. density-gradient, hydrodynamic simulation (QM HD).
Then, the Poisson and Schrdinger equations are solved self-consistently keeping the pseudo-potential Φ_n^0 found in the initial profile. This step has been introduced to provide a better initial guess for the potential QM correction (Λ^1 in Fig.13) than the one provided by QM HD (Λ^0), thus speeding up convergence.
Please notice in Fig. 13 that Λ^1 significantly deviates from Λ^0. Incidentally, this questions the accuracy of the standard density-gradient approach for 2D/3D cases.

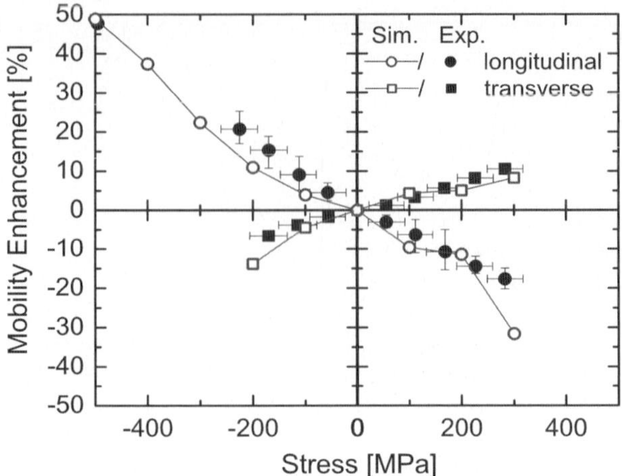

Fig. 10: Monte Carlo simulation of hole mobility enhancement in comparison with experimental data from wafer bending experiments of [Tho04] under uniaxial stress.

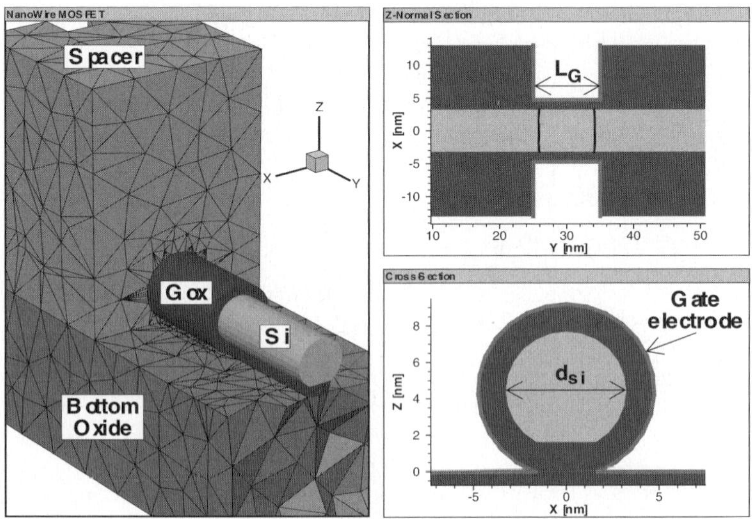

Fig. 11: Silicon NanoWire n-MOSFET reported in [Yan04] and simulated in this work. a) partial 3D view; b) horizontal (z-normal) section; c) channel cross-section. $L_G = 10nm$, $t_{ox} = 1.5nm$, $d_{si} = 6.5nm$. Not all dielectrics are shown.

Indeed, one observes that this density-gradient usually is calibrated to reproduce Poisson-Schrdinger results in 1D. Next, the real iteration loop is entered by performing a Monte Carlo-Poisson self-consistent simulation until a steady-state solution is reached. This is necessary to get a smooth solution for the potential and carrier pseudo-potential to be used in the Schrdinger Eq. solution to update Λ. Notice that any "noise" on $\Psi$and Φdirectly impacts Λ, and, if it is too large, may lead to unphysical results. The loop is then closed by solving the Schrödinger Equation as explained in the previous section. As it is possible to see in Fig. 13, only a couple of iterations are needed to get a stable solution for Λ. Finally, once a stable solution for Λ has

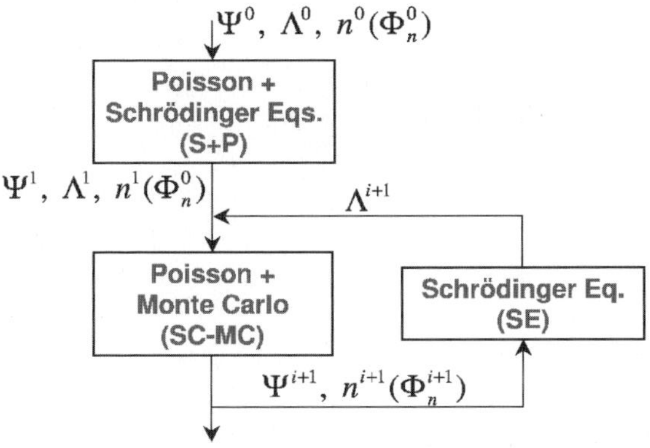

Fig. 12: Schematic representation of the iteration scheme. Convergence is reached after a few iterations. (Notice that $\Phi^1 = \Phi^0$)

been obtained, a longer Monte Carlo-Poisson loop is performed to collect smoother statistics data.

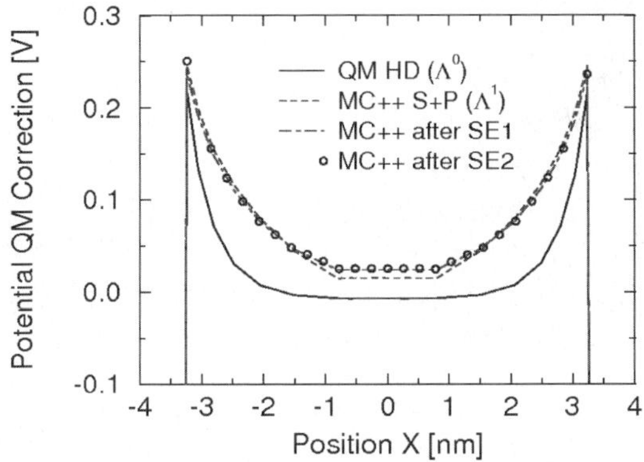

Fig. 13: Evolution of the potential QM correction during iterations. Λ^0 is the initial profile computed with conventional density-gradient hydrodynamic simulation (QM-HD). Λ^1 is the first guess provided by the self-consistent solution of the Schrödinger-Poisson Eq. (S+P).

All simulation results shown in the following have been obtained with the inclusion of strain. The strain tensor symmetry that can be inferred from the geometry of the device under investigation exhibits a biaxial compressive component in the plane perpendicular to the channel direction due to the gate all-around. Consequently, the current flows in the out-of-plane direction benefiting from the effect of mechanical strain (see Fig. 9).

Using the analytical model of [Kao88] and the process information available in [Yan04], the biaxial compressive strain is estimated to be 0.5%.

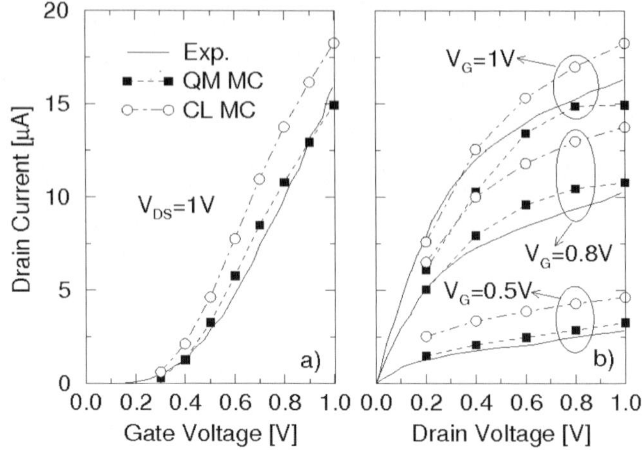

Fig. 14: Comparison of experimental (line) and simulated (symbols) drain current with (QM MC, ■) and without (Classic MC, ○) QM correction. a) trans-characteristics; b) output characteristics.

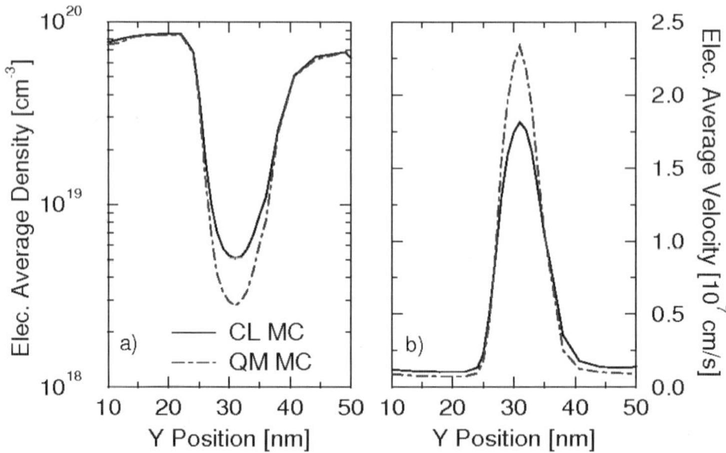

Fig. 15: Simulated electron density (a) and velocity (b) averaged on a channel cross-section as a function of the position for $V_G = 0.5V$, $V_{DS} = 1V$ with (QM MC, dot-dashed line) and without (CL MC, solid line) QM correction.

Simulated drain current with (QM MC) and without (Classic MC) QM correction is compared to experimental data in Fig. 14. A good agreement with experimental data is found only if QM effects are accounted for, while CL MC provides a higher current, as expected. This can be understood by looking at the electron concentration along the channel shown in Fig. 15.a. When QM effects are accounted for, there is a decrease of the free carrier density inside the channel, thus a smaller current, simply because quantization reduces the number of allowed states. This effect is of particular importance for small devices such as NW-MOSFET. However, this is not the only effect due to quantization. Fig. 15.b also reports the average velocity along the channel in the two cases. When QM effects are accounted for, electrons attain a larger average velocity while transiting in the channel ($\approx +25$ %), thus partially compensating the reduced charge concentration (≈ -50 %). This is due to the particular shape of the carrier space distribution resulting from the inclusion of QM correction. QM effects push electrons away from the interface providing the maximum concentration at the center of the NanoWire. On the contrary, without QM correction the maximum carrier concentration is attained at the gate oxide interface. Thus, in this latter case, electrons will experience more surface scattering (as confirmed by the larger average number of surface scattering per simulated particle), resulting in a smaller velocity.

5 Conclusions

The continuous technology shrinking mandates the need to account for more and more coupled physical effects. We showed how this is the case for two of the most important problems in the technology and device modeling area. These developments require further investment and collaboration between industry, research centers and software vendors, in order to provide accurate tools in time for effective usage.

References

[Abr00] A. Abramo, A. Cardin, L. Selmi, E. Sangiorgi: Two-Dimensional Quantum Mechanical Simulation of Charge Distribution in Silicon MOSFETs, IEEE Trans. on Electron Devices, **Vol. 47**, p. 1858–1863, 2000

[And99] E.Anderson, Z.Bai, C.Bischof, S.Blackford, J.Demmel, J.Dongarra, J.DuCroz, A.Greenbaum, S.Hammarling, A.McKenney, D.Sorensen: LAPACK User Guide, Philadelphia, PA: Society for Industrial and Applied Mathematics, 3rd edition, 1999

[Bud94] J. Bude, R.K.Smith: Phase-Space Simplex Monte Carlo for Semiconductor Transport, Semicond. Sci. Technol., **Vol. 9**, p. 840, 1994

[Buf00] F.Bufler, A.Schenk, W.Fichtner: Efficient Monte Carlo Device Modeling, IEEE Trans. on Electron Devices, **Vol.47, no.10**, p.1891–1897, 2000

[Eyr36] H. Eyring: Viscosity, plasticity and diffusion as examples of absolute reaction rate, Journal of Chemical Physics, pp. 283–291, 1936

[Fan05] P. Fantini, A. Ghetti, G.P. Carnevale, E. Bonera, D. Rideau: A full self-consistent methodology for strain-induced effects characterization in silicon devices, IEDM Tech. Digest, p. 1013–1016, 2005

[Fer06] M. Feraille, D. Rideau, A. Ghetti, A. Poncet, C. Tavernier, H. Jaouen: Low-field mobility in strained silicon with full-band Monte Carlo simulation using kp and EPM bandstructure, Proceedings SISPAD Conference, p. 264–266, 2006

[Fis96] M. Fischetti, S. Laux: Band structure, deformation potentials, and carrier mobility in strained Si, Ge, and SiGe alloys, journal of Applied Physics, **vol.80**, p. 2234–2252, 1996

[Fri05] M.Frigo, S.G.Johnson: The design and implementation of FFTW3, Proceedings of the IEEE, **vol.93, no.2**, pp. 216–231, 2005

[Ghe06] A.Ghetti, D.Rideau: 3D Monte Carlo Device Simulation of NanoWire MOSFETs including Quantum Mechanical and Strain Effects, Proceedings SISPAD Conference, p. 67–70, 2006

[Gil05] M.Gilbert, R.Akis, D.Ferry: Phonon-assisted ballistic to diffusive crossover in silicon nanowire transistors, Journal of Applied Physics, **vol.98**, p.094303.1–094303.8, 2005

[Hol05] C.Hollauer, H. Ceric, S. Selberherr: Three-Dimensional Simulation of stress dependent thermal oxidation, Proceedings SISPAD Conference, p. 183–186, 2005

[ITR05] International Technology Roadmap for Semiconductors 2005, *http://public.itrs.net*

[Kao88] D.-B. Kao, J.P. McVittie, W.D. Nix, K.C. Saraswat: Two-dimensional thermal oxidation of silicon II Modeling stress effects in wet oxides, IEEE Transaction on Electron Devices, **vol. 35, n. 1**, p. 25–37, 1988

[Kat03] G.Kathawala, U.Ravaioli: 3-D Monte Carlo Simulation of FinFETs, IEDM Tech. Digest, 2003

[Pir05] A.Pirovano, F.Pellizzer, A.Redaelli, I.Tortorelli, E.Varesi, F.Ottogalli, M.Tosi, P.Besana, R.Cecchini, R.Piva, M.Magistretti, M.Scaravaggi, G.Mazzone, P.Petruzza, F.Bedeschi, T.Marangon, A.Modelli, D.Ielmini, A.L.Lacaita, R.Bez: μ trench phase-change memory cell engineering and optimisation, Proc. ESSDERC, pp. 313–316, 2005

[Rid06] D. Rideau, M. Feraille, L. Ciampolini, M. Minondo, C. Tavernier, H. Jaouen: Strained Si, Ge and Si1-xGex alloys modeled with a first-principle-optimized full-zone kp method, Physical Review B, vol. 74, p. 195208:1–20, 2006

[Rim01] K.Kim, S.Koester, M.Hargrove, J.Chu, P.M.Mooney, J.Ott, T.Kanarsky, P.Ronsheim, M.Ieong, A.Grill, H.-S.P.WongK.Kim, S.Koester, M.Hargrove, J.Chu, P.M.Mooney, J.Ott, T.Kanarsky, P.Ronsheim, M.Ieong, A.Grill, H.-S.P.Wong

[Syn07] Synopsys User manuals, release Y-2006.06

[Tho04] S. E. Thompson, G. Sun, K. Wu, J. Kim, T. Nishida: Key differences for process-induced uniaxial vs. substrate-induced biaxial stressed Si and Ge channel MOS-FETs, Proceedings IEDM Conference 2004, p. 221–224

[ThoArm04] Thompson S.E., Armstrong M., Auth C., Alavi M., Buehler M., Chau R., Cea S., Ghani T., Glass G., Hoffman T., Jan C.-H., Kenyon C., Klaus J., Kuhn K., Zhiyong Ma., Mcintyre B., Mistry K., Murthy A., Obradovic B., Nagisetty R., Phi Nguyen., Sivakumar S., Shaheed R., Shifren L., Tufts B., Tyagi S., Bohr M., El-Mansy Y.: A 90-nm logic technology featuring strained silicon, IEEE Transaction on electron devices, **vol 51, n. 11**, p. 1790–1797, 2004

[Tuv06] *http://www.iue.tuwien.ac.at/99.0.html*

[Wan04] J.Wang, E.Polizzi, M.Lundstrom: A three-dimensional quantum simulation of silicon nanowire transistors with the effective-mass approxiamtion, Journal of Applied Physics, **vol.96**, pp.2192–2203, 2004

[Yan04] F.-L. Yang, D.-H. Lee, H.-Y. Chen, C.-Y. Chang, S.-D. Liu, C.-C. Huang, T.-H. Chung, H.-W. Chen, C.-C. Huang, Y.-H. Liu, C.-C. Wu, C.-C. Chen, S.-C. Chen, Y.-T. Chen, Y.-H. Chen, C.-J. Chen, B.-W. Chan, P.-F. Hsu, J.-H. Shieh, H.-J. Tao, Y.-C. Yeo, Y.Li, J.-W. Lee, P.Chen, M.-S. Liang, and C.Hu: 5-nm-Gate Nanowire FinFET, Proc. VLSI Technology Symposium, pp. 196–197, 2004

New Algorithm for the Retrieval of Aerosol's Optical Parameters by LIDAR Data Inversion

Camelia Talianu[1], Doina Nicolae[1], C. P. Cristescu[2], Jeni Ciuciu[1], Anca Nemuc[1], Emil Carstea[1], Livio Belegante[1], and Mircea Ciobanu[1]

[1] National Institute of R&D for Optoelectronic camelia@inoe.inoe.ro
[2] "Politechnica" University of Bucharest cpcris@physics.pub.ro

Atmospheric aerosols exhibit a high degree of variability in their properties and their spatial and temporal distribution. Laser remote sensing is now-days used to provide systematic monitoring of the temporal evolution of the aerosol in order to understand the radiative, physical, chemical and dynamic processes in the atmosphere. In this paper, we present a new method for solving the inverse problem in LIDAR sounding based on a hybrid regularization procedure.

1 Background of atmosphere investigation by LIDAR

Optical remote sensing techniques are used today for monitoring atmospheric characteristics due to the fact that, conformal to the diffraction theory, an obstacle can be "seen" by an electromagnetic wave having a wavelength of the same magnitude as the geometric dimension of the obstacle. The return signal of a LIDAR system (Light Detection And Ranging) contains information about the concentration and some physical characteristics of particles in the laser beam direction, in vertical profiles, but to extract this information complex data processing algorithms are necessary.
The description of the laser beam interaction with atmospheric constituents (i.e. molecules, particles, clouds) is based on the fundamental theory of electromagnetic wave propagation in various media, well represented in the scientific literature [MHT00]. But in fact, one must consider for the theoretical approach that the atmosphere contains a wide range of constituents extending from atoms and molecules (Angstrom range d$\sim 10^{-3} - 10^{-4}\mu m$) to aerosols (d$\sim 10^{-2} - 5\mu m$), cloud water droplets and ice crystals (d$\sim 1 - 15\mu m$ and even larger). In the last decades, atmospheric composition and properties were studied by various laboratories or in situ techniques can be "seen" by an electromagnetic wave having a wavelength of the same magnitude as the geometric dimension of the obstacle. The return signal of a LIDAR system (Light Detection And Ranging) contains information about the concentration and some physical characteristics of particles in the laser beam direction, in vertical profiles, but to extract this information complex data processing algorithms are necessary.
The description of the laser beam interaction with atmospheric constituents (i.e. molecules, particles, clouds) is based on the fundamental theory of electromagnetic

wave propagation in various media, well represented in the scientific literature [MHT00]. But in fact, one must consider for the theoretical approach that the atmosphere contains a wide range of constituents extending from atoms and molecules (Angstrom range d$\sim 10^{-3} - 10^{-4}\mu m$) to aerosols (d$\sim 10^{-2} - 5\mu m$), cloud water droplets and ice crystals (d$\sim 1 - 15\mu m$ and even larger). In the last decades, atmospheric composition and properties were studied by various laboratories or in situ techniques in different meteorological conditions and for different classifications [Hin99, KVP03, Jae93, Kop03]. From the laser remote sensing point of view, the basic information that can be used from these studies is that the mixture of these different components results in a series of complex atmospheric interactions that take place with a laser beam. All these phenomena can be used to derive information about the atmosphere. For that, the set-up of the system is different according to the selected phenomena. In all cases, the intensity of the light resulting from these processes is proportional with the initial intensity I_0, the number density of the active diffusers n and the differential angular cross - section σ and this can be used to derive some optical/microphysical parameters of the diffusers in the beam path.

1.1 LIDAR equation and inversion algorithm

Due to its relatively low cost, high reliability and easy operation, the backscatter LIDAR is commonly used for the study of aerosols and clouds. It measures the backscatter at the laser wavelength due to molecules (Rayleigh scattering) and particles (Mie scattering). In the presence of particles of size comparable to the excitation wavelength ($> 0.1\mu m$), Mie scattering processes become important (Mie theory). Thus the laser radiation is elastically scattered ($\lambda_D = \lambda_L$) by small atmospheric particles (i.e. aerosols) of size comparable to the radiation wavelength.
The magnitude of the received LIDAR signal is proportional to the number density of the atmospheric diffusers (molecules or aerosols), their intrinsic properties (i.e. probability of interaction with the electromagnetic radiation at the laser wavelengths, called cross-section value) and with the laser incident energy. In that sense, the direct problem in laser remote sensing is described by the so-called "LIDAR equation", which in the simplest case of elastic backscattering LIDAR can be written as follows [Mea92]:

$$\begin{aligned} RCS(\lambda, Z) = C_S(Z) \cdot [\beta_m(\lambda, Z) + \beta_a(\lambda, Z)] \\ \cdot exp\left[-2\int_{Z_0}^{Z}[\alpha_m(\lambda, Z) + \alpha_a(\lambda, Z)]dz\right] \end{aligned} \quad (1)$$

where: λ is the wavelength of sounding radiation, $RCS(\lambda, Z) = S(\lambda, Z) \cdot Z^2$ is the range corrected signal, S is the LIDAR signal, Z is the distance in the laser path from the transmitter, C_S is the system constant, β_m is the molecular backscatter coefficient, β_a is the aerosol backscatter coefficient, α_m s the molecular extinction coefficient, α_a is the aerosol extinction coefficient and Z_0 is the minimum relevant distance from the transmitter. In general, the inverse problem does not have unique solutions but finitely many branches of solutions. In the above equation, β_a and α_a are unknown parameters.

The atmospheric backscattering coefficient, $\beta_{atm}(\lambda, Z)$ is a key element of the LIDAR equation (1), and is proportional to the cross-section of the With these assumptions, the Ricatti solution of the backscatter LIDAR equation can be written [FHR72, Kle81]:

$$\beta(Z) = -\beta_m(Z) + RCS(Z) \cdot exp\left[-2(LR_a(Z) - LR_m) \cdot \int_{Z_C}^{Z} \beta_m(z)dz\right]$$

$$\left[-2LR_a(Z) \int_{Z_C}^{Z} RCS(z) exp\left[-2(LR_a(Z) - LR_m) \cdot \int_{Z_C}^{Z} \beta_m(z^{'})dz^{'}\right] dz + \right.$$

$$\left. + \frac{RCS(Z_C)}{\beta_a(Z_C)} + \beta_m(Z_C)\right]^{-1} \tag{2}$$

where LR_m is the molecular LIDAR ratio and has a constant value of $8\pi/3$. All molecular parameters can be calculated with sufficient accuracy from ground values of pressure and temperature using atmospheric model [KM94], for the reference value of backscattering coefficient a molecular assumed value at high altitude can be considered, but for solving the equation for the aerosol backscatter, the LIDAR ratio profile must be evaluated. LR_a depends on the aerosol microphysics and can vary between less than 10 sr (ice crystals) and more than 100 sr (heavily polluted air) [KVP03]. It depends on humidity and aerosol mixture and therefore, on height. One possibility is to measure LR_a using either high spectral resolution LIDAR, either Raman LIDAR. But even so, LR_a can only be measured for the lowest part of the profile, where the Raman signal is strong enough, and only when the background radiation is small enough to have a significant signal to noise ratio (nighttime generally). For this reason, additional methods to eliminate nondetermination in LIDAR equation were developed. If the backscattering coefficient in the calibration point can be measured by other methods or estimated from atmospheric model, the main parameter which can introduce significant errors remains the LIDAR ratio due to the fact that to know its values over the entire laser path is practically impossible.

2 Methodology

In order to overcome the nondetermination in the LIDAR equation and the lack of direct LIDAR ratio measurements, we improved the processing algorithm by using complementary data, such as those provided by the Ackermann model [Ack97].
LiSA system signal processing method (fig.1) is based on Fernald-Klett combined, atmospheric model and Mie algorithm for direct problem (theoretical calculation of optical parameters), all integrated in an iterative program to identify the proportions of aerosol components for which the best fit between theoretical and retrieved optical parameters is achieved.
The iterative hybrid regularization technique for lidar data processing and retrieval of the aerosols optical parameters is also used by EARLINET, but they consider as control parameter the pair of effective radius-refractive index of aerosol particle,set into a matrix of possible values. The aerosol is considered a spherical particle characterized by radius and refractive index. In LiSA hybrid algorithm (fig. 1), it is assumed

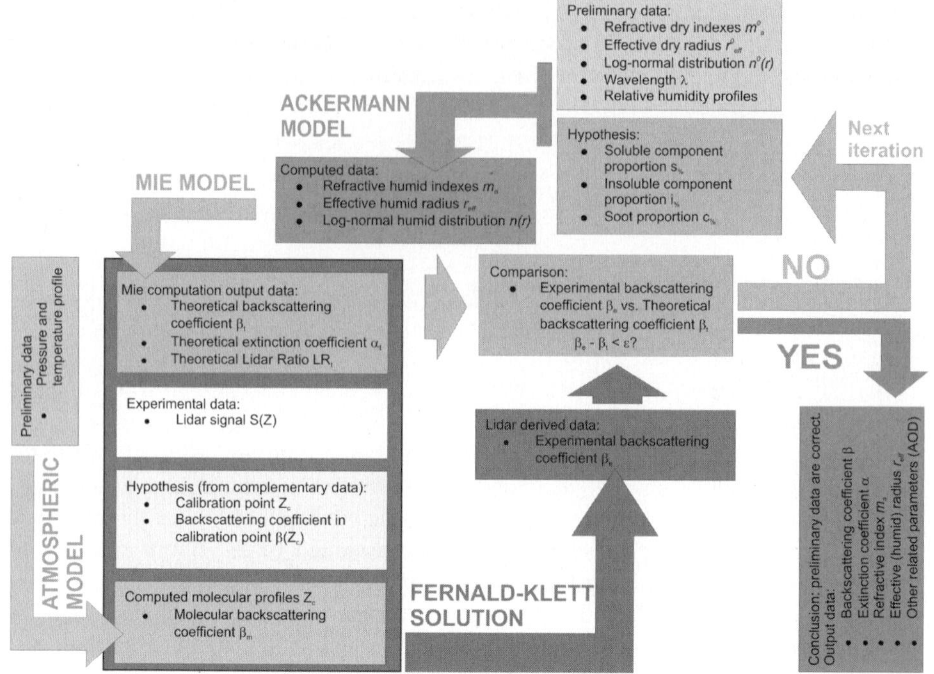

Fig. 1: LiSA hybrid algorithm

that the aerosol is an external mixture of internally mixed components. Each aerosol component is log-normally distributed with respect to the particle radius and representative to tropospheric continental aerosol type. Therefore we consider 3 types of aerosol particles [Ack97]: soluble(s), insoluble (i) and carbonic components (c) characterized by the number mixing ratio μ_s, μ_i, μ_c.

For given optical properties and a distinct relative humidity *RH*, the the variability of the LIDAR ratio is caused by different number mixing ratios μ_k. For given optical properties and a distinct relative humidity *RH*, the the variability of the LIDAR ratio is caused by different number mixing ratios μ_k. The iterative hybrid regularization technique for lidar data processing and retrieval of the aerosols optical parameters is also used by EARLINET, but they consider as control parameter the pair of effective radius-refractive index of aerosol particle,set into a matrix of possible values. The aerosol is considered a spherical particle characterized by radius and refractive index. In LiSA hybrid algorithm (fig. 1), it is assumed that the aerosol is an external mixture of internally mixed components. Each aerosol component is log-normally distributed with respect to the particle radius and representative to tropospheric continental aerosol type. Therefore we consider 3 types of aerosol particles [Ack97]: soluble(s), insoluble (i) and carbonic components (c) characterized by the number mixing ratio μ_s, μ_i, μ_c.

For given optical properties and a distinct relative humidity *RH*, the the variability of the LIDAR ratio is caused by different number mixing ratios μ_k. The water soluble component is the only component whose properties are affected by the relative humidity. The mixing ratio μ_s can be varied between 0.1 and 1 in steps of 0.1. Accordingly, since μ_i is about four orders less, μ_c, chosen to be chosen to be the controlling parameter in our algorithm, is iterated in 0.01 steps from 0.1 to 1. On each iteration we calculate humid log-normal distribution parameters and refractive indices for each aerosol component using Akermann's model [Ack97]:

$$n_k^{RH} = \frac{N_{tot} \cdot \mu_k}{(2\pi)^{1/2} \cdot r \cdot ln\sigma_k} \cdot exp\left[-\frac{ln(r/r_k^{RH})}{2ln^2\sigma_k}\right] \quad (3)$$

$$m_k^{RH} = m_a + (m_k^0 - m_a)\left(\frac{r_k^0}{r_k^{RH}}\right)$$

where *k* is referring to the soluble(s), insoluble (i) and carbonic components (c), the index *"0"* refers to the dry values of microphysical parameters and the *"RH"* indices refers to the corresponding value at relative humidity *RH*. These will be input in the Mie model for determination of the theoretical extinction and backscatter coefficients $\beta_t(Z)$ and theoretical LIDAR ratio.

$$LR_a^{RH} = \frac{\sum\limits_{k=s,i,c}^{f} Q_{ext.k}^{RH}(r, m_k^{RH}, \lambda)\pi r^2 n_k^{RH}(r)dr}{\sum\limits_{k=s,i,c} Q_{bksc.k}^{RH}(r, m_k^{RH}, \lambda)\pi r^2 n_k^{RH}(r)dr} = \frac{\sum\limits_{k=s,i,c} \alpha_k^{RH}}{\sum\limits_{k=s,i,c} \beta_k^{RH}} \quad (4)$$

Molecular backscattering coefficients calculated by the atmospheric model, the LIDAR signal and $\beta_a(Z_C)$ in the calibration point Z_c are used by the Fernald-Klett algorithm to derive the experimental backscattering coefficient $\beta_e(Z)$ conform to eq. (3). The control parameter μ_c will be varied until the difference between $\beta_t(Z)$ and $\beta_e(Z)$ is less then a threshold. At this point the conclusion is that the hypothesis made for the aerosols components is correct and microphysical aerosols parameters like AOD, effective radius, total volume concentration, can be derived. Also now we have the correct value for extinction and backscatter coefficients. For the next profile point the iteration will start with the controlling parameter determined. The method we are using is a regularization one because implies a regularization cycle for controlling parameter, μ_c until we are getting a theoretical profile of β_a almost equivalent to the measured one. This is a hybrid method because it on each iteration, for derivation of experimental β_a by LIDAR inversion we are using as an input the theoretical value of the LIDAR ratio LR_a obtained with the Akermann and Mie model.

3 Results

To illustrate the advantages of the algorithm, a dust intrusion episode was analysed, using both simple Fernald-Klett and our new algorithm for data processing. The intrusion was recorded on April 5th, 2006, during a LIDAR measurements campaign. Generally, no information about the origin of detected aerosols can be extracted from elastic backscatter LIDAR data, but running the iterative hybrid program we obtained the LIDAR ratio profile, being able in this way to discern between the local aerosol and the dust aerosols coming from the Iberic Peninsula. The phenomena is consistent with the prognosis of DREAM (The Dust REgional Atmospheric Model) developed by Dr. Slobodan Nickovic [NPK01] and running at the Supercomputing Center in Barcelona. An example which demonstrates the importance of considering a variable LIDAR ratio as input parameter in inversion algorithm is the case of cloudy sky. In the following figure some profiles of processed LIDAR data recorded during this event are presented.

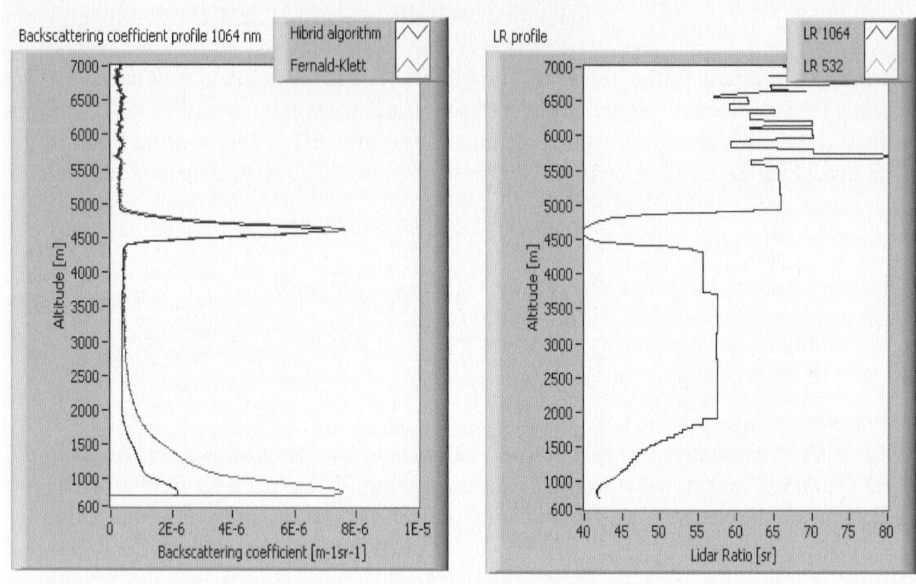

Fig. 2: The retrieval of backscattering coefficient (left) using Fernald-Klett (gray line) and LiSA (black line) algorithm 5th, and LIDAR ratio (right) in case of cloudy sky

Inside clouds the LIDAR ratio has an important variation due to the cloud particles composition and their refractive index. LIDAR ratio can no longer be considered constant with altitude. In the figure above only the profiles measured in the presence of cirrus clouds were considered and one can see that in this case the difference between the two methods becomes important. Inside a cloud, the dimension of hygroscopic aerosol increases, the refractive index varies inducing a stronger scattering and weaker absorption of light. For this reason, clouds are isibledirectly in LIDAR signal, unlike dust aerosols. The hygroscopic characteristics of cloud particles correspond to a incresing of soluble particles proportion and a decreasing of soot particles proportion.

It must be noted although that, in elaborating this algorithm, the multiple scattering was neglected, that's why the quantitative information referring to cloud particles's optical properties must be carrefully analyzed. Also, the calibration point must be properly chosen, above PBL and above the cloud, where the aerosol backscattering coefficient has a minimum value. This choice assures the convergence of the solution,

which is not always evident, especially if both foreward and backward integration is used.
For the presented case, the calibration point was chosen at 5500m altitude, above the cloud, and a 87 % soot proportion in that point was considered. This value was given by Ackermann [Ack97] as the average value for continental urban troposphere. The relative humidity profile was extracted from NOAA database. For the dry values of effective radius and refractive index of every aerosol component the Ackermann hypothesis were used. With these considerations, the backscattering coefficient obtained in calibration point was $4.5 \cdot e^{-7} m^{-1} m^{-1}$ for 1064 nm, respectively $1.16 \cdot e^{-6} m^{-1} m^{-1}$ for 532 nmand the corresponding LIDAR ratio was 51 sr for 1064 nm, respectively 50 sr for 532 nm, those values being used as input in Fernald-Klett algorithm.

4 Conclusions

Inherent assumptions made for the inversion of data, the quantitative results must be carefully analyzed and data validation must be done.

References

[MHT00] Mishchenco, M.I., Hovenier, J.W.,Travis, L.D.: Light Scattering by Nonspherical Particles, Academic Press, San Diego (2000)
[Hin99] Hinds, W.C: Aerosol technology, John Wiley & Sons, New York (1999)
[KVP03] Kulmala, M., Vehkamaki, H., Petaja T., dal Maso M., Lauri A., Kerminen V.M., Birmili W. and McMurry P.H.:Formation and growth rates of ultrafine atmospheric particles:A review of observations.In: J. Aerosol Sci., (2003)
[Jae93] Jaenicke, R.: Tropospheric aerosols in Aerosol-Cloud-Climate Interactions, P. V. Hobbs, Academic Press, San Diego, 1-31 (1993)
[Kop03] Koponen I.K.: Observations Of Tropospheric Aerosol Size Distributions. In Report Series In Aerosol Science 63 (2003)
[Ack98] Ackermann, J.: Two-frequency lidar inversion algorithm for a two-component atmosphere. In: Appl. Opt. 36, 5134-5143 (1998)
[Mea92] Measures R.M.: Laser Remote Sensing. Fundamentals and Applications, Krieger Publishing Company, Malabar, Florida (1992)
[FHR72] Fernald, F.G., Herman, B.M., and Reagan J.A.: Determination Of Aerosol Height Distribution By Lidar. In: J. Appl. Meteorol. 11, 482-489 (1972)
[Kle81] Klett J.D.: Stable Analytical Inversion Solution For Processing Lidar Returns. In: Appl. Opt. 20, 211-220 (1981)
[KM94] Kovalev V.A., Mossmuller H.: Distortion of particulate extinction profiles measured with lidar in a two-component atmosphere. In: Appl. Opt. 33, 6499-6507 (1994)
[Ack97] Ackermann J.: The extinction-to-backscatter ratio of tropospheric aerosol: a numerical study. In: Journal of Atmospheric and Oceanic Technology, 15, no. 4, 1043-1050 (1997)
[NPK01] Nickovic, S., Papadopoulos A., Kakaliagou O. and Kallos G.: Model for prediction of desert dust cycle in the atmosphere. In: J. Geophys. Res., 106, 18113-18129 (2001)

A Demonstrator Platform for Coupled Multiscale Simulation

Carlo de Falco[1,2], Georg Denk[2], and Reinhart Schultz[2]

[1] Bergische Universität Wuppertal
[2] Qimonda AG, München

Summary. In this communication we present the CoMSON Demonstrator Platform (DP), a software tool designed to help researchers in testing and validating models and algorithms for coupled simulation of nanoelectronic circuits and devices. The structure of the DP is presented with an explanation of the motivations behind the critical design choices. A multilevel simulation of a CMOS AND gate using two different coupling algorithms is provided as an application example. The example is intended to demonstrate the suitability of the DP as a flexible prototyping environment and its ability to cope with real life industrial problems. In the numerical simulations both the semi-classical Drift-Diffusion model (DD) and a Quantum Corrected DD model (QCDD) are employed and their predictions are compared.

1 Introduction

Currently, to design new integrated circuits or to port existing designs to a new technological platform, designers follow a path composed of different, almost independent, steps. At each stage of this path different software tools are used to support the design flow. Process simulators are used to predict geometries, doping profiles and other physical parameters of devices that can be produced in a given technological process. Device simulators are then used to predict electrical/thermal behavior of the new devices. Using physical considerations, often based on the drift-diffusion framework with simplifying assumptions on geometry, doping profiles, material parameters, one has to define compact models to describe the device behavior with simple, explicit analytical expressions. Very often a priori considerations lack predictiveness and accurate a posteriori calibration of model parameters based on numerical simulations and experimental data is needed. The compact device models are used in circuit simulations to predict the behavior of new circuit topologies or to evaluate the performance of existing topologies implemented with new technologies. Finally, an optimization step is used to maximize circuit performance by perturbing device parameters in the vicinity of the given values. This design flow presents some disadvantages that are becoming more relevant as CMOS technology is scaled down to its physical limits. To be as accurate as possible, compact models have grown to include several hundreds of parameters (see, for example, [7]) with little or no connection with physical characteristics of the devices. The lack of connection between model parameters and physical properties renders, on one hand, very delicate and cumbersome the parameter calibration stage and, on the other hand, it makes it

almost impossible to perform an optimization of the circuits based on the geometry and doping profiles of the devices. The latter effect is even more evident at the current stage of technological advancement where not only device dimensions are being scaled but completely new device geometries are being considered (DG, Tri-Gate, GAA, FinFET, nanotubes, ... see, for example, [12])
A possible approach to the solution of the problems described above is to create simulation tools where the behavior of the devices is represented not by evaluating the explicit analytical relations given by the compact models but by performing a direct simulation based on more accurate physical models taking the complete 2D/3D device geometry and realistic doping profiles as obtained by process simulation into account. This clearly comes at the cost of a great increase in computational effort, but the advantages are many-fold. First of all the use of few physically based design variables instead of many fitting parameters gives designers a much higher level of understanding which can lead faster to better design decisions and, furthermore, it can greatly help the construction of automatic optimization tools.

2 The CoMSON Demonstrator Platform

To achieve the above goal, many open problems still need to be solved. Apart from the computational cost (which will need to be reduced as much as possible, for example via Model Order Reduction techniques [5], or parallelization, but cannot be expected to be anywhere close to that of compact models) the coupling itself can lead to instability and convergence issues that need to be addressed properly by resorting to suitable numerical schemes. For this reason within the EU RTN project CoMSON (http://www.comson.org) a Demonstrator Platform (http://www.comson.org/dem) will be developed to connect numerical simulation tools available throughout the nodes of the CoMSON consortium through a common interface. In this way, researchers willing to be confronted with the problems arising in the framework of coupled simulation will be given the opportunity to abstract from the implementation of the basic tools (device simulator, circuit simulator, heat transfer simulator, ...) and to concentrate on the coupling itself. The architecture of the Demonstrator Platform will be the main focus of this communication.

2.1 Goals of the CoMSON Demonstrator Platform

The basic idea behind the Demonstrator Platform is to provide an integrated testing framework for researchers interested in new strategies for coupling simulation tools from different physical domains. Within this framework they will be able to implement, test and assess the applicability of their methods to real life problems without having to enter the details of the lower level tools. At the same time, researchers interested in new mathematical models for the basic physical phenomena can asses their relevance for overall system behavior taking advantage of the coupling with system level simulation tools.
It has has been designed to achieve the following objectives:

- providing a fast prototyping environment in which new and existing algorithms can be tested compared and assessed;

- allowing application of the algorithms, once assessed, to real life industrial problems.

2.2 The Structure of the CoMSON Demonstrator Platform

To achieve the results listed above, the structure depicted in Fig. 1 has been devised. The main components of the DP are:

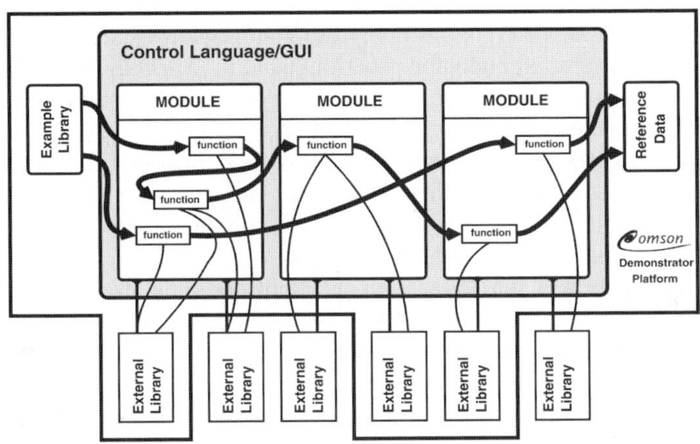

Fig. 1: The structure of the DP

1. a library of test examples and experimental measurements to be used as benchmarks for any new method,
2. a set of modules each consisting of a collection of functions providing the basic functionality of the single domain simulators,
3. a controlling programming language with which the aforementioned functions can be connected to form simulation algorithms. To separate the implementation of the basic functions from that of the coupling algorithms, the single domain simulators are organized as independent external libraries from which the DP functions are obtained via interfaces (bindings for the controlling language to the external libraries).

The test example library will contain both real-life industrial problems from the industrial nodes of the CoMSON consortium (NXP, Qimonda and STMicroelectronics) and simplified academic examples which display the same phenomena but without complications that are not essential for the understanding of the problem. This latter class of examples is especially fit for training purposes.
The initial set of functions in each module will be enriched if new algorithms will be studied that require lower level functions not initially available.
The programming language chosen as a controlling language is `Octave`. The main factors driving this choice were:

- the availability of a free language interpreter, and of a free API for building language extensions in `C, C++, Fortran`;

- the very high level of compatibility of the `Octave` interpreter with the `Matlab` programming language syntax which is the *de-facto* standard for teaching numerical algorithms;
- GPL licensing terms make it simple to distribute a fully functional system based on `Octave` including all needed software dependencies.

To better demonstrate the structure of the Demonstrator Platform and its use we will resort to a practical example. We will consider device/circuit coupling strategies belonging to two different classes:

- based on the extension of the device simulator by considering the network equations as general boundary conditions. Such an approach is used in [2] (in the case of stationary semiconductor equations) and in [1] (in the case of evolutionary semiconductor equations) to derive analytical results for the coupled system.
- Based on extension of the circuit simulator by adding the spatially discretized semiconductor equations to the system of network equations. This approach was applied in [11] for the numerical analysis of the coupled system and, together with a staggered solution approach, in [8, 9] for the simulation of the electrothermal behavior of an operational amplifier.

By implementing solvers based on such different coupling strategies, we demonstrate the flexibility of the Demonstrator Platform architecture. Moreover, we show how the abstraction layer provided by the Demonstrator Platform can be exploited for further generalization of the implemented algorithms by extending the coupling strategies considered to the case where more complex semiconductor models (like the Quantum-Corrected Drift-Diffusion class of models as described in [6]) are used for device simulation.

3 Two algorithms for coupled circuit-device simulation

In the current section we introduce two different strategies for simulating an electronic circuit where part of the composing elements is described through a full 2D Finite Element model and part is represented by lumped elements.

In Sec. (3.1) we introduce the system of equations stemming from the coupling of circuit and device equations.

The first algorithm is outlined in Sec. 3.2 and is referred to as *circuit-driven* algorithm because it is an approach that could be applied if one were to extend an existing circuit simulator to include distributed device models. The second algorithm, described in Sec. 3.3 is a viable option to extend a device simulation program based on the *Gummel Map* algorithm to include coupled simulation capabilities. In describing the algorithms we will point out which functionalities need to be exposed to the controlling language by the single domain simulators for their implementation.

For more details on implementation we invite the interested readers to download the software code and documentation which will be soon available at http://www.comson.com/dem.

3.1 The Circuit/Device Coupled Problem

Using charge/flux based MNA modeling for the network (see, for example, [10] for more details), we can write

$$\begin{array}{ll} A_q \mathbf{q}_{,t}(t) + \mathbf{f}(\mathbf{x}(t), t) = \mathbf{j}_N \\ \mathbf{q}(t) - g(\mathbf{x}(t)) \qquad = 0 \end{array} \tag{1}$$

where A_q is a constant *incidence matrix*, $\mathbf{f}(\mathbf{x}(t), t)$ and $g(\mathbf{x}(t))$ are non-linear functions, $\mathbf{x}$ is a vector formed by the values of the voltage nodes and of the currents through the inductors and voltage sources and $\mathbf{q}$ is the vector containing the values of the capacitor charges and the magnetic fluxes through the inductors. $\mathbf{j}_N$ represents the currents flowing from the circuit into the contacts of the distributed device. Furthermore note that the subscript $(\cdot)_{,t}$ indicates differentiation with respect to time.
Considering, for sake of brevity, the effect of charge transport due to electron carriers only, a very general form to express the equations for the distributed device which can fit the whole class of Quantum Corrected Drift Diffusion (QCDD, see [6]) is as follows

$$\begin{array}{ll} P(\Phi, n, p) = 0 \text{ in } \Omega & \Phi|_{\Gamma_i} = \phi_i \\ n_{,t} + C_n(\Phi, n) = 0 \text{ in } \Omega & n|_{\Gamma_i} = n_i \end{array} \tag{2}$$

In (2) Φ, n, p are the electric potential, electron density and hole density inside the device computational domain Ω respectively; P and C_n are non-linear differential operators for the Poisson equation and electron current continuity equation respectively; Γ_i is the i_{th} contact of the device and Φ_i and n_i are the values of the electric potential and electron density on each of the contacts.
From the values of Φ and n one can compute the charges q_{S_i} and currents j_{S_i} at the contacts of the device as

$$\begin{array}{ll} \int_{\Gamma_i} \varepsilon \nabla \Phi \cdot \nu \, d\gamma & = q_{s_i} \\ \int_{\Gamma_i} \mathbf{J}_n(\Phi, n) \cdot \nu \, d\gamma & = j_{s_i} \end{array}$$

where $\mathbf{J}_n$ represents the current density in the device, and ν being the unit outward normal to the boundary of the device. Finally the circuit and device can be coupled by enforcing charge conservation:

$$\begin{array}{ll} \mathbf{j}_N + A_s(\mathbf{j}_s + \mathbf{q}_{s,t}) = 0 \\ \alpha(\phi_N + \mathbf{V}_{BI}) \qquad = A_s^T \mathbf{x} \end{array} \tag{3}$$

In (3) A_s is an incidence matrix indicating to which nodes in the network the contacts of the distributed device are connected, Φ_N are the voltages of the network nodes connected to the device and $\mathbf{V}_{BI}$ are the corresponding built-in voltages, α is a scaling factor and the vectors $\mathbf{j}_s = [j_{s_1} \cdots j_{s_1}]^T$ and $\mathbf{q}_s = [q_{s_1} \cdots q_{s_1}]^T$ are the currents and charges flowing through the distributed device contacts.

3.2 The Circuit-Driven Algorithm

The basic idea behind this approach is to express the complete coupled system in a form as similar as possible to the MNA equations (1).
By using (1) and (3) and discretizing in time by applying Rothe's method and a BDF(m) formula, we can write the coupled problem as

$$\begin{array}{ll} \beta_0 \left(A_q \mathbf{q}(t_n) + A_s \mathbf{q}_s(t_n) \right) + \\ \quad + \mathbf{f}(\mathbf{x}(t_n), t_n) + A_s \mathbf{j}_s(A_s^T \mathbf{x}) = - \sum\limits_{k=1 \ldots m} \beta_k \left(A_q \mathbf{q}\left(t_{n-k}\right) + A_s \mathbf{q}_s(t_n) \right) \\ \mathbf{q}(t_n) - g(\mathbf{x}(t_n)) \qquad = 0 \\ \mathbf{q}_s(t_n) - g_s(A_s^T \mathbf{x}(t_n)) \qquad = 0 \end{array} \tag{4}$$

To solve this system with a Newton method we need a function to compute

- Currents and charges flowing through the distributed device contacts as a function of the node voltages
- Derivatives of such currents and charges with respect to the node voltages (local capacitance and conductance matrices)

Such function is implemented along the following lines.

1. Solve the DD equations with the Gummel map algorithm
2. Linearize the Poisson equation around the solution and compute the charges as the flux of $-\varepsilon\nabla\Phi$ through the contacts
3. Linearize the Continuity equation around the solution and compute the currents as the flux of $-\epsilon\mu_n(n\nabla\Phi - V_{th}\nabla n)$ through the contacts
4. Obtain the capacitance and conductance matrices via a Schur complement technique from the linearized Poisson (continuity) equation

The main requirement to implement this algorithm in the framework we described is that, to perform steps 2-3, we need the device simulation module to define functions that, given the contact potentials as input, produce as output the matrices for the linearized Poisson and continuity equation at each integration time point.
Once such matrices are available the computation of conductance and capacitance matrices is very straightforward.
Consider for example the Poisson equation for a device with two contacts. The discrete, linearized Poisson equation has the form

$$\begin{bmatrix} P_{11} & 0 & P_{1I} \\ 0 & P_{22} & P_{2I} \\ P_{I1} & P_{I2} & P_{II} \end{bmatrix} \begin{bmatrix} (\phi_1 + v_{BI_1})\mathbf{1}_{\Gamma_1} \\ (\phi_2 + v_{BI_2})\mathbf{1}_{\Gamma_2} \\ \Phi_I \end{bmatrix} = \begin{bmatrix} q'_{s_1} \\ q'_{s_2} \\ 0 \end{bmatrix} \tag{5}$$

where $\mathbf{1}_{\Gamma_i}$ represents a column vector of all ones with as many elements as there are on the mesh for the i-th contact, $\mathbf{\Phi}_I$ is the vector with the values of the electric potential at the internal mesh nodes, and q'_{s_1} is the vector of the charges at the mesh nodes on the i-th contact.
The total charge at the contacts can be expressed as

$$q_{s_1} = \mathbf{1}^T_{\Gamma_1} q'_{s_1}\,; \quad q_{s_2} = \mathbf{1}^T_{\Gamma_2} q'_{s_2}$$

and by eliminating $\mathbf{\Phi}_I$ one can get a relation for the charges in terms of the contact potentials of the form

$$\begin{pmatrix} q_{s_1} \\ q_{s_2} \end{pmatrix} = \begin{pmatrix} c_{11} & c_{12} \\ c_{21} & c_{22} \end{pmatrix} \begin{pmatrix} \phi_1 \\ \phi_2 \end{pmatrix} + \dots$$

where c_{ij} is the derivative of charge q_i with respect to node voltage Φ_j.

3.3 The Device-Driven Algorithm

The Device-Driven algorithm we present is a generalization of the well-known Gummel algorithm for the solution of the DD equations where the circuit equations are included as boundary conditions for each of the decoupled problems.
To set up such algorithm we need to decouple the problem into two subproblems, corresponding to the Poisson and continuity equations respectively:

Problem A (Poisson)

$$
\begin{aligned}
&P(A_s^T\mathbf{x}, \Phi_I, \mathbf{q}_s) &&= 0 \\
&\beta_0(A_q\mathbf{q}(t_n) + A_s\mathbf{q}_s(t_n)) + \mathbf{f}(\mathbf{x}(t_n), t_n) + A_s\mathbf{j}_s = \\
& &&= -\sum_{k=1\dots m} \beta_k(A_q\mathbf{q}(t_{n-k}) + A_s\mathbf{q}_s(t_{n-k})) \\
&\mathbf{q}(t_n) - \mathbf{g}(\mathbf{x}(t_n)) &&= 0
\end{aligned}
\tag{6}
$$

Problem B (Continuity)

$$
\begin{aligned}
&C_n\left(A_s^T\mathbf{x}, \Phi_{n_I}, \mathbf{j}_s\right) &&= 0 \\
&\beta_0(A_q\mathbf{q}(t_n) + A_s\mathbf{q}_s(t_n)) + \mathbf{f}(\mathbf{x}(t_n), t_n) + A_s\mathbf{j}_s = \\
& &&= -\sum_{k=1\dots m} \beta_k(A_q\mathbf{q}(t_{n-k}) + A_s\mathbf{q}_s(t_{n-k})) \\
&\mathbf{q}(t_n) - \mathbf{g}(\mathbf{x}(t_n)) &&= 0
\end{aligned}
\tag{7}
$$

In (6) Φ_I is the value of the electrical potential at the internal nodes of the device mesh $\mathbf{x}(t_n)$ is the vector of the network node voltages, the network and device node charges are $\mathbf{q}(t_n)$ and $\mathbf{q}_s(t_n)$ and the current through the device contacts is $\mathbf{j}_s$. In (7) Φ_{n_I} represents the vector of the values of the quasi-Fermi potentials at the internal nodes of the device mesh. As in Sec. 3.2 a BDF(m) formula has been applied for time discretization.

Having defined the two subproblems above, the procedure to be carried out at each time step can be described as follows:

Iterate through steps 1 and 2 below until consistency is reached:

1 Solve the non-linear Poisson equation [A] for the unknowns Φ_I, $\mathbf{x}(t_n)$, $\mathbf{q}(t_n)$, $\mathbf{q}_s(t_n)$ considering $\mathbf{j}_s$ a known quantity.

2 Solve the non-linear continuity equation with unknowns $\Phi_{n_I}, \mathbf{x}(t_n), \mathbf{q}(t_n), \mathbf{j}_s$ and considering $\mathbf{q}_s(t_n)$ fixed

Note that both step 1 and step 2 involve the solution of a system of non-linear equations so they require two more Newton loops to be nested within the iteration described above.

To be able to impose the appropriate boundary conditions we need the circuit simulation module to define a function that, given the values of the network unknowns as input, produces as output the matrices for the linearized MNA equations. This is the main requirement to be able to implement the Device-Driven method in our framework.

4 An Application Example

As an application for the algorithms described above, the test circuit in Fig.2(a), representing a CMOS AND gate, has been considered. For sake of simplicity only the n-type MOSFET in the output stage has been simulated using a full 2D simulation as shown in Fig. 2(b). The simulated device is a very aggressively scaled MOSFET

with a gate length of $15nm$. For such small devices, according to traditional scaling rules a V_{dd} voltage of $0.8V$ should be appropriate.
Fig. 3(a) displays the switching behavior of the AND gate computed with a DD Model for the distributed device and using both the Circuit and the Device-Driven algorithms. As stated in the previous section both coupling approaches can be applied

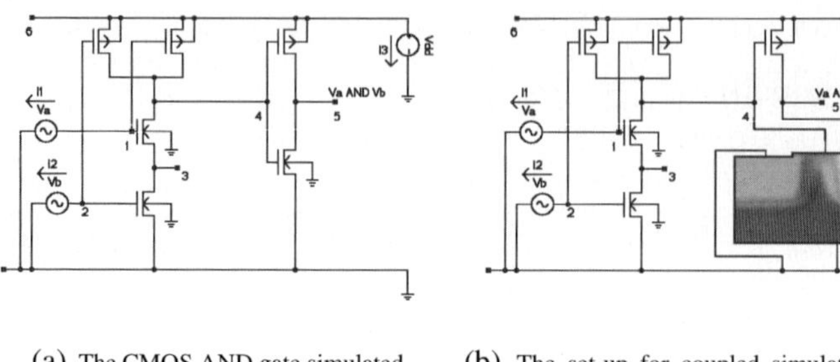

(a) The CMOS AND gate simulated

(b) The set-up for coupled simulation of the AND gate

Fig. 2:

with no effort to more advanced semiconductor models. To demonstrate this we repeated the simulation using a Density Gradient model (See [3], [4] for a description of the model). The impact of quantum correction on the performance of the circuit is shown in Fig. 2(b). Essentially the circuit does not behave as a digital gate at all. This is mainly due to a shift in the threshold voltage of the device connected to the increased *Equivalent Oxide Thickness* (see [6] for a description of this effect). Indeed, as shown in Fig. 3(c), if $V_{dd} = 1.6V$ is applied, the circuit displays better performance.

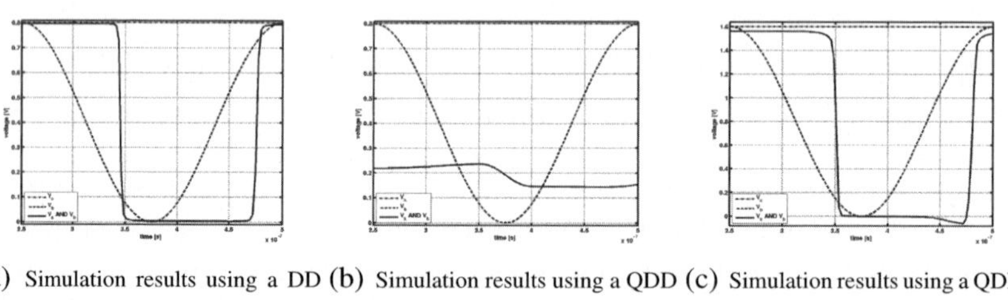

(a) Simulation results using a DD model with $V_{dd} = 0.8V$

(b) Simulation results using a QDD model with $V_{dd} = 0.8V$

(c) Simulation results using a QDD model with $V_{dd} = 1.6V$

Fig. 3:

Acknowledgment

The work described here is partly financially supported by the European Commission in the framework of the CoMSON RTN project, grant number MRTN-2005-019417.

References

1. G. Alì, A. Bartel, and M. Günther, *Parabolic differential-algebraic models in electrical network design*, SIAM J. Mult. Model. Sim. **4** (2005), no. 3, 813–838.
2. G. Alì, A. Bartel, M. Günther, and C. Tischendorf, *Elliptic partial differential-algebraic multiphysics models in electrical network design*, Mathematical Models and Methods in Applied Sciences **9** (2003), no. 13, 1261–1278.
3. M. G. Ancona and G. J. Iafrate, *Quantum Correction to the Equation of State of an Electron Gas in a Semiconductor*, Phys. Rev. B **39** (1989), 9536–9540.
4. M. G. Ancona and H. F. Tiersten, *Macroscopic physics of the silicon inversion layer*, Phys. Rev. B **35** (1987), no. 15, 7959–7965.
5. T. Bechtold, E. B. Rudnyi, and Jan G. Korvink, *Fast simulation of electro-thermal mems: Efficient dynamic compact models*, Springer, 2006.
6. C. de Falco, A. L. Lacaita, E. Gatti, and R. Sacco, *Quantum–corrected drift–diffusion models for transport in semiconductor devices*, J. Comp. Phys. **204** (2005), no. 2, 533–561.
7. G. Denk, *Circuit simulation for nanoelectronics*, Proceedings of Scientific Computing in Electrical Engineering (SCEE), Springer–Verlag, 2004, pp. 13–20.
8. T. Grasser, *Mixed-mode device simulation*, Ph.D. thesis, Institut für Mikroelektronik, TU/Wien, 1999.
9. T. Grasser and S. Selberherr, *Fully coupled electrothermal mixed-mode device simulation of SiGeHBT circuits*, IEEE Transactions on Electron Devices **48** (2001), no. 7, 1421–1427.
10. M. Günther, U. Feldmann, and J. ter Maten, *Modelling and discretization of circuit problems*, Handbook of Numerical Analysis (P.G. Ciarlet, W.H.A. Schilders, and E.J.W. ter Maten, eds.), vol. XIII, Elsevier North-Holland, 2005, pp. 523–659.
11. C. Tischendorf, *Coupled systems of differential algebraic and partial differential equations in circuit and device simulation. Modeling and numerical analysis*, Habilitationsschrift, Inst. für Math., Humboldt-Univ. zu Berlin, 2003.
12. H.-S. P. Wong, *Beyond the conventional transistor*, IBM J. Res. & Dev. **46** (2002), no. 2/3, 133–169.

Upon the Interaction between Magnetic Field and Electric Arc in Low Voltage Vacuum Circuit Breakers

Smaranda Nitu[1], Dan Pavelescu[1], Constantin Nitu[1], Gheorghe Dumitrescu[2], and Paula Anghelita[2]

[1] POLITEHNICA University of Bucharest - snitu@apel.apar.pub.ro
[2] Research and Development Institute for Electrical Industry - apel2@icpe.ro

Abstract - The paper presents an investigation of the magnetic field influence within low voltage switching process in vacuum, in the case of strong currents interrupting. The axial, transverse and radial magnetic field action upon the vacuum electric arc behavior is analyzed on a mono-phase model. The conclusions obtained by modeling the electromagnetic field in the vacuum quenching chamber are compared with the experimental results. The experimental set-up can reproduce the real switching conditions of the power vacuum circuit-breaker. The goal of the study is the improvement of the circuit-breaker switching capabilities.

1 Introduction

In spite of the very good dielectric properties of vacuum, the vacuum interrupters have to overcome some difficulties connected with the electric arc behavior. At low currents, up to 10 kA, the arc burns in diffuse mode, so that the contacts erosion and heating is acceptable. But currents up to 10 kA represent a range for which other interrupters, much cheaper, can be utilized with success. Thats why, larger currents, up to 100 kA, represent the goal for the vacuum interrupters performances.
At currents over 10 kA, owing to the Pinch effect, the arc column is concentrated, by the interaction between the current flowing through it and its own magnetic field. The result is a severe erosion of the contacts surfaces, caused by the intense heating and melting at the electric arc base.
Finally, the resulting contact surfaces rugosity diminishes the dielectric rigidity of the contact gap and the interrupters breaking capability.
A magnetic field (axial, transverse or radial) interacting with the current flowing through the arc has a beneficial action and increases the switchgear breaking performances. These magnetic fields are produced by the interrupted current itself, owing to a specific contact parts design.
The measurements were performed by using equipment able to provide asymmetric short-circuit currents up to 54 kA_{rms} (110 $kA_{\max}$ - for asymetrical short circuit current) at 1360 V_{rms}. It is provided also with the possibility of performing light intensity and spectroscopic measurements.

2 The axial Magnetic Field Action

The axial magnetic field (AMF) can modify the state of the electric arc in vacuum, by the interaction with the electric arc plasma. The result is the arc maintaining in the diffuse burning state up to large values of the current and a more uniform distribution of the electric arc power on the contact device surfaces. These consequences are more evident if the magnetic field is uniform in the area of interest.
The simplest solution to obtain a uniform magnetic field in the gap between the contact plates is an external coil placed around the quenching chamber (Fig.1). In the main electric circuit, the coil L is parallel connected to a resistor R_p which allows the adjustment of the coil current and, consequently, the one of the AMF. There are also specific contact parts design solutions capable to create an axial magnetic field, but in this case it is impossible to vary the magnetic field value in order to study its influence and establish the optimum value.

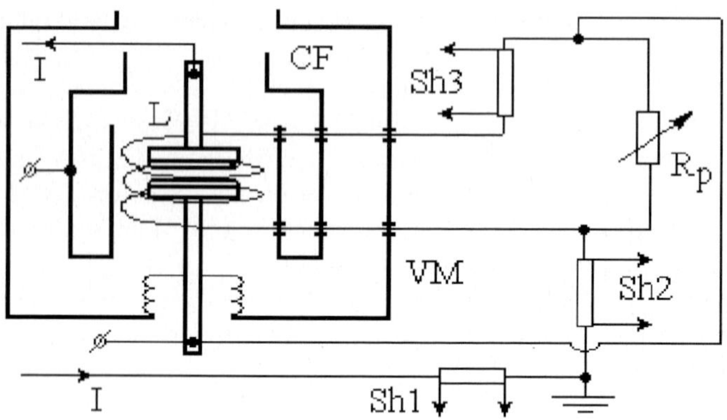

Fig. 1: Active functional part of the testing equipment

When the current passes through zero, the magnetic field phase is shifted behind the current, like in a real circuit breaker (Fig.2) [FL96]. The current limit for which breakdown failures appear is rising if an axial magnetic field is applied, up to a limit, at which the residual conductance at current zero moment, is stabilized [PDN99]. A too large residual magnetic field (B_0) can favor the arc reigniting, by maintaining the residual conductivity of the gap between the contacts, at large values.
A systematic analysis of the dependence on magnetic field variation (0...1 T) of the five parameters (Fig.2) which characterize the voltage drop on the large power electric arc developed in a vacuum chamber, with small gap between contacts and low voltage supply is presented.
Experimentally was observed the major influence of the magnetic field upon U_{a4} and the moderate one upon U_{a3}, the most important voltages as concerns the switching process.
As a consequence of the performed experimental determinations, it was found out that, for low voltage range and a variation of B_{max} between 0 and 1000 mT, U_{a3} and U_{a4} are power functions of B with under unit exponent. So, the average value of the arc voltages can be approximated by:

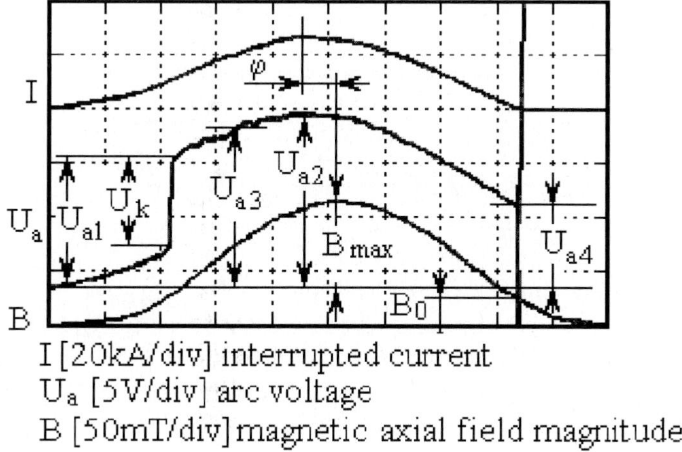

Fig. 2: Typical oscillogram of a successful interruption

$$U_{a3} = a + b \cdot B_{\max}^{\delta}, where : 0 < \delta < 1 \tag{1}$$

$$U_{a4} = U_{a4}^{*} + b \cdot B_{\max}^{\delta} \tag{2}$$

with $U^{*}_{/rma4}$, the value of U_{a4} for the case of no external magnetic field, dependent upon the current and approximated by an exponential law. Finally the variation of U_{a4} is found to be a curve family, with the current as parameter:

$$U_{a4} = \left[\left(c\hat{I} - d\right)\exp\left(\hat{I}/k\right) + b \cdot B_{\max}^{\delta}\right] \cdot \eta\left(\hat{I} - I_0\right) \tag{3}$$

where $\hat{I}$ is the hitting current, I_0 is the limit current for which the electric arc in vacuum exists, η is the unit step function and *b*, *c*, *d* and *k* are constants, determined from experimental values. This dependence of the final voltage U_{a4} is represented like a family of curves with the current as parameter, in Fig.3.
From these curves results the variation of maximum magnetic field magnitude versus the hitting value of the breaking current, when the 100% failure emerges. It is to be noticed a variation over 500 mT of the magnetic field magnitude has no significant effect upon breaking current raise.
So, the AMF produced by an outside coil is much stronger that the required one. In spite of the benefic action of an AMF, commercial applications are strongly affected by the Ohmic losses in the coil. The manufacturers found solutions to minimize the losses and optimize the axial magnetic field, to the minimum required.

3 Transverse and Radial Magnetic Field Action

The transverse (TMF) is obtained in a cup shaped contact (Fig.4) and the radial magnetic field (RMF) is created by the current flow through a spiral type contact (Fig.6). The TMF and RMF influence the arc behavior by yielding its motion on the contact plates surface, under the Laplaces force. Thus, the arc remains concentrate, but

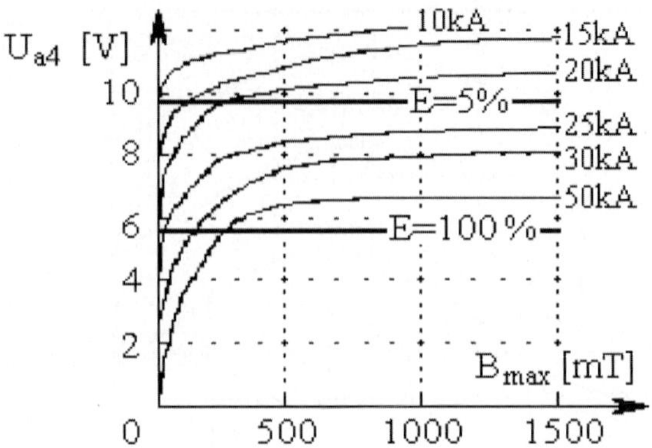

Fig. 3: Calculated values of U_{a4} and the probability of 5% and 100% (E=5%, E=100%) interruption failure (determined in a previous investigation [PTN96])

changing continuously its position, avoids the electrodes over heating and minimize the electrode surface erosion.

3.1 Transverse Magnetic Field

The TMF produces the column movement on the contact rim.
The graphics from Fig.4 represent the light intensity of the arc column distributed over the contact rim.

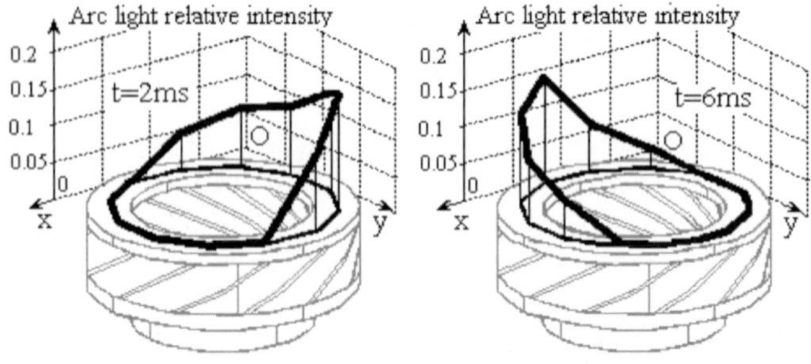

Fig. 4: Arc appearance for a 12 kA asymmetrical current (t=2ms and t=6ms)

In order to measure the light intensity emitted by the arc in points placed at 30° on the contact circumference were used 8 channels (Fig.5): Cxi, i=1..4 for recording the arc light intensity along the Ox direction and Cyj, j=1..4 for recording it along the Oy direction. The 8 signals are proportional to the line integral of the light emitted along the chosen direction. The intersection point A_{ij} of the observing directions of the channels Cxi and Cyj is assumed to be on the medium circle of the contact rim. The arc light intensity in each such point is proportional to the values measured through the channels Cxi and Cyj.
The arc evolution was evidenced in the case of an asymmetrical current successful interruption of 12 kA in Fig.4.

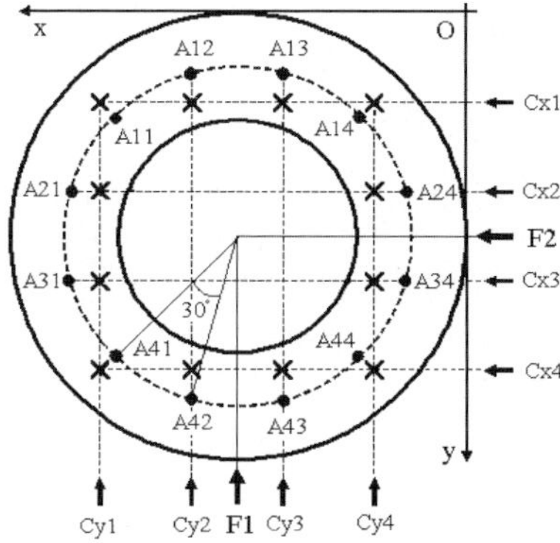

Fig. 5: The position of the 8 channels used for the electric arc investigation in the vacuum chamber

The arc velocity can be approximated by the maximum light point velocity, which is 5 - 8 m/s in the case of an interrupted current of 12 kA.

3.2 Radial Magnetic Field

The RMF is created by a specific geometry (Fig.6 - spiral type contact), at which the contact disc is divided in four curved petals, with variable transversal section. The electric arc is obliged by the Laplaces force to move from the base of the petal, where it is initiated, to the petals end, and than around the contact. The Laplaces force is calculated by a 2D FEM software, for a certain position of the arc, which is considered to have a cylindrical form, with a diameter of 10 mm. The difficulty is to appreciate the plasma mass of the electric arc and the resistant force, due to the viscosity.
The arc velocity is calculated using the relation deduced from [DSS02]:

$$v_{arc} = I^{5/6}\sqrt{j} \cdot \left[2 \cdot \left(\frac{L \cdot b_r \cdot h_{ev}}{m_N} \right)^2 \frac{1}{(T_s - T_0)\sqrt{\frac{\pi \kappa c \rho}{2}}} \right]^{1/3} \tag{4}$$

as function of the current intensity *I* and density *j*, with the same notations and in the same simplifying hypotheses.
The direction of the arc velocity is assumed to be the same as the tangential component of the electrodynamic force, numerically calculated from the electromagnetic field distribution. This calculation was performed for every arc position. In the v_{arc} relation, *I* (the interrupted current intensity), *j* (the current density), *L* (the arc length) and *b* (the magnetic field density) are calculated at each time step ($\Delta t = 0.05$ms), that is at each arc position. The other terms from the relation (4) are considered to be constant, like in the following relation:

$$v_{arc}(t) = K_1 \cdot K_2 \cdot I(t)^{5/6} \cdot j(t)^{1/2} \cdot [L(t) \cdot b(t)]^{2/3} \tag{5}$$

with the constants:

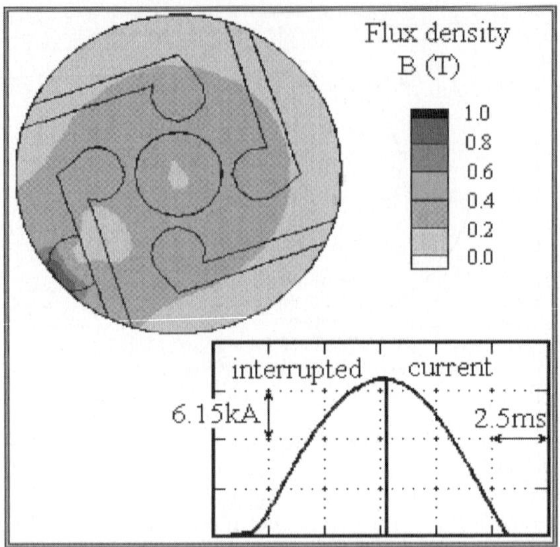

Fig. 6: Magnetic field distribution at t=6ms

$$K_1 = \left(\frac{h_{ev}}{m_N}\right)^{2/3} \quad ; \quad K_2 = \left[\frac{2}{(T_s - T_0)\sqrt{\frac{\pi\kappa c\rho}{2}}}\right]^{1/3} \tag{6}$$

The arc movement is considered to be uniform accelerated upon each time step. The calculated velocity values are presented in Fig.7, together with the interrupted current, the arc voltage and the light intensity measured by the F1 optical fiber from Fig.5. The arc velocity is also experimentally determined, from the light intensity measurements: between a maximum and a minimum of the light intensity, the arc is moving along a quarter of the contact circumference.
In Fig.7 point 1 represents the moment of the arc ignition; at moment 2 the current has 10 kA and the arc is constricted; at moment 3 the contact gap has 4 mm and the arc begins to move around.
So, the calculated results (dashed line in Fig.7) are much different from the experimental ones at the maximum current values, that is at current densities of 190...230 A/mm^2 (up to 40%). At current densities j = 100...120 A/mm^2, at the beginning and end of the arc movement, these differences are less, like in [DSS02]. The total arc displacement is 1.5 contact circumferences.

4 Conclusions

An efficient tool to predict arc behavior in the vacuum circuit breakers, under the magnetic field influence, is very necessary in order to optimize the apparatus design. The proposed model seems to be promising, but it will be improved by the computation of the eddy currents, caused by the magnetic flux variations, produced by the electric arc movement on the electrodes surface. This is necessary because the lack of the eddy currents influence in the contact model is the explanation for the great difference between the real and the calculated arc velocity values.

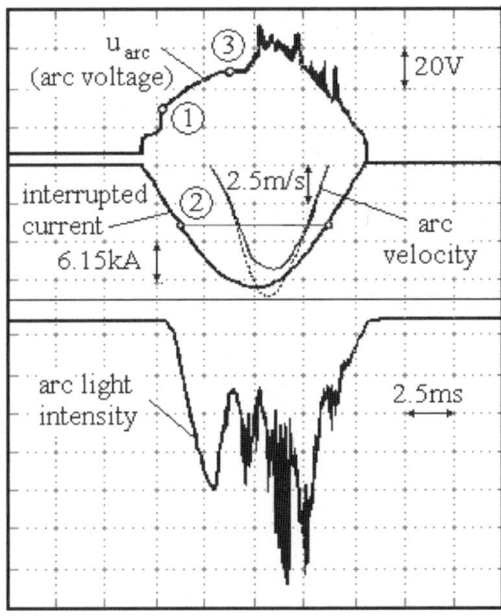

Fig. 7: Arc velocity, correlated with the interrupted current

References

[FL96] Fenski, B., Lindmayer, M.: Vacuum interrupters with axial field contacts - 3D finite element simulations and switching experiments, Proc. of the XVIIth ISDEIV, Berkeley, California, vol. I, 337-342 (1996)

[PDN99] Pavelescu, D., Dumitrescu, G., Nitu, S., Trusca, V., Pavelescu jr, D.: The Influence of the Axial Magnetic Field upon the Low Voltage Electric Arc in Vacuum, in IEEE Transactions on Power Delivery , vol.14, no.03, 948-954 (1999)

[PTN96] Pavelescu, D., Trusca, V., Nitu, S., Dumitrescu, G., Maricaru, M., Zoita, V.: Installation and equipment for the research of the electrical switching arc in advanced vacuum, Proceedings of the XVIIth ISDEIV, Berkeley, California, vol. I, 305-310 (1996)

[PPG05] Pavelescu, D., Pavelescu, G., Gherendi, F., Nitu, C., Dumitrescu, G., Nitu, S., Anghelita, P.: Investigation of the Rotating Arc Plasma Generated in a Vacuum Circuit Breaker, in IEEE Transactions on Plasma Science, vol.33, nr.5, part I, 1504-1510 (2005)

[FCS04] Fontchastagner, J., Chadebec, O., Schellekens, H., Meunier, G.: Model of a rotating vacuum arc by the coupling of a simple arc model with a commercial 3D FEM, Proceedings of the XXIth ISDEIV, Yalta, Ukraine, vol. I, 276-279 (2004)

[DSS02] Dullni, E., Schade, E., Shang, W.: Vacuum arcs driven by cross-magnetic Fields, Proc. of the XXth ISDEIV, Tours, France, 60-67(2002)

Accurate Modeling of Complete Functional RF Blocks: CHAMELEON RF *

H.H.J.M. Janssen, J. Niehof and W.H.A. Schilders

NXP Semiconductors Research, High Tech Campus 5, 5656 AE, Eindhoven, the Netherlands, {rick.janssen, jan.niehof, wil.schilders}@nxp.com

Abstract. Next-generation nano-scale RF-IC designs have an unprecedented complexity and performance that will inevitably lead to costly re-spins and loss of market opportunities. In order to cope with this, the aim of the European Framework 6 CHAMELEON RF project is to develop methodologies and prototype tools for a comprehensive and highly accurate analysis of complete functional IC blocks. These blocks will operate at RF frequencies of up to 60 GHz. In this paper an overview of the CHAMELEON RF project will be given and the first results achieved in the CHAMELEON RF project will be presented.

1 Introduction

IC design automation tools are indispensable for RF designers in the transition to the nano-scale era. These tools are needed to develop nano-scale designs of unprecedented complexity and performance and, in addition, enable the achievement of single-pass design success to avoid costly re-spins and the loss of market opportunities.

Next generation designs will be challenged by an increased number of trouble spots, many of which negligible at lower frequencies but representing a significant limitation for future designs. These trouble spots will have to be accounted for during the design phase in order to avoid costly mishaps that can originate potential failures and additional design and silicon iterations, and must be addressed in future design automation tools.

New coupling and loss mechanisms, including EM field coupling and substrate noise as well as process-induced variability, are becoming too strong and too relevant to be neglected, whereas more traditional coupling and loss mechanisms are more difficult to describe given the wide frequency range involved and the greater variety of structures to be modeled. All this will cause extra design iterations, over-dimensioning or complete failures, unless appropriate solutions are found to resolve these design issues.

* The CHAMELEON RF project is funded under the European Union IST 6th Framework Program: FP6/2004/IST/4-027378.

2 CHAMELEON RF

The key to these solutions is the recognition that devices, both active and passive, can no longer be treated in isolation. Complete RF blocks must be considered as one entity, and be treated as such by the design automation tools. Today, it is not possible to perform such analyses of complete RF blocks.
The CHAMELEON RF project will deliver the methodologies and prototype tools to make this possible.

Table 1: CHAMELEON RF Consortium

PARTNER NAME	COUNTRY
NXP Semiconductors Research Eindhoven	NL
austriamicrosystems	AT
MAGWEL	BE
Interuniversity Micro Electronics Centre	BE
INESC-ID	PT
Polytechnic University of Bucharest	RO
Delft University of Technology	NL

3 Objectives

Current state-of-the-art modeling techniques, such as developed in the FP5/IST project CODESTAR [1], include compact models for the active and passive components, as well as field solving and model order reduction procedures and tools for simulating such devices. Whereas the CODESTAR project aimed primarily at the creation of design tools suitable to study the coupling of electromagnetic effects by analyzing elementary design units such as spiral inductors, varactors, capacitors and interconnects, here, we will build on these results by actually combining them via the proposed 'hierarchical field solving using compact models with connectors' paradigm, with the goal of handling full circuit blocks instead of just one or a small number of devices in isolation. This requires revisiting modeling procedures and assumptions. The development of the design methods is further complicated by the fact that the behavior of these blocks is increasingly sensitive to external factors that are hard to control. Shrinking feature sizes lead to higher fabrication variability. Rising frequencies of operations also increase dependence on temperature and other operating conditions.
Therefore the general objective of the consortium is that of developing a methodology and prototype tools that take a layout description of typical RF functional blocks that will operate at RF frequencies up to 60 GHz and transform them into sufficiently accurate, reliable electrical simulation models taking variability into account.
The main goal of the project, against which the progress of the project work will be measured, is the silicon-accurate modeling of RF functional blocks (such as a VCO or an LNA) with up to 10 transistors, 10 passive devices and implemented in 90 and 180 nano-meter technology with a maximum of 10 levels of metal for frequencies up to 60 GHz.

4 Advance over state of the art

At the moment, no commercial RF simulation tools for large interconnect structures on semiconductor substrates exist. In the CHAMELEON RF project, prototype tools to accurately predict the behavior of complete RF functional blocks will be developed.
The functional requirements set for any future design framework allowing for the high-fidelity verification of RF blocks must include the following:

- Simulation tools need to allow for a multi-scale (MS) and a multi-resolution (MR) approach.
- Feedback effects should be estimated and approximations should be justified by sub-threshold feedback.
- Functional blocks should be described in terms of netlists (SPICE) to allow inclusion in the design flows.
- Extraction best practices need to be provided to actualize the SPICE models.
- Models of functional blocks must be manageable in size to allow accurate behavior verification. Such models must also account for the dependence on relevant design or operating parameters.

Currently, there is no standard framework that meets these requirements.
The contribution of CHAMELEON RF to advance the state-of-the-art in order to

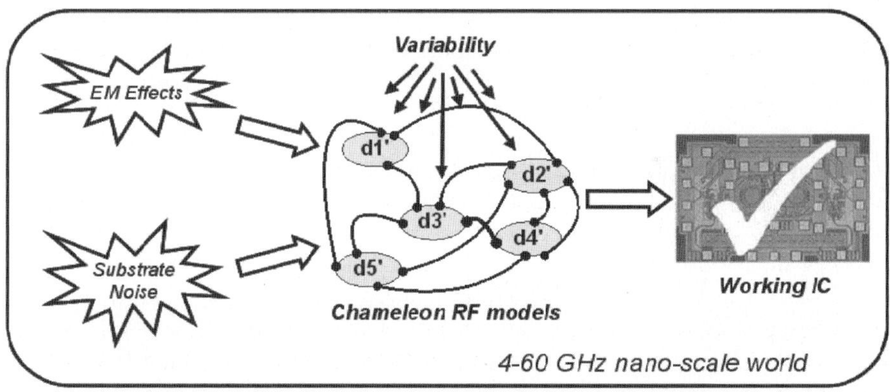

Fig. 1: Overview of the Chameleon RF system. For 4-60 GHz frequencies made possible by nano-scale integration technologies, electromagnetic and substrate noise effects require hierarchical connector-equipped models of full RF functional blocks in order to enable creation of working chips. The models will be variability-aware to account for relatively increasing effects of manufacturing tolerances.

fulfill the above requirements will come from the following:

- Hierarchical modeling procedures delivering compact models with connectors at the lower level(s).
- Top level modeling procedure for layout environment, working with connector-equipped compact models as gray boxes.
- Advanced Reduced Order Modeling (ROM) procedures including parametric support for variability and capable of dealing with the large number of inputs related to the compact model connectors and the variety of operating points.

5 Goals

The results will lead to design automation tools, in particular design verification tools, that can be used for comprehensive and highly accurate modeling of electromagnetic effects and other trouble spots in complete nano-scale RF blocks, thereby enabling designers to minimize turnaround time without compromising design quality and first-time-right requirements.

6 Work plan

In order to achieve the goals, the scientific and technical activities have been grouped in 5 work packages.
WP1 deals with the field solving aspects of coupling, at component level. The outcome will be a collection of method to build libraries of variable compact models with connectors of the typical parametric structures for the complete RF functional block.
WP2 aims at efficient and accurate methods of modeling the global interactions between the physical (on-chip) realizations of the circuit elements from the schematic. These elements include the active as well as the passive devices. Also process variability will be taken into account.
WP3 is aimed at developing efficient methods for generating compressed representations of parameterized inter-connected sets of compact models of interaction effects (ROM). WP4 is devoted to design and manufacturing of benchmark structures and WP5 is about validation of the solution against these benchmark results.

7 CODESTAR project results

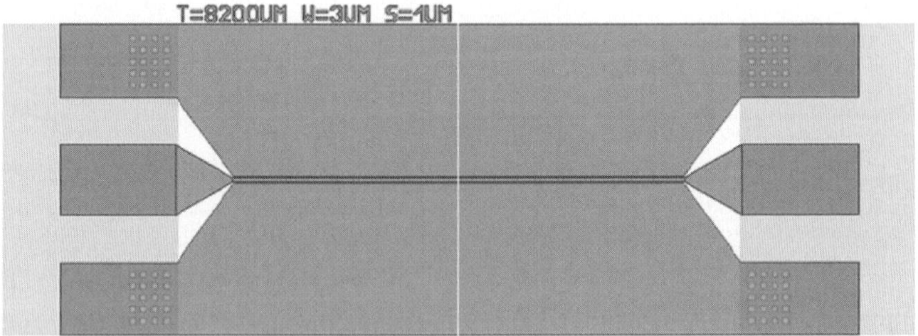

Fig. 2: Layout of coplanar line, 3 μm wide and 8.2 mm long, with 1 μm spacing between the line and ground.

As described in the CODESTAR final report [1], the project results were considered to be very successful. As an illustration we will describe an example of a coplanar line, of which the ground is situated next to the line in the same plane above the dielectric. The simulation of a coplanar line was successfully achieved for a number

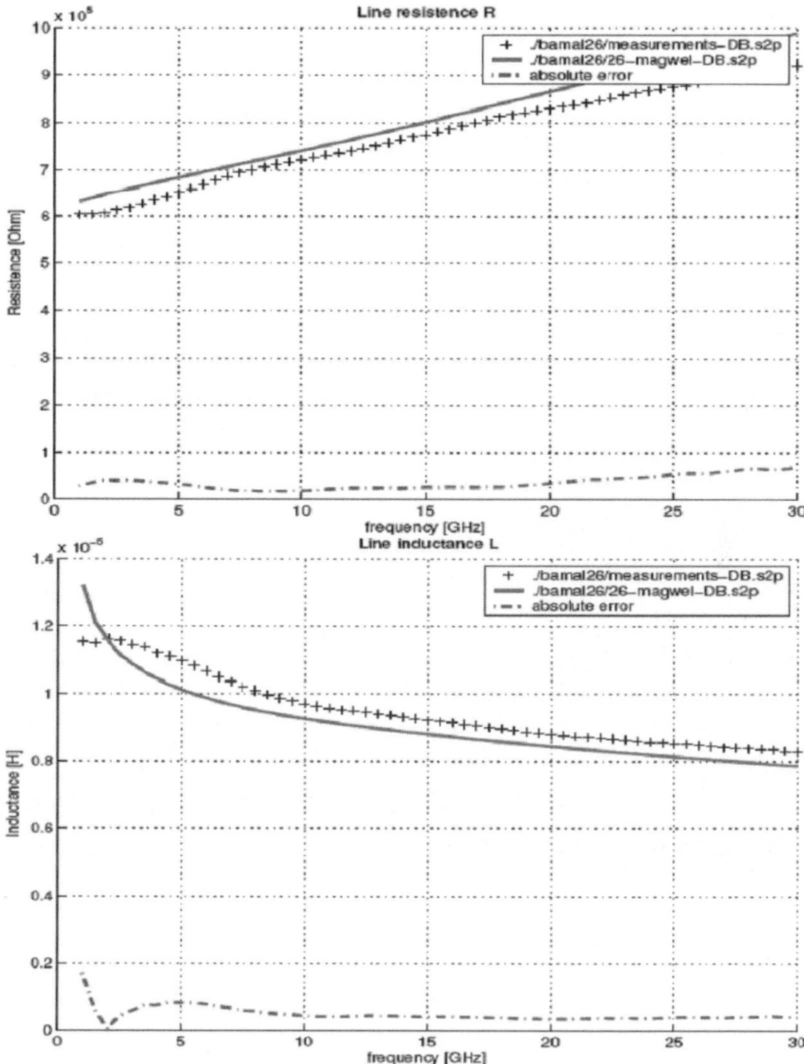

Fig. 3: Comparison of simulations with measurementsof the line parameters R and L (calculated from the impedance Z=R=jωL) for the coplanar line.

of EM solvers. In Fig. 2 the layout is shown (top view). The compact models obtained in CODESTAR cover a large frequency range: 0 - 30 GHz. Equivalent SPICE netlists of reduced order models were created and matching between measurement and (SPICE) simulation results was found in good agreement. The reduction was carried out using a vector fitting procedure [1] and a reduced model of order 10 was obtained.

8 Preliminary CHAMELEON RF project results

Preliminary simulations were performed on a substrate isolation structure, injecting noise into the substrate and analyzing the noise pick-up at the structure. A full-wave, 3D EM analysis is performed, where the substrate is modeled as a true semiconductor, using drift-diffusion equations with complex doping profiles [2]. The layout of the structure is shown in Fig. 4 and in Fig. 5 the S-parameters of simulations versus measurements are displayed, showing good agreement up to 10 GHz.

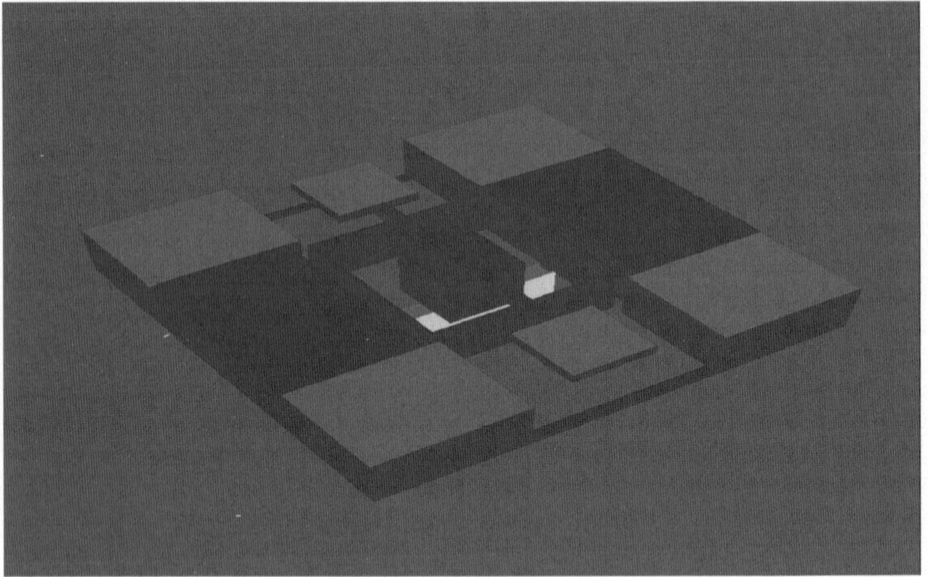

Fig. 4: Substrate isolation structure layout.

9 Conclusions

The aim of the CHAMELEON RF project is to develop methodologies and prototype tools for a comprehensive and highly accurate analysis of complete next-generation nano scale functional IC blocks that will operate at RF frequencies of up to 60 GHz. The project has started in November 2005, and will run for a period of two and a half years.

In the context of this project, attention is focused on the extension of the model order reduction work in two directions. First, the addition of parameters to handle geometric and process variability. Secondly, the two-level hierarchy pursued in the project asks for the possibility of hierarchical reduction of systems before assembling the full network.

Further information about the CHAMELEON RF project can be found at the project website [3].

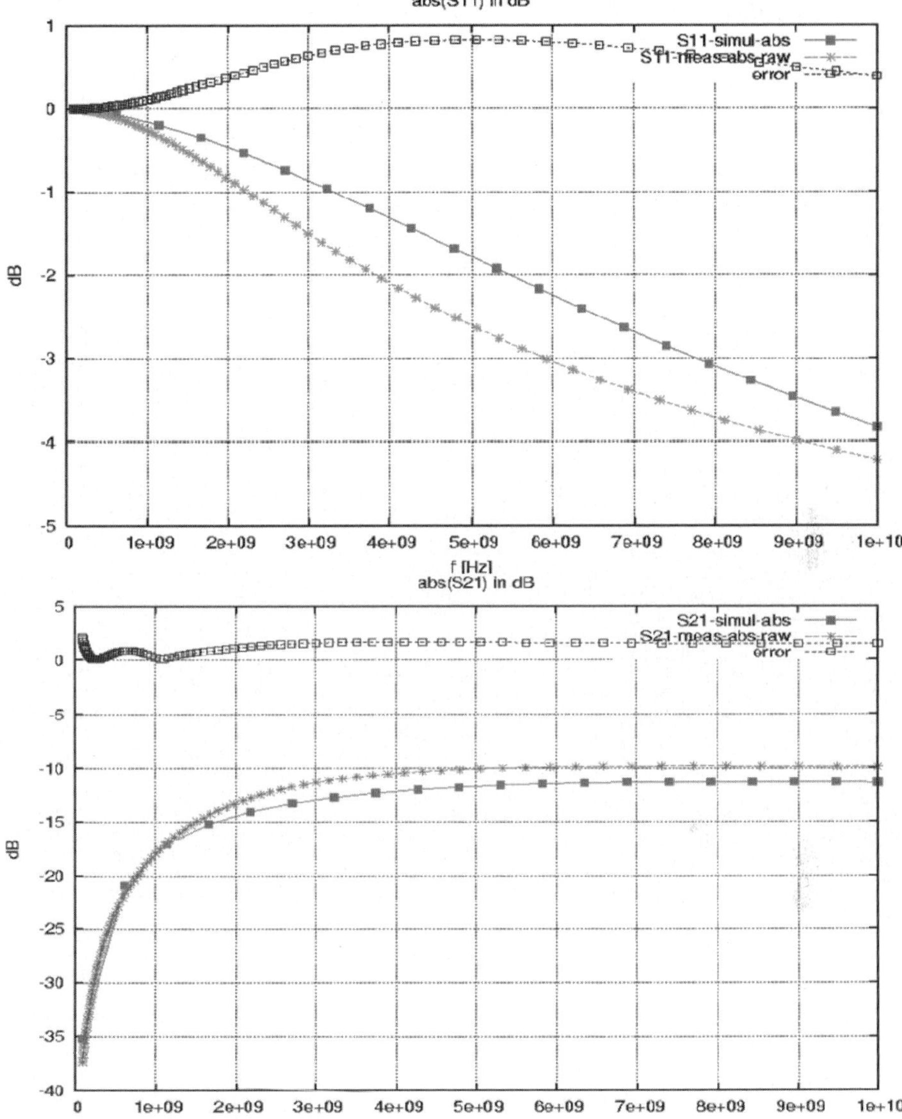

Fig. 5: S11 and S12 curves: the comparison between measurement and simulation shows good agreement (¡1 dB error) for frequencies up to 10 GHz.

References

[1] CODESTAR: http://www.magwel.com/codestar
[2] MAGWEL SolvEM: http://www.magwel.com.
[3] CHAMELEON RF: http://www.chameleon-rf.org.

Finite Element Analysis of Generation and Detection of Lamb Waves Using Piezoelectric Transducers

Sorohan St., Constantin N., Anghel V. and Gavan M.

Politehnica University of Bucharest, Spl. Independentei 313, 060042, Bucharest, Romania
sorohan@form.resist.pub.ro

Abstract – The paper reports the use of Finite Element (FE) simulation and experiments meant to explore the operation conditions of the PieZoelectric wafer Transducer (PZT). Piezoelectrics is the coupling of structural and electric fields and may be solved using the multi-physics approach. Accordingly, three different multi-physics models were developed to investigate a plane strain problem. The first one includes two PZTs mounted on an aluminium plate and is used to model both the emission and reception signals. The next two ones are developed to separately model the emission and detection processes, in order to decrease the computational effort. The wave displacements are generated by a PZT-like actuator and the output voltage is obtained at a PZT receiver both by a multi-physics approach. The analysis considered the transducer lengths, the effects of the finite pulse width, the pulse dispersion and the detailed interaction between the piezoelectric element and the transmitting medium. The transmitted and received signals for so-called A0 and S0 modes have maxima close to the frequencies predicted in other works. A series of sensitivity curves relating the generation and receiving of Lamb waves were also determined and plotted as a function of the pulse center frequency and of the PZT lengths.
Keywords – Lamb wave, multi-physics finite element models, piezoelectric effect, piezoelectric transducers.

1 Introduction

Many authors considered the use of Lamb waves for non-destructive testing. Lamb waves can propagate on long distances in plates and tubes, making it possible to detect flaws over a considerable area by a set of transducers [5], [6]. Complications that are encountered include the existence of multiple modes and their dispersive character. A partial solution to this complexity is the use of transducers that excite only a single mode, and various strategies have been employed to this end [6]. Recently there has been increasing interest in the use of PZT wafers as transducers, mainly due to the simplicity and low cost of such transducers [2]. PZT wafers have been excited with continuous sinusoidal or pulse signals for defect detection in plates and the influence of flaws on the Lamb wave transmission has been modelled in trial simulations.

In most papers, the mechanical interactions between the PZT wafer and the structure to be inspected are not directly included. Recently, it was observed that the wave emission and the reception using a PZT are physically distinct and both show a specific dependence on the pulse centre frequency [2], [3]. Taking the piezoelectric phenomena and the possible complicated geometry of the inspected items into account, a more accurate analysis of the whole coupled structural and electrical problem by the FE method is evaluated in this work.

2 Review of wave propagation and FE modelling

Piezoelectrics consists in the coupling of structural and electric fields, exploiting the natural material properties of quartz and of ceramics. A voltage difference applied to a piezoelectric material creates a displacement, and, reversely, vibrating a piezoelectric material generates a voltage difference. The electromechanical constitutive equations for linear material behaviour are [1], [7]:

$$\{T\} = [c]\{S\} - [e]\{E\}; \qquad \{D\} = [e]^T\{S\} + [\epsilon]\{E\}, \tag{1}$$

where $\{T\}$ is the stress vector, $\{D\}$ is the electric flux density vector, $\{S\}$ is the strain vector, $\{E\}$ is the electric field vector, $[c]$ is the elasticity matrix, $[e]$ is the piezoelectric stress matrix and $[\epsilon]$ is the dielectric matrix (evaluated at constant mechanical strain).
Using the variational principle, it is possible to derive a second order time-dependent system of equations that can be discretized using the FEM and that include the piezoelectric effect (for details see [1] and [7]):

$$\begin{bmatrix} [M] & [0] \\ [0] & [0] \end{bmatrix} \begin{Bmatrix} \{\ddot{u}\} \\ \{\ddot{V}\} \end{Bmatrix} + \begin{bmatrix} [C] & [0] \\ [0] & [0] \end{bmatrix} \begin{Bmatrix} \{\dot{u}\} \\ \{\dot{V}\} \end{Bmatrix} + \begin{bmatrix} [K] & [K_Z] \\ [K_Z{}^T] & [K_d] \end{bmatrix} \begin{Bmatrix} \{u\} \\ \{V\} \end{Bmatrix} = \begin{Bmatrix} \{F\} \\ \{L\} \end{Bmatrix} \tag{2}$$

where the submatrices and vectors used are: $[M]$-structural mass matrix; $[C]$-structural damping matrix; $[K]$-structural stiffness matrix; $[K_Z]$-piezoelectric stiffness matrix; $[K_d]$-dielectric coefficient matrix; $\{F\}$-applied nodal force vector; $\{L\}$-applied nodal charge vector; $\{u\}$-displacement vector; $\{V\}$-electric potential vector. The dot and double dots denote differentiation(s) with respect to time. To integrate this system, a full transient analysis using the Newmark method (with $\alpha = 0.25$, $\delta = 0.5$ and $\theta = 0.5$) was performed using ANSYS 7.0 code [7].
For a given PZT, the emitted and received wave modes depend on the applied signal frequency. This phenomenon is usually called mode selectivity and is particularly addressed. The obtained results provide more accurate predictions of the mode selectivity than previously reported ones using a simplified PZT model in the wave emission process [2]. Finally, this approach is readily adapted to explore the wave interactions with flaws and contour conditions.
In [3], for a similar study, a plate with thickness $h = 1.59$ mm has been used, together with a particular PZT. Here also several material properties have been made available. In our experiments we used the same input data for the numerical simulation, but allowed for different PZT dimensions. Simulations were made in the $100 - 600$ kHz range frequency. For the studied aluminium plate, only S0 and A0

Lamb modes (Fig. 1) exist at frequencies below 1 MHz, as it can be found on the dispersion curves in figure 2. Dispersion curves can be obtained by assuming a particular harmonic solution of the displacements into the Naviers equation for which the boundary conditions must be fulfilled. More detailed information about the dispersion curves can be found in [2], [4].

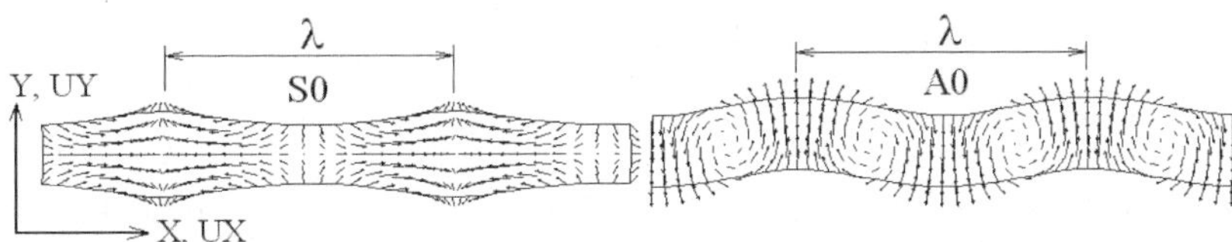

Fig. 1: Lamb modes in a plate. The wave propagates into X direction; λ denotes the wave length and the arrows show the instantaneous particle motion

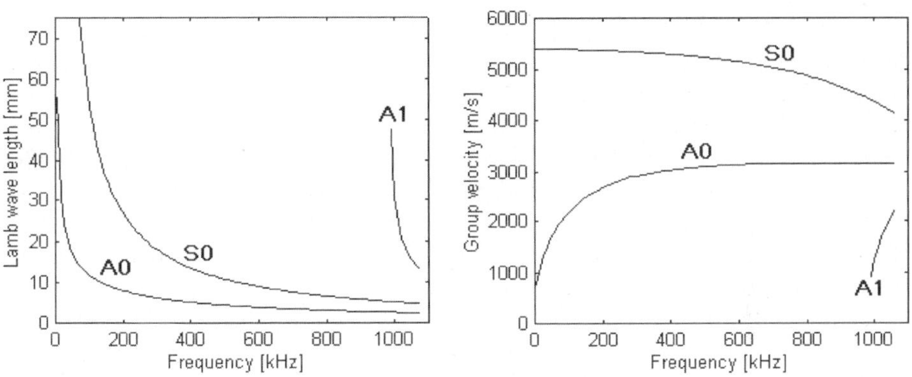

Fig. 2: Dispersion curves of an aluminium plate of 1.59 mm thickness

Figure 3 shows two similar wafer-type transducers bonded on an aluminium plate. The plate and the wafers are assumed to be of infinite extent in the Z direction (perpendicular to this paper), an ordinary plane strain assumption in structural mechanics. Only half of the plate was considered, with symmetry at the Y-axis. The wafer is a piezoelectric material with the poling direction normal to the surface of the plate and metallized on the top and bottom surfaces. The PZT actuation was excited with the pulse signal

$$V(t) = \begin{cases} 0.5V_0\left[1-\cos\left(\frac{2\pi f_0 t}{n_0}\right)\right]\cos(2\pi f_0 t);\ \text{if } t \leq \frac{n_0}{f_0} \\ 0;\ \text{if } t > \frac{n_0}{f_0} \end{cases} \quad (3)$$

where $V_0 = 10V$; the number of cycles $n_0 = 5$, and where the pulse centre frequency f_0 ranges between 100 and 600 kHz, is applied as input voltage between the metallized surfaces of the transducer (see Fig. 3).

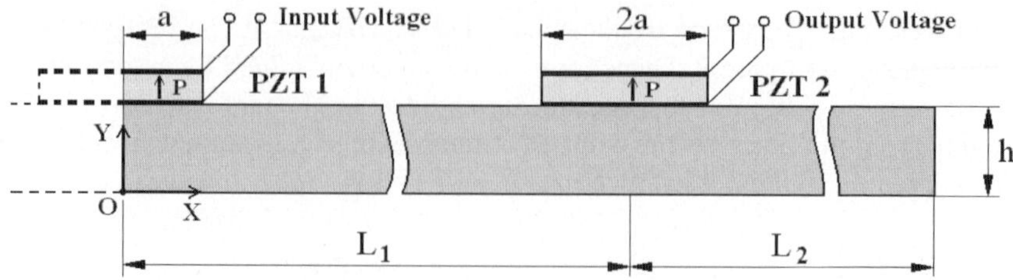

Fig. 3: The physical model used in finite element analysis

In FE simulations, the two PZTs and the plate were modelled individually using the PLANE13 element type within ANSYS. The Degrees Of Freedom (DOFs) of the two PZTs are the horizontal and the vertical displacements (UX and UY) and the electric potential (VOLT), while the DOFs of the plate are only UX and UY. The mesh parameters were chosen such that the element sizes were substantially smaller than a wavelength. Usually 10-20 elements per wavelength guarantee a good accuracy [7], in this case element size results between 0.2 and 0.4 mm (function of exciting frequency, see Fig.2a). The bottom surfaces of the PZTs were electrically grounded ($V = 0$) and an equipotential boundary condition was set on the top surfaces. Because the PZTs are bonded on the plate and the adhesive layer is neglected, the coincident nodes of the PZTs and the plate mesh were coupled both in X and Y direction. By symmetry, the X-displacement UX=0 at the origin. All other boundaries, except the input potential, were free. Simulations were performed in the time dependent mode with output time steps typically under one twentieth of a period $(0.1 - 0.25\mu s)$.
The above described approach more accurately matches with the actual PZT wafer /plate physical interaction. The generated 2D waves are propagating in the plate, inducing, by a similarly simulated interaction, an output voltage, collected at PZT 2. A simulation for a particular case ($L_1 = 200$ mm; $L_2 = 300$ mm; $2a = 6.4$ mm; average element size 0.4 mm - a total of 6333 nodes) shows the expected two propagating modes with S0 and A0 character (Fig. 4). The S0 mode has the highest group velocity (see also Fig.2) and shows particle displacements mostly in the X direction, and the slower A0 wave mode shows particle displacements mostly in the Y direction, asymmetric at the centre of the plate. It can be observed that the PZT 2 behaves like a reflecting flaw, converting the incident modes as physically must occur.

3 Emission of ultrasonic waves

In this section, simulations of the guided wave emission using PZT 1 as source are presented, putting in evidence the influence of the PZT length (2a in Fig. 3) upon the generated S0 and A0 Lamb modes. For reasons of computational effort, the model used in these simulations neglects the PZT 2 existing in figure 3, and considers a total plate length $L_1 + L_2 = 300$ mm. In order to quantify the variation of wave magnitudes with frequency, the maximum absolute value of UX for S0 mode and the maximum absolute value of UY for A0 mode were determined as maximum wave

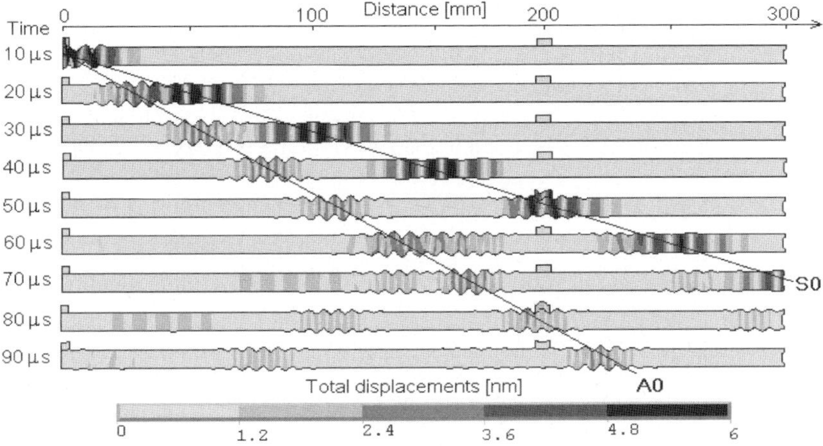

Fig. 4: Propagation of S0 and A0 modes at a central frequency of 300 kHz. The thickness of the plate has been exaggerated by a factor of 5 for improved visibility. The displacements have been scaled with a factor of 100000

magnitude when the two generated modes were clearly separated, for example at $t = 30\mu s$ in figure 4.

The sensitivity of generated waves, measured as maximum wave magnitude divided by the maximum input voltage, of about $V_0 = 10V$, is plotted in figure 5, separately for S0 and A0 mode, as a function of frequency bottom scale and Lamb wave length top scale, for different PZT lengths.

According to the simplified analytic model proposed by Giurgiutiu and Lyshevski [2], peak and null (or minimum) emission occur at wave length respectively, where n is an integer greater or equal to one. From relations (4) and dispersion curves (see Fig. 2a), the frequencies under 1 MHz, corresponding to maximum and minimum sensitivity, are determined and then summarized in the Tables 1 and 2, respectively.

$$\lambda^{(max)} = \frac{2a}{n - \frac{1}{2}}; \qquad \lambda^{(min)} = \frac{2a}{n}; \tag{4}$$

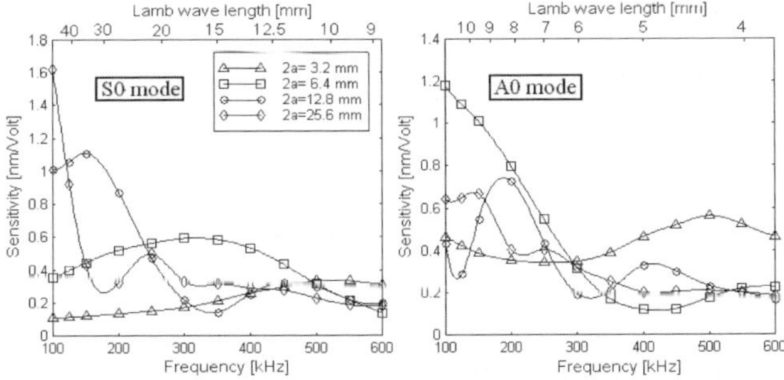

Fig. 5: The sensitivity to emission Lamb wave function of pulse centre frequency and PZT length

n	1		2		3		4	
2a[mm]	S0	A0	S0	A0	S0	A0	S0	A0
3.2	817.6	282.3	-	-	-	-	-	-
6.4	418.7	86.13	-	516.8	-	-	-	-
12.8	210.4	23.04	622.3	176.2	998.5	397.2	-	638.7
25.6	105.3	-	316.0	33.9	521.3	128.6	721.3	227.8

Table 1: Lamb wave fequencies [kHz] for maximum sensitivity as said by Eq. 4a

n	1		2		3		4	
2a[mm]	S0	A0	S0	A0	S0	A0	S0	A0
3.2	-	761.7	-	-	-	-	-	-
6.4	817.6	282.3	-	761.7	-	-	-	-
12.8	418.7	86.1	817.6	282.3	-	516.8	-	761.7
25.6	210.4	23.04	418.7	86.1	622.3	176.2	817.6	282.3

Table 2: Lamb wave fequencies [kHz] for minimum sensitivity as said by Eq. 4b

Comparing the results from figure 5 with the results from Table 1 and 2, a fair agreement between the Giurgiutiu & Lyshevski analytic results and the finite element results can be observed, especially for S0 mode.

4 Reception of Lamb waves

During the emission process, the complex motion of the plate and PZT 1 is symmetric about the Y axis and is later separated into S0 and A0 modes. During detection, the transducer interacts with a single mode which is propagating unidirectionally. Consequently, the displacements of PZT 2 will not be the same as during emission. Moreover, wave interaction with PZT generates additional reflected waves.
For computational effort reasons, the response of receiving PZT 2 was separately analysed for S0 and A0 modes. A particular S0 or A0 mode was selectively generated by adequately imposed displacements in interface nodes at X = 0. Therefore, PZT 1 was not included into the model and it was possible to consider $L_1 = 100$ mm and $L_2 = 100$ mm. The sensitivity of received waves (input voltage against maximum wave magnitude, measured by UX for S0 mode and UY for A0 mode) is plotted in figure 6.
It is clear that sensitivity increases as the PZT length decreases. However, especially for S0 mode, there are some optimal PZT lengths, function of the pulse centre frequency.

5 Global results and comparison with experiment

Figure 7 plots the ratio of the maximum received pulse amplitude to the exciting pulse amplitude as a function of frequency, namely the global sensitivity. Results are shown for four different transducer pairs, all separated by $L_1 = 200$ mm.
Measurements of the A0 and S0 wave amplitudes as a transfer functions have been

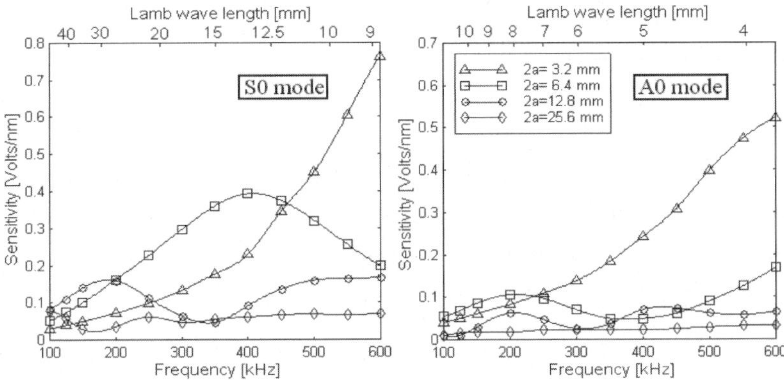

Fig. 6: The sensitivity to received Lamb wave function of pulse centre frequency and PZT length

reported in [3] only for $2a = 6.4$ mm and are in good agreement, in terms of frequencies for maximum sensitivity, with the results obtained during the numerical simulations presented in this work.

6 Conclusions

The operation of a PZT wafer transducer was analyzed for the generation and detection of guided Lamb waves using a multi-physics FE simulation and then compared with experiments reported in literature. The numerical simulations were intended to assess the possibility to follow a realistic mechanical interaction between the transducers and the transmitting inspected medium.

The results prove that multi-physics FE simulations are offering more accurate val-

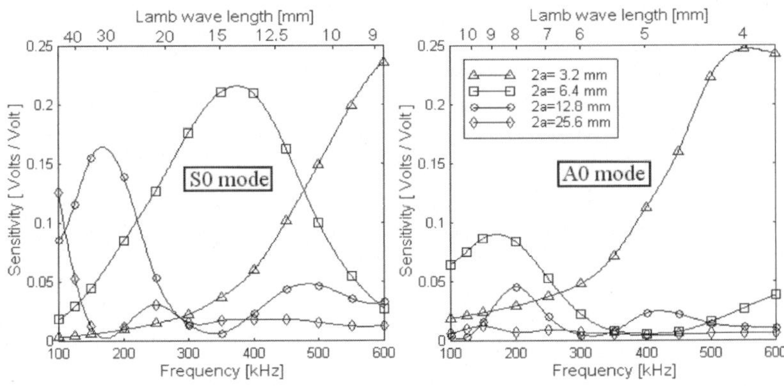

Fig. 7: The global sensitivity, or overall system transfer function $\frac{V_{out}}{V_{in}}$, as a function of pulse centre frequency and PZT length

ues for the peak frequencies and the mode selectivity. In addition, FE simulations make it possible to determine the optimum transducer length, which may be different for emission and reception.

Acknowledgement

The research work was financed by Sure2Grip CRAFT project, run in FP6 programme under COOP-CT-2004-513266 contract.

References

[1] Allik, H., and Hughes, J. R.:Finite Element for Piezoelectric Vibration, International Journal for Numerical Methods in Engineering, **No. 2**, 151–157 (1970)

[2] Giurgiutiu, V., Lyshevski, S., E.: Micromechatronics: Modeling, Analysis, and Design with Matlab, **CRC Press**, (2004)

[3] Nieuwenhuis, J. H., Neumann, J., Greve, D. W., and Oppenheim, I. J.: Generation and detection of guided waves using PZT wafer transducers, IEEE Trans. Ultrasonics, Ferroelectrics, and Frequency Control, **Vol.52**, 2103–2111, (2005)

[4] Rose, J. L.:Back to basics - Dispersion curves in guided wave testing, Material Evaluation, **Vol.60**, 20–23,(2003)

[5] Rose, J. L.:Guided wave nuances for ultrasonic nondestructive evaluation, IEEE Trans. Ultrasonics, Ferroelectrics, and Frequency Control, **Vol.47**, 575–583, (2000)

[6] Zhongquing S., Lin, Y., Ye, L.:Guided Lamb waves for identification of damage in composite structures: A review, Journal of Sound and Vibration, **Vol.295**, 753–780, (2006)

[7] ***:Ansys, Inc. Theory. Reference and User Guides (2000)

Optimization of a Switching Strategy for a Synchronous Motor Fed by a Current Inverter Using Finite Element Analysis

Vasile Manoliu

POLITEHNICA University of Bucharest, Electrical Engineering Faculty, Splaiul Independentei 313, 060042, Bucharest, Romania, vasilem@amotion.pub.ro

In a load-commutated synchronous motor, the torque is driven by the currents in the stator windings, and these currents depend on the rotor position. Since the stator phases are star-connected, the static torque generated by two stator phases connected in series and powered with a constant current offers sufficient information to design a suitable current control strategy. The synchronous machine studied has a reverse type of construction. The switching sequence for the current inverter is correlated with the rotor position by magnetostatic simulation in FLUX2D, using the maximization of static torque as optimization criteria. This study also covers a computational method for determining the operating parameters of a synchronous machine using simulation of the Standstill Frequency Response Test (SSFR).
Keywords synchronous machine, finite element, static torque, optimization.

1 Introduction

A synchronous machine with reverse type of construction has three identical phase-windings spaced around the internal periphery of the rotor magnetic core with a geometrical angle of $2\pi/3p$ between them (where p is the number of pole pairs). The rotor core is made of electrical-steel laminations, electrically insulated and slotted. The stator magnetic core (either with salient or non-salient poles) contains a d.c. (direct current) excitation winding.

The Park-Blondel equations are much used in studying the dynamic operation of a synchronous motor. In this reference system, the axis of a North pole of the stator is called the direct axis (denoted d) and the leading axis (with respect to the direction of the speed Ω) is called quadrature axis (denoted q).

In this paper a load-commutated synchronous machine model is studied. This model takes into account both the magnetic circuit configuration of the machine and the realistic construction of the damper winding.

These machines are robust and have a good weight/size to power ratio. Compared to a conventional brushed d.c. motor in which commutation is handled by carbon or copper brushes, the load-commutated synchronous motor commutation is controlled by electronics. In Fig.1 a schematic layout of the synchronous motor fed by current inverter is presented.

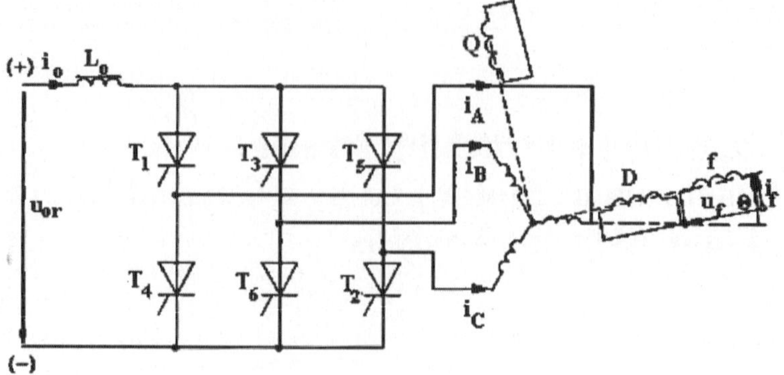

Fig. 1: Synchronous motor fed by a current inverter

An electronic supply by time-shaped rectangular currents uses a very simple and cheap rotor position sensor. The position of the rotor is needed to determine a proper timing of commutation.
Moreover, in using this kind of supply, the choice of a fractional number of slots per pole and phase ensures low torque ripple.
For this kind of synchronous motors, the drive of air compressors is an interesting application. In this case, since the starting (load) torque is very high, by replacing the asynchronous motor with a synchronous motor, the operating proves to be more economical.
Controlling without a position sensor is possible if the back electromotive force (emf) is measured; this control is recommended for applications where high starting torque is not required.
Since for the load-commutated synchronous motor the electromagnetic torque is affected by the back emf and the rotor (armature) currents, it is important to have a thorough knowledge of the static torques for designing a suitable current control strategy. For this reason, the static torque can be measured by connecting two rotor phases in series, using a constant current as source.
At the same time, the motor parameters affect the switching control. The useful characteristic parameters appear in equivalent circuits associated with the direct-axis and quadrature-axis of the synchronous machine.
By the FLUX 2D [FLU05] program 2D FEM field calculations can be made to determine these parameters as well as the static torque. This program offers a lot of useful features like partially automatic mesh generation, sliding air gap band for calculations in rotation, and external coupling circuits.
For determining the operating parameters of a synchronous machine, the simulation of the Standstill Frequency Response Test (SSFR) was used.
This approach can be used not only for the verification of the important characteristic quantities of an existing machine, but also allows optimization in designing a new machine.

2 Methods

The current supply of synchronous machine consists in imposing current amplitude in the machine windings and its phase with respect to the electromotive force E. To detect the phase angle of the electromotive force (leading angle) x_0, a rotor position sensor is generally used. In Fig.2 a simplified phasor diagram of a load-commutated synchronous motor is presented. The phasor diagram has been drawn using standard notations (X_d and X_q for direct-axis and quadrature-axis synchronous reactances, β for internal electrical angle) neglecting the winding resistance and commutation overlapping effect. In a natural commutation process, the current always leads the voltage and the machine power factor $cos\varphi^{(1)}$ is determined by the phase angle between the inverter triggering pulses and the machine voltage [Bos86].

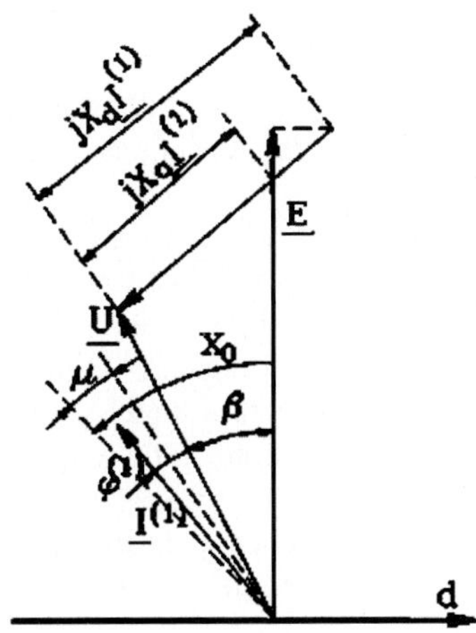

Fig. 2: The simplified phasor diagram of a synchronous motor with load commutation

The commutation angle, μ, is a function of commutation reactance X_c . At commutation limit (with negligible margin angle of commutation) the critical value of μ can be given as

$$\cos\mu_k = 1 - \frac{\pi}{3}\frac{\nu X_c}{U} I^{(1)}, \tag{1}$$

where ν is the frequency expressed in per-unit value, $I^{(1)}$ the r.m.s. (root mean squared) value of the fundamental wave of armature current and U the armature voltage.
Also, the commutation reactance can be calculated in terms of subtransient reactances $X_d^{"}$ and $X_q^{"}$ and the leading angle of commutation x_0, as follows:
A Finite Element two-dimensional analysis can be performed due to FLUX2D's unique features in simulating motion and allowing input from external electric circuits.
By maximizing the average torque calculated from the field analysis, the firing instants of the involved inverter thyristors can be determined.
The whole control imposes the leading angle x_0 determining the thyristors ignition instants with respect to the zero crossing instants of the electromotive force E. The

leading angle variation has the same meaning as the shift of brushes from the neutral axis in a direct current motor.

2.1 Static torque analysis using FLUX2D

In order to calculate the static torque with FLUX2D, a conventional circuit analysis method is used, to assemble the magnetic potential equations together with the current and voltage equations for each conductor. Non-linear functions like flux linkage-current relations are taken into account.
The nonlinear system is iteratively solved using the Newton-Raphson algorithm exploiting a conjugate gradient method to solve the intermediate linear systems.
The characteristics of the analyzed laboratory-used synchronous machine are: P_N = 3.2 kW, U_N = 220 V, 4 salient poles, in rotor: 27 slots, 198 turns/phase. The discretisation frame applied to the transverse section of the synchronous machine is shown in Fig.3.
Due to the fractional slot number (per pole and phase), (q = 9/4) the entire motor must be modeled. The discretisation frame contains 7675 triangular elements (second order) and 15405 nodes.
The static torque generated by two stator phases connected in series and powered with a constant current offers sufficient information to design a suitable current control strategy [TLW96]. This will be explained next.
First, energizing the same two rotor windings ($A-X$ and $B-Y$) connected in series, the static torque was computed for position angle (θ) values spaced at 15 electrical degrees, for a complete rotation of the rotor. The resulting curve of static torque T_{st} = $f(\theta)$ shows a quasi-sinusoidal variation [MK99].
Next, the positions corresponding to the maximum values of the static torque were quantisized in terms of the leading angle x_0 (as example, for x_0 = 60 electrical degrees the first maximum value of the static torque was obtained for a position angle θ = - 30 electrical degrees).

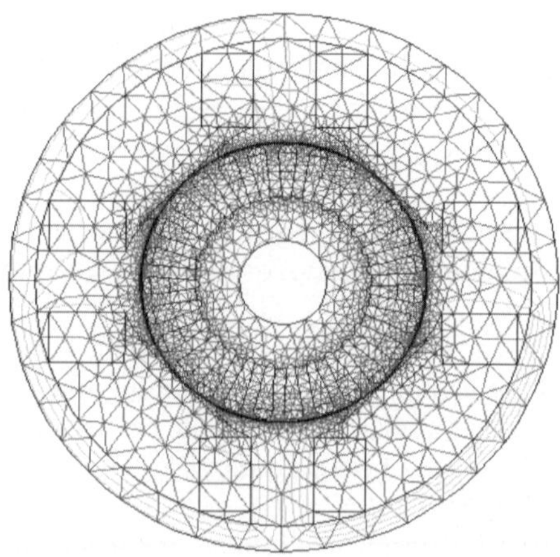

Fig. 3: Discretisation frame applied to the magnetic circuit of the synchronous motor

A magnetostatic simulation in FLUX2D was applied to the circuit structure of Fig.4. The *on* and *off* states of each thyristor (T_1 to T_6) are modeled using a resistance with a low value (1 mΩ for *on* state) or high value (100 kΩ for *off* state).

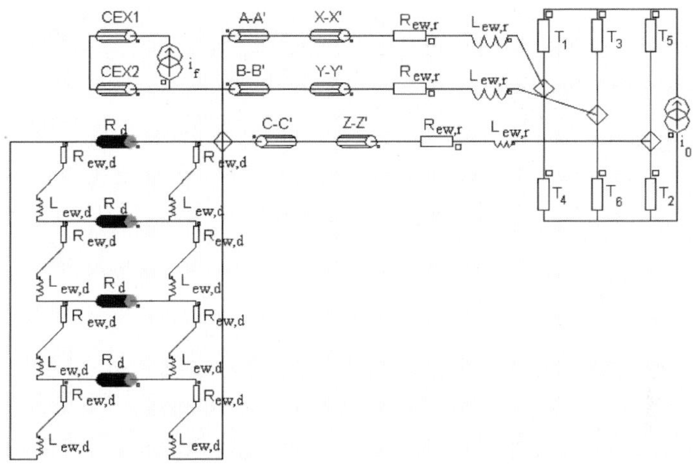

Fig. 4: Structure of the complete circuit for static torque simulation

The supply of the inverter is ensured by a direct current source (value of d.c. link current) i_0.
Each rotor winding is modeled by two stranded conductors (e.g. $A-A'$ and $X-X'$) plus a resistance and a inductance (lumped parameters). The resistance and inductance represent the end turns of winding which are not part of the finite element domain. The values used for modeling the end winding resistance and inductance are 0.1Ω and 1mH, respectively. The damper winding was modeled by four identical circuits, parallel-connected, each composed by a solid conductor resistance and end winding resistances ($R_{ew,d} = 2.87 \times 10^{-6}\Omega$) and inductances ($L_{ew,d} = 0.174mH$). The field winding circuit is modeled by two stranded conductors (CEX1 and CEX2) and a direct current source (value of field current) i_f.
The obtained variation of maximized static torque is shown in Fig.5, for per-unit (p.u.) values of field current (i_f = 1.52) and constant d.c. link current (i_0 = 0.81) [MK99].
Now, the switching moments of the inverter thyristors, can be determined; depending on the rotor position, the thyristors are turned on and off, for each conduction sequence (60 electrical degrees rotation), to produce a maximum average torque.

2.2 Simulation of the Standstill Frequency Response test

The Standstill Frequency Response test (SSFR) is carried out at standstill and consist in obtaining the Bode diagram for the operational inductance (in direct and quadrature axis) by measuring the armature voltage and current in a range of frequencies [NFC97].
The Finite Element simulation of the Standstill Frequency Response test (SSFR) was carried out by using FLUX2D in linear, quasi-static mode, with coupling to circuit

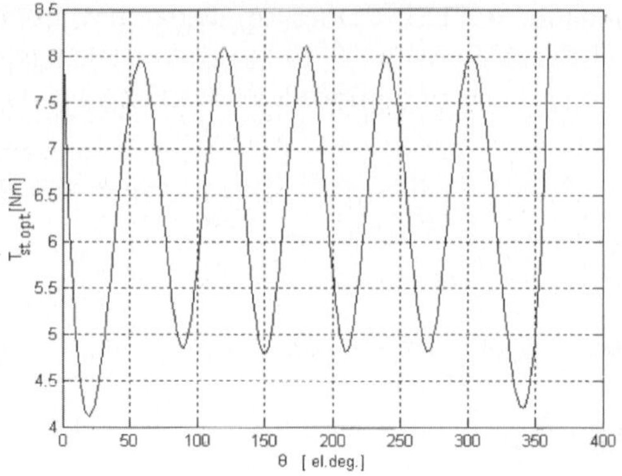

Fig. 5: The maximized static torque

equations. This approach allows to check the operating parameters (inductances and time constants) of a synchronous machine and to compare the simulated values of parameters with those obtained from the short-circuit and SSFR tests.
The circuit structure is shown in Fig.6, with the same notations as in Fig.4.

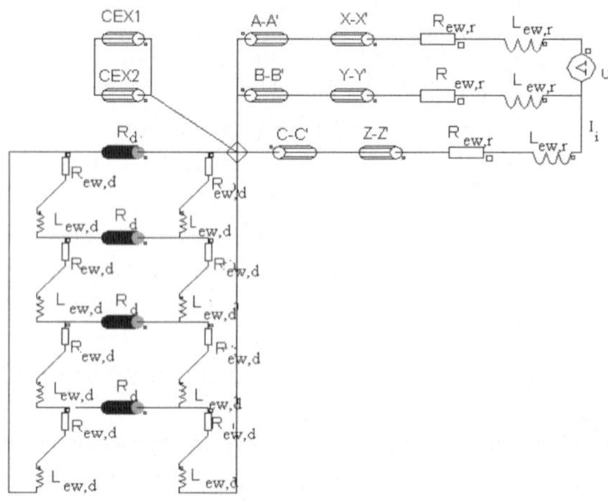

Fig. 6: Structure of the circuit for SSFR test simulation

The sinusoidal a.c. (alternating current) source of tension used in the armature circuit has the value: U_i = 31 mV and the simulation was carried out in a frequency range from 10 mHz to 1000 Hz.
The d-axis operational inductance was calculated by the following relation:

$$L_d(j\omega) = \frac{Z_d(j\omega) - R_a}{j\omega}, \tag{2}$$

with R_a - the armature resistance, $\omega = 2\pi f$, and:

$$Z_d(j\omega) = \frac{2}{3}\frac{U_i(j\omega)}{I_i(j\omega)} \tag{3}$$

The complex d-axis operational inductance depends on the applied frequency of the source. By studying the Bode plot of the amplitude, we can extrapolate linearly this to $f = 0$, (giving the direct-axis synchronous inductance L_d) and to f = 1000Hz, (giving the direct-axis subtransient inductance, $L_d^{"}$).

3 Results

3.1 Static torque analysis

For unsymmetrical supply (two-phase, without correlation with the rotor position) the flux lines shows a non-uniform distribution (Fig.7).

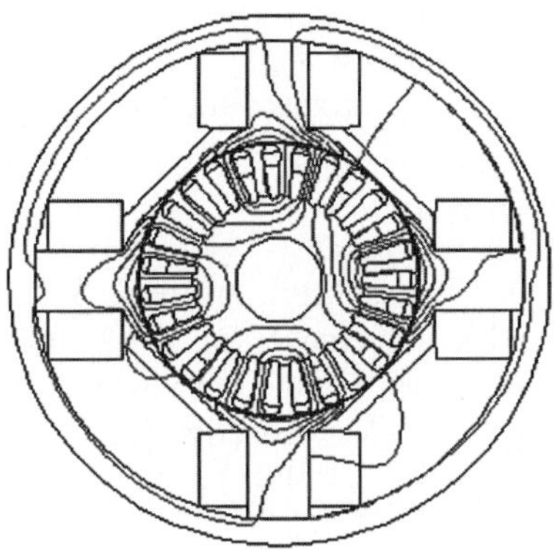

Fig. 7: Flux lines for one rotor position (unoptimized)

For optimized control (two-phase supply, correlated with the rotor position), flux lines becomes much more uniform (Fig.8).
Due to the fractional number of slots per pole and phase the harmonic content of electromotive forces is very small. Consequently, the electromagnetic torque is less affected by non-linearities due to the inverter operation.
For this control optimization was investigated for two cases of a synchronous motor: one with and one without damper winding. The difference between the maximum static torque for the two cases is quite small: $T_{st} = 8.138Nm$ - with damper winding and $T_{st} = 8.0Nm$ - without damper winding [MBM98].
By compensating variations of the current harmonics and by the reduction of overlapping commutation angle, the damper winding acts in reduction of the power factor angle, $\varphi^{(1)}$, thus determining an increase in the overload capability of the motor.
The copper damper winding takes over a part of the imposed magnetomotive force and thus reduces iron losses. The reduction becomes most notable after applying the optimization strategy.
As stated in [MK99], the static torque maximization allows for the correlation between the rotor position angle, θ, and the leading angle x_0 for the firing instants

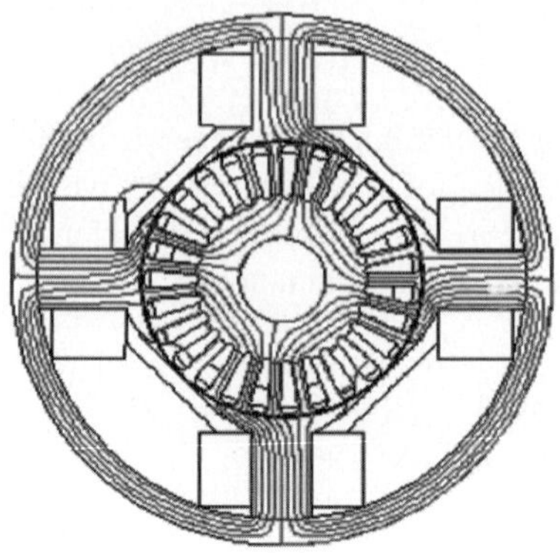

Fig. 8: Flux lines for one rotor position (optimized)

of every inverter thyristor. For example, for an arbitrary value of the leading angle $x_0 \in (0, \pi/2)$ rad., the rotor position angle has the value: $\theta = \pi/2 - x_0$ at the moment when the thyristor T_6 commutates with T_2.

3.2 Simulation of the SSFR test

By simulating the problem of the SSFR test with FLUX2D, for input voltage $U_i = 31mV$, a synchronous inductance $L_d = 34.385mH$ and a subtransient inductance, $L_d^{"} = 4.87mH$, were obtained. These values are in good agreement with experimental ones ($L_d = 36mH$, respectively, $L_d^{"} = 5.4mH$).
For an increased value of $U_i (U_i = 12.4V)$ the saturated values were calculated as well: $L_{d,sat} = 31.4mH$ and $L_{d,sat}^{"} \cong L_d^{"}$.
Also, for a configuration without a damper winding, an increased value of the subtransient inductance, $L_d^{"} = 5.67mH$ was obtained.
The simulation of the SSFR test was used, also, for computing the q-axis synchronous parameters: the synchronous q-axis inductance, L_q, and the sub-transient q-axis inductance, $L_q^{"}$.
The computed results of the machine parameters are compared to values obtained from experimental determinations. The results, expressed in per-unit values, are presented in Table 1.

Table 1: Parameters comparison

Method	l_d	l_q	$l_d^{'}$	$l_d^{"}$	$l_q^{"}$
Experimental	1.10	0.75	0.38	0.16	0.22
Computation FLUX2D	1.10	0.73	0.35	-	-
Simulation SSFR test	1.05	0.68	-	0.15	0.28

One can observe that the simulation of SSFR test gives satisfactory results, especially in the cases where the skin effect should be taken into account [MBM98].

4 Conclusions

For the synchronous motor studied, the harmonic content of emf's is very small due to the fractional number of slots per pole and phase. Consequently, the electromagnetic torque is less affected by non-linearities due to the inverter operation.
Maximum torque per unit current control strategy is the most widely studied approach in practice. For a given torque, this control strategy minimizes the current; thus, copper losses are minimized in the process.
An accurate determination of the parameters of synchronous machines at the design stage is very important to design engineers, allowing the determination of appropriate materials for machine manufactures by predicting electrical and mechanical overloads during transients.
The Finite Element analysis validates the parameters and evidentiates the need to consider the sub-subtransient reactances due to the skin effect.
By the Standstill Frequency Response test (SSFR) simulation using FLUX2D one can replace different tests performed by the manufacturer, making this tool useful for the design optimization of machines.

References

[FLU05] FLUX2D, version 9.2: Finite element software for electromagnetic applications; CEDRAT, France (2005)

[Bos86] Bose, B.K.: Power Electronics and AC Drives, Prentice-Hall (1996)

[TLW96] Taghezout, D., Lombard, P., Wendling, P.: Finite element prediction of a brushless DC motors dynamic behavior. In: Proc. Intelligent Motions Systems, pp. 59-67 (1996)

[MK99] Manoliu, V. Kisck, D.O: Modelling and simulation of self-controlled synchronous motor considering saturation. In: Proc. of Intern. Symp.ELECTROMOTION99, Patras, Greece, vol. 1, pp. 97-100 (1999)

[NFC97] Nabeta, S. Y., Foggia, A., Coulomb, J.-L., Reyne, G.: Finite element analysis of the skin-effect in damper bars of a synchronous machine. In: IEEE Transactions on Magnetics, vol. 33, no. 2, pp. 2065-2068 (1997)

[MBM98] Manoliu, V., Bl, C., Melcescu, L.: Effects of damper windings on the performances of self-controlled synchronous motor. In: Proc. of Intern. Symp. SPEEDAM98, Sorrento, Italy, pp. P5-57 - P5-61 (1998)

Finite Volume Method Applied to Symmetrical Structures in Coupled Problems

Ioana - Gabriela Sîrbu

University of Craiova, Electrical Engineering Faculty, Decebal Blv. No. 107, 200440-Craiova, Romania osirbu@elth.ucv.ro

This paper presents a possibility of modelling the electromagnetic field, by taking into account the symmetry of the domains. The domain discretisation and implicitly the finite volume shape will be derived from the type of the symmetry. Thus, a reduction of the problem size, as well as a guaranteed symmetry of the solution, and a better approximation of the borders and of the discontinuity surfaces are achieved.

1 Introduction

Cell complexes are basic tools of algebraic topology. A cell complex can be based on a coordinate system; in such a case the edges of the cells lie on the coordinate lines and the faces on the coordinate surfaces. The advantages of these complexes are the easy utilization, the amount of memory may be not too high and they are used to develop fast algorithms that are adapted to the vectorial architecture. Their limitations come from the domain borders and from the object discretisation, where special treatments are necessary, the main problem being the approximation of the solution near the borders and in the interface zones.
For numerical applications one prefers to drop out these cell complexes based on coordinates and to use non-structured cells of triangular or tetrahedral finite element type. These cells can easily be adapted to any complex geometry. Moreover finite element method was developed in great detail, the problem of the domain discretisation in triangular or tetrahedral cells being already solved. Almost all the software instruments for field modelling use the discretisation of triangular finite element type. These cell complexes have advantages on those that are based on coordinates: (1) the simplified cells can easily be adapted to the domain border; (2) can include regions of different materials; (3) the points can be taken on the separation surface; (4) the cells can have different dimensions in one zone or another. In order to model the electromagnetic field two cells complexes will be considered in this work: the Delaunay complex (primal complex) and associated Voronoi complex (dual complex) [Ton01].
In literature, the orthogonal grids were not completely abandoned. Some researchers try to find out solutions in order to eliminate the inconveniences of the method, for example by introducing a new consistent subgridding method, which allows an increased flexibility of mesh configurations with local refinement [PCW03].

This paper tries to find out a possibility of modelling the electromagnetic field in the symmetrical structures case, using curvilinear grids. This solution, even if it is not applicable to general problems, aims to simplify the problem size and to eliminate inconveniences near specific borders.

2 Symmetrical Structures Modelling

2.1 Electromagnetic Field Modelling

In rotationally symmetric structures, generation and adaptation of the grid may involve primal and dual cell complexes, as shown in Fig. 1.

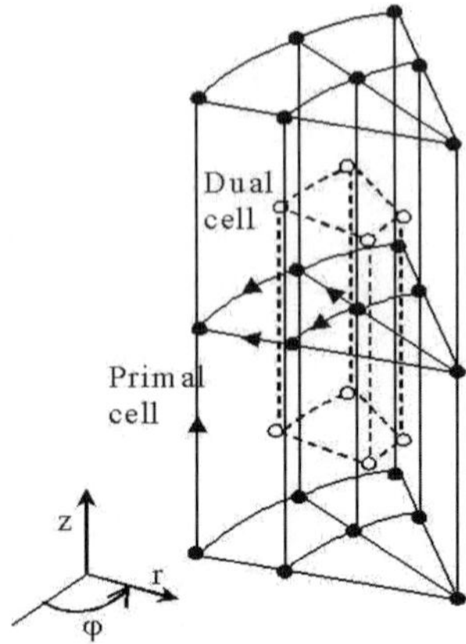

Fig. 1: Cells of the primal and dual complexes

A Yee-type scheme was chosen; every element (volume) of a cell complex includes a point of the other (dual) cell complex. Thus every surface of a grid will be intersected by a single line of the other grid. One associates the components of the electric field vector to the lines of the primal grid and the components of the magnetic field vector to the lines of the dual grid.
In order to model the electromagnetic field, the Maxwell's equations will be used:

$$\oint_{\Gamma} \overline{E} \cdot \overline{dr} = -\frac{d}{dt} \int_{S_{\Gamma}} \overline{B} \cdot \overline{dS} \tag{1}$$

$$\oint_{\Gamma} \overline{H} \cdot \overline{dr} = i_{S_{\Gamma}} + \frac{d}{dt} \int_{S_{\Gamma}} \overline{D} \cdot \overline{dS}$$

$$\oint_{\Sigma} \overline{D} \cdot \overline{dS} = q_{\Sigma}$$

$$\oint_{\Sigma} \overline{B} \cdot \overline{dS} = 0$$

where E, B, H and D are the electric field strength, the magnetic induction, the magnetic field strength and the electric induction respectively; q_{Σ} is the electric charge

inside Σ and i_{S_Γ} is the current through the surface delimited by Γ. These equations are completed by the constitutive equations.
With the notations:l- a line of the primal grid that belongs to the face f of this grid, $\widetilde{l}$- the face of the dual grid crossed by the line l and $\widetilde{f}$ - the line of the dual grid that crosses the face f, the first two relations of Maxwell's equations (1), applied to cells of the primal grid and of the dual grid respectively, becomes:

$$-\frac{\partial \varphi_f}{\partial t} = \sum_{l \in \partial f} \pm u_{el} \tag{2}$$

$$\frac{\partial \psi_{\widetilde{l}}}{\partial t} + i_{\widetilde{l}} = \sum_{\widetilde{f} \in \partial \widetilde{l}} \pm u_{m\widetilde{f}}$$

where φ_f is the magnetic flux through the face f, u_{el} is the electric voltage corresponding to the line l, $\psi_{\widetilde{l}}$ is the electric flux through the face $\widetilde{l}$ and $u_{m\widetilde{f}}$ is the magnetic voltage corresponding to the line $\widetilde{f}$; ∂f is the border of the face f and $\partial \widetilde{l}$ is the border of the face $\widetilde{l}$; the sign is "+" if the sign of the line l ($\widetilde{f}$) is associated to the sense of the face f ($\widetilde{l}$) by the rule "of the screw" and "-" if it is opposite.
Considering that on every face and on every line the electric or magnetic field value is constant and choosing as unknowns the magnetic fluxes and the electric voltages (corresponding to the primal grid), the system (2) becomes:

$$-\frac{\partial \varphi_f}{\partial t} = \sum_{l \in \partial f} \pm u_{el} \tag{3}$$

$$\frac{\partial}{\partial t}\left(\varepsilon \frac{A_{\widetilde{l}}}{L_l} u_{el}\right) + \sigma \frac{A_{\widetilde{l}}}{L_l} u_{el} = \sum_{\widetilde{f} \in \partial \widetilde{l}} \left(\pm \frac{1}{\mu} \cdot \frac{L_{\widetilde{f}}}{A_f} \cdot \varphi_f\right)$$

where ε is the permittivity of the medium, σ is the electrical conductivity, μ is the permeability; A_f ($A_{\widetilde{l}}$) is the area of the face f ($\widetilde{l}$) and L_l ($L_{\widetilde{f}}$) is the length of the line l ($\widetilde{f}$).
If the field variables do not depend on the z-coordinate and the domain has a cylindrical geometry, the problem is reduced to a 2-dimensional case (see Fig.2). In this case the first equation of (3) remains identical, while the second one is written as:

$$\frac{\partial}{\partial t}\left(\varepsilon \frac{\Delta r}{l_{arc}} u_{el}\right) + \sigma \frac{\Delta r}{l_{arc}} u_{el} = \sum_{\widetilde{f} \in \partial \widetilde{l}} \left(\pm \frac{1}{\mu} \cdot \frac{1}{A_{sector}} \cdot \varphi_f\right) \tag{4}$$

or

$$\frac{\partial}{\partial t}\left(\varepsilon \frac{l_{arc}}{\Delta_r} u_{el}\right) + \sigma \frac{l_{arc}}{\Delta r} u_{el} = \sum_{\widetilde{f} \in \partial \widetilde{l}} \left(\pm \frac{1}{\mu} \cdot \frac{1}{A_{sector}} \cdot \varphi_f\right) \tag{5}$$

in accordance to the line type that was considered (a part of an arc or of a radius).
For temporal discretisation the FDTD method was chosen.

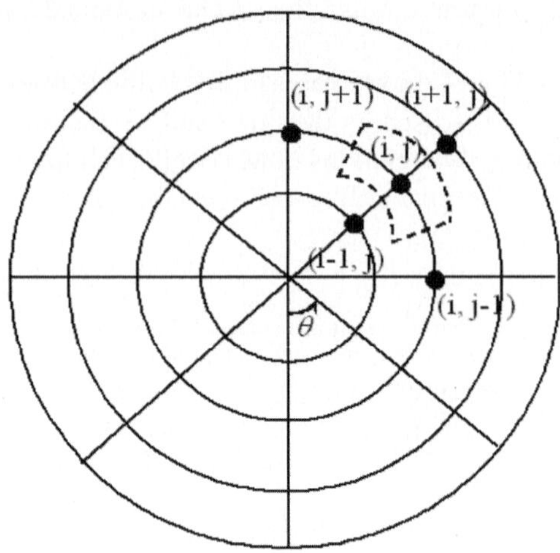

Fig. 2: Cylindrical symmetry structure - 2-dimensional case

2.2 Thermal Field Modelling

If the thermal field modelling is also wanted, the primal grid will be used. Thus coupled problems can be solved [MEU02].
The thermal conduction equation in a variable regime is considered. In a cylindrical symmetry case, one finds [POP02]:

$$\rho c_p \frac{\partial T}{\partial t} = \frac{1}{r} \cdot \frac{\partial}{\partial r}(r\lambda \frac{\partial T}{\partial r}) + \frac{1}{r^2} \cdot \frac{\partial}{\partial \theta}(\lambda \frac{\partial T}{\partial \theta}) + \frac{\partial}{\partial z}(\lambda \frac{\partial T}{\partial z}) + S \tag{6}$$

where ρ is the material density, c_p is the specific heat, T is the temperature, λ is the thermal conductivity and S is the power generated in a volume unit (function of electromagnetic field).
In the 2-dimensional case, if the field distribution in the section is not symmetrical, this equation becomes:

$$\rho c_p \frac{\partial T}{\partial t} = \frac{1}{r} \cdot \frac{\partial}{\partial r}(r\lambda \frac{\partial T}{\partial r}) + \frac{1}{r^2} \cdot \frac{\partial}{\partial \theta}(\lambda \frac{\partial T}{\partial \theta}) + S \tag{7}$$

For the primal grid represented in the Fig. 2, where the angle θ can be considered having the same value for all the sectors, this relation, integrated on a control volume built around a point (i,j) of the grid, becomes:

$$\int_t^{t+\Delta t} \int_{VC} \rho c_p \frac{\partial T}{\partial t} dV dt = \int_t^{t+\Delta t} \int_{VC} \frac{1}{r} \cdot \frac{\partial}{\partial r}(r\lambda \frac{\partial T}{\partial r}) dV dt + \\ + \int_t^{t+\Delta t} \int_{VC} \frac{1}{r^2} \cdot \frac{\partial}{\partial \theta}(\lambda \frac{\partial T}{\partial \theta}) dV dt + \int_t^{t+\Delta t} \int_{VC} S dV dt \tag{8}$$

where $dV = d\theta \cdot r \cdot dr \cdot 1$.
Applying Euler-backward time integration to (8), the following relation is obtained:

$$\rho c_p \theta (T_{i,j} - T^0_{i,j}) \cdot r_i \cdot \Delta r = r_{i+\frac{1}{2}} \cdot \lambda_{i+\frac{1}{2}} \cdot \theta \cdot \Delta t \frac{T_{i+1,j} - T_{i,j}}{\Delta r_i} - \tag{9}$$

$$-r_{i-\frac{1}{2}} \cdot \lambda_{i-\frac{1}{2}} \cdot \theta \cdot \Delta t \frac{T_{i,j} - T_{i-1,j}}{\Delta r_{i-1}} + \lambda_{j+\frac{1}{2}} \frac{2 r_i \Delta r \Delta t}{r^2_{i-\frac{1}{2}} \cdot r^2_{i+\frac{1}{2}}} \cdot \frac{T_{i,j+1} - T_{i,j}}{\Delta \theta} -$$

$$-\lambda_{j-\frac{1}{2}} \frac{2 r_i \Delta r \Delta t}{r^2_{i-\frac{1}{2}} \cdot r^2_{i+\frac{1}{2}}} \cdot \frac{T_{i,j} - T_{i,j-1}}{\Delta \theta} + S \cdot \theta \cdot \Delta r \cdot r_i \cdot \Delta t$$

where $T_{i+k,j+l}$ for $k = -1, 0, 1$, $l = -1, 0, 1$ are the (unknown) temperatures at the gridpoints at the current time level and $T^0_{i+k,j+l}$ is the (known) temperatures at the gridpoints at the previous time level.

2.3 Application

For the validation of the analysed discretisation technique, a numerical application was considered. The simple case of the conductor of copper, placed in air was studied, with radius a = 0.02 m. The r.m.s. value of the current through this conductor is I=200A. Two different frequencies are analysed: f = 50 Hz and 500 Hz.
A MATLAB program was developed for modelling both the electromagnetic field and the thermal field. The modelling results, obtained for the steady-state case, are presented in figures 3-8.

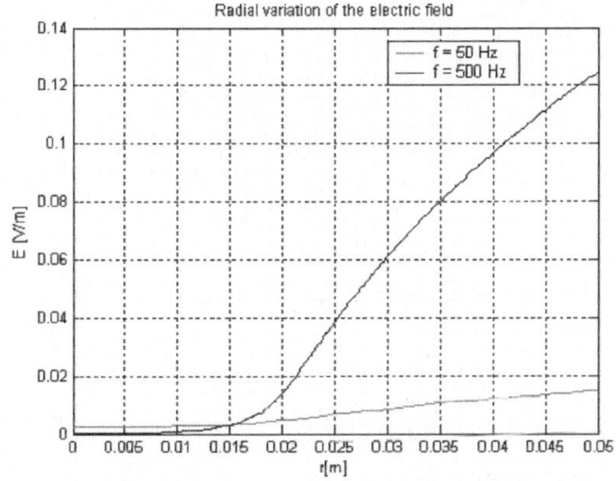

Fig. 3: The radial variation of the electric field

In figures 3 and 4 the electric field and the magnetic field are shown, for the two frequencies, in a comparative manner. Different types of radial variations in conductor and in air were observed. While the electric field increases from the conductor-air border to the domain border, the magnetic field maximum was obtained at the conductor-air border, having the same maximum value for the both values of the frequency.
In figures 5 and 6 the electrical and magnetic fields are presented as functions of r. In both cases f = 500 Hz was taken. We observe that near the z-axis both fields have

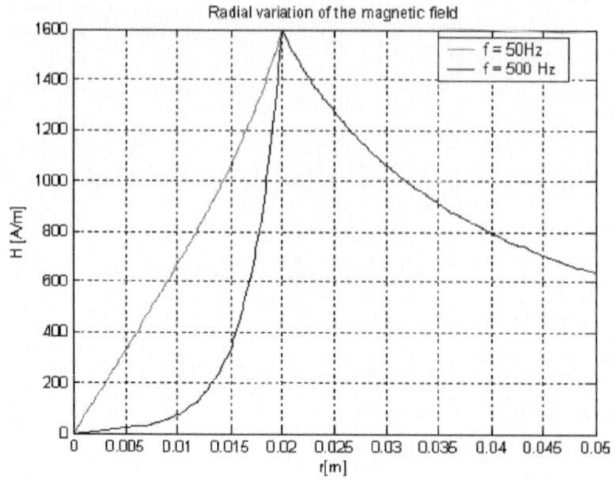

Fig. 4: The radial variation of the magnetic field

a different orientation when compared to the fields further away. This is a result of the high level of the frequency in the conductive medium.

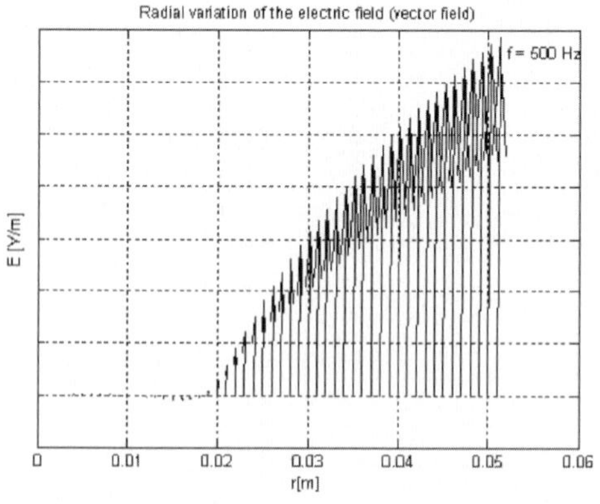

Fig. 5: The radial variation of the electric field (the vector field, for f = 500 Hz)

As for the thermal field, it was shown in figures 7 and 8 only for the conductive medium, for the same frequencies. A forced air cooling was considered. The boundary condition was of convection type. Because of the thinness of the conductor and of the cooling type that it was chosen, the variation of the temperature in the conductor resulted, in this case, very small.
Because of the symmetry of the resulted distributions, only the radial variations of the electric field, of the magnetic field and of the thermal field were represented here.

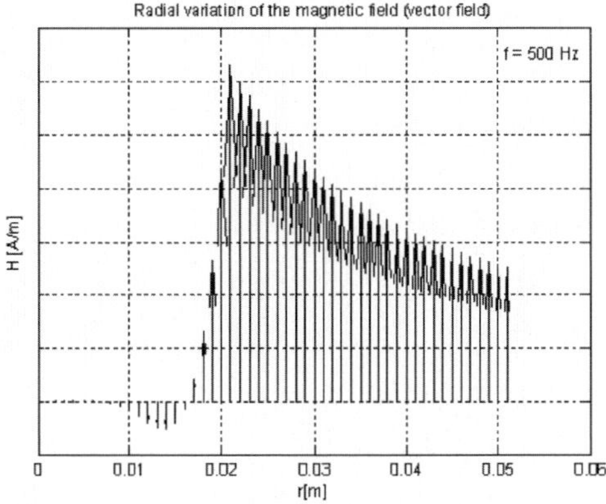

Fig. 6: The radial variation of the magnetic field (the vector field, for f = 500 Hz)

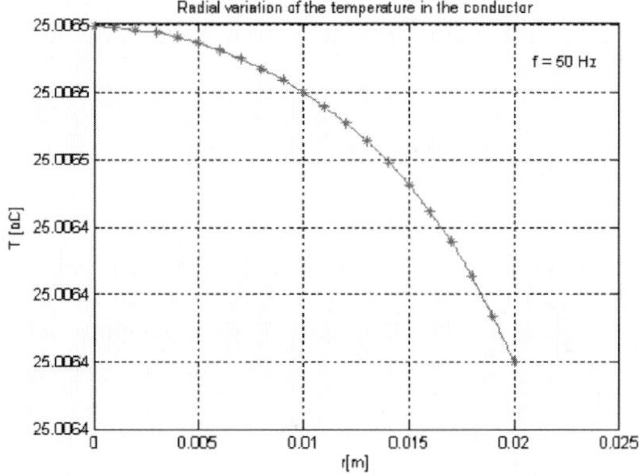

Fig. 7: The radial variation of the temperature for f =50 Hz

These results correspond to the electromagnetic field theory [MOC91]and to the thermal field theory [POP02] and they are similar to those obtained and published by the researchers.
The results were verified by using a dedicated software based on the discretisation of triangular finite element type (Quickfield). The Quickfield software cannot be used for coupled problems, but, using simplifications to accommodate for the couplings, results can be obtained that are approximative to those obtained by our MATLAB program.

3 Conclusion

The paper deals with the specific features of the electromagnetic - thermal coupled modelling in the particular case of the symmetrical distributions. This alternative method aims to increase the accuracy of the solution and to decrease the geometrical

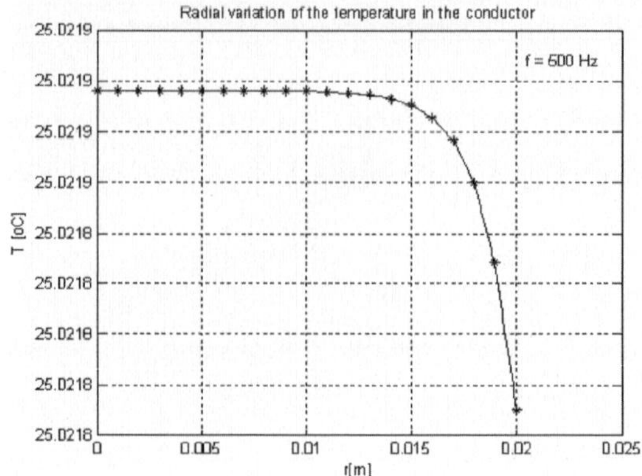

Fig. 8: The radial variation of the temperature for f =500Hz

complexity of the problem when compared to the case when the cartesian coordinates are used. This approach is an alternative to the discretisation of triangular finite element type.

References

[Ton01] Tonti, E.: Finite Formulation Of The Electromagnetic Field. In: Geometric methods for computational electromagnetics (ed) PIER Ser. vol 32 , 1– 44 (2001)

[PCW03] Poděbrad, O., Clemens, M., Weiland, T.: New Flexible Subgridding Scheme for the Finite Integration Technique. In: IEEE Transactions on Magnetics, **39**, no.3, may 2003, 1662–1665 (2003)

[MEU02] Meunier, G.: Electromagnétisme et problèmes couplés, vol.3. Hermes Science Publications, Paris (2002)

[POP02] Popa, I.C.: Modélisation numérique du transfert thermique - Méthode des volumes finis. Ed. Universitaria, Craiova (2002)

[MOC91] Mocanu, C.I.: Bazele electrotehnicii - Teoria câmpului electromagnetic. Ed. Didactică şi Pedagogică, Bucureşti (1991)

[GF03] Ghinea, M., Fireţeanu, V.: MATLAB - Calcul numeric. Grafică. Aplicaţii. Ed. Teora, Bucureşti (2003)

[***] ***: QuickField - User's Guide

Scattering Matrix Analysis of Cascaded Periodic Surfaces

Adriana Savin, Raimond Grimberg, Rozina Steigmann

National Institute of R&D for Technical Physics, 47 D.Mangeron Blvd, Iasi, 700050, Romania grimberg@phys-iasi.ro

1 Introduction

The design of microwave and millimeter wave devices requires more and more accurate synthesis procedures to satisfy the increasingly stringent specifications of modern communication systems [1]. Most of the available synthesis techniques are based on models that do not conveniently describe the physical behavior of circuits.
The goal of the present paper is to calculate the scattering matrix for the case of single and cascaded strip gratings.
The strip gratings allow a complete transmission of certain frequencies and a complete reflection at other frequencies and, then, exists a simple type of frequency selective surface [2]. A wide variety of filter characteristics can be obtained by cascading multiple layers of frequencies-selective surfaces. In this paper, a new method for scattered matrix analysis is presented. This method is based on Floquet harmonics analysis.

2 Floquet harmonics

The cross-sectional view of a strip grating is illustrated in Figure 1. This 2D geometry is an infinite periodic extension of a "unit cell" containing a single strip also illustrated in Figure 1. The precise location of the unit cell is arbitrary, and it may straddle two strips or even contain more than one complete strip, provided that its periodic repetition produces the original structure. We wish to determine the electromagnetic response of the grating due to a plane wave excitation of the form

$$E_z^{inc}(x,y) = E_0 e^{-jk(x\cos\theta + y\sin\theta)} \qquad (1)$$

The electromagnetic response to such kind of excitation can be expressed as a superposition of Floquet harmonics [3].
The Floquet harmonics form a complete orthogonal set over one period of the geometry in x and preserve the desired progressive phase shift imposed by the excitation for all x. From the infinite number of Floquet harmonics, generally only a few are associated with true propagation. The set of Floquet harmonics is entirely determined by the geometric period and the assumed phase progression along x and thus does

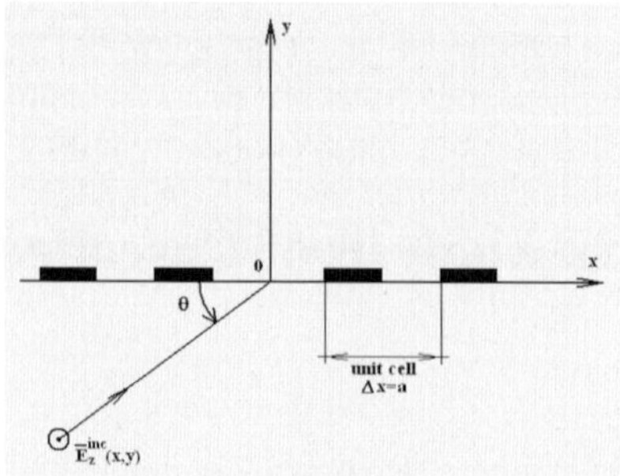

Fig. 1: Strip grating

not depend on the specific dimensions of the strip in the gratings [4].
The scattered electromagnetic field as the response of the plane wave excitation 1 can be expressed as

$$E_z^s(x,y) = \sum_{m=-\infty}^{\infty} e_m \Psi_m(x,y) \tag{2}$$

where

$$\Psi_m(x,y) = e^{-jk_{xm}x} e^{\pm j\sqrt{k^2-k_{xm}^2}y} \tag{3}$$

$$k_{xm} = k_x - \frac{2\pi}{a} m \tag{4}$$

a =unite cell length (defined in Figure 1)
When $k^2 > k_{xm}^2$, Floquet harmonics are waves that propagate away from the grating. When $k_{xm}^2 > k^2$, that sign of the square root is taken which ensures that the harmonics decay in a direction away from the grating. Because of the phase shift imposed by an incident field, the fields and currents of interest are not strictly periodic functions of x. Instead, they are modulated periodic functions of the general form

$$A_{mp}(x) = \sum_{l=-\infty}^{\infty} A(x-la) e^{-jk_x la} \tag{5}$$

The convolution theorem applied to the Fourier transform dictates that

$$\tilde{A}_{mp}(f) = \tilde{A}(f)\tilde{P}(f+f_0) = \Delta f \sum_{l=-\infty}^{\infty} A(l\Delta f - f_0)\delta(f - l\Delta f + f_0) \tag{6}$$

where

$$\Delta f = \frac{1}{a}; \quad f_0 = \frac{kx}{2\pi} \; and \; \tilde{P}(f+f_0) \tag{7}$$

is the Fourier transform of comb function P.
The comb function is defined as

$$P(f) = \sum_{l=-\infty}^{\infty} \delta(f - i\delta f).$$

As a consequence of the modulation, the transform of [5] is sampled at values of the transform variable f. Every periodic geometry can be associated with a spatial lattice at points in the original domain and a reciprocal lattice in the Fourier transform domain. The spatial lattice – reciprocal lattice concept is often useful in multidimensional applications, in the case in which the propagating harmonics can be identified. For the periodic structure depicted in Figure 1, the spatial lattice is defined by values of x

$$x = l \cdot a \tag{8}$$

in the set $\{.., -2, 0, 2, ..\}$ and the reciprocal lattice is defined by values of f

$$f = l\frac{l}{a} \tag{9}$$

in the set $\{.., -1, -\frac{1}{2}, 0, \frac{1}{2}, 1, ..\}$
The range $\left(\frac{k_x}{2p} - 1\right) < f < \left(\frac{k_x}{2p} + 1\right)$ defines the "visible region" of the spectrum containing the propagating Floquet harmonics [5].

3 The scattering from a conducting strip grating electric field integral equation approach

Consider a TM plane wave having the form of equation (1) incident on the infinite periodic structure grating illustrated in Figure 1.
The surface equivalence principle [2] can be used to replace the perfect conducting strip by an equivalent electric current density $J_z(x)$. Due to the phase progression imposed by the incident field, the equivalent currents must satisfy the Floquet conditions

$$J_z(x + a) = J_z(x)e^{-jk_x a} \tag{10}$$

A conventional electric field integral equation (EFIE) formulation requires the superposition of the field of each of the currents.
If $J_z(x)$ is considered nonzero only when x is located on the area of the conducting strips, the EFIE can be written as

$$E_z^{inc}(x, 0) = jk\eta \int_{-\infty}^{\infty} G_p(x, x')J_z(x')dx' \tag{11}$$

where $k = \sqrt{\omega^2 \mu_0 \varepsilon_0}$, $\eta = \frac{z}{\sqrt{\frac{\mu}{\varepsilon_0}}}$, and $G_p(\bar{x}, \bar{x}')$ is the Green's function. For our periodic distribution the Green's function can be write as a Fourier series.

$$\int G_p\left(x, x'\right) dx' = \frac{1}{4j} \sum_{l=-\infty}^{\infty} H_0^{(2)}\left(k \mid x - la \mid\right) e^{-jlk_x a} \tag{12}$$

where $x' = la$, $H_0^{(2)}$ is the two rank, 0–order, Hankel function [3].
By restricting the domain of the equation to a simple unit cell, the electrical size of the problem has been reduced to manageable proportions. The method of moments (MOM) discretization of the EFIE [6] follows in the usual manner. Suppose that the

conducting strip contained in the unit cell is divided into N intervals, over which pulse basis functions will reside. A Dirac delta testing function can be located in the center of each interval.
The resulting system has the general form

$$\begin{bmatrix} z_{11} & z_{12} & \cdots & z_{1N} \\ \cdot & \cdot & \cdot & \cdot \\ \cdot & \cdot & \cdot & \cdot \\ z_{N1} & z_{N2} & \cdots & z_{NN} \end{bmatrix} \begin{bmatrix} j_1 \\ \cdot \\ \cdot \\ j_N \end{bmatrix} = \begin{bmatrix} e_1 \\ \cdot \\ \cdot \\ e_N \end{bmatrix} \tag{13}$$

$$z_{mn} = jk\eta \int_{cell_n} G_p(x_m - x')dx' \tag{14}$$

$$e_m = E_0 e^{-jk_x x_m} \tag{15}$$

The current density can be expressed in terms of Floquet harmonics as:

$$J_z(x) = -\frac{2}{k\eta} \sum_{n=-\infty}^{\infty} e_n \sqrt{k^2 - k_{xn}^2} e^{-jk_{xn}x} \tag{16}$$

Using the orthogonality of the Floquet harmonics, the coefficients are

$$e_n = -\frac{k\eta}{2a\sqrt{k^2 - k_{xn}^2}} \int_0^2 J_z(x) e^{jk_{xn}x} dx \tag{17}$$

Reflection and transmission coefficients can be defined as

$$R_0 = \frac{e_0}{E_0}; \qquad T_0 = 1 - \frac{e_0}{E_0} \tag{18}$$

Figure 2 shows a plot of the reflection coefficient for a strip grating having conducting material occupying exactly half of the unit cell as a function of the unit cell size.

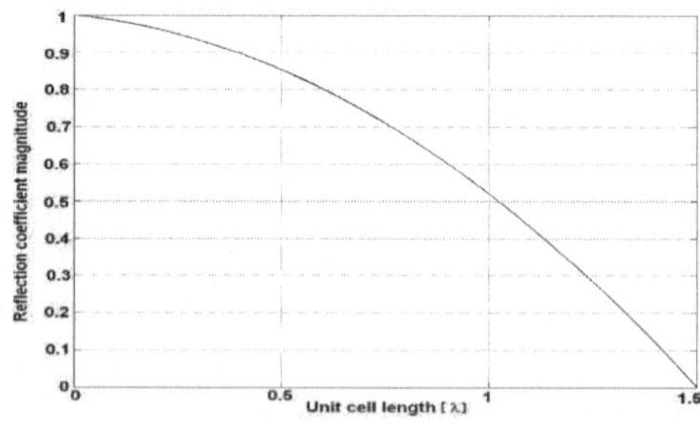

Fig. 2: Reflection coefficient magnitude vs. unit cell size

4 Acceleration procedure for the calculation of Green's function

The magnitude of the l-th term in the periodic Green's function (12) can be estimated from the asymptotic form of the Hankel function for large arguments [7].

$$H_0^{(2)}(kx) = \sqrt{\frac{2j}{\pi kx}} e^{-jkx} \tag{19}$$

Obviously, the summation in (12) would diverge if not for oscillating behavior of the exponential function. In practice, the direct summation required in (12) is prohibitively slow. We consider an acceleration technique based on the Fourier transform pair. This procedure is sometimes known as the "Poisson sum transformation" [1]. Using the Fourier transform of Hankel function

$$\tilde{H}_0^{(2)}(k \mid x \mid) = \frac{2}{\beta_y} \tag{20}$$

$$\beta_y = \begin{cases} \sqrt{k^2 - (2\pi f)^2} & \\ & if \begin{cases} k > (2\pi f) \\ k > (2\pi f) \end{cases} \\ -j\sqrt{(2\pi f)^2 - k^2} & \end{cases} \tag{21}$$

The Fourier transform of $G_p(x)$ can be found similar to (6) resulting to

$$\tilde{G}_p(f) = \frac{1}{2ja} \sum_{l=-\infty}^{\infty} \delta\left(f - \frac{l}{a} + \frac{kx}{2\pi}\right) \frac{1}{\beta_y} \tag{22}$$

Applying the inverse Fourier transform to (22) yields the result

$$G_p(x) = \frac{1}{2ja} \sum_{l=-\infty}^{\infty} \left[\frac{e^{j2\pi fx}}{\beta_y}\right]_{f=\frac{l}{a}-\frac{kx}{2\pi}} \tag{23}$$

The magnitude of the lth term behaves as

$$O\left(\frac{1}{l}\right) \quad as \ l \to \infty \tag{24}$$

5 Computing the Transversal Magnetic (TM) scattered field

The electromagnetic incident field is the TM plane wave with $\pi/2$ incident angle having the form presented in figure 3a and b.
The conducting strip grating is embedded into a dielectric slab having the electric permittivity $\epsilon_0\epsilon_r$ and located at $d = 0$, as shown in Figure 4.
In figure 5 is presented the scattered field (real and imaginary components) for the unit cell length $a = 1.5\lambda$
If the unit cell length satisfies the condition $a \leq 1.22\frac{\lambda}{2}$ [3], the diffraction of the electromagnetic incident wave is not possible; this situation is illustrated in figure 6, for $a = \lambda/2$.

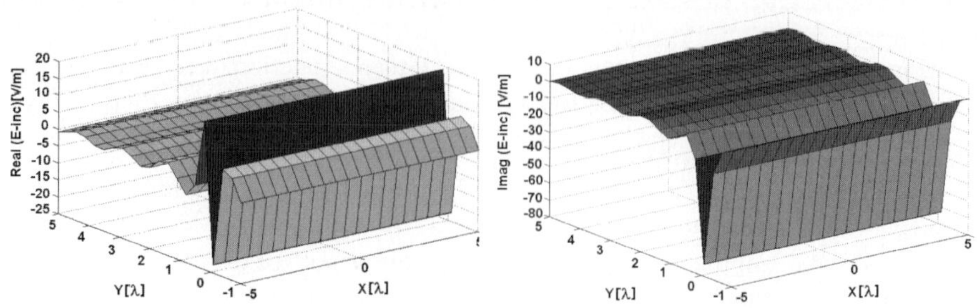

Fig. 3: The electromagnetic incident field a)Real component of TM electric field; b) Imaginary component of TM electric field

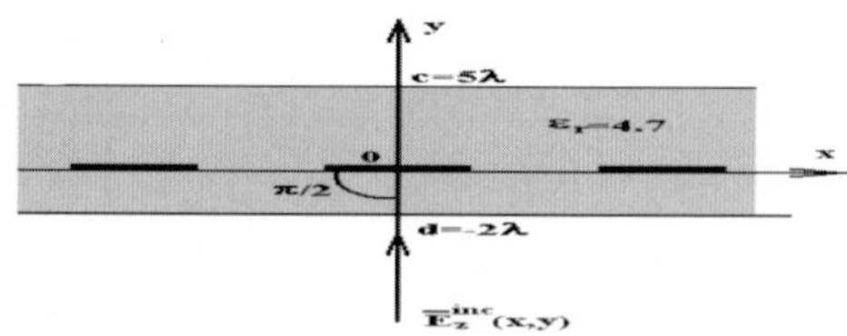

Fig. 4: The strip grating embedded into a dielectric slab, $\epsilon_r = 4.7$

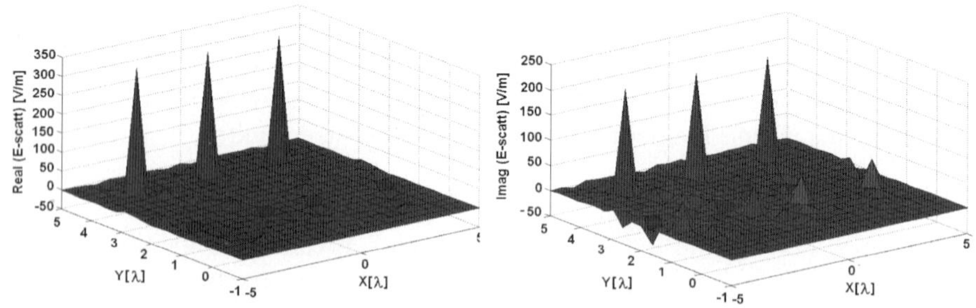

Fig. 5: The scattered field for $a = 1.5\lambda$. a)real component; b) imaginary component

6 Scattering matrix analysis of cascaded periodic surface

Because strip gratings allow complete transmission of certain frequencies and complete reflection at other frequencies, they are a simple type of frequency-selective surface.

To efficiently analyze multilayered structures, each periodic surface can be modeled as a multiport network with each port representing one Floquet harmonics determining the “far–field” reflection and transmission characteristics of periodic surfaces. Generalized scattering and transmission matrices representing each individual surface can subsequently be employed as a building block in the electromagnetic model of the multilayered structures.

The definition of the scattering matrix is presented in the context of the two port network in figure 7.

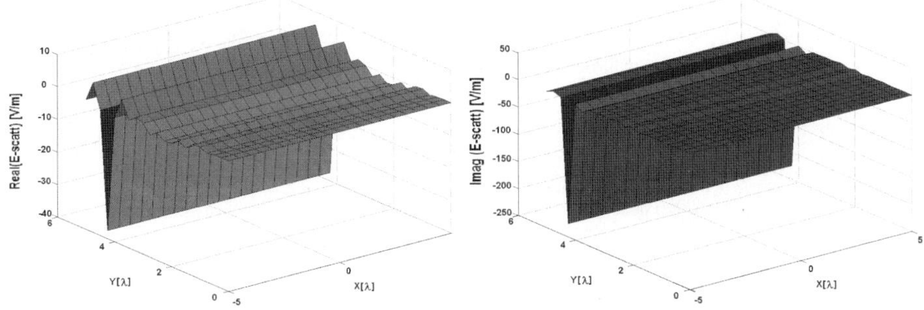

Fig. 6: The scattered field for $a = \lambda/2$. a)real component; b) imaginary component

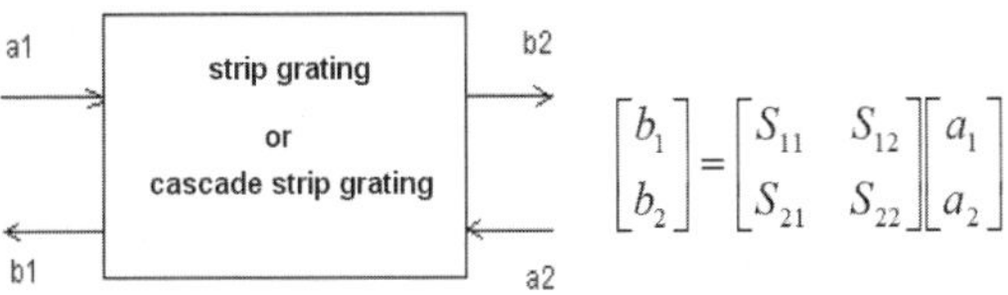

Fig. 7: Two port scattering parameters: strip grating or cascade strip gratings

The definition of S−parameters is given in Annex 1. For TM excitation, the scattering matrix arising from the strip grating problem is a 2×2 block matrix

$$S = \begin{bmatrix} S_{11} \; S_{12} \\ S_{21} \; S_{22} \end{bmatrix} \tag{25}$$

where each of the blocks is infinite dimensional matrix relating the coefficients of the scattered Floquet harmonics to those of the incident harmonics. The blocks take the form

$$S = \begin{bmatrix} S_{11}^{11} & S_{12}^{11} & \dots & S_{1n}^{11} \\ S_{21}^{11} & S_{22}^{11} & \dots & \dots \\ \vdots & \vdots & \vdots & \vdots \\ S_{m1}^{11} & S_{m2}^{11} & \dots & S_{mn}^{11} \end{bmatrix} \tag{26}$$

In practice, these matrices are truncated to finite dimensions. The entries of the generalized scattering matrices depend on the location of reference planes as indicated in Figure 8.

The (m,n) entry in the scattering matrix of equation (26) is proportional to the square root of the ratio of the power carried by the m−th reflected harmonic to the power carried by the n−th incident harmonic. In general, this is a complex quantity, having magnitude

$$mag\left(S_{mn}^{11}\right) = \sqrt{\frac{\int_0^a \mid \bar{E}_m^s \times \bar{H}_m^s \cdot \hat{y} \mid_{y=-d} dx}{\int_0^a \mid \bar{E}_n^{inc} \times \bar{H}_n^{inc} \cdot \hat{y} \mid_{y=-d} dx}} \tag{27}$$

and phase

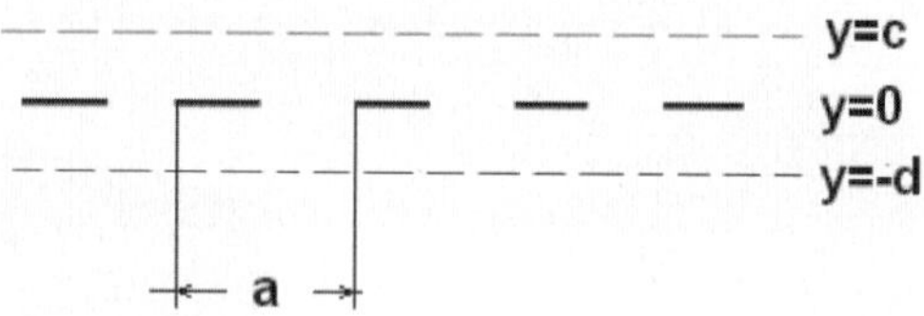

Fig. 8: Location of reference planes for the definition of S parameters

$$phase\left(S^{11}_{mn}\right) = phase\left\{E^{s}_{zm}\left(0,-d\right)\right\} - phase\left\{E^{inc}_{zn}\left(0,-d\right)\right\}. \tag{28}$$

The entries of the S_{21} matrix are defined in a similar manner, only in terms of the total transmitted fields instead of the scattered reflected field.

$$mag\left(S^{21}_{mn}\right) = \sqrt{\frac{\int_0^a \mid \bar{E}^{tot}_m \times \bar{H}^{tot}_m \cdot \hat{y} \mid_{y=c} dx}{\int_0^a \mid \bar{E}^{inc}_n \times \bar{H}^{inc}_n \cdot \hat{y} \mid_{y=-d} dx}} \tag{29}$$

$$phase\left(S^{21}_{mn}\right) = phase\left\{E^{tot}_{zm}\left(0,c\right)\right\} - phase\left\{E^{inc}_{zn}\left(0,-d\right)\right\} \tag{30}$$

The S_{12} and S_{22} matrices are defined in analogous manner.
For a single surface "illuminated" by a plane wave, generalized reflection and transmission coefficients are sometimes employed as an alternative to the scattering matrix description.
The magnitudes of these expressions simplify to

$$\mid R_n \mid = mag\left(S^{11}_{n0}\right) = \frac{\mid e_n \mid}{\mid E_0 \mid}\sqrt{\left|\frac{\sqrt{k^2 - k^2_{xn}}}{k^2 - k_x}\right|} \tag{31}$$

$$\mid T_n \mid = mag\left(S^{21}_{n0}\right) = \left|\delta^n_0 + \frac{e_n}{E_0}\sqrt{\frac{\sqrt{k^2 - k^2_{xn}}}{\sqrt{k^2 - k_x}}}\right| \tag{32}$$

where δ^n_0 is Kronecker symbol with properties $\delta^n_0 = \begin{cases} 1\ if\ n = 0 \\ 0 \quad otherwise \end{cases}$ and the phase defined in accordance with equations (28) and (30).
For cascading several layers, we use the transmission matrix representation

$$\begin{bmatrix} b_2 \\ a_2 \end{bmatrix} = \begin{bmatrix} T_{11}\ T_{12} \\ T_{21}\ T_{22} \end{bmatrix} \begin{bmatrix} a_1 \\ b_1 \end{bmatrix} \tag{33}$$

is more convenient than the scattering matrix description

$$\begin{aligned} T_{11} &= S_{21} - S_{22}S^{-1}_{12}S_{11} \\ T_{12} &= S_{22}S^{-1}_{12} \\ T_{21} &= -S^{-1}_{12}S_{11} \\ T_{22} &= S^{-1}_{12} \end{aligned} \tag{34}$$

The cascading of layers just maps onto multiplication of the transmission matrices.
In Table 1, we present the results obtained for one conducting strip grating with cell

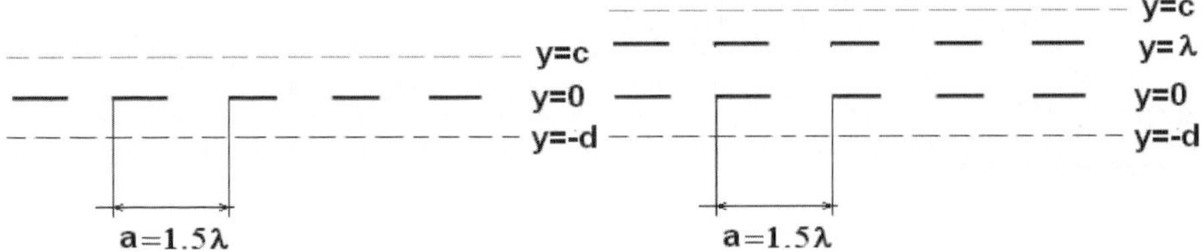

Fig. 9: a) One conducting strip grating; b) cascade of two strip gratings

length $a = 1.5\lambda$ and for a cascade of two similar conducting strip gratings located at $y = 0$ and $y = \lambda$ respectively, presented in Figure 9.
For normal illumination (inclination angle $\pi/2$), for the mode 0, one conducting

Inclination angle	mode	One conducting strip grating		Cascade of two conducting strip grating	
		R	T	R	T
$\pi/2$	0	0	1	0	1
	1	1	0	1	0
	2	1	0	1	0
$\pi/3$	0	0	1	1	0
	1	1	0	1	0
	2	1	0	0	1
$\pi/4$	0	0	1	0	1
	1	1	0	0	1
	2	0	1	0	1

Table 1:

strip grating and the cascade of two strip gratings present total transmission ($T = 1$) and reflection does not exist ($R = 0$). For modes 1 and 2 the strip grating and the cascade present total reflection ($R = 1, T = 0$).
For illumination at $\pi/3$, for mode 0, one strip grating performs a total transmission but the cascade performs a total reflection. For mode 1, one strip and the cascade perform total reflection.
For mode 2, one strip has total reflection, the cascade has total transmission.
For $\pi/4$, for modes 0 and 2, the reflection coefficient is 0 and the transmission is complete, and for mode 1, for one strip grating is total reflection and for two strip gratings the total transmission is performed.

7 Conclusions

The analysis of the scattering matrix for the case of single and multiple periodic surfaces is made using the Floquet harmonics method. TM scattering from a conducting strip grating is calculated using EFIE approach.
A very simple but efficient procedure for acceleration Green's function is presented. In this case, the use of only 25 basis functions for EFIE discretization has been absolutely enough. For TM excitation, the scattering matrix arising from the strip grating

problem is a 2×2 block matrix, where each of the blocks is a matrix of infinite-dimension. For this type of description, only 400 Floquet harmonics are sufficient, because the convergence in the sense of norm is smaller than 3×10^{-5} reported to the case in which 450 Floquet harmonics are used.

ACKNOWLEDGEMENT
This paper is supported by the Romanian Ministry of Education and Research, Excellence Research Program, under Projects nEDA and ROLIGHT.

References

[1] C.Kudisio, R.Cameron, W.C.Tang: Innovations in microwave filters and multiplexing networks for communication satellite system, *IEEE Trans. Microwave Theory Tech.*10.(1992), pp.1133–1149
[2] A. F. Peterson, S. L. Ray, and R. Mittra: *Computational Methods for Electromagnetics*. IEEE Press, New York, 1998
[3] M. Born , E. Wolf.*Principles of Optics*, 6th eds., Pergamon Press, Oxford, 1980.
[4] V.V.S. Prakash and Raj Mittra, An efficient technique for analyzing multiple frequency selective surface screens with dissimilar periods,*Microwave & Optical Technology Letters*, Vol. 35, No. 1, (2002), pp. 23–27
[5] Robert E. Collin,*Field Theory of Guided Waves*, 2nd Edition, IEEE Press, NY, 1991
[6] R.F. Harrington,*Field computation by moment methods*, Macmillan, New York, 1968
[7] M. Abramovitz, I. A. Stegun, eds.,*Handbook of Mathematical functions*, Dover Publications, New York, 1972

Annex 1. Definition of S–parameters

$$\begin{bmatrix} b_1 \\ b_2 \end{bmatrix} = \begin{bmatrix} S_{11} & S_{12} \\ S_{21} & S_{22} \end{bmatrix} \begin{bmatrix} a_1 \\ a_2 \end{bmatrix}$$

The independent variables a_1 and a_2 represent the amplitudes of incident waves at port 1 and 2, normalized to characteristic impedance of the two – port network viewed as a transmission line

$$a_1 = \frac{amplitude\ of\ the\ wave\ incident\ at\ port\ 1}{\sqrt{Z_0}}$$

$$a_2 = \frac{amplitude\ of\ the\ wave\ incident\ at\ port\ 2}{\sqrt{Z_0}}$$

The dependent variables b_1 and b_2 are normalized reflected amplitude wave

$$b_1 = \frac{amplitude\ of\ the\ wave\ reflected\ at\ port\ 1}{\sqrt{Z_0}}$$

$$b_2 = \frac{amplitude\ of\ the\ wave\ reflected\ at\ port\ 2}{\sqrt{Z_0}}$$

The S–parameters are
$S_{11} = \frac{b_1}{a_1} |_{a_2=0} =$ input reflection coefficient with the output port terminated by a matched load
$S_{22} = \frac{b_2}{a_2} |_{a_1=0} =$ output reflection coefficient with the input port terminated by a matched load
$S_{21} = \frac{b_2}{a_1} |_{a_2=0} =$ forward transmission gain with the output part terminated in a matching load
$S_{12} = \frac{b_1}{a_2} |_{a_1=0} =$ reverse transmission gain

Part II

Circuit Simulation and Design

Overview of Circuit-Simulation Activities at TKK CTL*

Janne Roos[1]

Helsinki University of Technology, Department of Electrical and Communications Engineering, Circuit Theory Laboratory, P.O.Box 3000, FI-02015 TKK, Finland.
janne@ct.tkk.fi

1 Introduction

This paper summarizes the recent circuit-simulation activities [Roo04]–[Sil06] at Helsinki University of Technology (TKK), Circuit Theory Laboratory (CTL). This paper is mostly based on the results of the national projects Advanced Radio Frequency SImulation and Modeling (ARFSIM 2002–2003) [Roo04], MOdeling and Simulation for Advanced Integrated Circuits and Systems (MOSAICS 2004–2005), and Accurate Models Aim for Zero Errors (AMAZE 2006–2008). All these projects have been funded by the National Technology Agency of Finland, Nokia Corporation, and AWR–APLAC Corporation; the annual volume at TKK CTL has been 4.0–5.5 man years. In these projects, APLAC circuit simulation and design tool [A06] has been used as a common platform for the circuit analysis and modeling methods developed.

This paper is organized as follows. Section 2 very briefly reviews transistor-model development. Section 3 lists our recent work with various analysis methods. In the following two sections, the current research interests of the author are treated in some more detail: Sections 4 and 5 discuss model-order reduction and behavioral modeling, respectively. Finally, Section 6 briefly summarizes that part of our recent research that has been carried out outside the ARFSIM, MOSAICS, and AMAZE projects.

2 Transistor models

During the ARFSIM, MOSAICS, and AMAZE projects, the C-code implementation of several BJT, MESFET [Kal02], and MOSFET semiconductor models has been improved. Also, an attempt has been made to make the transistor-model development more fluent: both a C-code model interface and a Philips SiMKit adapter have been implemented in APLAC. What comes to fundamental research, a new rule for MESFET gate-charge division based on the energy-conservation principle has been presented [KV04].

* Invited Paper at SCEE-2006

3 Analysis methods

During the ARFSIM, MOSAICS, and AMAZE projects, the following analysis methods have been studied and/or developed and/or implemented in APLAC:

- DC
  - speed/convergence improvements based on industrial feedback
  - piecewise-linear solution algorithm [RVV02], [Roo05], [Roo06]
  - nonmonotone norm-reduction method [Hon02a], [Hon02b]
  - nonlinear iteration/optimization methods [HRK06]
  - homotopy methods [Lin06]
  - parallel hierarchical analysis [Hon02c], [HK02], [KH02], [Hon03]
- AC
  - minor improvements
- Transient
  - event-based time-step control
  - truncation-error criteria
  - treatment of transmission lines
  - parallel hierarchical analysis [Hon03]
  - optimization of C-code implementation
- Multi-tone Harmonic Balance (HB)
  - reducing the memory consumption and increasing speed
  - efficient formulation of HB equations [Vir05]
  - inexact-Newton method with GMRES solver [Vir05]
  - nonmonotone norm-reduction method [Hon02b]
  - transient-assisted HB
  - multi-dimensional frequency mappings
  - sampling of nonlinear component-model functions [Vir05]
  - parallelization using threads [KH04]
  - oscillator analysis [Vir05]
  - frequency-divider analysis [Poh06]
- Multi-variate steady state time domain
  - GMRES preconditioners [LVV03], [Leh03]
  - multi-grid approach [Leh03]
- Large-signal–small-signal
  - GMRES preconditioners
  - amplitude/phase noise analysis [Vir05]
- Envelope
  - self-starting polynomial collocation/projection ODE-solver
  - MATLAB–APLAC prototype implementation
- Finite Difference Time Domain (FDTD)
  - FDTD–circuit/system co-simulation [Cos05]
  - optimization of C-code implementation [Cos05], [Cos06]

4 Model-order reduction for EM/circuit simulation

Let us divide the whole Model-Order Reduction (MOR) chain into three steps:

1. *Interconnect modeling*: model, using Electro-Magnetics (EM) simulation or other methods, the interconnect (e.g., layout parasitics) by a large RLC network.

2. *Linear MOR*: reduce the RLC network to obtain a reduced-order frequency-domain interconnect model (e.g., a set of poles and residues).
3. *Macromodel realization*: link the model obtained to transient simulation of the whole nonlinear circuit by generating an appropriate equivalent-circuit representation.

These three steps are treated in Sections 4.1, 4.2, and 4.3, respectively.

4.1 Interconnect modeling

Interconnects can be modeled using RC/RLC networks, (dispersive multi-conductor) transmission lines, (measurement or EM-simulation-based) tabulated frequency-domain scattering parameters, or even 3D full-wave models. The selection of a proper interconnect model depends on operation frequency, desired accuracy, available computational resources, etc.

In [Aal03], a dispersive inhomogenous two-conductor transmission line was treated in conjunction with the MOR method Padé-via-Lanczos (PVL). In [Pal04], in turn, a RLC lumped-element approximation for a dispersive multi-conductor transmission line was implemented in APLAC and treatment of tabulated frequency-domain scattering parameters was considered. Although we have studied interconnect modeling, it has not been the main focus area; in most cases, our starting point for MOR has been a given RLC netlist.

4.2 Linear MOR

During the last 15 years, various MOR methods have been proposed in the electrical-engineering literature. The first MOR methods were able to calculate single-input single-output transfer functions of linear circuits. The current MOR methods, in turn, are able to reduce large RLC networks such that the reduced-order models obtained can be consistently linked to the transient simulation of the whole nonlinear circuit (see Section 4.3).

In [Aal03], the following linear MOR methods were evaluated: Asymptotic Waveform Evaluation (AWE), Complex Frequency Hopping (CFH), Padé-via-Lanczos (PVL), reduction via split congruence transforms, coordinate-transformed Arnoldi algorithm, Passive Reduced-Order Interconnect Macromodeling Algorithm (PRIMA), and the PVL derivatives SyPVL, MPVL, and SyMPVL (see [Aal03] for all the references). According to [Aal03], "*Most of these methods were coded in a combination of C, MATLAB, and APLAC input language. Several test RLC networks were then reduced with the methods. In some cases, the result was a transfer function that was compared with that of the original circuit. In others, the result was a macromodel. Then, APLAC was used to run AC and transient analyses on both the original circuit and the reduced one. ... The best methods found were PRIMA and MPVL.*"

Based on this information, PRIMA [OCP98] was studied in more detail in [Pal04], where it was found that "*PRIMA provided passive reduced-order macromodels for interconnect circuits with excellent accuracy up to microwave frequencies. During this work, an attempt was made to develop a stopping criterion which would allow PRIMA iteration to be stopped right after numerical accuracy has been lost, thus allowing the easy generation of passive reduced-order models with the maximum available order. This attempt, however, proved futile: the instability of the reduced-order models could not be predicted from the properties of the matrices available during the iteration. In addition to this, an error estimate for PRIMA, presented in the literature, was evaluated. The results obtained with the error estimate were not always accurate enough, and the computation of the error estimate was too CPU-time consuming in some cases.*"

In [Aal03], the test RLC networks were quite small. While these small RLC networks were excellent in revealing the shortcomings of the reduction methods, they did not show all the potential of PRIMA (and MPVL). In [Pal04], in turn, much larger RLC networks (having nearly 1000 nodes) were reduced, and the overall impression of PRIMA was much more positive than in [Aal03], yet there were the problems mentioned above.

In [Pal04], PRIMA was implemented in APLAC, where it can be used as an off-line preprocessing tool for a large nonlinear circuit to be simulated. Namely, each large RLC block is treated as an N-port and reduced with PRIMA, the result being a file containing poles common for all the N-port Y-parameters, and the corresponding individual sets of residues for each Y-parameter.

Here, let us emphasize that PRIMA is by no means the last published linear MOR method; there are many newer methods that would be worth studying, too.

4.3 Macromodel realization

Since the reduced-order models are described in the frequency domain (or as differential equations), they have to be linked to the time-domain simulation of the total nonlinear circuit. This can be done by replacing the reduced-order models with appropriate macromodels.

Most of our scientific MOR activity has been just in this area: [Aal03], [Pal04] concentrate partially, and [AR02], [PR03], [PR04] fully on the macromodel realization. In particular, in [PR04] a comprehensive comparison of nine reduced-order interconnect macromodels for time-domain simulation is presented: the macromodels are reviewed, presented in a unified manner, and compared both theoretically and numerically.

The reduced-order macromodels can be divided into two groups:

- *Equivalent-circuit realizations*: a SPICE, APLAC, etc. netlist is synthesized using basic circuit elements. Nearly any time-domain circuit simulator can then be used.
- *Time-varying macromodels*: a macromodel with time explicitly present in the updating equations is generated. For most simulators, this method requires a modification of the simulator's source code.

In [PR04], we found the best macromodels for both categories. The overall fastest macromodel was the time-domain Differential-Equation Macromodel (DEM), which we proposed in [PR03]. In the course of the work of [Pal04], the time-domain DEM was implemented in APLAC. This time-varying macromodel is used as such in transient analysis. In other analysis methods (DC, AC, HB, etc.), another version of DEM is internally invoked; for example, in the case of AC analysis, $s = j\omega$ is inserted into relevant frequency-domain Y-parameter expressions.

The main message of [PR04] is that the macromodel realization has a great impact on the transient-simulation CPU time; in fact, the transient simulation of a poorly realized macromodel (along with the nonlinear circuit) may last longer than that of the original, unreduced, circuit.

5 Behavioral modeling of components and circuit blocks

In Behavioral Modeling (BM), there is no (fast and accurate) model available, only the input-output data; therefore, BM corresponds to nonlinear system identification. The BM methods can be classified in many ways, e.g., as follows:

- analog ↔ digital
- static ↔ dynamic
- linear ↔ nonlinear
- white box ↔ black box
- single component ↔ circuit block
- measurement/simulation based BM [WR05] ↔ nonlinear MOR [Vos05]

Sections 5.1–5.4 (that are partly based on the ongoing Ph.D. Thesis work of Tuomo Kujanpää) will concentrate on Artificial Neural Network (ANN) based "analog, static, nonlinear, black-box-like" BM of a single component. Then, Section 5.5 will very briefly discuss our most recent research on Dynamic Neural Network (DNN) based "analog, dynamic, nonlinear, black-box-like, simulation-based" BM of a circuit block.

5.1 Motivation

The modeling of RF/microwave components for computer-aided design continuously faces new challenges because of increasing operation frequencies, circuit complexity, integration density, and decreasing time to market. It is often impossible to derive analytical models for new devices. Conventional numerical methods like 3D EM simulation are accurate but CPU-expensive. Empirical models, in turn, are fast but inaccurate over a wide operation range. Recently, it has been shown that ANNs offer benefits to urgent modeling needs; fast and accurate ANN models have been created for a wide range of components [ZG00].

Our goal has been to develop and implement an easy-to-use ANN-model generator for industrial model developers and circuit designers, who are neither ANN experts nor willing to switch between various modeling tools and simulators. In order to reach this goal, we first implemented an ANN-model generator prototype [RSP03] using the flexible input language of APLAC. Later on, we implemented `ANNModelGenerator` in APLAC using C language. Thus, the trained ANN models can be readily used in the same simulation framework.

5.2 Multi-layer perceptron ANNs

The most widely used ANN in RF/microwave component modeling is the Multi-Layer Perceptron MLP [ZG00] (yet we have also studied radial basis function ANNs [Poh03]). In our `ANNModelGenerator`, any number of MLP layers can be specified, but let us, for simplicity and due to the universal approximation theorem [Hay99], concentrate on a three-layer MLP that realizes the following nonlinear mapping:

$$\tilde{y}_l(\mathbf{x}, \mathbf{w}) = w_{l0} + \sum_{j=1}^{N_\mathrm{h}} w_{lj} \tanh\left(w_{j0} + \sum_{i=1}^{N_\mathrm{i}} w_{ji} x_i \right), \tag{1}$$
$$l = 1, 2, \ldots, N_\mathrm{o},$$

where N_i, N_o, and N_h represent the number of inputs, outputs, and hidden-layer neurons, respectively; $\mathbf{x} = (x_1, x_2, \ldots, x_{N_\mathrm{i}})$, $\tilde{\mathbf{y}} = (\tilde{y}_1, \tilde{y}_2, \ldots, \tilde{y}_{N_\mathrm{o}})$, and $\mathbf{w} = (w_{10}, w_{11}, \ldots, w_{N_\mathrm{o}N_\mathrm{h}})$ represent ANN inputs, outputs, and weights, respectively. Let $\mathbf{y} = \mathbf{y}(\mathbf{x})$ be an unknown, nonlinear, multi-dimensional function to be approximated by the MLP mapping (1), that is, $\tilde{\mathbf{y}} = \tilde{\mathbf{y}}(\mathbf{x}, \mathbf{w})$. Let $\{(\mathbf{x}^k, \mathbf{y}^k), k = 1, 2, \ldots, N_\mathrm{tr}\}$ be an appropriate training set, N_tr being the number of samples, and the training-set inputs and outputs being scaled in the range $[-1, 1]$. Furthermore, let us define the normalized ANN training error as [ZG00]

$$E_\mathrm{tr}(\mathbf{w}) = \sqrt{\frac{1}{N_\mathrm{tr} N_\mathrm{o}} \sum_{k=1}^{N_\mathrm{tr}} \sum_{l=1}^{N_\mathrm{o}} \left(\frac{\tilde{y}_l(\mathbf{x}^k, \mathbf{w}) - y_l^k}{2} \right)^2}. \tag{2}$$

The training of the ANN means minimizing of $E_\mathrm{tr}(\mathbf{w})$ with respect to weights, $\mathbf{w}$, by optimization. The generalization capability of the trained ANN is evaluated by applying (2) to an independent test set, $\{(\mathbf{x}^k, \mathbf{y}^k), k = 1, 2, \ldots, N_\mathrm{te}\}$.

5.3 ANN-model generation

Overview

A block diagram of ANN-model generation (`ANNModelGenerator`) and ANN-model usage (`MODEL_FILE`, `ANNModel`, and `ANNFunc`) is shown in Fig. 1. The obligatory and optional blocks are drawn with solid and dashed lines, respectively. The operation of (most of) these blocks is explained in the following five subsections.

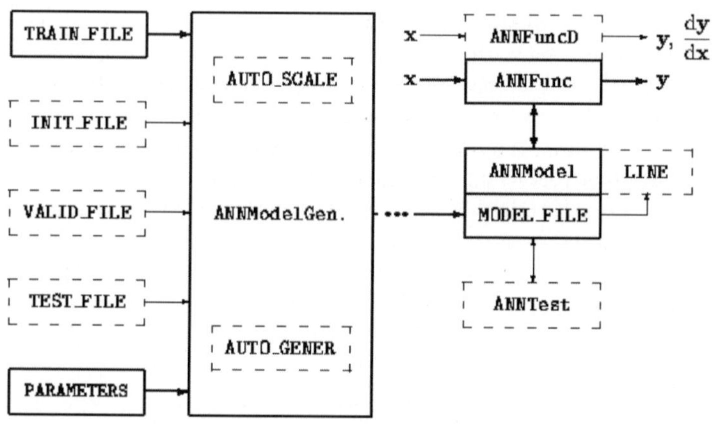

Fig. 1: ANN-model generation and usage.

ANN-structure selection

One challenge in ANN-based modeling is the determination of the number of hidden layers and their neurons; a very simple MLP with few weights may not offer enough degrees of freedom for the approximation problem, while a complex MLP structure results in many weights to be optimized. In `ANNModelGenerator`, the default number of hidden layers is one, which should be enough in many cases. The (optimistic) default number of neurons in that hidden layer is $N_\mathrm{h} = N_\mathrm{i} + N_\mathrm{o}$. Naturally, the user can increase the number of layers and/or neurons (by adjusting '`PARAMETERS`', See Fig. 1), if needed.

Training/validation/test-set generation

The very first step in ANN training is to generate a training-set file. The training-set file (obligatory `ANNModelGenerator` input `TRAIN_FILE` in Fig. 1) includes a collection of data samples, each consisting of relevant inputs (e.g., MOSFET DC-bias voltages V_gs and V_ds) and desired outputs (e.g., MOSFET drain current $I_\mathrm{ds}(V_\mathrm{gs}, V_\mathrm{ds})$), obtained from measurements or simulations.
If ANN training continues for too long, overlearning, or oscillatory overfitting to (noisy) training data, may occur [Hay99]. Therefore, an independent validation set may be used to avoid overlearning by early stopping [Hay99] at the lowest validation error obtained. Moreover, the generalization ability of the ANN should be tested with an independent test set after ANN training. The optional validation-set file (`VALID_FILE`) and test-set file (`TEST_FILE`) should be constructed such that all their inputs, $\mathbf{x}^k = (x_1^k, x_2^k, \ldots, x_{N_\mathrm{i}}^k)$, $k = 1, 2, \ldots, \{N_\mathrm{va}, N_\mathrm{te}\}$, are located inside the region defined by the training-set inputs; otherwise, this data will be also used for validating/testing the (non-guaranteed) extrapolation capability of the ANN.

Data scaling

In the literature, many heuristic methods have been suggested for improving ANN training [Hay99]. One of these methods is scaling: since in typical RF/microwave modeling applications the orders of magnitude of input/output parameter values are very different from one another, scaling of training data is desirable for ANN training [ZG00].
In `ANNModelGenerator`, each ANN input and output, that is, each column of the training (and validation/test) set, is scaled in the range $[-1, 1]$ before the actual ANN training. Linear scaling is used by default, but there is also another option, namely automatic scaling (`AUTO_SCALE`) that is based on our work in [RP04]. The automatic scaling first finds a suitable logarithmic scaling function [ZG00] for each column, after which it optimizes the shape parameter of the function such that the scaled values are spread as equally as possible in the range $[-1, 1]$.

ANN training

APLAC contains 10 optimization methods:
- Global methods:
  - Genetic algorithm

  - Simulated annealing
- Gradient-based methods:
  - Steepest descent
  - Conjugate gradient
- Direct-search methods:
  - Hooke–Jeeves
  - Nelder–Mead
  - Multi-directional search
- Other methods:
  - MinMax
  - Random
  - Exhaustive search

All these methods can be used to optimize virtually any (circuit) variable with respect to any design goal. Thanks to the internal 'ANNModelGenerator $\leftrightarrow$ optimization methods' C-code interface, all these 10 methods are readily available for ANN training, too. Until now, we have only modified the gradient-based methods for ANN training by adding the Error Back Propagation (EBP) algorithm [Hay99] for fast and accurate gradient evaluation. In ANNModelGenerator, the default optimization method is conjugate gradient with EBP and Hestenes–Stiefel search-direction determination [KRH05].

One important factor in ANN training is weight initialization [Hay99]. Currently, we initialize the weights randomly in the range $[-0.25, 0.25]$; this simple scheme will be improved in the future [KR06].

ANNModelGenerator calculates the normalized training error, E_{tr}, from (2). If, say, $E_{\mathrm{tr}} \leq 0.5\ \%$, ANN training was succesful.

ANN validation/testing

If VALID_FILE is specified, ANNModelGenerator calculates the normalized validation error, E_{va}, by applying (2) at every 10th optimization cycle. If E_{va} is smaller than the previous one, the current ANN model (only) is saved (the memory requirement being mainly determined by the, say, $50 \ldots 500$ ANN weights, $w_{10}, w_{11}, \ldots, w_{N_{\mathrm{o}}N_{\mathrm{h}}}$). When ANN training terminates (e.g., at the maximum number of optimization cycles), the best ANN model, with the lowest E_{va} at the early-stopping point, is fetched. If TEST_FILE has been specified, ANNModelGenerator calculates, at the end of ANN training, the normalized test error, E_{te}. If, say, $E_{\mathrm{te}} \leq 1\ \%$, the generalization capability of the ANN is very good.

5.4 Connection to circuit simulation

The end result of ANN training is the ANN-model file (MODEL_FILE) that contains, e.g., the ANN structure, the values of ANN weights, and relevant comment lines for ANN training/validation/test error, ANN-training CPU-time, etc.

The trained ANN model can be connected to APLAC circuit simulation by ANNModel (see Fig. 1), which reads in MODEL_FILE and stores the parameters. The actual on-line calculation of ANN outputs (during circuit simulation) is done using ANN-evaluation functions ANNFunc and ANNFuncD, which use the parameters stored by ANNModel. In nonlinear modeling applications (e.g., MOSFET drain current

$I_{ds}(V_{gs}, V_{ds})$), the use of `ANNFuncD` is recommended, since it also returns the analytically calculated derivatives [AO00] (e.g., $\partial I_{ds}/\partial V_{gs}$ and $\partial I_{ds}/\partial V_{ds}$) that are needed in circuit simulation for the Newton–Raphson iteration method.
To summarize, in our approach both ANN-model generation and usage can be seamlessly done inside the same circuit-simulation framework: APLAC.

5.5 Dynamic behavioral modeling using DNNs

Very recently, we have also started to study behavioral modeling of dynamic nonlinear circuit blocks like Power Amplifiers (PAs) using Dynamic Neural Networks (DNNs) [WR05] (Ch. 6 and 7), [Mei96], [PAS01]. Untill now, we have developed and implemented a prototype version of `DNNModelGenerator` using a combination of `ANNModelGenerator` and APLAC input language. We have tested this tool by generating a DNN model for an audio amplifier and for a 5 GHz PA. These DNN models could be used to replace the original circuit block in a HB simulation.

6 Other research activities

Finally, it is worth mentioning that during the years 2002–2006, we have also carried out research outside the ARFSIM, MOSAICS, and AMAZE projects. In [Kuj02], digital components were modeled (for mixed-mode simulation with the in-house development version of APLAC). In [Sun04], APLAC simulation models for striplines with conductor surface roughness were generated. In [Vei06], in turn, several APLAC simulation models for Micro-Electro-Mechanical Systems (MEMS) are documented and the related MEMS publications of TKK CTL are listed. Finally, it is worth mentioning (outside the topic 'circuit simulation') that [Sil06] contains a list of publications on network analyzer calibration methods.

References

[Roo04] Roos, J., Aaltonen, S., Honkala, M., Karanko, V., Kujanpää, T., Lehtovuori, A., Palenius, T., Pohjala, A., Valtonen, M., Virtanen, J.: Advanced radio frequency simulation and modeling of electronic circuits. ECMI Newsletter, 10–11 (2004)

[Kal02] Kallio, A.: MESFET models in the APLAC circuit simulator. M.Sc. Thesis, Helsinki University of Technology, Finland (2002)

[KV04] Kallio, A., Valtonen, M.: A new rule for MESFET gate charge division based on the energy conservation principle. Int. Journal of Circ. Theory and Appl., **32**, 139–165 (2004)

[RVV02] Roos, J., Valtonen, M., Virtanen, J.: Implementation of piecewise-linear DC analysis in APLAC. In: Proc. ICECS 2002, **3**, 1139–1142 (2002)

[Roo05] Roos, J.: On simplex-based piecewise-linear approximations of nonlinear mappings. Int. Journal of Circ. Theory and Appl., **33**, 109–134 (2005)

[Roo06] Roos, J.: Speed-up and performance evaluation of piecewise-linear DC analysis. Int. Journal of Circ. Theory and Appl., accepted to be published (2006)

[Hon02a] Honkala, M.: Nonmonotone norm-reduction method for circuit simulation. IEE Electronics Letters, **38**, 22, 1316–1317 (2002)

[Hon02b] Honkala, M.: Nonmonotone norm-reduction method in numerical circuit analysis. In: TKK CTL report series, **46** (2002)

[HRK06] Honkala, M., Roos, J., Karanko, V.: On nonlinear iteration methods for DC analysis of industrial circuits. In: Di Bucchianico, A., Mattheij, R.M.M., Peletier, M.A. (ed) Mathematics in Industry 8 — Progress in Industrial Mathematics at ECMI 2004. Springer, Berlin Heidelberg. 144–148 (2006)

[Lin06] Linja-aho, V.: Homotopy methods in circuit DC analysis. M.Sc. Thesis, Helsinki University of Technology, Finland (2006)

[Hon02c] Honkala, M.: Parallel hierarchical DC analysis. Lic.Sc. Thesis, Helsinki University of Technology, Finland (2002)

[HK02] Honkala, M., Karanko, V.: Improving the convergence of combined Newton–Raphson and Gauss–Newton multilevel iteration method. In: Proc. ISCAS 2002, **2**, 229–232 (2002)

[KH02] Karanko, V., Honkala, M.: Least squares solution of nearly square overdetermined sparse linear systems. In: Proc. ISCAS 2002, **4**, 830–833 (2002)

[Hon03] Honkala, M.: Parallel Processing in APLAC — Part A: DC and Transient Analyses. In: TKK CTL report series, **47** (2003)

[Vir05] Virtanen, J.: Harmonic balance and phase-noise analysis methods in the APLAC circuit simulator. Lic.Sc. Thesis, Helsinki University of Technology, Finland (2005)

[KH04] Karanko, V., Honkala, M.: A parallel harmonic balance simulator for shared memory multicomputers. In: Proc. EUMW 2004 (2004)

[Poh06] Pohjala, A.: Harmonic balance analysis for frequency-divider circuit (in Finish). Lic.Sc. Thesis, Helsinki University of Technology, Finland (2006)

[LVV03] Lehtovuori, A., Virtanen, J., Valtonen, M.: GMRES preconditioners for multivariate steady-state time-domain method. In: Proc. IMS 2003, 2129–2132 (2003)

[Leh03] Lehtovuori, A.: Multivariate steady-state time-domain analysis method. Lic.Sc. Thesis, Helsinki University of Technology, Finland (2003)

[Cos05] Costa, L.: Incorporating lumped elements into an efficiently implemented FDTD-based electromagnetic simulator. Lic.Sc. Thesis, Helsinki University of Technology, Finland (2005)

[Cos06] Costa, L.: Implementing efficient array traversing for FDTD-lumped element cosimulation. In: Di Bucchianico, A., Mattheij, R.M.M., Peletier, M.A. (ed) Mathematics in Industry 8 — Progress in Industrial Mathematics at ECMI 2004. Springer, Berlin Heidelberg. 149–153 (2006)

[Aal03] Aaltonen, S.: Order reduction of interconnect circuits. Lic.Sc. Thesis, Helsinki University of Technology, Finland (2003)

[Pal04] Palenius, T.: Efficient time-domain simulation of interconnects characterized by large RLC circuits or tabulated S parameters. Lic.Sc. Thesis, Helsinki University of Technology, Finland (2004)

[AR02] Aaltonen, S., Roos, J.: Simple reduced-order macromodels with PRIMA. In: Proc. ICECS 2002, **1**, 367–360 (2002)

[PR03] Palenius, T., Roos, J.: An efficient reduced-order interconnect macromodel for time-domain simulation. In: Proc. ISCAS 2003, **4**, 628–631 (2003)

[PR04] Palenius, T., Roos, J.: Comparison of reduced-order interconnect macromodels for time-domain simulation. IEEE Trans. Microwave Theory and Techniques, **52**, 9, 2240–2250 (2004)

[RSP03] Roos, J., Şengör, N.S., Pohjala, A.: Artificial neural network based RF-model generator — version 0.2. In: TKK CTL report series, **48** (2003)

[Poh03] Pohjala, A.: Generation of simulation models from measurement data using interpolation and radial basis function networks. M.Sc. Thesis, Helsinki University of Technology, Finland (2003)

[RP04] Roos, J., Pohjala, A.: Development and comparison of formulas for scaling ANN inputs and outputs in RF-modeling applications. In: Buikis, A., Ciegis, R., Fitt, A.D. (ed) Mathematics in Industry 5 — Progress in Industrial Mathematics at ECMI 2002. Springer, Berlin Heidelberg. 197–201 (2004)

[KRH05] Kujanpää, T., Roos, J., Honkala, M.: Experimental comparison of optimization methods in ANN training. In: Proc. PRIME 2005, **2**, 430–433 (2005)

[KR06] Kujanpää, T., Roos, J.: Efficient initialization of artificial neural network weights for electrical component models. In: Book of Abstracts of SCEE 2006, 47–48 (2006)

[Kuj02] Kujanpää, T.: Modeling of digital components in the APLAC circuit simulator. M.Sc. Thesis, Helsinki University of Technology, Finland (2002)

[Sun04] Sundström, S.: Stripline simulation models with conductor surface roughness, M.Sc. Thesis, Helsinki University of Technology, Finland (2004)

[Vei06] Veijola, T., *et. al.*: Modeling of micromechanical devices (2006) `http://www.ct.tkk.fi/research/mems/main.html`

[Sil06] Silvonen, K.: Network analyzer calibration methods (2006) `http://www.ct.tkk.fi/research/calibration.html`

[A06] APLAC 8.2 Manuals. AWR–APLAC Corporation, Finland (2006) `http://www.aplac.com`

[OCP98] Odabasioglu, A., Celik, M., Pileggi, L.T.: PRIMA: passive reduced-order interconnect macromodeling algorithm. IEEE Trans. Computer-Aided Design of Integrated Circuits and Systems, **17**, 88, 645–654 (1998)

[WR05] Wood, J., Root, D.E. (ed): Fundamentals of Nonlinear Behavioral Modeling for RF and Microwave Design. Artech House, Boston London (2005)

[Vos05] Voss, T.: Model reduction for nonlinear differential algebraic equations. M.Sc. Thesis (Philips Nat.Lab. Unclassified Report PR-TN-2005/00919), University of Wuppertal, Germany (2005)

[ZG00] Zhang, Q.J., Gupta, K.C.: Neural Networks for RF and Microwave Design. Artech House, Boston London (2000)

[Hay99] Haykin, S.: Neural Networks — A Comprehensive Foundation. Prentice Hall, New Jersey (1999)

[AO00] Antonini, G., Orlandi, A.: Gradient evaluation for neural-networks-based electromagnetic optimization procedures. IEEE Trans. Microwave Theory and Techniques, **48**, 5, 874–876 (2000)

[Mei96] Meijer, P.B.L.: Neural network applications in device and circuit modelling for circuit simulation, Ph.D. Thesis, Eindhoven University of Technology, The Netherlands (1996)

[PAS01] Plebe, A., Anile, A.M., Rinaudo, S.: Sub-micrometer bipolar transistor modeling using neural networks. In: van Riene, U., Günther, M., Hecht, D. (ed): Scientific Computing in Electrical Engineering, Lect. Notes in Comput. Science and Engineering, **18**, Springer, Berlin Heidelberg. 259–266 (2001)

Outstanding Issues in Model Order Reduction*

João M. S. Silva[1], Jorge Fernández Villena[1], Paulo Flores[1], and L. Miguel Silveira[1,2]

[1] INESC ID / Instituto Superior Técnico
Technical University of Lisbon
Rua Alves Redol, 9
1000-029 Lisboa, Portugal
{jmss, jorge, lms}@algos.inesc-id.pt
[2] Cadence Laboratories
Cadence Design Systems

Summary. With roots dating back to many years ago and applications in a wide variety of areas, model order reduction has emerged in the last few decades as a crucial step in the simulation, control, and optimization of complex physical systems. Reducing the order or dimension of models of such systems, is paramount to enabling their simulation and verification. While much progress has been achieved in the last few years regarding the robustness, efficiency and applicability of these techniques, certain problems of relevance still pose difficulties or renewed challenges that are not satisfactorily solved with the existing approaches. Furthermore, new applications for which dimension reduction is crucial, are becoming increasingly relevant, raising new issues in the quest for increased performance.

Keywords—Model order reduction, massively coupled systems, orthogonal projection, parametric systems, circuit simulation.

1 Introduction

Model reduction algorithms are standard techniques nowadays in many areas, including the microelectronics design community. The goal of model order reduction is to replace a large-scale model of a physical system by a model of lower dimension which exhibits similar behavior, typically measured in terms of its frequency or time response characteristics. Such techniques are commonly used for analysis, approximation, and simulation of models arising from electromagnetic formulation of physical structures. The need to accurately account for all relevant physical effects implies that the mathematical formulation used to describe such structures often results in very large models. Reducing the order or dimension of these models is crucial to enabling the simulation and verification of such systems [2, 1].
An area to which extensive research has been devoted in the last few years is the problem of order reduction of nonlinear systems [20, 18, 4]. A discussion of such methods is however beyond the scope of this paper. Due to space constraints we will restrict the discussion to issues arising from linear systems reduction. Nevertheless this discussion is still relevant in the nonlinear case as most existing nonlinear reduction algorithms are based on extensions of linear methods or the solution of carefully selected sequences of linear problems. While enormous progress has been achieved in the last decades in this field, both from a theoretical as

* Invited Paper at SCEE-2006

well as a practical standpoint, still greater challenges lie ahead as new and exciting applications are being researched for which order reduction is again a crucial step.
Existing methods for linear model reduction can be broadly characterized into two types: those that are based on projection methods, and those based on balancing techniques (sometimes also referred to as SVD[3]-based [1]). Among the first, Krylov subspace projection methods such as PVL [6] and PRIMA [15] have been the most widely studied over the past decade. They are very appealing because of their simplicity and performance in terms of efficiency and accuracy, despite the fact that they exhibit several known shortcomings. The lack of a general strategy for error control and order selection, as well as a dependence on the original model's structure if passivity is to be guaranteed after the reduction are among the more obvious such shortcomings. The alternative methods, those in the truncated balanced realization (TBR) family [14], perform reduction based on the concept of controllability and observability of the system states and are purported to produce nearly optimal models and have easy to compute *a-posteriori* error bounds. However, they are awkward to implement and expensive to apply, which limits their applicability to small and medium sized problems. Hybrid techniques that combine some of the features of each type of methods have also been presented [11, 9, 10]. Recently, a new technique was also proposed that attempts to establish a bridge between the two techniques. The Poor Man's TBR [19] is based on a projection scheme where the projection matrix approximately spans the dominant eigenspaces of the controllability and observability matrices and provides an interesting platform for bridging between the two types of techniques. Still the technique is not without drawbacks, as it relies on proper choice of sampling points, a non-trivial task in general.
In spite of their shortcomings, all of the mentioned methods are in widespread use nowadays. Still, there are situations that challenge the existing knowledge in the field. For instance, consider the problem of reducing systems with a large number of ports, also known as massively coupled systems. Such systems typically occur in substrate, power grid and package parasitic networks. Furthermore, the trend to nano-scale dimensions together with the increasing frequencies of operation implies that non-neglectable electromagnetic effects have to be accounted for in the models, which will also give rise to these massively coupled problems. Projection-based algorithms are inefficient for such systems as they rely on block iterations, where the size of the block equals the number of ports. Therefore, each block iteration increases the size of the model by an amount equal to the number of ports, leading to large models even for moderate reduction order. This trend is particularly troublesome when simulation with such models is necessary. TBR is intrinsically somewhat less sensitive to the number of input ports. Unfortunately such systems are typically very large, which makes reduction based on balancing techniques impractical.
Additionally, new challenges are being posed that require further research. As an example, consider the problem of order reduction of parametrized systems. Parameter-based descriptions are now starting to be used as the basis for variability-aware design models. For high frequencies, at nano-scale feature sizes, process variability effects, as well as dependence on operating conditions become extremely relevant and should be accounted for in the models. Existing techniques for handling such systems are, for the most part, straightforward extensions of the basic order reduction algorithms [3, 12]. Projection-based techniques match Taylor-series coefficients, which in parameter-based descriptions are multidimensional moments. Unfortunately this technique has exponential cost increase with the number of parameters and is thus expensive except for small size and small number of parameters. Building a projection space assuming small perturbations around the nominal operating point is also problematic: it is hard to do anything beyond first-order and thus it is not clear how to dial in accuracy. Sampling the parameter space also presents a challenge, as it is not clear where to place sample point in such a multidimensional space. Still if some information regarding the statistical distribution of the parameter values is available, this can be used to guide the sampling and to build the model accordingly.

[3] SVD – Singular value decomposition.

In this paper we review some of these current and future challenges for which much research is still needed in model order reduction. In Section 2 we discuss the problem of reducing massively coupled problems, and in Section 3 we discuss the reduction of parametrized systems, a recent topic of much research work. Finally in Section 4, we present some conclusions.

2 Massively Coupled Systems

As an illustration of the problems pertaining to massively coupled systems, results from the study of the reduction of power distribution networks, also known as power grids, will be presented. Power grids are fairly regular structures which must cover the whole area of the chip for power delivery purposes. Since all devices, wells and substrate plugs, are connected to the power grid, the total number of ports of such circuits can be as high as hundreds of thousands, or millions. This unfortunately brings added difficulty to the reduction process.

2.1 Background

Modeling a power grid as an RC network and using the nodal analysis formulation leads to:

$$\begin{aligned} C\dot{v} + Gv &= Mu \\ y &= N^T v \end{aligned} \tag{1}$$

where $C, G \in \mathbb{R}^{n\times n}$ are the capacitance and conductance matrices, respectively, $M \in \mathbb{R}^{n\times p}$ is a matrix that relates the inputs $u \in \mathbb{R}^p$ to the states $v \in \mathbb{R}^n$ that describe the node voltages, $N \in \mathbb{R}^{n\times q}$ being its counterpart with respect to the outputs $y \in \mathbb{R}^q$, n is the number of states, p the number of inputs and q the number of outputs. The $p \times q$ matrix transfer function of the network is then given by $H(s) = N^T(G + sC)^{-1}M$. Typically, matrices C and G are very sparse but also very large. For a typical power grid, the number of nodes will be in the order of several millions but the number of ports, input and output, is also quite large. Solving Eqn. (1) directly or using it inside a circuit simulator is therefore too expensive.
The goal of model-order reduction is, generically, to determine a reduced model,

$$H_k(s) = \hat{N}^T(\hat{G} + s\hat{C})^{-1}\hat{M} \tag{2}$$

of size $k \ll n$, that closely matches the input-output behavior of the original model, and where the state description is given by $z = V^T v \in \mathbb{R}^k$. However, even if $k \ll n$, the reduced-order model may fail to provide relevant compression. This may happen because, for large networks, the matrices C and G are sparse, having a number of non-zeros entries of order $\mathcal{O}(n)$. If the number of non-zero entries in the reduced-order model increases with the number of ports, the benefits of reduction may vanish with increasingly large p and q.

Projection-based framework

Projection-based Krylov subspace algorithms, such as PRIMA [15], provide a general-purpose, rigorous framework for deriving interconnect modeling algorithms and have been shown to produce excellent compression in many scenarios involving on- and off-chip interconnect and packaging structures. In its simplest form, they can be used to compute individual approximations to each of the $p \times q$ matrix transfer function entries. However, more commonly, they are used to generate a single approximation to the full system transfer function. The PRIMA algorithm [15], for instance, reduces a state-space model in the form of (1) by use of a projection matrix V, through the operations:

$$\hat{G} = V^T G V, \quad \hat{M} = V^T M, \quad \hat{C} = V^T C V, \quad \hat{N} = V^T N \tag{3}$$

to obtain a reduced model in the form of (2). In the standard approach, the projection matrix V is chosen as an orthogonal basis of a block Krylov subspace, $\mathcal{K}_m(A, b) = span\{b, Ab, \ldots, A^{m-1}b\}$, a typical choice being $A = G^{-1}C$ and $b = G^{-1}M$. The construction of the projection matrix V is done iteratively by blocks, with each block being generated through a back-orthogonalizing procedure. When the projection matrix is constructed in this way, the moments of the reduced model can be shown to match the moments of the original model to some order. Consequently, the reduced model size is proportional to the number of matched moments multiplied by the number of ports. Furthermore, the reduced system matrices will be dense. Therefore, these methods present two problems when dealing with networks with a large number of ports. First, the cost associated with model computation is directly proportional to the number of inputs, p, i.e. to the number of columns in the matrices defining the inputs. This is easy to see by noting that the number of columns in the projection matrix V in (3) is directly proportional to p (a direct result of the block construction procedure described). This implies that model construction for systems with large number of ports is costly. Second, the size of the reduced model is also proportional to p, as was discussed earlier and can directly be seen from (3). While the cost of model construction can perhaps be amortized in later simulations, the large size of the model is more problematic since it directly affects simulation cost.

Truncated balanced realizations

An alternative class of reduction algorithms are based on Truncated Balanced Realization (TBR). The TBR algorithm first computes the observability and controllability Gramians, X and Y, by solving the Lyapunov equations:

$$GXC^T + CXG^T = MM^T, \tag{4}$$
$$G^TYC + C^TYG = N^TN \tag{5}$$

and then reduces the model by projection onto the space associated with the dominant eigenvalues of the product XY [14]. Model size selection and error control in TBR is based on the eigenvalues of XY, also known as the Hankel singular values. In the proper case, there is an *a-posteriori* theoretical bound on the frequency-domain error for the TBR model given by [14]:

$$\|H - H_k\| \leq 2 \sum_{i=k+1}^{n} \sigma_i \tag{6}$$

The existence of such an error bound is an important advantage of the TBR class of algorithms as there is no counterpart in the projection-based algorithms. Theoretically, the model selection criteria, and therefore the size of the generated model, can be done independently of the number of inputs. However, there is an indirect dependence in most problems and in particular for networks such as power grids, that exhibit a large number of inputs (see [19] for additional discussion on the topic). In this case, useful reductions are not achievable. Furthermore, the solution of the Lyapunov equations required to obtain X and Y is computationally intensive for large systems and as such the technique is only of theoretical interest in this context. A variety of approximate methods have been proposed that attempt to circumvent this problem (see [19] and references therein).

2.2 Methods

As stated previously, the difficulty with standard projection algorithms like PRIMA or multipoint projection schemes, is that the models produced have size proportional to the number of ports. This limits their applicability to problems such as power grids, where the number of network ports is likely to be very large. An interesting question that might be raised is whether this restriction is inherent to the system, given the number of ports, or an artifact of

the computation scheme chosen. In other words, one might ask whether accurate modeling and analysis of a power grid, modeled as a large RC mesh, does indeed require so much dynamic information. This question is all the more relevant as there is a common popular belief that only a few poles are required to accurately model an RC circuit. It is now widely accepted that in certain settings that is indeed the case, but this conclusion is emphatically not general (see [22]).
In the following, two recently proposed methods for overcoming the difficulties faced by standard MOR methods are presented. The first method is based on the analysis of singular values of the system moments while the second one is a "cheaper" version of a TBR class method previously mentioned [19], also based on projection.

Singular Value Decomposition MOR (SVDMOR)

The SVDMOR [5] algorithm was developed to address the reduction of systems with a large number of ports, like power grids. While the size of a reduced model produced via PRIMA is directly proportional to the number of ports in the circuit, SVDMOR theoretically overcomes this problem using singular value decomposition (SVD) analysis in order to truncate the system to any desired order.
The main idea behind SVDMOR is to assume that there is a large degree of correlation between the various inputs and outputs. SVDMOR further assumes that such input-output correlation can be captured from observation of structural system properties, evidenced in matrices M and N. The method can, for instance, use an input-output correlation matrix, like the one given by the zero order moment matrix $S_{DC} = N^T G^{-1} M$, which contains only DC information. Alternatively, more complicated response correlations can be used such as frequency, s_j-shifted moments, $S_{DC}^{(s_j)} = N^T(G + s_jC)^{-1}M$, a more generic k-order moment, $S_k = N^T(G^{-1}C)^k G^{-1}M$, or even combinations of these. Let K be the appropriate correlation matrix. If the basic correlation hypothesis holds true, then K can be approximated by a low-rank matrix. This low rank property can be revealed by computing the SVD of K, $K = U\Sigma W^T$, where U and W are orthogonal matrices and Σ is the diagonal matrix containing the ordered singular values. Assuming correlation, there will be only a small number, $m \ll p + q$, of dominant singular values. Therefore, we can approximate $K \approx U_m \Sigma_m V_m^T$, where truncation is performed keeping the m most significant singular values. The method further approximates:

$$\begin{aligned} M &\approx b_M V_m^T = M V_m (V_m^T V_m)^{-1} V_m^T \\ N &\approx b_N U_m^T = N U_m (U_m^T U_m)^{-1} U_m^T \end{aligned} \tag{7}$$

where b_M and b_N are obtained using the Moore-Penrose pseudo-inverse, resulting in:

$$H(s) \approx U_m \underbrace{b_N^T (G + sC)^{-1} b_M}_{H_m(s)} V_m^T \tag{8}$$

Standard MOR methods, like PVL or PRIMA, can now be applied to $H_m(s)$, leading to $\tilde{H}_m(s)$, an r-th order model, from which a final model approximation $H(s) \approx H_r(s) = U_m \tilde{H}_m(s) V_m^T$ is computed. The reduced system is $p \times q$ with a number of nonzero elements of order $\mathcal{O}(r^2)$.

Input-Correlated Poor Man's TBR (PMTBR)

The PMTBR algorithm [19, 22] was motivated by a connection between frequency-domain projection methods and approximation to truncated balanced realization. The method is less expensive in terms of computation, but tends to TBR when the order of the approximation increases. The actual mechanics of the algorithm are akin to multi-point projection. In a multi-point rational approximation the projection matrix columns are computed by sampling at several frequency points along a desired frequency interval. The samples are given by

$z_i = (G + s_i C)^{-1} M$, where $s_i = j\omega_i$ (with $i = 1, 2, \ldots, P$) are P frequency sample points. The frequency sampled matrix thus obtained can then be used to project the original system in order to obtain a reduced model. In the PMTBR algorithm, a similar procedure is used. The connection to TBR methods is made by noting that an approximation $\hat{X}$ to the Gramian X can be can be computed as:

$$\hat{X} = \sum_i w_i z_i z_i^H \tag{9}$$

where the ω_i which defines each sample, and the w_i can be interpreted as nodes and weights of a quadrature scheme applied to a frequency-domain interpretation of the Gramian matrix (see [19] for details). Let Z be a matrix whose columns are the z_i, and W the diagonal matrix of the square root of the weights. Eqn. (9) can be written more compactly as:

$$\hat{X} = ZW^2 Z^H \tag{10}$$

If the quadrature rule applied is accurate, $\hat{X}$ will converge to X, which implies the dominant eigenspace of $\hat{X}$ converges to the dominant eigenspace of X. Computing the singular value decomposition of ZW, $ZW = V_Z S_Z U_Z$ (with S_Z real diagonal, and V_Z, U_Z unitary matrices), it is easy to see that V_Z converges to the eigenspaces of X, and the Hankel singular values are obtained directly from the entries of S_Z. V_Z can then be used as the projection matrix in a model order reduction scheme. The method was shown to perform quite well in a wide variety of settings [19].
An interesting additional interpretation was more recently presented [22] which is of relevance in our context. It has been shown that if further information revealing time-domain correlation between the ports is available, a variant of PMTBR can be used that can lead to significant efficiency improvement. This idea is akin to the basic assumptions in SVDMOR and relate to exploiting correlation between the inputs. Unlike SVDMOR, however, it is assumed that the correlation information is not contained in the circuit information directly, but rather in its inputs. In this variant of PMTBR, a correlation matrix K is formed by columns which are samples of port values along the time-steps of some interval. Those samples should characterize as well as possible the values expected at the inputs of the system, i.e. K should be a suitably representative model of the possible inputs. An SVD is then performed over K in order to retain only the r most significant components of the input correlation information, $K \approx U_r \Sigma_r V_r^T$. With this additional correlation information, the samples relative to multipoint approximation become $z_i = (G + s_i C)^{-1} M U_r \Sigma_r$. Using these z_i as columns of the Z matrix in (10), leads to the input-correlated TBR algorithm (ICTBR). See [19] for more details and a more thorough description of the probabilistic interpretation of both PMTBR as well as ICTBR.

2.3 Results

Both the standard model order reduction as well as the methods described in the previous section can be applied to massively coupled systems. Methods like SVDMOR are reported to provide significant advantages over the standard algorithms if certain conditions are met, namely that significant port correlation exists and can be ascertained in a practical way. PMTBR is a more general algorithm for model reduction, which can nonetheless be applied to large systems, given its reduced computational complexity.
In this section, results are presented for two types of topologies: a first mesh, grid A, with voltage inputs on the left side and current outputs on the right one, and a second mesh, grid B, with voltage ports along the left side and current ports randomly distributed over the remaining nodes. For practical reasons, we have kept the mesh sizes smaller than they would be in realistic applications but scaling of all appropriate dimensions and sizes would produce qualitatively the same results. There are two main differences between the two setups described. The first one concerns formulation. While in grid A matrices M and N in Eqn. (1) are distinct (M yields input information and N yields output information), in grid B, $M = N$, thus all

ports are controllable and observable. The second main difference consists in the separation between ports. In grid A the separation between inputs and outputs is maximal, while in grid B not only every port is both input and output, but also the geometric proximity between ports is reduced. Grid A is thus expected to be fairly compressible, but smaller reductions are expected for grid B. Grid A is similar to the one used in [5], while grid B was created in order to illustrate a more realistic setup. The electrical model of all grids is as follows: every connection between nodes is purely resistive and at every node there is a capacitance to ground. While this is not necessary, it simplifies the ensuing description (furthermore, a parasitic capacitance is usually extracted at all nodes). Resistance and capacitance values were randomly generated in the interval $(0.9, 1.1)$. In the following set of experiments the size of the reduced model is the same for all methods and was pre-determined. The correlation matrix of SVDMOR is the DC moment matrix. For this method, after computing the SVD and choosing how many singular values to keep, a number of PRIMA iterations is performed in order to generate a model of the required size. The number of frequency samples of PMTBR was set such that a model of the same size can be drawn from matrix Z. Samples were chosen uniformly in the frequency range shown in the plots, with an additional sample added at DC.

Highly-correlated ports

The previously discussed methods were first used to reduce grid A. The Bode plot of an arbitrarily selected transfer function is presented in Figure 1 (left). The number of retained states was forced at $r = 1200$. In the case of SVDMOR, 15 singular values were kept and 80 PRIMA iterations were run, yielding the reduced model of $15 \times 80 = 1200$ states. One observes that SVDMOR shows good results, better than PRIMA and PMTBR. In order to understand the reason for these results the plot of the singular values of SVDMOR and PMTBR methods is presented in Figure 1 (right). The singular values (s.v.) of the DC moment, used by SVDMOR to guide the reduction, decay quite fast. Therefore keeping just the first 15 yields a good approximation. On the other hand the PMTBR s.v. decay very slowly. Table 1 shows the

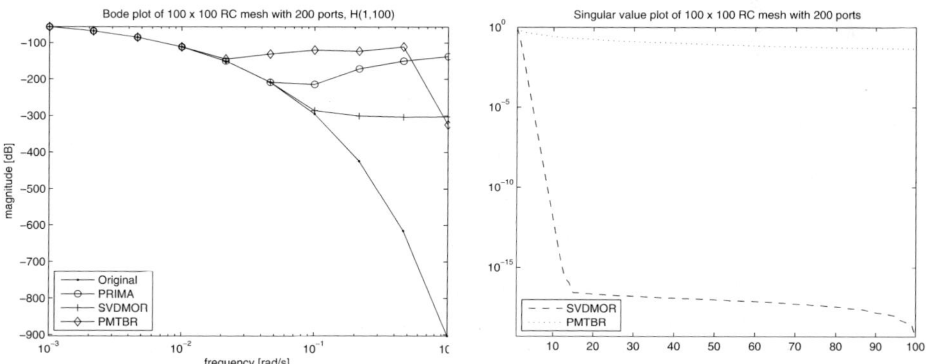

Fig. 1: Results for grid A ($r = 1200$): Bode plot of arbitrarily selected entry of 100×100 transfer function matrix (left); normalized plot of singular values: SVDMOR moment matrix and PMTBR samples matrix (right).

maximum absolute error of the transfer matrix, $max\{|H(s) - H_r(s)|\}$. Analysis of the table indicates that in the overall model, SVDMOR shows the smallest error as expected for this grid setup.

Table 1: Maximum absolute error of $|H(s) - H_r(s)|$ for 100×100 mesh with 100 inputs on the left side and 100 outputs on the right side. SVDMOR used 15 singular values.

$r = 1200$	PRIMA	SVDMOR	PMTBR
$max\{\lvert H - H_r\rvert\}$	1.443×10^{-6}	1.406×10^{-7}	1.160×10^{-5}

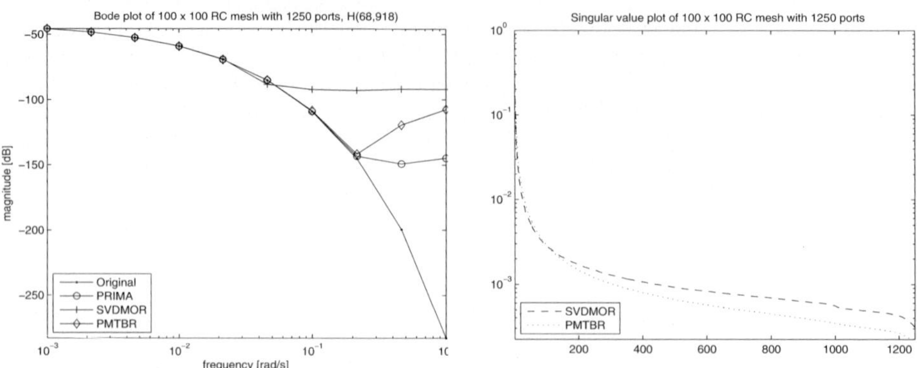

Fig. 2: Results for grid B ($r = 2500$): Bode plot of arbitrarily selected entry of 100×100 transfer function matrix (left); normalized plot of singular values: SVDMOR moment matrix and PMTBR samples matrix (right).

Table 2: Maximum absolute error of $|H - H_r|$ for 100×100 mesh with 100 ports on the left side and 1150 randomly distributed ports over the mesh.

$r = 2500$	PRIMA	SVDMOR	PMTBR
$max\{\lvert H - H_r\rvert\}$	$1.284e \times 10^{-2}$	2.533×10^{-1}	1.545×10^{-3}

Weakly-correlated ports

In grid B the objective was to emulate a more realistic situation whereby potentially many devices, modeled as current sources, are attached to the power grid and can draw or sink current from/to it when switching. The number of current sources was chosen to be $1/8$ of the number of nodes. There are 1150 current sources and 100 voltage sources (for a total of 10000 nodes). This is a harder problem to reduce, due to port proximity, and thus interaction, and the results show it. Again the Bode plot of an arbitrarily selected transfer function is presented in Figure 2 (left). The number of retained states was now forced at $r = 2500$ already showing smaller reduction than for grid A. In this case, the approximation produced by SVDMOR is less accurate. This is expected from inspection of Figure 2 (right), where one observes that the s.v. of SVDMOR decay slower than in the previous case. Clearly, the assumption of highly correlated ports is not valid here. The results concerning the error of the transfer matrix are in Table 2. PMTBR produces the most accurate model, while PRIMA shows a reasonable approximation.

Note that while the Bode plots show large errors for higher (normalized) frequencies, concerning to higher order moments which are harder to match, these frequencies are uninteresting in practical simulations. Note also that the matrices in the reduced models for all methods in both experiences are full, which has drastic consequences for usage of these models in a simulation environment.

3 Parametrized System Descriptions

In any manufacturing process there is always a certain degree of uncertainty involved given our limited control over the environment and other physical conditions. For the most part this uncertainty was previously ignored when analyzing or simulating systems, but as we step towards the nano-scale and higher frequency eras, such environmental, geometrical and electromagnetic fluctuations become more significant. Nowadays, parameter variability can no longer be disregarded, and its effect must be accounted for in early design stages so that unwanted consequences can be minimized. This leads to parametric descriptions of systems, including the effects of the manufacturing variability, which further increases the complexity of such models. When model reduction is required, these parametric representations must be addressed and the resulting reduced models must retain the ability to model the effects of small random fluctuations, in order to accurately predict behavior and optimize designs. This is the aim of the Parametric Model Order Reduction (pMOR).

3.1 Background

Actual fabrication of physical devices is prone to the variation of certain circuit parameters due to deliberate adjustment of the process or from random deviations inherent to this manufacturing. This variability leads to a dependence of the extracted circuit elements on several parameters, of electrical or geometrical origin. This dependence results in a parametric state-space system representation, which in descriptor form can be written as

$$\begin{aligned} &C(\lambda_1,\ldots,\lambda_L)\dot{v}(\lambda_1,\ldots,\lambda_L) + G(\lambda_1,\ldots,\lambda_L)v(\lambda_1,\ldots,\lambda_L) = Mu \\ &y = N^T v(\lambda_1,\ldots,\lambda_L) \end{aligned} \tag{11}$$

where $C, G \in \mathbb{R}^{n\times n}$ are again, respectively, the capacitance and conductance matrices, $M \in \mathbb{R}^{n\times p}$ is the matrix that relates the input vector $u \in \mathbb{R}^p$ to the inner states $v \in \mathbb{R}^n$ and $N \in \mathbb{R}^{n\times q}$ is the matrix that links those inner states to the outputs $y \in \mathbb{R}^q$. The elements of the matrices C and G, as well as the states of the system v, depend on a set of L parameters $\lambda = [\lambda_1, \lambda_2, \ldots, \lambda_L]$ which model the effects of the mentioned uncertainty. Usually the system is formulated so that the matrices related to the inputs and outputs (M and N) do not depend on the parameters. This time-domain descriptor yields a parametric dependent frequency response modeled via the transfer function

$$H(s,\lambda_1,\ldots,\lambda_L) = N^T(sC(\lambda_1,\ldots,\lambda_L) + G(\lambda_1,\ldots,\lambda_L))^{-1}M \tag{12}$$

for which we seek to generate a reduced order approximation, able to accurately capture the input-output behavior of the system for any point in the parameter space.

$$\hat{H}(s,\lambda_1,\ldots,\lambda_L) = \hat{N}^T(s\hat{C}(\lambda_1,\ldots,\lambda_L) + \hat{G}(\lambda_1,\ldots,\lambda_L))^{-1}\hat{M} \tag{13}$$

In general, one attempts to generate a reduced order model whose structure is, as much as possible, similar to the original, i.e. exhibiting a similar parametric dependence.

3.2 Methods

In the following we summarize the main methods presented for dealing with this problem.

Perturbation-Based Techniques

One of the earliest attempts to address this variational issue was to combine perturbation theory with moment matching MOR algorithms [13]. To model the variational effects of the interconnects, an affine model can be built for the capacitance and conductance matrices, so that

$$\begin{aligned} G(\lambda_1,\ldots,\lambda_L) &= G_0 + \lambda_1 G_1 + \ldots + \lambda_L G_L \\ C(\lambda_1,\ldots,\lambda_L) &= C_0 + \lambda_1 C_1 + \ldots + \lambda_L C_L \end{aligned} \tag{14}$$

where now C_0 and G_0 are the nominal matrix values, i.e. the value of the matrices under no parameter variation, and C_i and G_i, $i = 1, \cdots, L$, are its sensitivities with respect to those parameters. For small parameter variations, the projection matrix obtained via a moment-matching type algorithm such as PRIMA also suffers small perturbations. Therefore, the idea was to draw several samples in the parameter space for the system matrices $G(\lambda_1,\ldots,\lambda_L)$ and $C(\lambda_1,\ldots,\lambda_L)$, and for each sample PRIMA was applied so a projection matrix is obtained. Fitting is later applied over all the computed projectors in order to determine the coefficients of a parameter dependent projection matrix

$$V(\lambda_1,\ldots,\lambda_L) = V_0 + \lambda_1 V_1 + \ldots + \lambda_L V_L \tag{15}$$

which is in turn applied in a congruence-like transformation to the parametric system in (11), yielding a reduced system parametrized with respect to the set $[\lambda_1, \lambda_2, \ldots, \lambda_L]$.
Another approach also based on perturbation theory arguments was applied to the Truncate Balanced Realization (TBR) [14, 17] framework, so that a theoretically based perturbation matrix was obtained starting from the affine models shown in (14) [8]. This matrix was then applied via a congruence transformation over the Gramians to address the variability, and yield the perturbed Gramians. These in turn were used inside a balancing truncation procedure. As with most TBR-inspired methods, this one is also expensive to compute and hard to implement.
The above methods have obvious drawbacks, perhaps the most glaring of which is the heavy computation cost required for obtaining the reduced models and the limitation that comes from first order approximations possibly leading to inaccuracy in certain cases.

Multi-Dimensional Moment Matching

These techniques appear as extensions to nominal moment-matching techniques [15, 6, 21]. Moment matching algorithms have gained a well deserved fame in nominal MOR due to their simplicity and efficiency. The extensions of these techniques to the parametric case are usually based in the implicit or explicit moment matching of the parametric transfer function (12).
This type of algorithms assumes small fluctuations of the parameters, so that a model based on the Taylor Series expansion can be used for approximating the behavior of the conductance and capacitance, $G(\lambda)$ and $C(\lambda)$, expressed as a function of the parameters

$$\begin{aligned} G(\lambda_1,\ldots,\lambda_L) &= \textstyle\sum_{i_1=0}^{\infty} \cdots \sum_{i_L=0}^{\infty} G_{i_1,\ldots,i_L} \lambda_1^{i_1} \ldots \lambda_L^{i_L} \\ C(\lambda_1,\ldots,\lambda_L) &= \textstyle\sum_{i_1=0}^{\infty} \cdots \sum_{i_L=0}^{\infty} C_{i_1,\ldots,i_L} \lambda_1^{i_1} \ldots \lambda_L^{i_L} \end{aligned} \tag{16}$$

where $G_0, C_0, G_{i_1,\ldots,i_L}$ and $C_{i_1,\ldots,i_L}$ are the multidimensional Taylor series coefficients. This Taylor series can be extended up to the desired (or required) order, including cross derivatives, for the sake of accuracy. If this formulation is used, the structure for parameter dependence may be maintained if the projection is not only applied to the nominal matrices, but to the sensitivities as well.
The Multi-Parameter Moment Matching method is a single-point expansion of the transfer function (12) in the joint space of the frequency s and the parameters $\lambda_i, i = 1, \cdots, L$, in order to obtain a power series in several variables $s, \lambda_1, \ldots, \lambda_L$ [3],

$$v(s,\lambda_1,\ldots,\lambda_L) = \sum_{k=0}^{\infty} \sum_{k_s=0}^{k} \sum_{k_1=0}^{k-k_s} \cdots \sum_{k_L=0}^{k-k_s-k_1\ldots-k_{L-1}} M_{k,k_s,k_1,\ldots,k_L} s^{k_s} \lambda_1^{k_1} \ldots \lambda_L^{k_L} \tag{17}$$

where $M_{k,k_s,k_1,\ldots,k_L}$ is a k-th ($k = k_s + k_1 + \ldots + k_L$) order multi-parameter moment corresponding to the coefficient term $s^{k_s}\lambda_1^{k_1}\ldots\lambda_L^{k_L}$. Following the same idea used in the nominal moment matching techniques, a basis for the subspace formed from these moments can be built and the resulting matrix V can be used as a projection matrix for reducing the original system. It has been shown that this parametrized reduced model matches up to the k-th order multi-parameter moment of the original system. The main inefficiency of this method is that process parameters fluctuate in a small range around their nominal value, whereas the frequency range is much larger, and a higher number of moments are necessary in order to capture the global response for the whole frequency range. For this reason, the reduced model size grows exponentially with the number of parameters and the moments to match. A similar idea but more efficient, is to rely in a two-step moment matching scheme [12]. In this method, one first matches in an explicit way the multi-parameter moments for the process variability parameters (by expanding the state space vector v and the matrices G and C in its Taylor Series only w.r.t. the parameters), and in a second stage implicitly match moments with respect to the frequency via Krylov projection. This two-step approach avoids the exponential growth of model size with the number of moments matched, suffered by the multi-parameter moment matching. This method allows a certain degree of flexibility as the number of moments matched with respect to the frequency and to the parameters can be different. In principle, in spite of the larger size of the augmented model, the order of the reduced system can be much smaller than in the previous cases. On the other hand, the structure of the dependence with respect to the parameters is lost since the parametric dependence is shifted to the later projected output related N matrix.
A different multi-dimensional moment matching approach was also presented [7], which relies on the computation of several subspaces, built separately for each dimension, i.e. the frequency s and the parameter set λ. So given a parametric system (11), the first step of the algorithm is to obtain the k_s block moments of the transfer function with respect to the frequency when the parameters take their nominal value (for example, via PRIMA). This block moments will be denoted as Q_s. The next step is to obtain the subspaces which match k_{λ_i} block moments of v with respect to each of the parameter λ_i, and will be denoted by Q_{λ_i}. Once all the subspaces have been computed, an orthonormal basis can be obtained so that its columns spans the joint of all subspaces. Applying the resulting matrix in a projection scheme ensures that the parametric ROM[4] matches k_s moments of the original system with respect to the frequency, and k_{λ_i} moments with respect to the parameter λ_i. If the cross-term moments are needed for accuracy reasons, the subspace that spans these moments can be also included by following the same scheme.

Variational PMTBR

A novel approach was recently proposed that extends the PMTBR algorithm to include variability [16]. This approach is based on the statistical interpretation of the algorithm (see [19] for details) and enhances its applicability. In this interpretation, the Gramian is seen as a covariance matrix for a Gaussian variable, $v(0)$, obtained by exciting the (presumed stable) system with white noise. Rewriting the Gramian as

$$X_\lambda = \int_{S_\lambda}\int_{-\infty}^{\infty}(sC_\lambda + G_\lambda)^{-1}MM^T(sC_\lambda + G_\lambda)^{-H}p(\lambda)dwd\lambda \tag{18}$$

where $p(\lambda)$ is the probability density of λ in the parameter space, S_λ. Just as in PMTBR, a quadrature rule can be applied in the overall parameter plus frequency space to approximate the Gramian via numerical computation. But in this case the weights are chosen taking into account the PDF[5] of λ_i and the frequency constraints. This can be generalized to a set of

[4] Reduced Order Model
[5] PDF – Probability density function.

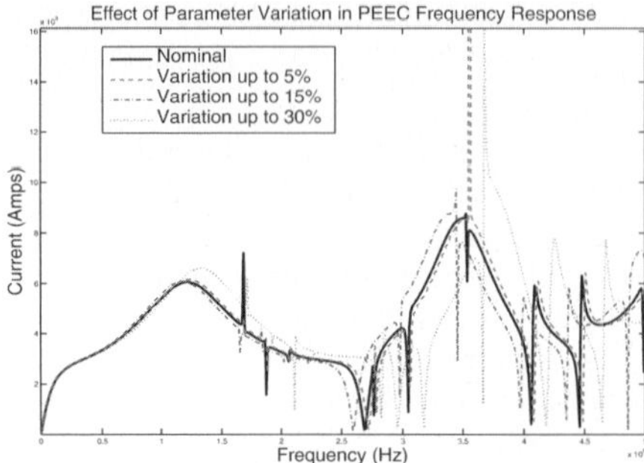

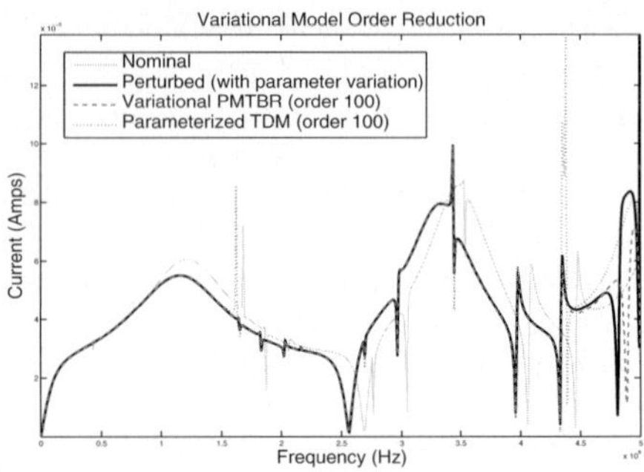

Fig. 3: Variational PEEC: effects on the frequency response (left) and performance of parametric MOR methods (right).

parameters, where a joint PDF of all the parameters can be applied to the joint parameter space, or the individual PDF of each parameter can be used. The ability to do this represents an interesting advantage, since *a-priori* knowledge of the parameters and the frequency can be included in order to constrain the sampling and yield a more accurate reduced model. As in the deterministic case, an error analysis and control can be included, via the eigenvalues of the SVD, but in this variational case only an expected error bound can be given:

$$E\{\|\hat{v}_O - v_O\|_2^2\} \leq \sum_{i=r+1}^{n} \sigma_i^2 \tag{19}$$

where r is the reduced order and n the original number of states. In this method, the issue of sample selection, already an important one in the deterministic version, becomes even more relevant, since the sampling must now be done in a potentially much higher-dimensional space.

3.3 Results

To illustrate (for a qualitative analysis mostly) the effect of parameter variability on the response of a circuit we resort to a simple example of a partial equivalent electric circuit (PEEC) model. The system under analysis is an RLC model of a connector of order 304. In this example we consider the effect of five geometric parameters, each having a different effect on the conductance and capacitance matrices. Figure 3(left) shows the effect of random variations on each parameter up to a limit of $5\%, 15\%$ and 30%. It can be seen that even small range variations in the parameters can result in large deviations from nominal. An important effect of the parameter variation is that those deviations not only can change the overall shape of the frequency response but also cause frequency shifts in the pole location. Figure 3(right) shows a comparison of the reduction of the variational system with two different methods: variational PMTBR and parametrized time-domain macromodels [7], all of the same order, versus the nominal response and the system response under parameter variation (Perturbed). As can be seen, the parametric MOR algorithms are able to maintain an acceptable accuracy up to high frequencies in the presence of strong variations.

4 Conclusions

Model order reduction is a crucial enabling technique for simulation, control, and optimization of complex physical systems. In this paper we discussed how, in spite of the progress achieved in the area in the last few years, certain types of problems such as those derived from massively coupled systems, still pose difficulties to the existing approaches. We also discussed new challenges in the field, brought by new applications such as the reduction of parametric systems, that are becoming increasingly relevant, raising new issues in the quest for increased performance. Clearly, we have but scratched the surface of the relevant issues facing us. Other challenging problems exist, like the reduction on nonlinear systems, which has also been subject to extensive research.

Acknowledgements

Jorge Fernández Villena, Paulo Flores and L. Miguel Silveira acknowledge the financial support from the FP6/IST/027378 Chameleon-RF project (`http://www.chameleon-rf.org`). João M. S. Silva was supported by a PhD fellowship from Fundação para a Ciência e Tecnologia (FCT), Portugal, with the reference SFRH/BD/10586/2002.

References

1. A. C. Antoulas. *Approximation of Large-Scale Dynamical Systems.* Society for Industrial and Applied Mathematics, Philadelphia, PA, USA, 2005.
2. P. Benner, V. Mehrmann, and D. Sorensen, editors. *Dimension Reduction of Large-Scale Systems*, volume 45 of *Lecture Notes in Computational Science and Engineering.* Springer-Verlag, Berlin/Heidelberg, Germany, 2005.
3. L. Daniel, O. C. Siong, S. C. Low, K. H. Lee, and J. K. White. A multiparameter moment-matching model-reduction approach for generating geometrically parametrized interconnect performance models. *IEEE Trans. Computer-Aided Design*, 23:678–693, May 2004.
4. N. Dong and J. Roychowdhury. Piecewise polynomial nonlinear model reduction. In 40th *ACM/IEEE Design Automation Conference*, pages 484–489, Anaheim, CA, June 2003.
5. P. Feldmann. Model order reduction techniques for linear systems with large number of terminals. In *DATE'2004 - Design, Automation and Test in Europe, Exhibition and Conference*, volume 2, pages 944–947, Paris, France, February 2004.
6. P. Feldmann and R. W. Freund. Efficient linear circuit analysis by Padé approximation via the Lanczos process. *IEEE Transactions on Computer-Aided Design of Integrated Circuits and Systems*, 14(5):639–649, May 1995.
7. P. Gunupudi, R. Khazaka, M. Nakhla, T. Smy, and D. Celo. Passive parameterized time-domain macromodels for high-speed transmission-line networks. *IEEE Trans. On Microwave Theory and Techniques*, 51(12):2347–2354, December 2003.
8. P. Heydari and M. Pedram. Model reduction of variable-geometry interconnects using variational spectrally-weighted balanced truncation. In *International Conference on Computer Aided-Design*, pages 586–591, San Jose, CA, USA, November 2001.
9. I. M. Jaimoukha and E. M. Kasenally. Krylov subspace methods for solving large Lyapunov equations. *SIAM Journal on Numerical Analysis*, 31:227–251, 1994.
10. M. Kamon, F. Wang, and J. White. Generating nearly optimally compact models from Krylov-subspace based reduced-order models. *IEEE Transactions on Circuits and Systems II: Analog and Digital Signal Processing*, 47(4):239–248, April 2000.
11. J.-R. Li, F. Wang, and J. White. Efficient model reduction of interconnect via approximate system grammians. In *International Conference on Computer Aided-Design*, pages 380–383, San Jose, CA, November 1999.
12. X. Li, P. Li, and L. Pileggi. Parameterized interconnect order reduction with Explicit-and-Implicit multi-Parameter moment matching for Inter/Intra-Die variations. In *International Conference on Computer Aided-Design*, pages 806–812, San Jose, CA, November 2005.
13. Y. Liu, L. T. Pileggi, and A. J. Strojwas. Model order reduction of RC(L) interconnect including variational analysis. In 36th *ACM/IEEE Design Automation Conference*, pages 201–206, June 1999.

14. B. Moore. Principal Component Analysis in Linear Systems: Controllability, Observability, and Model Reduction. *IEEE Transactions on Automatic Control*, AC-26(1):17–32, February 1981.
15. A. Odabasioglu, M. Celik, and L. T. Pileggi. PRIMA: passive reduced-order interconnect macromodeling algorithm. *IEEE Trans. Computer-Aided Design*, 17(8):645–654, August 1998.
16. J. Phillips. Variational interconnect analysis via PMTBR. In *International Conference on Computer Aided-Design*, pages 872–879, San Jose, CA, USA, November 2004.
17. J. Phillips, L. Daniel, and L. M. Silveira. Guaranteed passive balancing transformations for model order reduction. *IEEE Trans. Computer-Aided Design*, 22(8):1027–1041, August 2003.
18. J. R. Phillips. Projection-based approaches for model reduction of weakly nonlinear, time-varying systems. *IEEE Trans. Computer-Aided Design*, 22:171–187, 2003.
19. J. R. Phillips and L. M. Silveira. Poor Man's TBR: A simple model reduction scheme. *IEEE Trans. Computer-Aided Design*, 24(1):43–55, Jan. 2005.
20. M. Rewienski and J. White. A trajectory piecewise-linear approach to model order reduction and fast simulation of nonlinear circuits and micromachined devices. *IEEE Transactions on Computer-Aided Design of Integrated Circuits and Systems*, 22(2):155–170, Feb. 2003.
21. L. M. Silveira, M. Kamon, I. Elfadel, and J. K. White. A coordinate-transformed Arnoldi algorithm for generating guaranteed stable reduced-order models of rlc circuits. In *International Conference on Computer Aided-Design*, pages 288–294, San Jose, California, November 1996.
22. L. M. Silveira and J. Phillips. Exploiting input information in a model reduction algorithm for massively coupled parasitic networks. In 41st *ACM/IEEE Design Automation Conference*, pages 385–388, San Diego, CA, USA, June 2004.

Positive Real Balancing for Nonlinear Systems

Tudor C. Ionescu[1] and Jacquelien M. A. Scherpen[1]

Rijksuniversiteit Groningen, t.c.ionescu, j.m.a.scherpen@rug.nl

Summary. We extend the positive real balancing procedure for passive linear systems to the nonlinear systems case. We show that, just like in the linear case, model reduction based on this technique preserves passivity.

Keywords—positive real, passive, energy functions, Hamilton-Jacobi equations, nonlinear balancing, truncation.

1 Introduction

Positive Real Balancing for linear systems is an attractive tool for passivity preserving model reduction [Ant05]. The method deals with the class of passive linear systems. It combines the useful properties of the balancing technique with the passivity theory. The latter provides a particular pair of energy functions to be balanced. The balanced form of the energy functions reveal the positive real singular values. They measure the energetic importance of the states. The less important states are omitted to obtain a reduced order system. If the full order system were passive then the reduced model would be passive too [Ant05].
The idea in this paper is to extend this method to the case of passive nonlinear systems. It is motivated by the wide range of applications such as power systems stability analysis and controller design, see e.g. [Giu05]. We use the nonlinear balancing method developed in [Sch93, Sch94] in combination with the passivity theory in [Wil72, vdS00]. In this case, the positive real singular values are nonlinear positive functions of the state, having the same significance as in the linear case, i.e. measure the energetic importance of the states.
In Section 2, a brief overview of the passivity and positive realness properties is given and the energy functions, the available storage and the required supply, will be defined. Section 3 shortly reviews the positive real balancing procedure for linear systems and the properties of the reduced model. Section 4 presents the energy functions as the solutions of a Hamilton Jacobi equation. Section 5 is an adaptation of the nonlinear balancing procedure to the positive real systems case. We define the positive real singular value functions. The outcome of it is used in Section 6, where the truncation itself is done and the reduced system will pe proved to be passive. Some conclusions and future work make up Section 7.
The nonlinear systems we treat are given in the state space representation as:

$$x = f(x) + g(x)u, \; y = h(x) + d(x)u, \tag{1}$$

where $x \in \mathbb{R}^n, u \in \mathbb{R}^m, y \in \mathbb{R}^p$, with $m = p$. x is called the state vector, u is the input and y is the output of the system. f, g, h are smooth nonlinear vectorfields depending on the state vector x. n is called the dimension of system (1). The input u will be considered to have finite energy, i.e. $u \in L_2(\mathbb{R}^p)$.

2 Passivity, Energy Functions and Positive Realness

In this section, we give a brief overview on the dissipativity theory as in [Wil72, Wil71, vdS00]. A function $w : \mathbb{R}^p \times \mathbb{R}^p \to \mathbb{R}$ will be called the supply rate. The dissipativity property is defined with respect to the supply rate w.

Definition 1 *[Wil72, vdS00] A system (1) is called dissipative with respect to the supply rate $w(u, y)$, if there exists a storage function $S : \mathbb{R}^n \to \mathbb{R}$, with the following properties:*
1. $S(x) \geq 0$
2. $S(x_0) + \int_{t_0}^{t_1} w(u, y)dt \geq S(x_1)$,

where $x_0 = x(t_0)$ and $x_1 = x(t_1)$. A particular case is when the supply rate represents the energy supplied at the terminals of the system, that is $w(u, y) = u^T y$. In this case the system is called passive. □

Remark 2 *If the inequality is strict, we will call the system strictly passive, that is the internal energy of the system is decreasing even when supplied at the terminals. In case of equality the system is called lossless. It means that the internal energy of the system is not changing. Property 2. can also be written in a differential form as:*

$$\frac{\partial S(x)}{\partial x}(f(x) + g(x)u) \leq u^T h(x) + u^T d(x)u \tag{2}$$

□

For our purpose, from the set of storage functions satisfying the definition or (2), two particular types of storage functions are of interest: the available storage and the required supply.

Definition 3 *[Wil72, vdS00] The available storage function of a system (1) is the energy function:*

$$S_a(x_0) = -\min_u \int_0^{\infty} u^T y\, dt,\ x(0) = x_0,\ x(\infty) = 0 \tag{3}$$

The required supply function of system (1) is the energy function:

$$S_r(x_0) = \min_u \int_{-\infty}^{0} u^T y\, dt,\ x(0) = x_0,\ x(-\infty) = 0 \tag{4}$$

□

$S_a(x)$ represents the maximal amount of energy that can be extracted from the terminals of the system when starting at the initial state x_0. $S_r(x)$ represents the minimal amount of energy required to be supplied to the system in order to reach x_0 from the equilibrium.
The property of the system being reachable from x_0 is a condition for the existence and non-negativity of the energy functions defined above.

Lemma 4 *[Wil71] Let system (1) be passive as in Definition 1 and reachable from the state x_0. Then, the energy functions S_a and S_r as in Definition 3 exist and are nonnegative. Moreover, $S_a \leq S_r$.* □

Definition 5 *[BIW91] A system (1) is called positive real if, for all $u \in L_2(R^p)$,*

$$\int_0^t u(\tau)^T y(\tau) d\tau \geq 0. \tag{5}$$

□

Combined with Lemma 4, we obtain:

Proposition 6 *[BIW91] A passive system (1) is positive real. Conversely, a positive real system (1), that is reachable from the state x_0, is passive.* □

Remark 7 *If the inequality is strict, the system is strictly positive real.* □

3 Linear Systems Case

A linear system is a particular case of system (1), given as: $\dot{x} = Ax + Bu,\ y = Cx + Du$, where A, B, C, D are constant matrices of appropriate dimensions. The system is assumed to be reachable and observable (minimal) and $R = D + D^T > 0$. Then, strict positive realness, can be studied with the Kalman-Yakubovitch-Popov lemma, see e.g. [Ant05]. The energy functions are quadratic and related to a pair of matrices called the positive real Gramians of the system.

Theorem 8 *[Wil72] Assume that the linear system is strictly passive. Then* $S_a(x) = \frac{1}{2}x^T K_{min} x$ *and* $S_r(x) = \frac{1}{2}x^T K_{max} x$*, where* K_{min} *and* K_{max} *are the minimal, respectively maximal solution of the Positive Real Algebraic Riccati equation:*

$$KA + A^T K + (KB - C^T)R^{-1}(B^T K - C) = 0 \tag{6}$$

□

Definition 9 *[Ant05] A positive real linear system is called positive real balanced if* $K_{min} = (K_{max})^{-1} = \mathrm{diag}(\pi_1 I_{s_1}, \pi_2 I_{s_2}, ..., \pi_q I_{s_q})$*, where* $1 \geq \pi_1 > \pi_2 > ... > \pi_q > 0$*,* $s_1 + s_2 + ... s_q = n$*.* □

The positive real singular value π_k, $k = 1, ..., q$ represents the energetic measure of the state components $x_{s_1+...+s_{k-1}+1}, ..., x_{s_1+...s_k}$. If π_l is much larger than π_{l+1}, then the state vector can be truncated from $w = s_1 + ... + s_l + 1$ to n, i.e. $x_{s_1+...+s_l+1} = 0, ..., x_n = 0$. A reduced model of dimension $\hat{n} = s_1 + ... + s_l < n$ is obtained. Then:

Theorem 10 *Let the passive linear system be brought into the positive real balanced form* (A_b, B_b, C_b, D_b)*. The reduced system obtained after truncation with dimension* l*, i.e. dim* $\hat{x} = \hat{n}$*, is minimal and passive.* □

4 Nonlinear Systems Case

In this section we consider a system (1), under the following assumptions:

1. 0 is an equilibrium point of the system and $h(0) = 0$;
2. the system is strictly positive real, i.e. $r(x) = d(x) + d^T(x) > 0$, and reachable from x_0;
3. $x \in Y$, where Y is a neghbourhood of 0.

Assumption 1 is made for the sake of simplicity, but generality is not lost. Assumption 2 is in accordance with the nonlinear version of the Kalman-Yakubovitch-Popov lemma which characterizes the property of (strict) positive realness [Moy74, HilMoy76]. We mention that the smoothness assumed in the definition of system (1) guarantees the existence of solutions to be introduced. This condition could be relaxed, but it is kept for convenience.
Denote by $||v||_M^2 = v^T M v$, $(\forall) v \in \mathbb{R}^n$, $M \in \mathbb{R}^{n \times n}$.
The energy functions are computed as the stabilizing and antistabilizing solution, respectively, of a Hamilton-Jacobi equation, which is the nonlinear generalization of the Positive Real Algebraic Riccati equation, (6) from the previous section.

Theorem 11 *Let system (1) be, satisfying Assumptions 1-3. Then the Hamilton-Jacobi equation:*

$$\frac{\partial S(x)}{\partial x} f(x) + \frac{1}{2}\left(\frac{\partial S(x)}{\partial x} g(x) - h^T(x)\right) r^{-1}(x) \left(g^T(x)\frac{\partial S^T(x)}{\partial x} - h(x)\right) = 0 \tag{7}$$

has the smooth solution $S_a(x)$*,* $S_a(0) = 0$*, such that*

$$f(x) + g(x) r^{-1}(x) \left(g^T(x) \frac{\partial S_a{}^T}{\partial x} - h(x)\right) \tag{8}$$

is asymptotically stable and the smooth solution $S_r(x)$*,* $S_r(0) = 0$*, such that*

$$-\left(f(x) + g(x) r^{-1}(x) \left(g^T(x) \frac{\partial S_r{}^T}{\partial x} - h(x)\right)\right) \tag{9}$$

is asymptotically stable. □

Proof: Because system (1) is passive and reachable, according to Lemma 4, $S_a(x(t))$ and $S_r(x(t))$ exist and are nonnegative. We develop the proof for $S_r(x)$. The sequel follows the idea in Scherpen [Sch94], Section 3, Theorem 3.1.3. By definition, $S_r(x) = \min_{u, x(-\infty)=0}$

$\int_{-\infty}^{t} u^T(s)y(s)ds$. Because $S_r(x)$ exists, there exists an optimal input u^*, i.e. $S_r(x(t)) = \int_{-\infty}^{t} u^{*T}(s)y^*(s)ds$, where $y^*(s)$ is the output of the system with the input u^*. Differentiating $S_r(x(t))$ with respect to time we get:

$$\dot{S}_r(x(t)) = u^{*^T} y^* \Rightarrow \frac{\partial S(x)}{\partial x}(f(x) + g(x)u^*) - u^{*T} y^* = 0. \tag{10}$$

On the other hand, using completion of squares and (7), we have that

$$u^T y - \frac{1}{2}\left\| u - r^{-1}\left(g^T \frac{\partial S^T}{\partial x} - h\right)\right\|_r^2 = \frac{\partial S_r}{\partial x} gu - \frac{1}{2}\left(\frac{\partial S_r}{\partial x} g - h^T\right) r^{-1}\left(g^T \frac{\partial S_r^T}{\partial x} - h\right)$$
$$= \frac{\partial S_r}{\partial x}(f + gu) = \dot{S}_r. \tag{11}$$

Relation (10), can be written as $\frac{\partial S_r}{\partial x} f + \frac{\partial S_r}{\partial x} gu^* - u^{*^T} y^* = 0$. Relations (11), (7) give: $u^{*^T} y^* - \frac{\partial S_r^T}{\partial x} gu^* = -\frac{1}{2}\left(\frac{\partial S_r}{\partial x} g - h^T\right) r^{-1}\left(g^T \frac{\partial S_r^T}{\partial x} - h\right) + \frac{1}{2}\left\| u - r^{-1}\left(g^T \frac{\partial S^T}{\partial x} - h\right)\right\|_r^2$

$$\Rightarrow \frac{\partial S_r}{\partial x} f + \frac{1}{2}\left(\frac{\partial S_r}{\partial x} g - h^T\right) r^{-1}\left(g^T \frac{\partial S_r^T}{\partial x} - h\right) - \frac{1}{2}\left\| u^* - r^{-1}\left(g^T \frac{\partial S_r^T}{\partial x} - h\right)\right\|_r^2 = 0 \tag{12}$$

Now we show that $u^* = r^{-1}\left(g^T \frac{\partial S^T}{\partial x} - h\right)$. Let u be any continuous admissible control that steers the state from $x(t)$ to $x(-\infty) = 0$ (as the system is considered reachable). Let

$$\hat{u}(\bar{t}) = \begin{cases} u(\bar{t}), & t - \delta \leq \bar{t} \leq t \\ u^*(\bar{t}), & -\infty \leq \bar{t} \leq t - \delta \end{cases}$$

Denoting by $\hat{y}(s)$ the output of system (1) with input $\hat{u}(s)$ and by $J(\hat{u}) = \int_{-\infty}^{t} \hat{u}^T(s)\hat{y}(s)ds$, we have: $J(\hat{u}) = \int_{-\infty}^{t-\delta} u^{*T}(\bar{t})y^*(\bar{t})d\bar{t} + \int_{t-\delta}^{t} \hat{u}^T(\bar{t})\hat{y}(\bar{t})d\bar{t} = S_r(x(t-\delta)) + \int_{t-\delta}^{t} \hat{u}^T(\bar{t})\hat{y}(\bar{t})d\bar{t}$. The integral can be approximated as follows: $\int_{t-\delta}^{t} \hat{u}^T(\bar{t})\hat{y}(\bar{t})d\bar{t} = \delta\hat{u}^T(t)\hat{y}(t) + o(\delta)$, where $o(\delta)/\delta \to 0$, as $\delta \to 0$. By the smoothness of $S_r(x)$ we have that: $S_r(x(t)) = S_r(x(t-\delta)) + \delta\frac{dS_r(x(t))}{dt} + o(\delta) = S_r(x(t-\delta)) + \delta\frac{\partial S_r(x)}{\partial x}(f(x) + g(x)u) + o(\delta)$.
At the same time we know that $S_r(x) \leq J(\hat{u})$ which leads to: $\frac{\partial S_r(x)}{\partial x}(f(x)+g(x)u) - u^T y \leq 0$. Using relation (12) we conclude that $\frac{\partial S_r}{\partial x} f + \frac{1}{2}\left(\frac{\partial S_r}{\partial x} g - h^T\right) r^{-1}\left(g^T \frac{\partial S_r^T}{\partial x} - h\right)$
$-\frac{1}{2}\left\| u - r^{-1}\left(g^T \frac{\partial S_r^T}{\partial x} - h\right)\right\|_r^2 \leq 0$ Taking into account equation (12) the equality holds for $u = u^* = r^{-1}\left(g^T \frac{\partial S_r^T}{\partial x} - h\right)$. Hence, because $S_r(0) = 0$ and u^* steers the state from t to $-\infty$ in 0, we conclude that $S_r(x)$ satisfies (7) such that (9) is asymptotically stable. The proof for $S_a(x)$ follows the exact same line. □
S_a and S_r are the minimal, respectively maximal solution of (7).

Remark 12 *If S_a and S_r were quadratic as in the linear systems case, everything would boil down to the Positive Real Algebraic Riccati equation and the positive real Gramians from Theorem 8, in Section 3.* □

Proposition 13 *If $S(x) \geq 0$ is a solution of (7), then $0 \leq S_a \leq S \leq S_r$.* □
Proof: Follows the ideas of [Moy74]. □

Remark 14 *This result is in accordance with the ideas in [Wil72, vdS00].* □
The energy functions can be used according to [Wil72] as Lyapunov functions for system (1).

Lemma 15 *If the system (1) is passive and zero-state observable then any solution $S(x)$ of (7) is positive definite $(\forall)$ $x \neq 0$.* □
Proof: This follows the line in [Moy74, Lemma 2]. □

Corollary 16 *A system (1) that is passive, with $r(x) > 0$ and zero state observable is asymptotically stable.* □

5 Nonlinear Balancing

In this section a system (1) is considered, under assumptions 1, 2, 3, and:

4. it is zero-state observable on Y;
5. $S_a(x)$ and $S_r(x)$ exist and are smooth on Y;

According to the previous section, the Assumptions 1-5 insure that S_a and S_r are the minimal, respectively the maximal positive definite solutions of the equation (7), for all $x \in Y$. The sequel follows the procedure in Scherpen [Sch93]. The goal is to find the coordinate transformation $\overline{z} = \xi(x)$ which brings the system into the positive real balanced form.

Theorem 17 *[Sch93] There exists a coordinate transformation $x = \phi(\overline{x}), \phi(0) = 0$ s.t., in the new coordinates $S_a(\phi(\overline{x})) = \frac{1}{2}\overline{x}^T\overline{x}$ and $S_r(\phi(\overline{x})) = \frac{1}{2}\overline{x}^T M(\overline{x})\overline{x}$, where $M(\overline{x})$ is an $n \times n$ symmetric matrix whose entries are smooth functions of $\overline{x}$.* □

Proof: See Lemma 3.2.2 in [Sch93], Chapter 3. □

For the sequel an extra assumption is needed

6. on Y, $M(\overline{x})$ has a constant number of distinct eigenvalues ([Sch94], Lemma 3.2.3)

According to Kato's result, [Ka82, Theorem 5.13a], Assumption 6 insures that $M(\overline{x})$ can be brought into a diagonal form, while leaving S_a in the same form.

Theorem 18 *[Sch93] Under assumptions 1-6, there exists a coordinate transformation $\overline{x} = \psi(z)$, s.t.*

$$S_a(z) = S_a(\psi(z)) = \frac{1}{2}z^T z, \tag{13}$$

and

$$S_r(z) = S_r(\psi(z)) = \frac{1}{2}z^T \mathrm{diag}(v_1(z), ..., v_n(z))z \tag{14}$$

□

The nonlinear system (1) is brought in positive real balanced form using the following coordinate transformation $\overline{z} = \eta(z) = [\eta_1(z_1) ... \eta_n(z_n)]^T$, where $\eta_i(z_i) = v_i(0, ..., z_i, ..., 0)^{\frac{1}{4}} z_i > 0$. Applying the transformation we get:

$$S_a(\overline{z}) = \frac{1}{2}\overline{z}^T \mathrm{diag}(\pi_1(\overline{z}_1)^{-1}, ..., \pi_n(\overline{z}_n)^{-1})\overline{z} \tag{15}$$

$$S_r = \frac{1}{2}\overline{z}^T \mathrm{diag}(\pi_1(\overline{z}_1)^{-1}v_1(\eta^{-1}(\overline{z})), ..., \pi_n(\overline{z}_n)^{-1}v_n(\eta^{-1}(\overline{z})))\overline{z} \tag{16}$$

$v_k(\eta^{-1}(\overline{z})) > 0$ for all k, can be called the positive real singular value functions of (1) and $\pi_k(\overline{z}_k) = \sqrt{v_k(0, ..., \eta_k^{-1}(\overline{z}_k), ..., 0)}$. Applying this coordinate transformation to (1), it becomes:

$$\dot{\overline{z}} = \overline{f}(\overline{z}) + \overline{g}(\overline{z})u, \quad \overline{y} = \overline{h}(\overline{z}) + \overline{d}(\overline{z}).$$

A system having the available storage and required supply of the form (15) and (16) is in the positive real balanced form.

So, given a system (1), by directly applying the coordinate change $\overline{z} = \xi(x) = (\eta \circ \psi^{-1} \circ \phi^{-1})(x)$, it is brought into positive real balanced form.

The available energy extracted at component $\overline{z}_k$ is given by the quantity $S_a(0, ..., \overline{z}_k, ...0) = \frac{1}{2}\overline{z}_k^2\pi_k^{-1}(\overline{z}_k)$ and the energy supply required to reach component $\overline{z}_k$ is measured as $S_r(0, ..., \overline{z}_k, ..., 0) = \frac{1}{2}\overline{z}_k^2\pi_k(\overline{z}_k)$. So, if $v_k(\overline{z}) \gg v_{k+1}(\overline{z})$, then $\pi_k^{-1}(\overline{z})v_k(\overline{z}) \gg \pi_{k+1}^{-1}(\overline{z})v_{k+1}(\overline{z})$. This means that to reach state component $\overline{z}_k$ less supply of energy is required than for the component $\overline{z}_{k+1}$ and at state component $\overline{z}_k$ is stored more energy available than at state component $\overline{z}_{k+1}$. This makes components $\overline{z}_1, ..., \overline{z}_k$ more important from energetic point of view than state components $\overline{z}_{k+1}, ..., \overline{z}_n$. Thus, model truncation can be applied, meaning that the $\overline{z}_{k+1}, ..., \overline{z}_n$ components can be made 0.

6 Model Reduction - Truncation

Partition the state vector $\overline{z}$ into $[\overline{z}^{1^T}, \overline{z}^{2^T}]^T$, where $\overline{z}^1 = [\overline{z}_1 ..., \overline{z}_k]^T$ and $\overline{z}^2 = [\overline{z}_{k+1} ... \overline{z}_n]^T$. Accordingly, the system can be partitioned into:

$$\overline{f}(\overline{z}) = \begin{bmatrix} \overline{f}_1(\overline{z}^1, \overline{z}^2) \\ \overline{f}_2(\overline{z}^1, \overline{z}^2) \end{bmatrix}, \overline{g}(\overline{z}) = \begin{bmatrix} \overline{g}_1(\overline{z}^1, \overline{z}^2) \\ \overline{g}_2(\overline{z}^1, \overline{z}^2) \end{bmatrix}, \overline{h}(\overline{z}) = \overline{h}(\overline{z}^1, \overline{z}^2), \overline{d}(\overline{z}) = \overline{d}(\overline{z}^1, \overline{z}^2).$$

According to the previous section, the energetic analysis of the state components tells that $\overline{z}^2$ is less important than $\overline{z}^1$. Hence, to reduce the system, we truncate i.e. we set $\overline{z}^2 = 0$. The reduced system is described by:

$$\dot{\overline{z}}^1 = \overline{f}_1(\overline{z}^1, 0) + \overline{g}_1(\overline{z}^1, 0)u, \quad \overline{y} = \overline{h}(\overline{z}^1, 0) + \overline{d}(\overline{z}^1, 0)u \tag{17}$$

The available storage of the reduced system is: $S_a(\overline{z}^1, 0)$. Because of the form in (15) we have that $\frac{\partial S_a}{\partial \overline{z}^2}(\overline{z}^1, 0) = 0$.
The Hamilton-Jacobi equation (7) is satisfied as follows:

$$\frac{\partial S_a}{\partial \overline{z}^1}(\overline{z}^1, 0)\overline{f}_1(\overline{z}^1, 0) + \frac{1}{2}\left(\frac{\partial S_a}{\partial \overline{z}^1}(\overline{z}^1, 0)\overline{g}_1(\overline{z}^1, 0) - \overline{h}^T(\overline{z}^1, 0)\right)\overline{r}^{-1}(\overline{z}^1, 0) \cdot$$
$$\left(\overline{g}_1^T(\overline{z}^1, 0)\frac{\partial^T S_a}{\partial \overline{z}^1}(\overline{z}^1, 0) - \overline{h}(\overline{z}^1, 0)\right) = 0$$

Substituting the required supply $S_r(\overline{z}^1, 0)$ from relation (16) it is obtained that:

$$\frac{\partial S_r}{\partial \overline{z}^1}(\overline{z}^1, 0)\overline{f}_1(\overline{z}^1, 0) + \frac{1}{2}\left(\frac{\partial S_r}{\partial \overline{z}^1}(\overline{z}^1, 0)\overline{g}_1(\overline{z}^1, 0) - \overline{h}^T(\overline{z}^1, 0)\right)\overline{r}^{-1}(\overline{z}^1, 0) \cdot$$
$$\left(\overline{g}_1^T(\overline{z}^1, 0)\frac{\partial^T S_r}{\partial \overline{z}^1}(\overline{z}^1, 0) - \overline{h}(\overline{z}^1, 0)\right) + F\left(\frac{\partial S_r}{\partial \overline{z}^2}(\overline{z}^1, 0), \overline{g}_2(\overline{z}^1, 0), \overline{h}(\overline{z}^1, 0)\right) = 0$$

where

$$F = \frac{\partial S_r}{\partial \overline{z}^1}\overline{f}_2 + \left(\frac{\partial S_r}{\partial \overline{z}^1}\overline{g}_1 - \overline{h}^T\right)\overline{r}^{-1}\overline{g}_2^T\frac{\partial^T S_r}{\partial z^2} + \frac{\partial S_r}{\partial \overline{z}^2}\overline{g}_2\overline{r}^{-1}\overline{g}_2^T\frac{\partial^T S_r}{\partial \overline{z}^2}$$

The required supply of the reduced system does not equal the reduced required supply, unless an extra condition is fulfilled, i.e. $F = 0$.

Remark 19 *Being an input-output property, (strict) passivity is not affected by the coordinate transformation which brings the original system into (strictly) positive real balanced form. It means that the (strictly) positive real balanced system is again (strictly) passive.* □

Theorem 20 *The reduced order system is strictly passive.* □
Proof: We check if the strict passivity property in (2) is satisfied by the reduced system. We can write (2) for the full order strictly positive real balanced system:

$$\begin{bmatrix} \frac{\partial S_a}{\partial \overline{z}^1} & \frac{\partial S_a}{\partial \overline{z}^2} \end{bmatrix}\left(\begin{bmatrix} \overline{f}_1(\overline{z}^1, \overline{z}^2) \\ \overline{f}_2(\overline{z}^1, \overline{z}^2) \end{bmatrix} + \begin{bmatrix} \overline{g}_1(\overline{z}^1, \overline{z}^2) \\ \overline{g}_2(\overline{z}^1, \overline{z}^2) \end{bmatrix} u\right) < u^T\overline{h}(\overline{z}^1, \overline{z}^2) + u^T\overline{d}(\overline{z}^1, \overline{z}^2)u.$$

Setting $\overline{z}^2 = 0$ we have that $\frac{\partial S_a}{\partial \overline{z}^2}(\overline{z}^1, 0) = 0$, $S_a(\overline{z}^1, 0) > 0$. Substituting in the above inequality we get:

$$\frac{\partial S_a}{\partial \overline{z}^1}(\overline{z}^1, 0)(\overline{f}_1(\overline{z}^1, 0) + \overline{g}_1(\overline{z}^1, 0)u) < u^T\overline{h}(\overline{z}^1, 0) + u^T\overline{d}(\overline{z}^1, 0)u.$$

It means that the reduced order system satisfies inequality (2), hence the reduced system is strictly passive. □

Theorem 21 *If* $F\left(\frac{\partial S_r}{\partial \overline{z}^2}(\overline{z}^1, 0), \overline{g}_2(\overline{z}^1, 0), \overline{h}(\overline{z}^1, 0)\right) = 0$ *for all* $\overline{z}^1$ *around* 0*, then the reduced system is in strictly positive real balanced form having the singular value functions:* $v_1(z^1, 0) \geq ... \geq v_k(z^1, 0)$*, for* $z^1 = \eta^{-1}(\overline{z}^1, 0)$*.* □
Proof: If the condition on F is satisfied, then $S_r(\overline{z}^1, 0)$ as in (16) is the required supply of the reduced system. $S_a(\overline{z}^1, 0)$ as in (15) satisfies directly the Hamilton-Jacobi equation (7), so it is the available storage of the reduced system. Thus, the system is in positive real balanced form with the positive real singular value functions $v_1(z^1, 0) \geq \ldots \geq v_k(z^1, 0)$, where $z^1 = [\eta^{-1}(\overline{z}_1) \ldots \eta^{-1}(\overline{z}_k)]^T$. □

Remark 22 *If the singular value functions are independent of* $\overline{z}^2$*, then* $\frac{\partial S_r}{\partial \overline{z}^2}(\overline{z}^1, 0) = 0$*. Then immediately* $F\left(\frac{\partial S_r}{\partial \overline{z}^2}(\overline{z}^1, 0), \overline{g}_2(\overline{z}^1, 0), \overline{h}(\overline{z}^1, 0)\right) = 0$ *follows.* □

7 Future Work

We present a passivity preserving model reduction technique, based on positive real balanced truncation. The results in Section 5 are coordinate dependent, leading to the fact that the balanced representation and the singular value functions are not unique, i.e. the choice of different sets of singular value functions gives different reduced systems. For future research, developments such as in [FujSch05], is to be taken into account for the nonlinear positive real balancing case.
If the system is not strictly positive real, but is positive real, there is no Hamilton-Jacobi equation to solved. However, if one can compute S_a and S_r in a different way, the balancing procedure and the results of this paper can still be applied. Additionally, for physical systems, such as port-Hamiltonian systems (see [vdS00]) it may be useful to preserve besides passivity, an additional energy/power-based structure in the model for control purposes. This is also a topic for future research.

Acknowledgment

We thank NWO (Dutch Science Foundation) for supporting this research. We would also like to thank prof. Romeo Ortega (LSS, Supelec, France) and Alvaro Giusto (Departamento de Control y Electrnica Industrial, IIE, Uruguay) for presenting the problem in the power systems area, which led to the idea of this paper.

References

[Ant05] A. C. Antoulas: Approximation of Large-Scale Dynamical Systems. SIAM (2005).

[BIW91] C. I. Byrnes, A. Isidori, J. C. Willems: Passivity, feedbackequivalence, and the global stabilization of minimum phase nonlinear systems. IEEE Trans. Automat. Contr., **36**, 1228-1240 (1991).

[FujSch05] K. Fujimoto and J. M. A. Scherpen: Nonlinear input-normal realizations based on the differential eigenstructure of Hankel operators. IEEE Trans. on Aut. Contr., **50**, 2-18 (2005).

[Giu05] A. Giusto: On Transient Stabilization of Power Systems: A Power-Shaping Solution for Structure-Preserving Models , submitted to IEEE Conf on Decision and Control, Dec (2006).

[HilMoy76] D. Hill and P. J. Moylan: The stability of nonlinear dissipative systems, IEEE Trans. on Aut. Contr., **21**, 708-711 (1976).

[Ka82] T. Kato: A Short Introduction to Perturbation Theory for Linear Operators. Springer-Verlag (1982).

[Moy74] P. J. Moylan, Implications of passivity in a class of nonlinear systems, IEEE Trans. on Aut. Contr., **19**, 373-381 (1974).

[Sch93] J. M. A. Scherpen: Balancing for nonlincar systems. System & Control Letters, **21**, 143-153 (1993).

[Sch94] J. M. A. Scherpen: Balancing for nonlinear systems. PhD Thesis, Univ. of Twente (1994).

[vdS00] A. J. van der Schaft, L_2-gain and Passivity Techniques in Nonlinear Control. Springer-Verlag (2000).

[Wil71] J. C. Willems: The generation of Lyapunov functions for input-output stable systems. J. SIAM Contr., **9**, 105-134 (1971).

[Wil72] J. C. Willems: Dissipative Dynamical Systems - Part I: General theory, - Part II: Linear systems with quadratic supply rates. Archive for Rational Mechanics and Analysis, **45**, 321-393 (1972).

Efficient Initialization of Artificial Neural Network Weights for Electrical Component Models

Tuomo Kujanpää and Janne Roos

Helsinki University of Technology, Department of Electrical and Communications Engineering, Circuit Theory Laboratory, P.O.Box 3000, FI-02015 TKK, Finland.
tuomo.kujanpaa@tkk.fi
janne@ct.tkk.fi

1 Introduction

The modeling of RF/microwave components for computer-aided design is facing new challenges because of increasing operation frequencies, circuit complexity, integration density, and decreasing time to market. Recently, it has been shown that Artificial Neural Networks (ANNs) offer solutions to urgent modeling problems encountered with conventional numerical methods (e.g., 3-D EM simulation) and empirical models. Fast and accurate models based on ANNs have been created for a wide range of components [ZG00k], [PAR01].
The crucial part in ANN-based modeling is ANN training, that is, optimization of ANN weights with given measurements or, say, 3-D EM simulation data. In [TF97] several ANN weight-initialization methods were introduced and compared mainly by means of classification problems. It was shown how the choice of an initialization method influences the convergence of the optimization and the optimal initial weights are, by some means, determined by the measurement/simulation data set. However, weight-initialization methods have not previously been systematically evaluated for electrical component modeling problems and the nature of the problems — the functions to be approximated — differs significantly from, e.g., classification problems with discrete/Boolean input/target values.
In this paper, three methods for an initialization of ANN weights are experimentally evaluated for electrical component modeling applications. The third method, a special modification of the second method, is not found in literature. The methods are evaluated with respect to average ANN training error, ANN test error, and ANN training CPU time. Also, the standard deviations of ANN training and test errors are calculated for robustness analysis of the methods.

2 Artificial neural networks

The most widely used ANN in the field of RF/microwave component modeling is the Multi-Layer Perceptron (MLP) [ZG00k]. The three-layer MLP used in this work realizes the nonlinear mapping

$$\tilde{y}_l(\mathbf{x}, \mathbf{w}) = w_{l0} + \sum_{j=1}^{N_\mathrm{h}} w_{lj} a \tanh\left(b \cdot (w_{j0} + \sum_{i=1}^{N_\mathrm{i}} w_{ji} x_i) \right), \tag{1}$$

$$l = 1, 2, \ldots, N_\mathrm{o},$$

where N_i, N_h, and N_o represent the number of inputs, hidden-layer neurons, and outputs, respectively; $\mathbf{x} = (x_1, x_2, \ldots, x_{N_\mathrm{i}})$, $\tilde{\mathbf{y}} = (\tilde{y}_1, \tilde{y}_2, \ldots, \tilde{y}_{N_\mathrm{o}})$, and $\mathbf{w} = (w_{10}, w_{11}, \ldots, w_{N_\mathrm{o} N_\mathrm{h}})$

represents ANN inputs, outputs, and weights, respectively. The function $a\tanh(bv_j)$ is called the Activation Function (AF), where the parameters a and b determine the maxima and the steepness, respectively, and $v_j = w_{j0} + \sum_{i=1}^{N_\mathrm{i}} w_{ji}x_i$ is the induced local field of the function. Let $\mathbf{y} = \mathbf{y}(\mathbf{x})$ be an unknown, nonlinear, multidimensional function to be approximated by the MLP mapping (1): $\tilde{\mathbf{y}} = \tilde{\mathbf{y}}(\mathbf{x}, \mathbf{w})$. Let $\{(\mathbf{x}^k, \mathbf{y}^k), k = 1, 2, \ldots, N_\mathrm{tr}\}$ be an appropriate training set, N_tr being the number of samples, and the training-set inputs and targets being scaled linearly in the range $[-1, 1]$. Furthermore, let us define the normalized training error as

$$E_\mathrm{tr}(\mathbf{w}) = \sqrt{\frac{1}{N_\mathrm{tr}N_\mathrm{o}} \sum_{k=1}^{N_\mathrm{tr}} \sum_{l=1}^{N_\mathrm{o}} \left(\frac{\tilde{y}_l(\mathbf{x}^k, \mathbf{w}) - y_l^k}{2} \right)^2}. \tag{2}$$

The training of the ANN means minimizing $E_\mathrm{tr}(\mathbf{w})$ with respect to the weights, $\mathbf{w}$, using a suitable optimization method — in this work, Hestenes–Stiefel conjugate-gradient with Error Back Propagation (EBP) [KRH05k]. The generalization capability of the trained ANN is evaluated by applying Eq. (2) to an independent test set, $\{(\mathbf{x}^k, \mathbf{y}^k), k = 1, 2, \ldots, N_\mathrm{te}\}$, to obtain the normalized test error $E_\mathrm{te}(\mathbf{w})$.

3 Weight-initialization methods

Weight initialization tries to provide initial weight values close to the global minimum of $E_\mathrm{tr}(\mathbf{w})$, in the hope of avoiding local minima. There are several strategies for initializing the MLP weights; the most developed strategies can also be regarded as training methods [EFP05]. However, the most widely utilized strategy for ANN-based RF/microwave component modeling is, still, initializing the weights as random real numbers from a Uniform Distribution (UD) with fixed or variable range. The weight-initialization Methods (Ms) evaluated in this work include: M1. random initialization from UD with fixed range [ZG00k], M2. random initialization from UD with variable range and special input data scaling [Hay99k], and M3. random initialization from UD with variable range and special input and target training data scaling.
Utilizing M1 [ZG00k], one sets $a = b = 1$ and $w_{ji}, w_{lj} \in [-c, c]$, where, e.g., $c = 1.0$. This heuristic initialization tries to ensure the local field (v_j) of the AFs to be such that it forces the AFs to operate in an approximately linear transition region determined by maxima of the second derivative, $\max(\partial^2 \tanh(v_j)/\partial v_j^2)$. This would be desirable for the convergence of optimization because, when using EBP [Hay99k], $\partial E_\mathrm{tr}^2/\partial w_{ji} \sim \partial \tanh(v_j)/\partial v_j$ and the latter has its maximum value in the transition region. However, the heuristic weight initialization does not take into account the mean, $\bar{x}_i$, and the standard deviation of input data, σ_{x_i}, and, therefore, AFs may operate in saturation regions slowing down the optimization [Hay99k].
M2 [Hay99k] forces the specific AFs to operate in the transition region (between (–1,–1) and (1,1) for $a = 1.7159$ and $b = 2/3$) with $w_{ji} \in [-\sqrt{3/N_\mathrm{i}}, \sqrt{3/N_\mathrm{i}}]$, $w_{lj} \in [-\sqrt{3/N_\mathrm{h}}, \sqrt{3/N_\mathrm{h}}]$, $w_{j0} = 0$, and $w_{l0} = 0$. This initialization is based on a special input data scaling, with $\bar{x}_i = 0$ and $\sigma_{x_i} = 1$.
When one utilizes M2 and approximates the transition region of AFs as a straight line going through the origin with slope 1, the distribution parameters of the MLP outputs, $\tilde{y}_l$, are $\bar{\tilde{y}}_l = 0$ and $\sigma_{\tilde{y}_l} = 1$ as for MLP inputs x_i. A hypothesis to be tested is presented (M3): scaling of the target training data, y_l, such that $\bar{y}_l = 0$ and $\sigma_{y_l} = 1$, improves the convergence of optimization. The idea of M3 is to equalize the distribution parameters of the MLP outputs and the target training data, possibly aiding the convergence.

4 Experimental setup

In the evaluation, we had eight representative modeling problems: 1. approximation of a modulated sinusoidal function, 2. the same problem with additive normal-distributed noise, 3.

MEMS gas-damper behavior, 4. rounded-stripline-bend parallel capacitance and series inductance vs. device geometries, 5. JFET DC characteristics, 6. spiral-inductor S-parameters vs. geometries, 7. power amplifier output power vs. supply voltage and frequency, and 8. MESFET drain and gate currents vs. bias voltages and temperature.
For each problem, three appropriately sized MLPs (N_h and N_w get three different values as given in Table 1) were utilized. The modeling-problem characterization and corresponding MLPs are shown in Table 1, where N_w is the resulting number of ANN weights, i.e., optimization variables, N_{tr} is the number of training-set samples, and $N_g = N_{tr}N_o$ is the resulting number of optimization goals.

Table 1: Modeling-problem characterization

problem	N_i	N_h	N_o	N_{tr}	N_w	N_g
1	1	{5,10,15}	1	20	{16,31,46}	20
2	1	{5,10,15}	1	20	{16,31,46}	20
3	3	{5,10,15}	1	40	{26,51,76}	40
4	3	{5,10,15}	2	50	{32,62,92}	100
5	2	{5,10,20}	3	306	{33,63,123}	918
6	5	{10,15,25}	5	486	{115,170,280}	2430
7	2	{10,20,30}	1	4667	{41,81,121}	4667
8	3	{10,15,25}	2	37597	{62,92,152}	75194

Each MLP was trained 30 times with each weight-initialization method — M1 with $c = 0.001, 0.005, 0.01, 0.1, 0.5, 1.0, 5.0$ — and E_{tr}, E_{te}, and training CPU time noted in hundred-step increments. The results obtained for each method were averaged over all runs at each value of the optimization cycles. In addition, the standard deviations of the training and test errors were calculated for each problem and method. Finally, the standard deviations were averaged over all the problems at each value of the optimization cycles.
A total number of 6480 runs were carried out by semi-automatic scripts using APLAC 8.2 `ANNModelGenerator` [A06k] on an Ia64 HP Server rx5670 with a 1.3 GHz processor and 4 Gbyte memory.

5 Analysis of results

A set of representative results is shown in Figs. 1–5. The convergence of M1 degraded rapidly with increasing or decreasing c (as in [TF97]), and therefore only the best results (obtained with $c = 0.5$) for M1 are shown.
According to the results obtained, the hypothesis presented is true; comparing the new M3 to M1 (with $c = 0.5$), the training and test errors decreased by 13.6 % and 1.4 %, respectively. The smallest standard deviations for training and test errors show that M3 is also more robust than other methods (41.6 % and 2.1 % improvement, respectively, compared to M1 with $c = 0.5$). The performance improvement is obtained with a slight increase in the training CPU time (7.0 % increase compared to M1 with $c = 0.5$).
M2 forces the AFs to operate in the transition region and improves the convergence when compared to heuristic M1 with other values of c. Thus, one can conclude that when $a = b = 1$, $c = 0.5$, and the training-set inputs and targets are scaled linearly in the range $[-1, 1]$, the AFs are forced, on the average, to operate in the transition region. However, this may not be true with a single modeling problem.

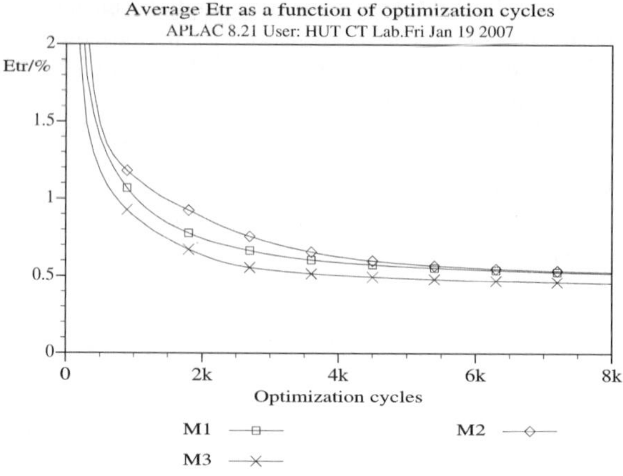

Fig. 1: Average training error vs. optimization cycles

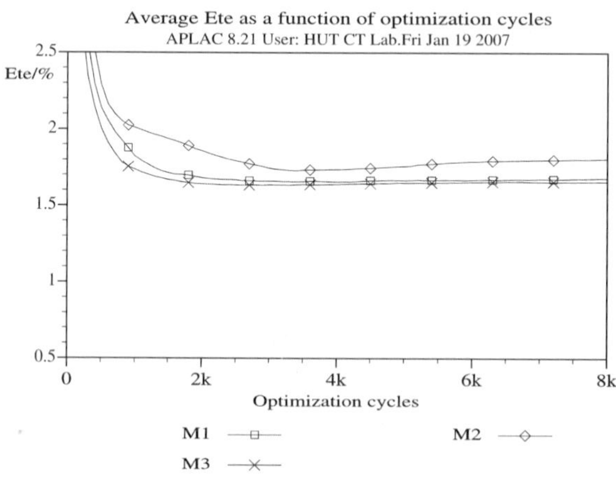

Fig. 2: Average test error vs. optimization cycles

6 Conclusions

Three methods for the initialization of MLP-ANN weights were experimentally evaluated for electrical component modeling applications. A new weight-initialization method was also presented. The methods were evaluated with respect to average training error, test error and training CPU time. Also, the standard deviations of training and test errors were calculated and utilized to analyze robustness of the methods.

According to the results obtained, the hypothesis presented is true: the new method proposed (M3) improves the convergence and robustness of MLP-ANN training for electrical component modeling problems. The performance is improved because the AFs are forced to operate in the transition region and the target training data is scaled so that its distribution parameters correspond to the ones of the MLP outputs. This is not true with the heuristic weight initialization (M1), even though it is possible to find empirically a good value of c for a specific modeling problem.

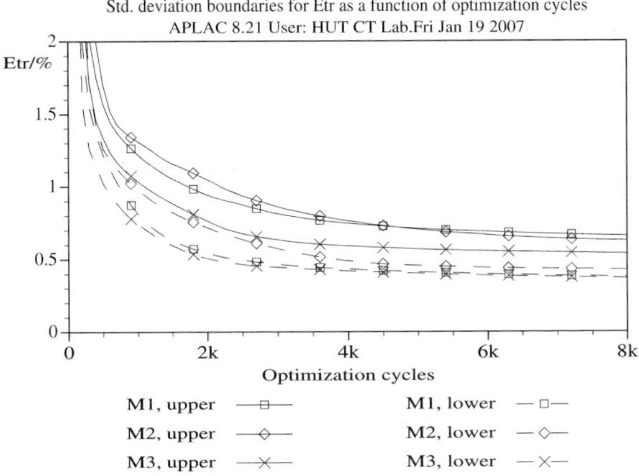

Fig. 3: Standard deviation boundaries for training error vs. optimization cycles

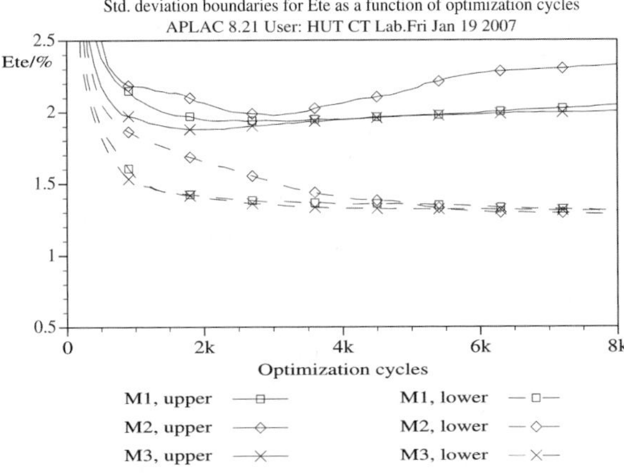

Fig. 4: Standard deviation boundaries for test error vs. optimization cycles

Acknowledgment

This work was funded by Nokia Corporation and AWR-APLAC Corporation through projects TEKES/ELMO/MOSAICS (grants 2078/31/03 and 2440/31/03) and TEKES/MASI/AMAZE (grant 3239/31/05).

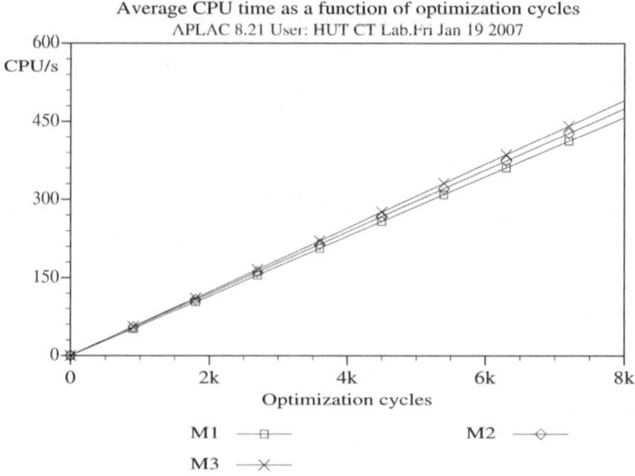

Fig. 5: Average training CPU time vs. optimization cycles

References

[ZG00k] Zhang, Q.J., Gupta, K.C.: Neural Networks for RF and Microwave Design. Artech House, Boston London (2000)

[PAR01] Plebe, A., Anile, A.M., Rinaudo, S.: Sub-micrometer bipolar transistor modeling using neural networks. In: van Rienen, U., Günther, M., Hecht, D. (Eds.): Scientific Computing in Electrical Engineering, Lecture Notes in Computational Science and Engineering, **18**, Springer, Berlin Heidelberg, 259–266 (2001)

[TF97] Thimm, G., Fiesler, E.: High-order and multilayer perceptron initialization. IEEE Trans. on Neural Networks, **2**, 349–359 (1997)

[KRH05k] Kujanpää, T., Roos, J., Honkala, M.: Experimental comparison of optimization methods in ANN training. In: Proc. PRIME 2005, **2**, 430–433 (2005)

[EFP05] Erdogmus, D., Fontenla-Romero, O., Principe, J. C., Alonso-Betanzos, A.: Linear-least-squares initialization of multilayer perceptrons through backpropagation of the desired response. IEEE Trans. on Neural Networks, **2**, 325–337 (2005)

[Hay99k] Haykin, S.: Neural Networks — A Comprehensive Foundation. Prentice Hall, New Jersey (1999)

[A06k] APLAC 8.2 Manuals. AWR-APLAC Corporation (2006)

Trajectory Piecewise Linear Approach for Nonlinear Differential-Algebraic Equations in circuit simulation

T. Voß[1], R. Pulch[2], E.J.W. ter Maten[3], A. El Guennouni[4]

[1]University of Groningen,Faculty of Mathematics and Natural Sciences, Nijenborgh 4, 9747 AG Groningen, The Netherlands, t.voss@rug.nl
[2]Bergische Universitt Wuppertal, Fachbereich Mathematik, Gaustr. 20, 42119 Wuppertal, Germany, pulch@math.uni-wuppertal.de
[3]Philips Semiconductors, High Tech Campus 48, 5656 AE Eindhoven, The Netherlands, jan.ter.maten@philips.com
[4]Magma Design Automation, Eindhoven, The Netherlands

Summary. In this paper we extend the Trajectory Piecewise Linear (TPWL) model order reduction (MOR) method for nonlinear differential algebraic equations (DAE). The TPWL method is based on combining several linear reduced models at different time points, which are created along a typical trajectory, to approximate the full nonlinear model.
We discuss how to select the linearization tuples for linearization and the choice of linear MOR method. Then we study how to combine the local linearized reduced systems to create a global TPWL model. Finally, we show a numerical result.

1 Introduction

Nowadays a lot of circuits which are used in many fields are not only purely digital or analogue. These circuits are a mixture of analogue and digital and are called mixed-signal circuits. For developing these large circuits there is a need of tools which can simulate these circuits efficiently during the design phase as well as during the verification phases. The digital part in mixed-signal designs contains also several sub-circuits that are reused several times. So, simplifying these circuits could give a speed-up for the transient analysis.
To do this we could use MOR methods, which are based on linear or quadratic reduction [PH00] or nonlinear methods, e.g. proper orthogonal decomposition (POD) [VOL99]. However, these methods are mostly developed for weakly nonlinear systems. Therefore these methods are not so useful in circuit simulation, which often deals with highly nonlinear circuits. To overcome this issue, a TPWL [REW03] approach for ordinary differential equations (ODE) was developed. We will show how we can adapt this method to DAEs. We handle here only DAEs of index 1, because a large number of circuits can be modeled with an index 1 DAE.
In the next section we present our TPWL approach for nonlinear DAEs. In Section 3 we show how the method performs in practice. Finally in Section 4 we draw our conclusions.

2 Trajectory Piecewise Linear Model Order Reduction

In this section we discuss how we can apply the TPWL method to a nonlinear DAE which is used to describe the dynamical behavior of a circuit. The DAE system we want to reduce is

$$\frac{d}{dt}\mathbf{q}(t,\mathbf{x}) + \mathbf{j}(t,\mathbf{x}) + \tilde{B}\tilde{\mathbf{u}}(t) = \mathbf{0}, \mathbf{x}(0) = \mathbf{x}_0$$

where $\mathbf{q}, \mathbf{j} : \mathbb{R} \times \mathbb{R}^n \rightarrow \mathbb{R}^n$, $\tilde{B} \in \mathbb{R}^{n \times m}$ and $\tilde{\mathbf{u}} : \mathbb{R} \rightarrow \mathbb{R}^m$. Here $\mathbf{q}$ represents contributions from capacitances and the inductances, $\mathbf{j}$ represents contributions from resistances while $\tilde{B}$ is the input distribution matrix and $\tilde{\mathbf{u}}$ is the given input for the circuit.
The idea behind the TPWL method is to linearize the system at special time points t_i along a typical trajectory. The trajectory itself should represent the full nonlinear behavior of the system. Then we reduce each locally linearized system with a linear model reduction technique and store the basis of each locally reduced subspace S_i. With the help of the S_i we compute a globally reduced subspace S. S is then used as the subspace for all locally linearized systems. The final TPWL model is a weighted sum of all locally linearized reduced systems. The TPWL model can then be solved by a standard DAE time integrator. In the following subsection we show how we apply the described steps.

2.1 Creating the local linearized models

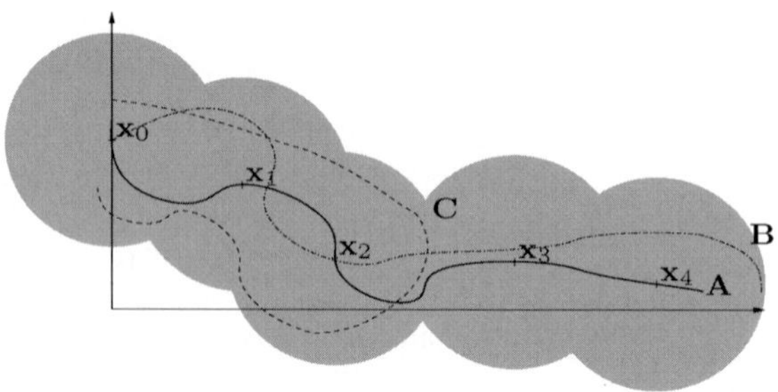

Fig. 1: Creating the linearization tuples, and curves with different sources and/or initial values

The disadvantage of the standard linearization methods is that we can only trust in the results if the solutions stays close to the linearization tuple (LT), time and solution space, around which we have created the linearized model[1]. To overcome this drawback the idea is to take several linearized models to create the TPWL model. These LTs will be taken along a trajectory which represents the typical behavior of that system. If we do this we can trust in the results as long as the solution stays close to one of the LT. Figure 1 illustrates a typical situation. In this situation we have 5 LTs $(x_0, \ldots, x_4)$, which are created along trajectory **A**, and their related accuracy region (the gray region). We can see that trajectory **B** and **C** stay in the accuracy region even if they have different inputs (**B**) or different initial values (**C**). And as long as the trajectories stay in the accuracy region we can also be sure that we have a good approximation of the original system.
We will discuss an approach which will choose as many LTs as needed to reach a given accuracy and as few as possible to get the maximum speed-up in the TPWL model. In [REW03] they propose more accurate methods for selecting the LTs, but these methods are only applicable to ODEs, because they assume that $\frac{\partial \mathbf{q}}{\partial \mathbf{x}}(\mathbf{x}_0, t_0)$ is regular, but since we handle DAEs so $\frac{\partial \mathbf{q}}{\partial \mathbf{x}}(\mathbf{x}_0, t_0)$ is singular.
To get the LTs we need a solution of the original model. But we only need low accuracy of this solution because it is enough if the LTs are close to the exact trajectory. The reason why we need only low accuracy is that we just need the LTs to stay close to the exact trajectory. With this in mind, we see that it is a good idea to include the selection of the LTs directly in a solver of a nonlinear DAE.
Similarly to a step size controller, the accuracy of the actual local reduced model depends on the selection of the future LTs. The reason for this is that for calculating the global reduced subspace we use *all* local subspaces we have created along the trajectory, also future ones. In consequence, we can only make local accuracy assumptions and so, we use a quite simple strategy for selecting a new LT.
From the overview we know that the final TPWL model consists of a weighting of several reduced linearized systems. We create the reduced linearized systems by projecting all of them into the same reduced global subspace. The basis for the global reduced subspace is created by merging all locally reduced subspaces which we got during the creation of the linearized

[1] For simplicity we sometimes omit the time dependency of the LT

models. This is done in such a way that the global reduced subspace represents the most dominant parts of the locally reduced subspaces. So a good approximation for the globally reduced subspace is then just the actual locally reduced subspace. This means that we create the local reduced subspace with a linear model reduction technique and we use this subspace to create the local linearized subspace. Next we simulate both systems, the original and the local linearized reduced system, until the distance between the solutions of both systems is bigger then a given bound. At this point we set a new LT. Algorithm 1 shows the procedure to find LT $i+1$. We continue with this procedure until we have reached the end of the given

---

Algorithm 1 Linearization tuple controller

1. Set an accuracy factor $\varepsilon > 0$
2. Linearize the system around the last (i-th) LT (x_i, t_i). So we get
$$C_i\dot{\mathbf{x}} + G_i\mathbf{x} + B\mathbf{u}(t) = 0$$
where $C_i = \frac{\partial \mathbf{q}}{\partial \mathbf{x}}(\mathbf{x}_i, t_i) \in \mathbb{R}^{n\times n}$, $\det C_i = 0$ and $G_i = \frac{\partial \mathbf{j}}{\partial \mathbf{x}}(\mathbf{x}_i, t_i) \in \mathbb{R}^{n\times n}$. Save C_i, G_i and B.
3. Reduce the i-th linearized system to dimension $r_i \ll n$ with a linear model reduction method, e.g. 'Poor Man's TBR (PMTBR) [PS05], and project the system to this locally reduced subspace which is spanned by P_i. Be aware of that r_i can be different for all different LTs
$$C_i^r\dot{\mathbf{y}} + G_i^r\mathbf{y} + B^r\mathbf{u}(t) = 0$$
where $C_i^r = P_i^\top C_i P_i$, $G_i^r = P_i^\top G_i P_i$ and $B^r = P_i^\top B$ with $P_i \in \mathbb{R}^{n\times r_i}$. $\mathbf{y} \in \mathbb{R}^{r_i}$ is the approximation to $\mathbf{x}$ with $\mathbf{x} \approx P_i\mathbf{y}$. Save P_i.
4. Simulate both the locally linearized reduced system with $\mathbf{y}_0^i = P_i^\top \mathbf{x}_i$ and the original system with a step size determined from the original nonlinear system. If at t the relative distance between the two solutions $\frac{||P_i\mathbf{y}-\mathbf{x}||}{||\mathbf{x}||}$ becomes bigger than ε we set the $(i+1)$-th LT to $(\mathbf{x}, t)$ and go to step 2.

---

trajectory.
An extension to this approach is to calculate several typical trajectories to create a bigger accuracy region. However the more LTs we have, the more memory for saving the TPWL model we need, and the more involving the weighting procedure will be.

2.2 Creating the global reduced subspace

After we have created p linearized systems and p related local reduced subspaces we have to construct the global reduced subspace. We need a global reduced subspace because we want a smooth transition from one accuracy region to another accuracy region while solving the TPWL model. If we had for each local subspace a separate reduced subspace the transition from one to another subspace would be way too difficult.
Let us assume we have p local reduced subspaces which are spanned by $P_i \in \mathbb{R}^{n\times r_i}, i = 0, \ldots, p-1$. The columns P_i span the optimal reduced subspace for the i-th local linearized system. So one idea is to create a new matrix $\tilde{P}$ which contains all columns of the P_i's. So $\tilde{P} := [P_1, \ldots, P_p]$ spans then the union of all reduced subspaces.Of course the columns of $\tilde{P}$ are in general linearly dependent and also the number of columns is in general larger than n, so $\tilde{P}$ is not a good global projection matrix. It is even high likely that several P_i are quite similar because the linearized systems are also. Hence the columns of these P_i are quite dominant in the matrix $\tilde{P}$. To extract the most dominant columns of $\tilde{P}$, and so the most dominant part of the union of the local reduced subspaces, we use a singular value decomposition (SVD) of $\tilde{P} = U\Sigma V^\top$. Then U contains the most dominant columns of $\tilde{P}$, and so from all P_i, ordered

by their importance. As the global reduced subspace we take the one which is spanned by the first r columns of U. With this global reduced subspace we can establish a smooth transition from one local system to another one. Summarizing we obtain Algorithm 2.

Algorithm 2 Creating the global reduced subspace

1. Define $\tilde{P} = [P_1, \ldots, P_p]$.
2. Calculate the SVD of $\tilde{P}$. So $\tilde{P} = U\Sigma V^\top$ with $U = [u_1, \ldots, u_n] \in \mathbb{R}^{n \times n}$,$\Sigma \in \mathbb{R}^{n \times rp}$ and $V \in \mathbb{R}^{rp \times rp}$.
3. Define P as $[u_1, \ldots, u_r]$.
4. Create the p local linearized reduced systems given as $C_{ir}\dot{\mathbf{y}} + G_{ir}\mathbf{y} + B_{ir}\mathbf{u}(t) = \mathbf{0}$ with $C_{ir} = P^\top C_i P$, $G_{ir} = P^\top G_i P$ and $B_{ir} = P^\top B$

2.3 Creating the TPWL reduced order model by weighting

Now we have p locally linearized reduced systems which are all lying in the same global reduced subspace, but we still need to combine them to get a global TPWL model. We do this by calculating a weighted sum of local models

$$\sum_{i=0}^{p-1} w_i(\mathbf{y}) \left(C_{ir}\dot{\mathbf{y}} + G_{ir}\mathbf{y} + B_{ir}\mathbf{u}(t)\right) = 0.$$

To see how we should choose the weights we take a look to a simple example, see Figure 2. In this example we have 3 LTs $\mathbf{x}_0, \mathbf{x}_1$ and $\mathbf{x}_2$ and the related accuracy region, shown as circles. We also have 3 possible trajectory points of the TPWL model. $\mathbf{y}_0$ lies only in the accuracy region of $\mathbf{x}_1$ so the related local system should have the biggest influence to the TPWL model. Hence we should choose $w_1 \approx 1$ and w_0, $w_2 \approx 0$. If we look to $\mathbf{y}_1$ we see that this point lies in the accuracy region related to $\mathbf{x}_1$ and $\mathbf{x}_2$, so we should take a combination of both local models this means that we should choose the weights as following: $w_1 + w_2 \approx 1$ and $w_3 \approx 0$. For $\mathbf{y}_2$ we have the following situation: the solution has left *all* accuracy regions so we should stop the simulation at this point or give at least a warning. A template for a weighting procedure is described in Algorithm 3.

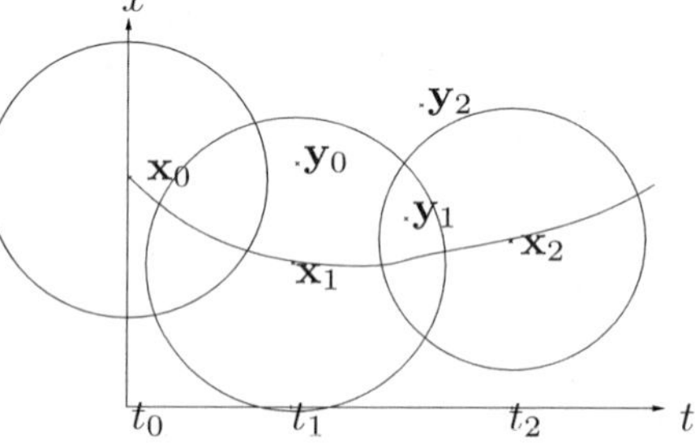

Fig. 2: Simple TPWL model

After calculating the weights we normalize them to get a convex combination of the local linearized reduced systems. If we choose the weight in the way described in the example we get a distance depending weighting scheme as shown in Algorithm 4.

There also is an extended approach which uses instead of the distance an approximation of the linearization error to calculate the weights [VO05]. This approach is more complex because we have to compute an estimate of the Hessian's of $\mathbf{q}$ and $\mathbf{j}$ but in this way we obtain an even better TPWL model.

To summarize we have changed, compared to [REW03], the way how the LTs are chosen. We also tried several linear model order reduction techniques to see which performs the best. Additionally we tried to improve the weighting procedure.

Algorithm 3 Weighting template

given p LTs $(t_{l_i}, \mathbf{y}_{l_i})$, $i = 0, \ldots, p-1$ and $b = 0$
for $i = 0$ to $p - 1$
    if $\mathbf{y}$ lies in the accuracy region of the i-th LT
        $0 \ll w_i \leq 1, b = 1$
    else
        $0 \leq w_i \ll 1$
    end
end
if $b = 0$
    Create warning
end
Such that $\sum_{i=0}^{p-1} w_i = 1$

Algorithm 4 Distance dependent weights

Given actual state $\mathbf{y}$, actual time t, p LTs $(t_{l_i}, \mathbf{y}_{l_i})$ and $\alpha_{\mathbf{y}}, \alpha_t \geq 0$ with $\alpha_{\mathbf{y}} + \alpha_t = 1$

1. For $i = 0, \ldots, p-1$ compute $d_i = \alpha_{\mathbf{y}} \|\mathbf{y} - \mathbf{y}_{l_i}\| + \alpha_t |t - t_{l_i}|$
2. For $i = 0, \ldots, p-1$ calculate $\tilde{w}_i = e^{-\frac{d_i \beta}{m}}$ with $m = \min_{i=0,\ldots,p-1} d_i$, $\beta > 0$
3. Normalize the weights such that the given constraints hold
   $w_i = \frac{\tilde{w}_i}{s}$ with $s = \sum_{i=0}^{p-1} \tilde{w}_i$

3 Example

We show how the TPWL method performs on a practical example. As a test circuit we have chosen a chain of inverters, which consists of 100 inverters, connected in series. The circuit behaves nonlinearly so it is a good test for the TPWL method. Additionally, we have dependencies between all nodes which is also not an optimal behavior for a model reduction process. The DAE which is describing the dynamics of the circuit has 104 states. For selecting the LTs we have used Algorithm 4. For linear model reduction technique we used PMTBR [PS05], that was adapted to deal with our DAE.
In Figure 3 we illustrate the results of our test setup. The upper picture shows the relative error of the TPWL model compared to a highly accurate solution, calculated with a backward differential formula (BDF) method. The lower one compares the solution for the voltage of node 50 of the TPWL method to the solution of the BDF method. As input for both figures we used a slightly different input than for training the TPWL model. We used an input signal with a added sinus wave and/or delayed input signal.
In Table 1 we sum-up the speed-up for the simulation with the same input as the training input. *Extr. time + BDF* is the time it needs to create a TPWL model including the time for BDF method. *Simul. time* is the time needed to simulate the final TPWL model.

	r	# LTs	Extr. time + BDF	Simul. time	max. error	Speed-up	mean error
PMTBR	50	62	240s (220s BDF)	41s	0.037484	5.4	0.012051
PMTBR	40	62	236s (220s BDF)	31s	0.033233	7.2	0.016672
PMTBR	35	62	233s (220s BDF)	27s	0.057046	8.3	0.026128

Table 1: Final LT controller

It can be seen that the relative error is most of the time lower then the given error bound, see Figure 3. For all orders we have to use the same number of LTs (62), which comes from the fact that the local systems only need relatively small subspaces to get the desired accuracy.

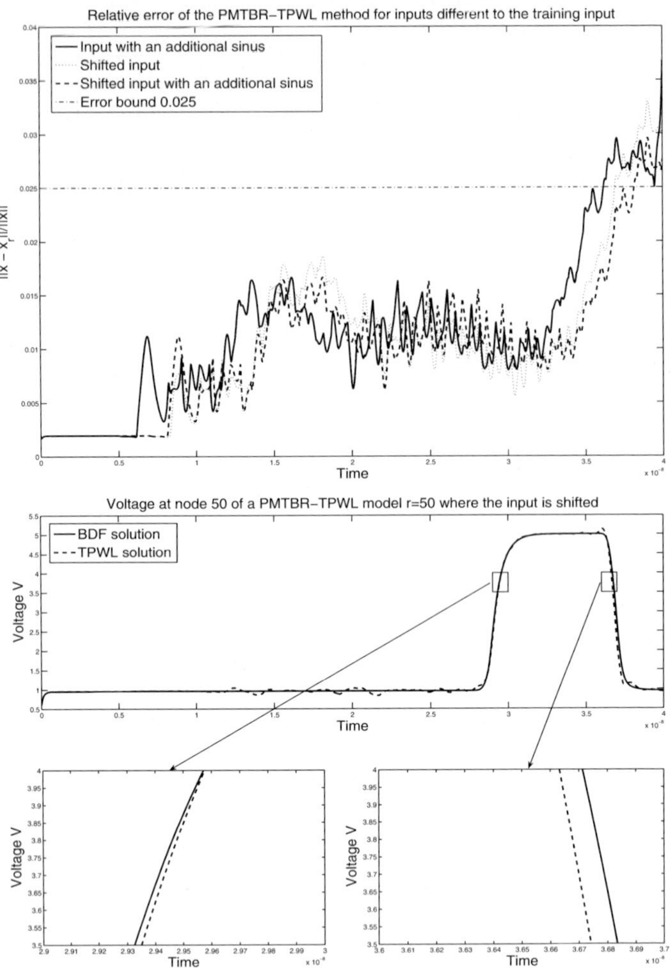

Fig. 3: Relative error of the PMTBR-TPWL method compared to high accurate BDF method (top) Voltage at node 50 compared to a high accurate BDF solution (bottom)

The resulting speed-up is between 5.4 and 8.3 compared to a BDF method, which needs 220s. This speed-up is related to the fact that we are solving linear reduced systems instead of the full nonlinear DAEs, which require a nonlinear solver including Newton iterations. The speed-up would be even bigger if we used a circuit with a lot of devices which have all to be evaluated. We can also notice that in our example the extraction time is quite high, since we have used a high accuracy for the BDF-method during the extraction. We can reduce this time if we use a lower accuracy during the extraction because we only need a rough approximation of the solution to create the TPWL model.
Even if we can reduce the extraction time using a lower accuracy during the extraction time, one can argue that a full simulation of the circuit is just a bit more expensive. This is of course correct but if we think about that we reduce parts of the circuit which are reused a lot, it is easy to see that we still get a quite large speed-up. So we propose to use the TPWL method in a kind of library approach.

4 Conclusion

The TPWL method, applied to nonlinear DAEs, is a promising technique to reduce the simulation time. It has several advantages compared to other methods. First of all we can get a big speed-up in simulation time. We can also use the well-developed linear model reduction techniques. And we are able to create a linearization tuple controller that can be used directly in a BDF method. In future we plan to apply this method to a more practical example, then we will also address open questions, e.g. what happens if the circuit contains a feed back. We also aim to improve the method for selecting the LTs and the weighting procedure and to quantify the speed-up obtained a priori.

Acknowledgment

Thanks go to Ir. Arie Verhoeven, University of Technology Eindhoven, The Netherlands and Prof. Dr. Michael Gnther, University of Wuppertal, Germany.

References

[PS05] Phillips, J., Silveira, L.M.: Poor Man's TBR: A simple model reduction scheme, IEEE transactions on computer-aided design of integrated circuits and systems. Vol. 14 No. 1, 2005

[PH00] Phillips, J.R.: Projection frameworks for model reduction of weakly nonlinear systems, DAC 2000, Los Angeles, California, 2000

[REW03] Rewieński, M.J.: A trajectory piecewise-linear approch to model order reduction of nonlinear dynamical systems, PhD Thesis, Massachusetts Institute of Technology, 2003

[VOL99] Volkwein, S.: Proper Orthogonal Decomposition and Singular Value Decomposition, SFB-Preprint No. 153, University of Graz, 1999

[VO05] Voß, T.: Model reduction for nonlinear differential algebraic equations, MSc. Thesis University of Wuppertal, 2005; Unclassified Report PR-TN-2005/00919, Philips Research Laboratories, 2005

Model Order Reduction of Large Scale ODE Systems: MOR for ANSYS versus ROM Workbench

A.J. Vollebregt[1], T. Bechtold[2], A. Verhoeven[3], E.J.W. ter Maten[2,3]

[1] Bergische Universität Wuppertal
[2] Philips Semiconductors - NXP, Eindhoven
[3] Technical University of Eindhoven

Summary. In this paper we compare the numerical results obtained by different model order reduction software tools, in order to test their scalability for relevant problems of the microelectronic-industry. MOR for ANSYS is implemented in C++ and ROM Workbench is a MATLAB code. We further compare two Arnoldi-based reduction algorithms, which seems to be the most promising for microsystem design applications. The chosen benchmarks are large scale linear ODE systems, which arise from the finite element discretisation of electro-thermal MEMS models.

1 Introduction

Decreasing size of silicon chips and their increasing integration density require permanently new and more powerful simulation tools and strategies in microelectronics and microsystem technology. Model order reduction (MOR) approaches [1] are successfully used to considerably reduce both the computational time and the resources. Mathematical development of MOR is an active area of research, which is growing from the reduction of linear ordinary differential equation systems (ODEs) towards the reduction of parameterized and nonlinear differential-algebraic equations (DAEs) and partial differential-algebraic equations (PDAEs). The implementation aspects of model order reduction are advancing as well. Practical MOR has developed from academic prototyping environments to several strong tools that can be easily used as an extension of the commercial simulators like e. g. ANSYS [2].
In today's age of fast computers it is possible to use quick prototyping tools like MATLAB or Mathematica for convenient implementation and testing of new MOR methods. However, the run time for the usually large-scale industry relevant problems enforces the use of programming languages, like for example C++. Such implementations offer better performances, but also demand more time and programming skills from the developer.
The goal of this paper is to numerically compare two MOR tools, which belong to the mentioned streams: MOR for ANSYS (M4A) [2] and ROM Workbench (RW) [3]. The first was developed at the university of Freiburg, Germany, as an extension to the commercial finite element simulator ANSYS. However, it can be easily coupled to an arbitrary circuit simulator, provided the matrices of the linear dynamical system are exported in the Matrix Market format [4]. Back coupling of the reduced model with the rest of the circuitry might be done by either converting a reduced ODE/DAE system into equivalent electrical circuit, or by enabling a simulator to incorporate a reduced model as a black-box. M4A implements block Arnoldi algorithm from [5] and SOAR from [6]. RW is a MATLAB library of different MOR methods, which has been developed at the University Politehnica of Bucharest, Romania, within the European project CODESTAR. It implements a PRIMA version of block-Arnoldi based on [7]. Both tools are planned for use in the European project COMSON [8], which joins the

efforts of the major European semiconductor companies and academic nodes to develop a demonstrator platform in a software code, that could fulfill the demands of the modern microelectronic industry. Such a comparison will give us a clear understanding up to which size and for what structure of the industrial problem the MATLAB code can be used and at which point one should switch to the compiled language implementation.
In section 2 we prove the equivalence of algorithms [5] and [7]. In section 3 we comment on the implementations within two codes and describe two electro-thermal MEMS (micro-electro-mechanical-systems) devices used as case studies for model order reduction. In section 4, the numerical results for block Arnoldi-based order reduction with both tools are presented. In section 5, we conclude the paper.

2 Block Arnoldi Algorithms

In microsystem simulation, the spatial discretization of computational domain often results in a linear multiple-input multiple-output ODE systems of the form

$$\begin{aligned} C \cdot \dot{\mathbf{x}} + G \cdot \mathbf{x} &= B \cdot \mathbf{u}(t) \\ \mathbf{y} &= L^T \cdot \mathbf{x}\,, \end{aligned} \tag{1}$$

with initial condition $\mathbf{x}(\mathbf{0}) = \mathbf{x}_0$. Here, t is the time variable, $\mathbf{x}(t) \in R^n$ the state vector, $\mathbf{u}(t) \in R^m$ the input excitation vector and $\mathbf{y}(t) \in R^p$ the output measurement vector. G, $C \in R^{n\times n}$ are linear (not depending on $\mathbf{x}$ and t) symmetric and sparse system matrices, $B \in R^{n\times m}$ and $L \in R^{n\times p}$ are (constant) input and output distribution arrays, respectively. n is the dimension of the system and m and p are the number of inputs and outputs.
Model order reduction is based on the projection of (1) onto some low-dimensional subspace. Most MOR methods generate two projection matrices $V, W \in R^{n\times\nu}$, to construct a reduced system of the order ν as

$$\begin{aligned} C_r \cdot \mathbf{z} + G_r \cdot \mathbf{z} &= B_r \cdot \mathbf{u}(t) \\ \mathbf{y}_r &= L_r^T \cdot \mathbf{z}\,, \end{aligned} \tag{2}$$

with $C_r = V^T CW$, $G_r = V^T GW$, $B_r = V^T B$, and $L_r = W^T L$. The ultimate goal of MOR is to find matrices V and W in such a way that $\nu \ll n$, while minimizing the error between the full and the reduced system in either time domain $min||y - y_r||$ or Laplace domain. Furthermore, the stability and passivity of the original system should be preserved in (2).
The basic idea behind the Krylov-subspace based block-Arnoldi algorithm is to transfer (1) into the implicit (left-hand side) formulation

$$\begin{aligned} A\dot{\mathbf{x}} &= \mathbf{x} + R\mathbf{u} \\ \mathbf{y} &= L^T\mathbf{x}\,, \end{aligned} \tag{3}$$

with $A = -(G + s_0C)^{-1}C$, and $R = -(G + s_0C)^{-1}B$, and to write down the transfer function of (3) in the frequency domain, using a Taylor series in s_0 as

$$H(s) = -L^T(I - (s - s_0)A)^{-1}R = \sum_{i=0}^{\infty} m_i(s - s_0)^i, \tag{4}$$

where $m_i = -L^T A^i B$ is called the i-th moment around s_0. One aims to find a reduced system whose transfer function $H_r(s)$ will have the same moments as $H(s)$ up to a degree ν. However, due to numerical instabilities, the moments are not computed explicitly, but via the right-sided Krylov subspace $Kr(A, R, \rho) := span(R, AR, A^2R, \ldots, A^{\rho-1}R)$. Block Arnoldi algorithm generates a single orthonormal basis W for $Kr(A, R, \rho)$ and the system (3) is reduced by projection to

$$\begin{aligned} A_r\dot{\mathbf{z}} &= \mathbf{z} + R_r\mathbf{u} \\ \mathbf{y}_r &= L_r^T\mathbf{z}\,, \end{aligned} \tag{5}$$

with $A_r = W^TAW$, $R_r = W^TR$ and $L_r = W^TL$. The order of (5) is $\nu = \rho \cdot m$. The property of the Krylov subspace is such that the first ν moments of $H_r(s) = -L_r^T(I - (s - s_0)A_r)^{-1}R_r$ and of $H(s)$ are identical.
As the reduced system (5) is not necessarily passive (this means that the system generates no energy, which property is important for applications in circuit simulation), two alternatives to "classical" block-Arnoldi have been suggested: PRIMA algorithm [7] and Freund's Arnoldi [5]. Both are described and compared below.

2.1 PRIMA

The PRIMA algorithm was designed in 1998 to guarantee the passivity of the reduced system. PRIMA [7] stands for Passive Reduced-order Interconnect Macromodeling Algorithm. An orthonormal basis, X, is generated such that span(X) = $Kr(A, R, \rho)$, but X is used for an explicit projection of (1), which means that $C_r = X^TCX$, $G_r = X^TGX$, $B_r = X^TB$ and $L_r = X^TL$. In [7] is proven that for this reduced system the passivity is preserved if C is positively semi-definite and that the first n moments of the transfer function of the original and the reduced system are matched. Introducing the notation $X_k = [x_{km+1}| \ldots |x_{(k+1)m}]$, an implementation of PRIMA can be found in Algorithm 1.

2.2 Freund's Arnoldi

Freund suggests in [5] that vectors which are almost linearly dependent with other vectors in the span of the orthonormal matrix should be eliminated. He calls this method of eliminating vectors *deflation*. His algorithm is vector based, although there is a block structure visible for multiple input multiple output systems. Instead of generating orthonormal blocks the algorithm generates candidate vectors, and each vector $\hat{\mathbf{v}}_k$ that satisfies

$$||\hat{\mathbf{v}}_k|| < DTOL, \tag{6}$$

for some appropriate threshold $DTOL$, is removed. Therefore, the number of vectors per orthonormal block m can be ither smaller then, or equal to, the number of vectors of block $m-1$. If the deflation is omitted we get Algorithm 2.

Algorithm 1
Block-Arnoldi as in PRIMA [7]

1: $\hat{X}_0 = R$
2: **for** $j = 1, \ldots, m$ **do**
3: $\quad \mathbf{x}_j = \hat{\mathbf{x}}_j / ||\hat{\mathbf{x}}_j||$
4: $\quad$ **for** $i = j+1, \ldots, m$ **do**
5: $\quad\quad \hat{\mathbf{x}}_i = \hat{\mathbf{x}}_i - \mathbf{x}_j \mathbf{x}_j^T \hat{\mathbf{x}}_i$
6: $\quad$ **end for**
7: **end for**
8: **for** $k = 1, 2, \ldots, \rho - 1$ **do**
9: $\quad$ Determine $\hat{X}_k = A X_{k-1}$
10: $\quad$ **for** $j = 1, \ldots, k$ **do**
11: $\quad\quad \hat{X}_k = \hat{X}_k - X_{k-j} X_{k-j}^T \hat{X}_k$
12: $\quad$ **end for**
13: $\quad$ **for** $j = km+1, \ldots, (k+1)m$ **do**
14: $\quad\quad \mathbf{x}_j = \hat{\mathbf{x}}_j / ||\hat{\mathbf{x}}_j||$
15: $\quad\quad$ **for** $i = j+1, \ldots, (k+1)m$ **do**
16: $\quad\quad\quad \hat{\mathbf{x}}_i = \hat{\mathbf{x}}_i - \mathbf{x}_j \mathbf{x}_j^T \hat{\mathbf{x}}_i$
17: $\quad\quad$ **end for**
18: $\quad$ **end for**
19: **end for**

Algorithm 2
Freund's Arnoldi ignoring deflation

1: **for** $i = 1, \ldots, m$ **do**
2: $\quad \hat{\mathbf{v}}_i = \mathbf{r}_i$, $\mathbf{r}_i$ being the i-th column of R
3: **end for**
4: **for** $k = 1, 2, \ldots, \nu$ **do**
5: $\quad \mathbf{v}_k = \hat{\mathbf{v}}_k / ||\hat{\mathbf{v}}_k||$
6: $\quad$ Determine $\hat{\mathbf{v}}_{k+m} = A\mathbf{v}_k$
7: $\quad$ **for** $i = 1, \ldots, k$ **do**
8: $\quad\quad \hat{\mathbf{v}}_{k+m} = \hat{\mathbf{v}}_{k+m} - \mathbf{v}_i \mathbf{v}_i^T \hat{\mathbf{v}}_{k+m}$
9: $\quad$ **end for**
10: $\quad$ **for** $i = k-1, \ldots, k-m+1$ **do**
11: $\quad\quad \hat{\mathbf{v}}_{i+m} = \hat{\mathbf{v}}_{i+m} - \mathbf{v}_k \mathbf{v}_k^T \hat{\mathbf{v}}_{i+m}$
12: $\quad$ **end for**
13: **end for**

2.3 Comparison between both Algorithms

We state that the exact results of Algorithm 1 and Algorithm 2 are the same. In other words, Freund's Arnoldi is PRIMA with deflation. We are now going to prove that this is indeed the case.
Proving that the exact results of both algorithms are the same, is equivalent to proving that $X = V$ (V is the orthonormal projection basis of Algorithm 2), i. e. that both methods produce the same space and that for V holds that

$$\forall j,\ 0 < j \leq \nu: \ \mathbf{v}_j = \frac{\tilde{\mathbf{v}}_j}{||\tilde{\mathbf{v}}_j||},\ \tilde{\mathbf{v}}_j = (1 - \sum_{i=1}^{j-1} \mathbf{v}_i \mathbf{v}_i^T)\mathbf{g}(\mathbf{v}_{j-m}), \tag{7}$$

with $\mathbf{g}(\mathbf{v}_{j-m}) = A\mathbf{v}_{j-m}$ if $j > m$ and $\mathbf{g}(\mathbf{v}_{j-m}) = \mathbf{r}_j$ if $0 < j \leq m$, $\mathbf{g} : R^m \to R^m$. (7) states that the columns of V indeed form the wanted Krylov space with orthogonalization, which is the wanted result of the original Arnoldi algorithm for multiple starting vectors. In the following we will call the projection matrix for both algorithms V. Introducing the notation

$$\mathbf{q}(\mathbf{v}_j, \tau, \varphi) = (1 - \sum_{i=\tau}^{\varphi} \mathbf{v}_i \mathbf{v}_i^T)\mathbf{g}(\mathbf{v}_j), \tag{8}$$

we need to prove that for all j between 0 and q we have $\mathbf{v}_j = \frac{\mathbf{q}(v_{j-m},1,j-1)}{||\mathbf{q}(v_{j-m},1,j-1)||}$. Also, introduce the set of invariants

- $\Gamma(p) \equiv \{\forall j,\ 0 < j \leq p-1: \ \mathbf{v}_j = \frac{\mathbf{q}(v_{j-m},1,j-1)}{||\mathbf{q}(v_{j-m},1,j-1)||}\}$, $\Gamma : N \to B = \{false, true\}$.
- $\Lambda(p, \tau, \varphi) \equiv \{\hat{\mathbf{v}}_p = \mathbf{q}(\mathbf{v}_{p-m}, \tau, \varphi - 1)\}$, $\Lambda : N^3 \to B$.
- $\Omega(p, \tau, \varphi, \omega) \equiv \Lambda(p, \tau, \varphi) \wedge \ldots \wedge \Lambda(p + \omega - 1, \tau, \varphi)$, $\Omega : N^4 \to B$.

If we can prove that $\Gamma(\nu+1)$ holds at the end of Algorithm 1 and Algorithm 2 we can conclude that $V = X$. We use Lemma 1 in the proof.

Lemma 1. *If for the set parameters (p, τ, φ) the invariant $\Lambda(p, \tau, \varphi)$ holds, then after*

$$\hat{\mathbf{v}}_p = \hat{\mathbf{v}}_p - \mathbf{v}_\varphi \mathbf{v}_\varphi^T \hat{\mathbf{v}}_p, \tag{9}$$

is executed, $\Lambda(p, \tau, \varphi + 1)$ holds.

Proof. Assume $\Lambda(p, \tau, \varphi)$ holds. Then

$$\hat{\mathbf{v}}_p = \mathbf{q}(\mathbf{v}_{p-m}, \tau, \varphi - 1). \tag{10}$$

Substitute this into (9) we get

$$\begin{aligned}\hat{\mathbf{v}}_p &:= (1 - \mathbf{v}_\varphi \mathbf{v}_\varphi^T)\mathbf{q}(\mathbf{v}_{p-m}, \tau, \varphi - 1) \\ &= (1 - \sum_{j=\tau}^{\varphi-1} \mathbf{v}_j \mathbf{v}_j^T)\mathbf{g}(\mathbf{v}_{p-m}) - \mathbf{v}_\varphi \mathbf{v}_\varphi^T [(1 - \sum_{j=\tau}^{\varphi-1} \mathbf{v}_j \mathbf{v}_j^T)\mathbf{g}(\mathbf{v}_{p-m})] \\ &= (1 - \sum_{j=\tau}^{\varphi} \mathbf{v}_j \mathbf{v}_j^T)\mathbf{g}(\mathbf{v}_{p-m}), \end{aligned} \tag{11}$$

since $v_i^T v_j = 0$ for all i and j, $i \neq j$. □

Theorem 1. *$\Gamma(\nu + 1)$ holds at the end of both algorithms.*

Proof. We start with the observation that $\Gamma(1) \wedge \Omega(1, 1, 1, m)$ holds after line 3 in Algorithm 2 and after line 1 in Algorithm 1. Afterwards, we prove that $\Gamma(k) \wedge \Omega(k, 1, k, m)$ implies $\Gamma(k + 1) \wedge \Omega(k + 1, 1, k + 1, m)$ within the loop. This can be done by using Lemma 1. Then it indeed follows by induction that $\Gamma(\nu + 1)$ holds at the end of both algorithms. The complete proof can be found in [9]. □

3 Implementation and Case Studies

In the previous section we have proved the mathematical equivalence of the generated subspaces, while neglecting the numerical errors. In this section we will comment on the implementation of both algorithms within the software tools MOR for ANSYS and RW and will point out what adjustments we have made to the RW function in order to improve the performance for chosen case-studies.

3.1 MOR for ANSYS

MOR for ANSYS is an extension to the commercial finite element simulator ANSYS. It takes as input a linear ANSYS model (file.full), reduces it and gives as output the matrices of the reduced system (2) in MatrixMarket format. However, for the purpose of the COMSON project it has been adjusted to also take as input the matrices of the arbitrary linear dynamical system (1). The code is a C++ implementation of Algorithm 2. The solve step in line 6 can be done with several forward-backward substitution methods (like LU- and Cholesky decomposition), which are available via the TAUCS-library. The reordering is done with METIS [10].

3.2 ROM Workbench (RW) and its Modification (symRW)

Rom Workbench is written in MATLAB. It implements several MOR methods including the Algorithm 1. The PRIMA function takes as input the matrices of the linear system (1) in MATLAB-(sparse)array format. Unfortunately, the efficiency of this implementation is limited because it has no special treatment for symmetric C and G, as the only available factorization is the LU decomposition of MATLAB with *colamd* as re-ordering scheme. However, the dynamical systems which arise from technical applications, as MEMS or electrical circuits, usually do have symmetric system matrices. As the COMSON Demonstrator Platform [8] should be able to handle a wide variation of industry-relevant problems, we have adjusted the PRIMA function of RW in such a way, that for symmetric matrices the performance is increased. We have implemented Cholesky-decomposition (as Cholesky is at least two times faster than LU decomposition) with *symamd* as re-ordering scheme. In the following we will call our adjusted version symRW. Please note, that it is further possible to generate C code directly from MATLAB function. This would speed up the loops, but one would still need the quick solvers from the TAUCS-library.

3.3 Case Studies

In order to test the presented MOR tools on industry-relevant problems, we have chosen two electro-thermal MEMS devices [11]. The pyrotechnical microthruster is based on the integration of the solid fuel with a silicon micro-machined structure. The thermally tunable optical filter is a Fabry-Perot interferometer fabricated as a free-standing membrane. Both models have been made and meshed in ANSYS, using low and high-order finite elements.

4 Numerical Results

We have reduced the described case studies using MOR for ANSYS, RW and its adjusted version symRW. We have limited opurselfs to the single-input case, so that deflation had no impact and no error control has been used. In Fig. 1 a good match between the step response of the full-scale and that of the reduced order model at a single output node of the pyrotechnical microthruster are displayed. The difference between the reduced model computed with MOR for ANSYS and RW/symRW is neglectable, as expected. In Tables 2 and 1 we compare the reduction time (down to order 30) of M4A, RW and symRW.To analyze the bottlenecks of different implementations we divide the algorithm courses into several phases. These are

- Phase 1: Reading the original matrices into the memory from file and writing the reduced matrices to file.
- Phase 2: Reordering the matrix G.
- Phase 3: Factoring G and constructing the first basis vector.
- Phase 4: Constructing the rest basis vectors via the back substitution in each iteration.

CPU time of RW is up to 60 times longer than the CPU time of MOR for ANSYS. Due to our improvement, this difference has been reduced to 12 times for the largest case study. As expected the main speed up was achieved by introducing Cholesky factorization for the symmetric G and a more effective ordering. The remaining CPU time difference is mainly due to the interpretation overhead in MATLAB.

5 Conclusion

We have compared two software tools (a single MOR routine), which are meant to be integrated into the COMSON demonstrator platform. They belong to the two main implementation streams, fast prototyping in the interpreter environment and the compiled language implementation in C++. We have proven that both algorithms generate the same reduced basis and that the most important bottleneck for MATLAB is the decomposition phase. We have implemented Cholesky factorization for the symmetric problems in RW and have switched to a *symamd* re-ordering. Hence, the present run times in MATLAB allow for testing moderate-size industry-relevant problems within the COMSON demonstrator platform.

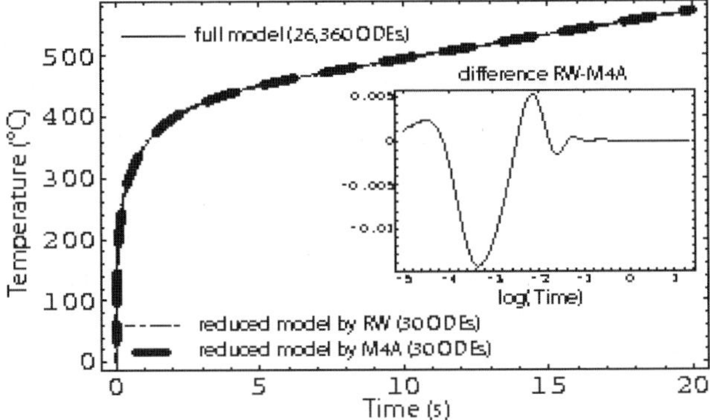

Fig. 1: Step response of the full scale and reduced models (computed with MOR for ANSYS and RW) in a single output node of microthruster (model C).

Table 1: Complete reduction times in s for all the case studies on AMD Opteron with 2.4 GHz and 16 Gb RAM. nnz is the number of nonzero matrix elements of G and its factor l (from the LU decomposition).

Model			M4A		RW		symRW	
	n	nnz(G)	time	nnz(l)	time	nnz(l)	time	nnz(l)
A	1668	6.21e3	0.19	2.46e4	0.2	3.39e4	0.2	2.32e4
B	106437	1.41e6	45.5	1.89e7	402	4.82e7	114	2.84e7
C	26360	2.65e5	7.34	5.00e6	216	1.68e7	53.8	1.06e7
D	79171	2.22e6	98.5	4.56e7	6600	1.88e8	1660	1.24e8

Table 2: Computational times in seconds on AMD Opteron with 2.4 GHz and 16 Gb RAM.

Model A	M4A	RW	symRW
Phase 1	0.08	0.11	0.06
Phase 2	0.02	0.01	0.03
Phase 3	0.05	0.02	0.01
Phase 4	0.02	0.06	0.01

Model B	M4A	RW	symRW
Phase 1	22.4	11	10.4
Phase 2	3.42	2.35	2.21
Phase 3	10.5	349	73
Phase 4	9.22	39.8	28.5

Model C	M4A	RW	symRW
Phase 1	2.34	1.09	1.07
Phase 2	0.465	1.31	0.28
Phase 3	2.52	202	44.8
Phase 4	2.02	11.3	7.65

Model D	M4A	RW	symRW
Phase 1	3.54	16.8	16.6
Phase 2	4.04	7.63	2.50
Phase 3	41.9	6460	1560
Phase 4	17.1	113	76.8

Acknowledgment

We would like to thank Dr. Evgenii B. Rudnyi from the University of Freiburg for helping us with MOR for ANSYS, Prof. Dr. Gabriela Ciuprina from the University Politehnica of Bucharest for her help with ROM Workbench and to acknowledge the EU support through the COMSON project.

References

1. A.C. Antoulas: *Approximation of Large-Scale Dynamical Systems*, SIAM, 2005.
2. http://www.imtek.de/simulation/mor4ansys
3. http://www.imek.be/codestar
4. http://math.nist.gov/MatrixMarket/
5. R.W. Freund: *Krylov-subspace methods for reduced order modeling in circuit simulation*, Journal of Comp. and Appl. Math., Vol. 123, pp. 395–421, 2000.
6. Z.J. Bai, K. Meerbergen, Y. F. Su: *Arnoldi methods for structure-preserving dimension reduction of second-order dynamical systems, in P. Benner, V. Mehrmann, D. Sorensen (eds), Dimension Reduction of Large-Scale Systems, Lecture Notes in Computational Science and Engineering*, Springer-Verlag, Berlin/Heidelberg, Germany, 2005
7. A. Odabasioglu, M. Celik, T. Pileggi: *PRIMA: Passive Reduced-order Interconnect Macromodeling Algorithm*, IEEE Trans. Comp. Aid. Design Int. Circ. Syst., Vol. 17, pp. 645-654, 1998.
8. http://www.comson.org
9. S. Vollebregt, T. Bechtold, A. Verhoeven, E.J.W. ter Maten: *Model Order Reduction of Large Scale ODE Systems: MOR for ANSYS versus ROM Workbench*, CASA-Report 06-38, 2006, ftp://ftp.win.tue.nl/pub/rana/rana06-38.pdf.
10. G. Karypis, V. Kumar: *A Fast and High Quality Multilevel Scheme for Partitioning Irregular Graphs*, Technical Report TR 95-035, Department of Computer Science, University of Minnesota, 1995.
11. Oberwolfach Model Reduction Benchmark Collection, http://www.imtek.de/simulation

Adjoint Transient Sensitivity Analysis in Circuit Simulation

Z. Ilievski[1], H. Xu[1], A. Verhoeven[1], E.J.W. ter Maten[1,2], W.H.A. Schilders[1,2] and R.M.M. Mattheij[1]

[1] Technische Universiteit Eindhoven; e-mail: Z.Ilievski@tue.nl
[2] NXP Semiconductors Eindhoven

Summary. Sensitivity analysis is an important tool that can be used to assess and improve the design and accuracy of a model describing an electronic circuit. Given a model description in the form of a set of differential-algebraic equations it is possible to observe how a circuit's output reacts to varying input parameters, which are introduced at the requirements stage of design. In this paper we consider the adjoint method more closely. This method is efficient when the number of parameters is large. We extend the transient sensitivity work of Petzold et al., in particular we take into account the parameter dependency of the dynamic term. We also compare the complexity of the direct and adjoint sensitivity and derive some error estimates. Finally we sketch out how Model Order Reduction techniques could be used to improve the efficiency of adjoint sensitivity analysis.

Keywords – Sensitivity Analysis; Transient Analysis; Adjoint Method; Model Order Reduction

1 Introduction

A typical quantity in circuit analysis is the product of voltage difference times the current through an electronic component (power) and, when integrated of time, this reflects the total power that is dissipated. Another time domain problem is the determination of the time moment when a certain unknown, or an expression, crosses a particular value. Such a moment can be the moment at which synchronization is required in co-simulation between a circuit simulator and another simulation tool.
In transient analysis, the adjoint method can be formulated as a convolution of the circuit equations with a carefully constructed function, that, by its nature, requires a backward integration in time of a related DAE (and for which a proper initial value has to be determined). The method has been popularized in [8, 12] for linear problems. For more general DAEs the method has been studied in [7] in a more mathematical way.
In [16] the application to the nonlinear DAEs of circuit equations was studied more closely. Nice applications can be derived for the problem of finding optimal sources in detecting faults in analog circuits [5]. However, in studying sizing problems (in which for instance the physical area of a capacitor has to be taken into account), it appears that especially parameters of capacitors give rise to terms that require additional investigation. Here the effect of the index of the related DAE shows up.
Apart from purposes of optimization, adjoint systems are of interest in determining optimal reduced order models [3, 10], in which case a large number of parameters occurs. Because the adjoint systems are linear the equations themselves can be made subject to a reduced order modeling process.
We will describe ways how to calculate sensitivities in a stable and an efficient way.

2 Transient sensitivity analysis

Equation (1) is a general Differential-Algebraic Equation (DAE) that can be used to describe how any circuit behaves over a period of time. In Modified Nodal Analysis [12], $\mathbf{x}(t) \in \mathbb{R}^N$ is the state vector and represents the node voltages and the currents through voltage sources and inductors, $\mathbf{j}$ and $\mathbf{q}$ are vector functions that describe the current and charge (capacitors) or flux (inductors) behavior. All source values are comprised in $\mathbf{s}(t)$

$$\frac{\mathrm{d}}{\mathrm{d}t}[\mathbf{q}(\mathbf{x}(t))] + \mathbf{j}(\mathbf{x}(t)) = \mathbf{s}(t). \tag{1}$$

The initial solution at $t = 0$, the DC-solution $\mathbf{x}_{\mathrm{DC}}$, satisfies

$$\mathbf{j}(\mathbf{x}_{\mathrm{DC}}) = \mathbf{s}(0). \tag{2}$$

Applying Euler-Backward time integration between time points t_n and $t_{n+1} = t_n + \Delta t$ enables to calculate $\mathbf{x}^{n+1}$ as approximation at t_{n+1}:

$$\frac{1}{\Delta t}[\mathbf{q}(\mathbf{x}^{n+1}) - \mathbf{q}(\mathbf{x}^{n})] + \mathbf{j}(\mathbf{x}^{n+1}) - \mathbf{s}(t_{n+1}) = 0 \tag{3}$$

A Newton-Raphson procedure involves the coefficient matrix $\mathbf{Y} = \frac{1}{\Delta t}\mathbf{C} + \mathbf{G}$, in which $\mathbf{C} = \partial \mathbf{q}/\partial \mathbf{x}$ and $\mathbf{G} = \partial \mathbf{j}/\partial \mathbf{x}$. Making explicit that the equations and their solution depend on a parameter $\mathbf{p} \in \mathbb{R}^P$ we will write

$$\frac{\mathrm{d}}{\mathrm{d}t}[\mathbf{q}(\mathbf{x}(t,\mathbf{p}),\mathbf{p})] + \mathbf{j}(\mathbf{x}(t,\mathbf{p}),\mathbf{p}) = \mathbf{s}(t,\mathbf{p}). \tag{4}$$

By adjusting these parameters it is possible to optimize the behavior of a required functionality. The sensitivity of $\mathbf{x}(t,\mathbf{p})$ with respect to $\mathbf{p}$ is denoted by $\hat{\mathbf{x}}(t,\mathbf{p}) \equiv \partial \mathbf{x}(t,\mathbf{p})/\partial \mathbf{p} = (\partial x_i(t,\mathbf{p})/\partial p_j) \in \mathbb{R}^{N\times P}$, and similarly for $\hat{\mathbf{x}}_{\mathrm{DC}}(\mathbf{p})$. After solving (3), and saving of the LU-decomposition of the matrix $\mathbf{Y} = \mathbf{LU}$, the sensitivity $\hat{\mathbf{x}}^{n+1}(\mathbf{p}) \approx \hat{\mathbf{x}}(t_{n+1},\mathbf{p})$ may be calculated by recursion [9, 11]

$$\hat{\mathbf{x}}^{n+1}(\mathbf{p}) = \mathbf{Y}^{-1}\mathbf{f}, \quad \text{in which} \tag{5}$$

$$\mathbf{f} = -\frac{1}{\Delta t}\left[\frac{\partial \mathbf{q}}{\partial \mathbf{p}}^{n+1} - \frac{\partial \mathbf{q}}{\partial \mathbf{p}}^{n}\right] - \frac{\partial \mathbf{j}}{\partial \mathbf{p}}^{n+1} + \frac{\partial \mathbf{s}}{\partial \mathbf{p}}^{n+1} + \frac{1}{\Delta t}\mathbf{C}\hat{\mathbf{x}}^{n}(\mathbf{p}). \tag{6}$$

The vector $\mathbf{f}$ requires $\mathcal{O}(PN^2)$ operations for the last term in addition to $\mathcal{O}(PN)$ evaluations for a term like $\frac{\partial \mathbf{q}}{\partial \mathbf{p}}$ etc... For simplicity we assume full matrices. Solving the system requires an additional $\mathcal{O}(PN^2)$ operations.
A more general basic observation function is denoted by $\mathbf{F}(\mathbf{x}(t,\mathbf{p}),\mathbf{p}) \in \mathbb{R}^F$ from which other observation functions can be obtained, like

$$\mathbf{G}(\mathbf{x}(\mathbf{p}),\mathbf{p}) = \int_0^T \mathbf{F}(\mathbf{x}(t,\mathbf{p}),\mathbf{p})\mathrm{d}t. \tag{7}$$

From (7) we derive

$$\frac{\mathrm{d}}{\mathrm{d}\mathbf{p}}\mathbf{G}(\mathbf{x}(\mathbf{p}),\mathbf{p}) = \int_0^T \left(\frac{\partial \mathbf{F}}{\partial \mathbf{x}} \cdot \hat{\mathbf{x}} + \frac{\partial \mathbf{F}}{\partial \mathbf{p}}\right)\mathrm{d}t. \tag{8}$$

If $\frac{\partial \mathbf{F}}{\partial \mathbf{x}}$ can be determined rather cheaply in (8), the main emphasis in sensitivity analysis is in the efficient calculation of $\hat{\mathbf{x}}$, or even in efficiently calculating the inner-product $\frac{\partial \mathbf{F}}{\partial \mathbf{x}} \cdot \hat{\mathbf{x}}$. Note that, from (5), we derive $\frac{\partial \mathbf{F}}{\partial \mathbf{x}} \cdot \hat{\mathbf{x}} = \frac{\partial \mathbf{F}}{\partial \mathbf{x}} \cdot \mathbf{Y}^{-1}\mathbf{f} = [\mathbf{Y}^{-T}[\frac{\partial \mathbf{F}}{\partial \mathbf{x}}]^T]^T\mathbf{f}$, which can be calculated in $\mathcal{O}(\min(F,P)N^2 + FPN)$ operations (in addition to those already mentioned above: the overall leading N^2-term still has coefficient P). This is a direct, forward, analysis.

When, additionally, some library for evaluating $\mathbf{q}$, $\mathbf{j}$, or $\mathbf{s}$, does not allow symbolic differentiation, here also a symmetric finite difference will be made (at the cost of two additional evaluations for each quantity $\frac{\mathbf{dq}}{\mathbf{dp}} \approx \frac{\mathbf{q}(\mathbf{p}+\mathbf{\Delta p})-\mathbf{q}(\mathbf{p}-\mathbf{\Delta p})}{2\mathbf{\Delta p}}$). This means that at each interior time point of (8) the integrand will have an error $\mathcal{O}(||\Delta\mathbf{p}||^2)$ (assuming this discretization error is dominant). A quadrature rule like the Trapezoidal Rule adds up to $\mathcal{O}(||\Delta\mathbf{p}||^2/\Delta t)$ leading to $||\Delta\mathbf{p}|| = o(\sqrt{\Delta t})$ if $\Delta t \to 0$ and no persistent errors in sensitivities are wanted. In the sequel, we now consider an approach based on (backward) adjoint integration [7]. We differentiate (4) w.r.t. $\mathbf{p}$ and multiply the result with a function $\lambda^\star(t) \in \mathbb{R}^{F\times N}$ (in which the * means transpose), yielding

$$0 = \int_0^T \lambda^\star(t)\Big[\frac{\mathrm{d}}{\mathrm{d}t}\frac{\mathbf{dq}}{\mathbf{dp}} + \frac{\mathbf{dj}}{\mathbf{dp}} - \frac{\partial\mathbf{s}}{\partial\mathbf{p}}\Big]\mathrm{d}t \tag{9}$$

$$= \Big[\lambda^\star(t)\frac{\mathbf{dq}}{\mathbf{dp}}\Big]|_0^T + \int_0^T \Big[-\frac{\mathrm{d}\lambda^\star}{\mathrm{d}t}\frac{\mathbf{dq}}{\mathbf{dp}} + \lambda^\star\Big(\frac{\mathbf{dj}}{\mathbf{dp}} - \frac{\partial\mathbf{s}}{\partial\mathbf{p}}\Big)\Big]\mathrm{d}t$$

$$= \Big[\lambda^\star(t)\frac{\mathbf{dq}}{\mathbf{dp}}\Big]|_0^T + \int_0^T \Big[-\frac{\mathrm{d}\lambda^\star}{\mathrm{d}t}\Big(\frac{\partial\mathbf{q}}{\partial\mathbf{x}}\cdot\frac{\partial\mathbf{x}}{\partial\mathbf{p}} + \frac{\partial\mathbf{q}}{\partial\mathbf{p}}\Big) + \lambda^\star\Big(\frac{\partial\mathbf{j}}{\partial\mathbf{x}}\cdot\frac{\partial\mathbf{x}}{\partial\mathbf{p}} + \frac{\partial\mathbf{j}}{\partial\mathbf{p}} - \frac{\partial\mathbf{s}}{\partial\mathbf{p}}\Big)\Big]\mathrm{d}t$$

$$= \Big[\lambda^\star(t)\frac{\mathbf{dq}}{\mathbf{dp}}\Big]|_0^T + \int_0^T \Big[-\frac{\mathrm{d}\lambda^\star}{\mathrm{d}t}\Big(\mathbf{C}\hat{\mathbf{x}} + \frac{\partial\mathbf{q}}{\partial\mathbf{p}}\Big) + \lambda^\star\Big(\mathbf{G}\hat{\mathbf{x}} + \frac{\partial\mathbf{j}}{\partial\mathbf{p}} - \frac{\partial\mathbf{s}}{\partial\mathbf{p}}\Big)\Big]\mathrm{d}t$$

$$= \Big[\lambda^\star(t)\frac{\mathbf{dq}}{\mathbf{dp}}\Big]|_0^T + \int_0^T \Big[-\Big(\frac{\mathrm{d}\lambda^\star}{\mathrm{d}t}\mathbf{C} - \lambda^\star\mathbf{G}\Big)\hat{\mathbf{x}} - \frac{\mathrm{d}\lambda^\star}{\mathrm{d}t}\frac{\partial\mathbf{q}}{\partial\mathbf{p}} + \lambda^\star\Big(\frac{\partial\mathbf{j}}{\partial\mathbf{p}} - \frac{\partial\mathbf{s}}{\partial\mathbf{p}}\Big)\Big]\mathrm{d}t \tag{10}$$

This result holds for any $\lambda^\star$. We now consider some choices.

2.1 Backward, adjoint sensitivity for $\frac{\mathbf{d}}{\mathbf{dp}}\mathbf{G}(\mathbf{x}(\mathbf{p}),\mathbf{p})$

In (8) we encounter the product $\frac{\partial\mathbf{F}}{\partial\mathbf{x}}\cdot\hat{\mathbf{x}}$. Equation (10) will enable us to get rid of the $\hat{\mathbf{x}}$, which does not need to be calculated explicitly. We choose $\lambda(t) \in \mathbb{R}^{N\times F}$ appropriately and require that $\lambda(t)$ satisfies the linear 'adjoint' DAE (we assume the index 1 case, which is not trivial [4]).

$$\mathbf{C}^\star\frac{\mathrm{d}\lambda}{\mathrm{d}t} - \mathbf{G}^\star\lambda = -\Big(\frac{\partial\mathbf{F}}{\partial\mathbf{x}}\Big)^\star. \tag{11}$$

This does not yet make $\lambda(t)$ unique, because we did not specify the initial value yet. But we are now able to express $\frac{\partial\mathbf{F}}{\partial\mathbf{x}}\cdot\hat{\mathbf{x}}$ in terms of $[\mathbf{C}^\star\frac{\mathrm{d}\lambda}{\mathrm{d}t} - \mathbf{G}^\star\lambda]\cdot\hat{\mathbf{x}}$, after which we can apply (10).

$$\frac{\mathrm{d}}{\mathbf{dp}}\mathbf{G}(\mathbf{x}(\mathbf{p}),\mathbf{p}) = \int_0^T \Big(\frac{\partial\mathbf{F}}{\partial\mathbf{x}}\cdot\hat{\mathbf{x}} + \frac{\partial\mathbf{F}}{\partial\mathbf{p}}\Big)\mathrm{d}t$$

$$= \int_0^T \Big(\Big[-\frac{\mathrm{d}\lambda^\star}{\mathrm{d}t}\mathbf{C} + \lambda^\star\mathbf{G}\Big]\cdot\hat{\mathbf{x}} + \frac{\partial\mathbf{F}}{\partial\mathbf{p}}\Big)\mathrm{d}t$$

$$= -\Big[\lambda^\star(t)\frac{\mathbf{dq}}{\mathbf{dp}}\Big]|_{t=0}^T + \int_0^T \Big(\frac{\mathrm{d}\lambda^\star}{\mathrm{d}t}\frac{\partial\mathbf{q}}{\partial\mathbf{p}} - \lambda^\star\Big(\frac{\partial\mathbf{j}}{\partial\mathbf{p}} - \frac{\partial\mathbf{s}}{\partial\mathbf{p}}\Big) + \frac{\partial\mathbf{F}}{\partial\mathbf{p}}\Big)\mathrm{d}t. \tag{12}$$

The first term involves $\hat{\mathbf{x}}(0) = \hat{\mathbf{x}}_{\mathrm{DC}}$, which we know. However, if $\lambda^\star(T)\frac{\partial\mathbf{q}}{\partial\mathbf{x}}(T) \neq 0$, one still needs $\hat{\mathbf{x}}(T)$, which we did not want to determine explicitly. Hence good choices to define the initial conditions for $\lambda(t)$ are

$$\mathbf{C}(T) = \frac{\partial\mathbf{q}}{\partial\mathbf{x}}(T) \neq 0 \Longrightarrow \lambda(T) = 0, \quad \text{or if} \tag{13}$$

$$\mathbf{C}(t) = \frac{\partial\mathbf{q}}{\partial\mathbf{x}}(t) \equiv 0 \Longrightarrow \lambda(T) \equiv \lambda_{\mathrm{DC}}, \quad \text{with} \quad -\mathbf{G}^\star\lambda_{\mathrm{DC}} = -\Big(\frac{\partial\mathbf{F}}{\partial\mathbf{x}}\Big)^\star\Big|_{t=0} \tag{14}$$

Note that (13) is only allowed for DAEs of index up to 1. With this choice (12) simplifies to

$$\frac{\mathrm{d}}{\mathrm{d}\mathbf{p}}\mathbf{G}(\mathbf{x}(\mathbf{p}),\mathbf{p}) = \lambda^\star(0)\left[\frac{\partial \mathbf{q}}{\partial \mathbf{x}}(0)\cdot\hat{\mathbf{x}}_{\mathrm{DC}} + \frac{\partial \mathbf{q}}{\partial \mathbf{p}}(0)\right] + \int_0^T \left(\frac{\mathrm{d}\lambda^\star}{\mathrm{d}t}\frac{\partial \mathbf{q}}{\partial \mathbf{p}} - \lambda^\star\left(\frac{\partial \mathbf{j}}{\partial \mathbf{p}} - \frac{\partial \mathbf{s}}{\partial \mathbf{p}}\right) + \frac{\partial \mathbf{F}}{\partial \mathbf{p}}\right)\mathrm{d}t. \tag{15}$$

The circuit that describes the adjoint system (11) is described in more details in [8, 12]. There the above expression is derived by conservation properties based on the Kirchhoff Laws which imply Tellegen's Theorem. Note that (9) is just the time integral of inner-products of "branch-currents$_{\mathrm{AdjointCircuit}}$ * branch-voltages$_{\mathrm{OriginalCircuit}}$".
The formulation used above follows [7], but reveals more closely the effect of a non-trivial function $\mathbf{q}$ in (12) that explicitly depends on $\mathbf{p}$. By this, also the velocity $\frac{\mathrm{d}\lambda^\star}{\mathrm{d}t}$ will be needed. Already a linear function $\mathbf{q}(\mathbf{x},\mathbf{p}) = \mathbf{C}(\mathbf{p})\mathbf{x}$ requires this. Because of the DAE-nature of the problem (1), in [16], $\frac{\mathrm{d}\lambda^\star}{\mathrm{d}t}$ was estimated by symmetric finite differences in the interior of the interval and with one-sided approximations at the boundaries. When, as before, a library for evaluating $\mathbf{q}$, $\mathbf{j}$, or $\mathbf{s}$, does not allow symbolic differentiation, here also a symmetric finite difference will be made (at the cost of two additional evaluations for each quantity). This means that at each interior time point the integrand will have an error $\mathcal{O}(\Delta t^2 + ||\Delta\mathbf{p}||^2)$. Any quadrature rule (like the Trapezoidal Rule) may add these errors up to an error $\mathcal{O}(\Delta t + ||\Delta\mathbf{p}||^2/\Delta t)$ which is additional to the error of the quadrature rule, which means that one will require $||\Delta\mathbf{p}|| = \mathcal{O}(\Delta t)$ (note that a one-sided difference will even need $||\Delta\mathbf{p}|| = \mathcal{O}(\Delta t^2)$). It also shows that the Trapezoidal Rule may not give better results than simple, first order, Euler integration.
System (11) can be determined by integrating backwards in time after $\mathbf{x}(t,\mathbf{p})$ has been determined in the nominal analysis. This backward time integration of (11) requires Jacobian matrices $\frac{1}{\Delta t}\mathbf{C}+\mathbf{G}$ (assuming Euler backwards), similar as in the forward analysis. When the same step sizes are used and the LU-decompositions of the Jacobian matrices at the converged values have been saved from the nominal analysis, the transposed decompositions can be reused for the Jacobians when integrating (11) (one may also save approximative inverse matrices, or preconditioning matrices). However we will assume that one will re-decompose them during the backward integration.
Note that in [6] it is remarked that if the same time step is used in the forward (for $\mathbf{x}$) and backward analysis (for λ) this may give rise to very inaccurate solutions for λ. This general step size approach introduces effects due to interpolation (effects which we have not yet studied).
Let $\mathcal{W} = \mathcal{O}(N^\alpha)$ represent the number of operations for the LU-decompositions with $1 \leqslant \alpha \leqslant 2$ for sparse systems and $\alpha = 3$ for full systems. Note that $\lambda^\star(0)\frac{\partial \mathbf{q}}{\partial \mathbf{x}}(0)\cdot\hat{\mathbf{x}}_{\mathrm{DC}} = [\mathbf{C}^\star\lambda(0)]^\star\hat{\mathbf{x}}_{\mathrm{DC}}$, which can be handled similarly as in adjoint sensitivity analysis in the DC-problem, without explicitly calculating $\hat{\mathbf{x}}_{\mathrm{DC}}$, in $\mathcal{W} + \mathcal{O}(\min(F,P)N^2 + FPN)$ operations (assuming that $\mathbf{G}$ has been decomposed again at $t = 0$). Each time integration step of (11) requires $\mathcal{W} + \mathcal{O}(FN^2)$ operations, after which the integrand in (15) at each time point requires $\mathcal{O}(PN + FP)$ evaluations and $\mathcal{O}(FPN)$ additional operations. In practise $F \ll P$, which (apart from the $\mathcal{W}$ term) makes (15) more efficient as the direct, forward method for (5)-(6).
Reduction of the $\mathcal{W}$ term is discussed in Section 2.2.

2.2 MOR applied to the global adjoint sensitivity equations

The main burden of the backward adjoint sensitivity equations still is the $\mathcal{W} = \mathcal{O}(N^\alpha)$ work needed for the LU-decompositions in the case $\alpha = 3$ when integrating backwards in time for the adjoint equations for $\lambda(t)$. In order to reduce this we observe that in the interior we only need to know $\lambda(t)$ for coordinates where $\frac{\partial \mathbf{q}}{\partial \mathbf{p}}$, $\frac{\partial \mathbf{j}}{\partial \mathbf{p}}$, and $\frac{\partial \mathbf{s}}{\partial \mathbf{p}}$ are non trivial. However, at

$t = 0$ also for the nontrivial rows of $\frac{\partial \mathbf{q}}{\partial \mathbf{x}}(0)$ coordinates of $\lambda(t)$ should be known (which may significantly increase the number of needed coordinates). More precisely, in (15) a term like $\lambda^\star(0)\frac{\partial \mathbf{q}}{\partial \mathbf{x}}(0) \cdot \hat{\mathbf{x}}_{\text{DC}} \in \mathbb{R}^{F\times P}$ shows that the total number of output is $r' = F \times P$, and when $r' \ll N$, one may think to apply MOR.

We observe that there is a tentative opportunity to apply Proper Orthogonal Decomposition (POD) [1, 13], since the forward time integration to determine $\mathbf{x}(t)$ delivered a nice series of snapshots $\{\mathbf{x}(t_0), \ldots, \mathbf{x}(t_N)\}$ (and, even cheaply, also of $\frac{\partial \mathbf{F}}{\partial \mathbf{x}}(t_k)$). With POD a matrix $\mathbf{V} \in \mathbb{R}^{N\times r}$ is found such that $\mathbf{x} \approx \mathbf{V}\tilde{\mathbf{x}}$, $\tilde{\mathbf{x}} \in \mathbb{R}^r, r \ll N$, in which $\mathbf{V}$ is time independent and similar snapshots could have been obtained from a modified problem

$$\frac{\mathrm{d}}{\mathrm{d}t}[\mathbf{V}^T\mathbf{q}(\mathbf{V}\tilde{\mathbf{x}}(t))] + \mathbf{V}^T\mathbf{j}(\mathbf{V}\tilde{\mathbf{x}}(t)) = \mathbf{V}^T\mathbf{s}(t). \tag{16}$$

For (16) the matrices are of size $r \times r$, which indicates a nice option for MOR. In this paper we apply the $\mathbf{V^T}, \mathbf{V}$ matrices from the POD directly to the adjoint problem (11). Note that in this approach we can neglect the dependency of $\mathbf{V}$ on $\mathbf{p}$. For each $\mathbf{p}$ one integrates (1) and calculates the snapshots, resulting in a POD Model Order reduction projection matrix $\mathbf{V}$. With $\mathbf{V}$ we can reduce the system of DAEs for λ (in which only the error due to POD matters) resulting in $\lambda_{POD}(t) \sim \lambda(t)$ (please see [13] for error estimates). Note that POD can also be used to reduce (1) itself. In this case the dependency of $\mathbf{V}$ on $\mathbf{p}$ introduces an additional error in the procedure that can not be neglected. This last approach could further reduce costs and has to be studied further. In [2] it was shown that (in general) POD not directly applies to DAEs. Here a Least-Square POD remedy was introduced that can be applied to the linear DAE (11).

Alternatively to POD, during the forward integration one could additionally determine projection matrices for the Trajectory Piecewise-Linear Method [14, 15]. Next, similarly as to the POD case, a global matrix $\mathbf{V}$ is determined that allows for Piecewise-Linear MOR. In this case the reduced DAE problems can directly be solved [15].

3 Results

p=$l_R(m)$	**dG/dp**	**dG/dp** $withPOD$
0.02	-0.235 $\times 10^{-10}$	-0.235 $\times 10^{-10}$
0.021	-0.247 $\times 10^{-10}$	-0.247 $\times 10^{-10}$
0.022	-0.256 $\times 10^{-10}$	-0.256 $\times 10^{-10}$

Table 1. Sensitivities, where **p** is the parameter, in this case the parameter is the length of the second resistor. Units are $(Wattsec.m^{-1})$

Equation (12) was implemented in Matlab for a simple circuit and the sensitivities of the energy $\mathbf{G}$ dissipated for a resistor $\mathbf{R} = \mathbf{R}(l_R)$ were observed while changing the length parameter l_R

$$\mathbf{G} = \int_0^T \mathbf{I}(\mathbf{R}) * \mathbf{V}(\mathbf{R})dt. \tag{17}$$

To see the dependency of $\mathbf{V_R}$ for different values of $\mathbf{p}$, in Fig.1. we plotted,
A: $(\mathbf{V_R}(\mathbf{p} = 0.021) - \mathbf{V_R}(\mathbf{p} = 0.020))(t)$
B: $(\mathbf{V_R}(\mathbf{p} = 0.022) - \mathbf{V_R}(\mathbf{p} = 0.021))(t)$.

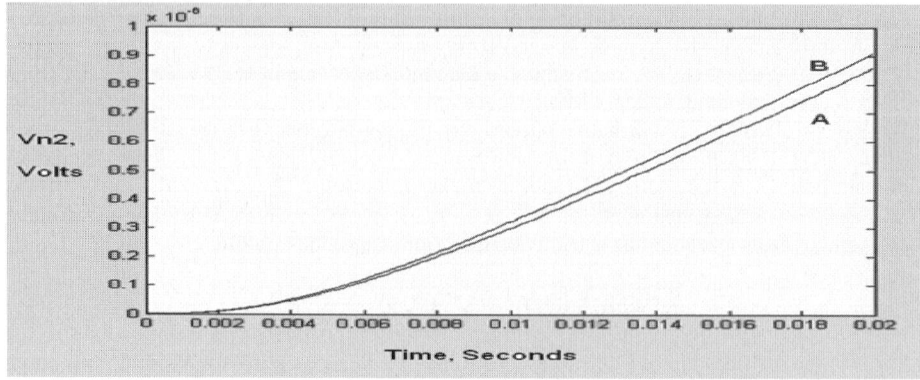

Fig. 1: Voltage differences for successive sensitivity values, node 2

In Table 1 the sensitivity $\mathbf{dG/dp}$, calculated by the backward adjoint method, is shown for 3 different values of $\mathbf{p}$. To test POD we created a larger, equivalent system by splitting one resistor in to a number of smaller resistors. Doing this enabled us to compare the sensitivity of our observed resistor in the reduced large system with the sensitivity in the original system. We generated our projection matrices by applying the singular value decomposition on the new state snapshot matrix for the enlarged problem. Indeed the number of nontrivial singular values for both systems was the same. We applied the projection matrices to reduce only the backward adjoint calculation steps. The size of the matrices was considerably reduced and the calculated sensitivities from the reduced system were exact to at least 3 significant figures, please see Table 1 for sensitivity results for the POD and for the non POD approach.

4 Conclusions

The backward, adjoint sensitivity methods are immediately attractive when the original DAE (1) is linear and when the number of parameters $P \gg 1$. Direct forward and backward adjoint approaches impose different accuracy conditions to finite difference approximations. The direct forward method exploits the re-use of LU-decompositions. The backward adjoint methods becomes more of interest when MOR can be applied, or when otherwise approximate LU-decompositions could have been saved during the forward time integration. In these cases they can outperform the direct forward method when the number of parameters P is large (but still smaller than N, usually $F \ll P$).
We have shown that applying POD MOR to the backward adjoint step is possible and works very well. MOR techniques can also be used to reduce the effort in sensitivity calculation in the forward analysis. In future we want to study more closely the application of our sensitivity calculations to larger circuits and with more industrial relevance. We also want to consider the effect of different time stepping in forward and backward analysis. And, finally, we want to study the sensitivity of the reduced DAE (16) in which $\mathbf{V} = \mathbf{V}(\mathbf{p})$ depends on $\mathbf{p}$ as well.

References

1. P. Astrid: *Reduction of process simulation models: a proper orthogonal decomposition approach*, PhD-Thesis, Department of Electrical Engineering, Eindhoven University of Technology, 2004.
2. P. Astrid, A. Verhoeven: *Application of least squares MPE technique in the reduced order modeling of electrical circuits*, TU Eindhoven, Center for Analysis, Scientific Computing and Applications, CASA Report 11, 2006.
3. O. Balima, Y. Favennec, M. Girault, D. Petit: *Comparison between the modal identification method and the POD-Galerkin method for model reduction in nonlinear diffusion systems*, Int. J.Numer. Meth. Engng., Vol. 67, pp. 895–915, 2006.
4. K. Balla, R. März : *Linear differential-algebraic equations of index-1 and their adjoint equations*, Results in Maths, Vol.37, pp.12-35, 2000.
5. B. Burdiek: *Generation of optimal test stimuli for nonlinear analog circuits using nonlinear programming and time-domain sensitivities*, Proc. DATE 2001, Munchen, pp. 603-608.
6. B. Burdiek: *Zur Berechnung von Testsignalen für nichtlineare analoge Schaltungen unter Verwendung von Methoden der Optimalsteurungstheorie*, Shaker, Achen, 2005 (PhD Thesis Univ. of Hannover, 2005).
7. Y. Cao, S. Li, L. Petzold, R. Serban: *Adjoint sensitivity for differential-algebraic equations: the adjoint DAE system and its numerical solution*, SIAM J. Sci. Comput., Vol. 24-3, pp. 1076–1089, 2002.
8. A.R. Conn, P.K. Coulman, R.A. Haring, G.L. Morrill, C. Visweswariah, C.W. Wu: *JiffyTune: circuit optimization using time-domain sensitivities*, IEEE Trans. on CAD of ICs and Systems, Vol. 17-12, pp. 1292–1309, 1998.
9. L. Daldoss, P. Gubian, M. Quarantelli: *Multiparameter time-domain sensitivity computation*, IEEE Trans. on Circuits and Systems - I: Fund. Theory and Applics, Vol. 48-11, pp. 1296–1307, 2001.
10. Y. Favennec, M. Girault, D. Petit: *The adjoint method coupled with the modal identification method for nonlinear model reduction*, Inverse Probl. in Science and Engng., Vol. 14, No. 3, 153–170, 2006.
11. D.E. Hocevar, P. Yang, T.N. Trick, B.D. Epler: *Transient sensitivity computation for MOSFET circuits*, IEEE Trans. on CAD of Integr. Circuits and Systems, Vol. CAD-4, Nr. 4, pp. 609–620, 1985.
12. L.T. Pillage, R.A. Rohrer, C. Visweswariah: *Electronic circuit and system simulation methods*, McGraw-Hill, Inc, New York, USA, ISBN 0070501696, 1994.
13. M. Rathinam, L.R. Petzold: *A new look at proper orthogonal decomposition*, SIAM J. Numer. Analysis, Vol. 41-5, pp. 1893–1925, 2003.
14. M.J. Rewienski: *A trajectory piecewise-linear approach to model order reduction of nonlinear dynamical systems*, PhD-Thesis, Massachutes Institute of Technology, 2003.
15. T. Voss: *Model reduction for nonlinear differential algebraic equations*, MSc. Thesis University of Wuppertal, 2005; Unclassified Report PR-TN-2005/00919, Philips Research Laboratories, 2005.
16. H. Xu: *Transient Sensitivity Analysis in Circuit Simulation*, MSc-Thesis, Department of Mathematics and Computing Science, Eindhoven University of Technology, 2004. [Online via author search at Library at http://w3.win.tue.nl/en/]

Index Reduction by Element-Replacement for Electrical Circuits

Simone Bächle and Falk Ebert[1 2]

[1] Institute of Mathematics, Technical University of Berlin, MA 4-5 Straße des 17. Juni 136, 10623 Berlin, Germany
email: {baechle,ebert}@math.tu-berlin.de
[2] Supported by the DFG Research Center MATHEON "Mathematics for key technologies" in Berlin.

Summary. In this paper we will discuss certain aspects of the transient simulation of electrical circuits. It is a well known problem that DAEs in circuit simulation may possess a higher index (e.g. 2) and thus exhibit undesirable numerical behaviour. While methods for the reduction of the higher index exist, they are usually algebraic in nature. The large size of the systems in VLSI circuit simulation prohibits the use of algebraic methods for index reduction. We will present a topological approach to index reduction that changes certain elements of the circuit netlist to obtain a circuit DAE with usually improved numerical behaviour with respect to workload or accuracy.

1 Introduction

Simulation of electrical circuits is a commonly used tool to test new electrical circuits prior to producing an actual prototype. Especially in chip design it is important to be able to have a quick and reliable method for simulating the behaviour of a circuit. But, in this context, the respective circuits tend to contain millions of elements. Thus, numerical simulation may become difficult, just because of the sheer size of the problem.
The main methods for the simulation of circuits are the Modified Nodal Analysis (MNA), the charge-/flux-oriented MNA (MNA c/f) and the Sparse Tableau Approach, cf. [9]. Kirchhoff's Laws and branch constitutive relations are set up to form a system of equations describing the important properties of the circuit, e.g., voltages and currents.
As this system contains differential relations as well as algebraic ones, it is called 'Differential-Algebraic-Equation' (DAE). A well known problem of DAEs is that besides obvious algebraic relations, they may contain so-called hidden constraints that are revealed only by differentiating certain equations or linear combinations of equations. In circuit simulation, these DAEs are known to have index 2, given some topological properties of the network, cf. [6]. This higher index leads to several undesirable effects in the numerical solution of the DAEs, e.g. loss of accuracy, unnecessary small stepsizes and difficulties when computing consistent initial values. Recent approaches try to lower the index of DAEs to improve the numerical behaviour, see [13] and references therein. These methods usually involve numerical rank determinations and costly algebraic transformations of the equations. Especially, for large scale circuit equations, these transformations become too costly to be efficient.
The hidden constraints in the case of the MNA and the MNA c/f can been determined, and in [8] it has been shown, how they can be obtained without algebraic transformations of the

circuit equations by only using information contained in the topology of the circuit. The information gained this way has until now mainly been used to determine consistent initial values, fulfilling the hidden constraints as well as the explicit constraints, for numerical integration of the circuit equations [6]. Recently, a concept called *minimal extension*, see [12], has been used to include the hidden constraints into the process of integration, see [2]. The DAE obtained in this way is of index 1, while many systems arising in practice can be of index 2. The problem with this approach is that in order to include hidden constraints into the system, the number of variables increases. We will take a slightly different approach to the topological analysis, one that focuses on index reduction. It will be shown that it is possible to incorporate the hidden constraints into the network equations while retaining the MNA or MNA c/f structure of the equations. The method proposed here will use the results of the topological index analysis on a circuit element level. The main advantage will be, that no algebraic transformations of the full circuit equations will have to be performed, but the method will change the circuit itself. Hence, prior to the actual simulation, a preprocessing step will have to be performed, analyzing the structure of the circuit and exchanging certain elements by newly defined elements, in order to obtain a circuit that is described by a DAE of index 1. The advantage of this approach is that after the preprocessing, no more algebraic transformations have to be performed, but the netlist, i.e., the list containing all structure- and element related information, itself is changed. The changed netlist can then be processed by the same simulation tools as the original netlist, provided they are able to handle the introduced new elements. As the new netlist produces circuit equations of index 1, the integration process is often faster and more accurate than for the index 2 case. The linear systems arising from discretization are usually better conditioned. Additionally, it is much easier to obtain a set of consistent initial values for the index 1 case. The only extra cost is a one-time preprocessing step, while existing software can be used for integration of the equations.
Here, we will give an overview of the element replacement technique and point out some restrictions for its applicability. We present a numerical example where the method is applied to a NAND gate test circuit.

2 Element replacement

We will subsequently assume that we are dealing with circuits consisting of general resistances and two-term capacitances and inductances. These capacitances and inductances are assumed to be locally controlled and independent of time, which means

$$j_C = \frac{d}{dt} q_C(u_C) = C(u_C)\frac{d}{dt}u_C, \quad u_L = \frac{d}{dt}\phi_L(j_L) = L(j_L)\frac{d}{dt}j_L.$$

Here, u_* and j_*, $* \in \{C, L\}$ denote voltage and current across the respective element, q_C denotes charge of a capacitance and ϕ_L flux in an inductance. The charge-voltage- and flux-current-relations $q_C(u_C)$ and $\phi_L(j_L)$ are assumed to be strictly monotonous such that $C(u_C) = \frac{\partial}{\partial u_C} q(u_C)$ and $L(j_L) = \frac{\partial}{\partial j_L}\phi(j_L)$ are positive for all arguments u_C and j_L. The main topological properties of a circuit that are responsible for a higher-index behaviour are CV loops, i.e. loops formed by branches that are occupied by capacitances or voltage sources, and LI cutsets, i.e. cutsets formed by branches containing inductances and current sources only. For capacitances and inductances not appearing in such structures the strict 'locally controlled two-term element' restriction above may be dropped. We will not consider hidden constraints that arise due to special choices of circuit parameters and controlled sources that do not fulfill the requirements in [6]. Under these assumptions the index of the circuit equations is guaranteed to be not higher than 2.
For every loop and every cutset, an orientation can be chosen. A more detailed description of these structures is given in [4]. Loops of voltage sources and cutsets of current sources are generally not allowed, because they might lead to inconsistent equations.
We will derive expressions for the hidden constraints in such a way that the steps needed to

perform an index reduction become clearly visible. Consider one specific CV loop. Let n_C^{loop} and n_V^{loop} be the number of capacitances and voltage sources in the loop, respectively. The hidden constraint arising from that CV loop can be interpreted as follows:

- Denote the voltages across capacitances in the loop by $u_{C,k}^{loop},\ k = 1 \dots n_C^{loop}$,
- denote the source voltages by $v_k^{loop},\ k = 1 \dots n_V^{loop}$.
- Set $\alpha_{*,k} = \pm 1$, where $* \in \{C, V\}$. For every element in the loop, the constant $\alpha_{*,k}$ is 1 if the element is oriented in the same way as the loop and -1 otherwise.
- Kirchhoff's voltage law over that loop states that

$$\sum_{k=1}^{n_C^{loop}} \alpha_{C,k} u_{C,k}^{loop} + \sum_{k=1}^{n_V^{loop}} \alpha_{V,k} v_k^{loop} = 0. \tag{1}$$

- The derivative of (1) holds as well,

$$\sum_{k=1}^{n_C^{loop}} \alpha_{C,k} \frac{d}{dt} u_{C,k}^{loop} + \sum_{k=1}^{n_V^{loop}} \alpha_{V,k} \frac{d}{dt} v_k^{loop} = 0. \tag{2}$$

As the currents through capacitances depend on the derivatives of the respective capacitance voltages, Equation (2) imposes a constraint on the branch currents as well. This constraint is not originally visible in the system and has been obtained by differentiation, thus representing a hidden constraint.
We want condition (2) to be fulfilled, hence, it has to appear explicitly in the circuit equations. For this purpose, we choose one of the involved capacitances. Without loss of generality, we assume that this is C_1^{loop} and that the direction of the loop is identical to the one of the capacitance. The corresponding voltage is $u_{C,1}^{loop}$. Then, we split (2) as follows

$$\frac{d}{dt} u_{C,1}^{loop} = -\sum_{k=2}^{n_C^{loop}} \alpha_{C,k} \frac{d}{dt} u_{C,k}^{loop} - \sum_{k=1}^{n_V^{loop}} \alpha_{V,k} \frac{d}{dt} v_k^{loop}. \tag{3}$$

We multiply this equation by C_1^{loop} to obtain

$$j_{C,1}^{loop} = C_1^{loop} \frac{d}{dt} u_{C,1}^{loop} = -\sum_{k=2}^{n_C^{loop}} \alpha_{C,k} C_1^{loop} \frac{d}{dt} u_{C,k}^{loop} - \sum_{k=1}^{n_V^{loop}} \alpha_{V,k} C_1^{loop} \frac{d}{dt} v_k^{loop}. \tag{4}$$

By definition, the term $C_1^{loop} u_{C,1}^{loop}$ describes the charge q_1^{loop} of the capacitance and $C_1^{loop} \frac{d}{dt} u_{C,1}^{loop} = \frac{d}{dt} q_1^{loop}$ is the current through this capacitance. In this way, we have expressed one branch current explicitly. Hence, we can remove the capacitance C_1^{loop} and replace it by a current source $i_{C,1}^{loop}$ that provides the current given by the right hand side of (4). Here, i is used instead of j to characterize the difference between a current as an unknown of the system and a current source. This differentially controlled current source (DCS) is an element not easily available in many common circuit simulators, e.g. SPICE [11]. For this reason it has to be emulated using common elements such as current-controlled current sources (CCCS). To this end, we again use the relation $C_k^{loop} \frac{d}{dt} u_{C,k}^{loop} = \frac{d}{dt} q_k^{loop} = j_k^{loop}$, or to be more precise $\frac{d}{dt} u_{C,k}^{loop} = \frac{1}{C_k^{loop}} \frac{d}{dt} q_k^{loop} = \frac{1}{C_k^{loop}} j_k^{loop}$. Using this expression for $\frac{d}{dt} u_{C,k}^{loop}$, Equation (4) can be rewritten as

$$i_{C,1}^{loop} = -\sum_{k=2}^{n_C^{loop}} \alpha_{C,k} \frac{C_1^{loop}}{C_k^{loop}} j_{C,k}^{loop} - \sum_{k=1}^{n_V^{loop}} \alpha_{V,k} C_1^{loop} \frac{d}{dt} v_k^{loop}. \tag{5}$$

In this way, the current $i_{C,1}^{loop}$ can be expressed as a linear combination of the other capacitance currents of the loop and some terms generated by the voltage sources of the loop. It also allows, to model the current source $i_{C,1}^{loop}$ by means of standard current sources and CCCS. However, since capacitance currents usually do not appear as variables in MNA and charge-oriented MNA equations, these currents have to be introduced artificially. This is commonly done by inserting a zero-voltage voltage source in series with the current to measure. The current through the voltage source appears as a variable in the equations and can thus be used as controlling current for the CCCS.
The approach in the case of LI cutsets is similar. Using an analogous notation as in the case of CV loops, we can select one inductance in an LI cutset (here L_1^{cut}) and using Kirchhoff's current law, we can express the voltage across the element as follows

$$v_{L,1}^{cut} = L_1^{cut} \frac{d}{dt} j_{L,1}^{cut} = -\sum_{k=2}^{n_L^{cut}} \alpha_{L,k} L_1^{cut} \frac{d}{dt} j_{L,k}^{cut} - \sum_{k=1}^{n_I^{cut}} \alpha_{I,k} L_1^{cut} \frac{d}{dt} i_k^{cut}. \tag{6}$$

Here, v has been used instead of the original u to underline the role of the element as controlled voltage source instead of an inductance. Again, for the sake of emulation of this differentially controlled voltage source (DVS) by common sources, we use the relationship $L_k^{cut} \frac{d}{dt} j_{L,k}^{cut} = \frac{d}{dt} \phi_k^{cut} = u_{L,k}^{cut}$.

$$v_{L,1}^{cut} = -\sum_{k=2}^{n_L^{cut}} \alpha_{L,k} \frac{L_1^{cut}}{L_k^{cut}} u_{L,k}^{cut} - \sum_{k=1}^{n_I^{cut}} \alpha_{I,k} L_1^{cut} \frac{d}{dt} i_k^{cut}. \tag{7}$$

Remark 1. It has to be pointed out that the choice of the capacitance or inductance to be replaced is not completely arbitrary. If for instance one capacitance appears in several loops at once, it cannot be replaced in neither. This is due to the fact that once a capacitance is replaced in one loop it cannot be used as controlling element in the other. Similar restrictions limit the choice of inductances in overlapping cutsets. However, in [3] it has been proven, that with an adequate choice of loops and cutsets a correct selection and consecutive element replacement is always feasible. For the relevant graph algorithms, we refer to the same source.

Remark 2. In the previous considerations it has always been assumed that all considered capacitances and inductances are only locally controlled two-term elements. Admittedly, this imposes a severe restriction on the applicability of the method. Currently, an extension of the method is in preparation that also allows multi-port capacitances and inductances.
Until now, no in-depth investigations with respect to charge- and flux conservation of the reduced-index network have been undertaken and will not be undertaken in this paper.

Remark 3. While it is relatively simple to implement emulated DVS in LI cutsets such as in (7) in a SPICE type simulator, this becomes more difficult for the emulated DCS in CV loops. Conversely to the DVS case, the controlling currents are not explicitly present as variables. It has been suggested to put zero-voltage voltage sources in series with capacitances in order to measure the respective currents. It has to be noted, that each such voltage source adds two unknowns to the system, one for the current through the source and one for the extra node between capacitance and voltage source. This has to be done for every capacitance that in some way is part of a CV loop. For networks with many CV loops and especially with many capacitances mounted in parallel, this may become infeasible and increase the size of the DAE system considerably.

3 Numerical Example

The following example has been taken from [3]. It represents a NAND gate test circuit originally found at [15]. The circuit equations have been set up with QPSIM [3], an extension of PSIM, cf. [14], to the charge/flux oriented MNA formulation.

The MOSFETs have been simulated using LEVEL B replacement circuits. These replacement models include nonlinear but locally controlled capacitances. For details and specifications we refer to [15, 3]. Initial values have been taken from [7]. The used input signals and a reference solution for the output signal are depicted in 2. The reference solution has been computed with RADAU5, [10], with index reduction for tolerance settings of 10^{-15}, as this was the setting with the best possible tolerance. The element replacement procedure has been applied to the circuit to obtain a DAE of index 1. In the process 7 voltage sources have to be introduced in order to measure capacitance currents. The size of the MNA c/f system increased from 29 to 36 unknowns. QPSIM exploits the structure of the MNA c/f equations for the inserted voltage sources such that only the currents through these sources are added as unknowns while the node potentials of the inserted nodes need not be considered, hence only 7 additional variables have to be introduced instead of 14 as asserted in Remark 3. Numerical solutions have

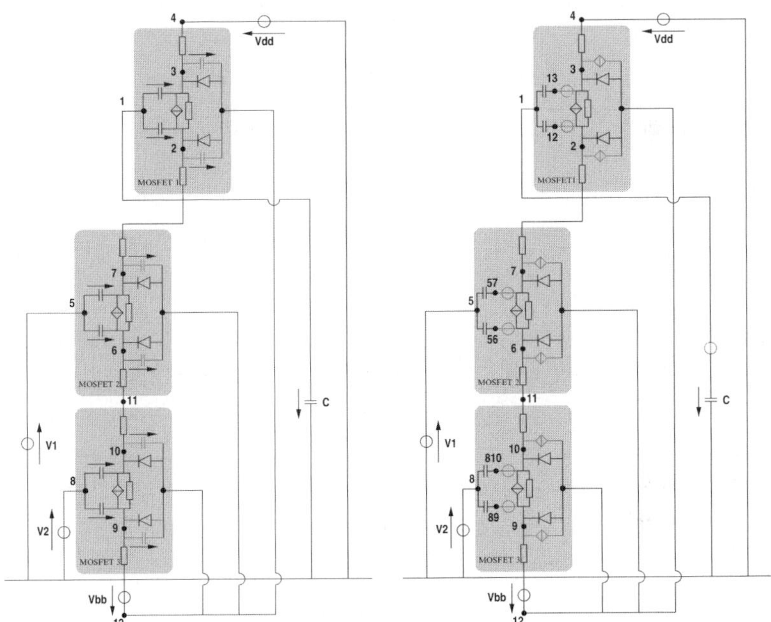

Fig. 1: NAND gate replacement circuit, index 2 circuit (left), index 1 circuit with additional voltage sources (right)

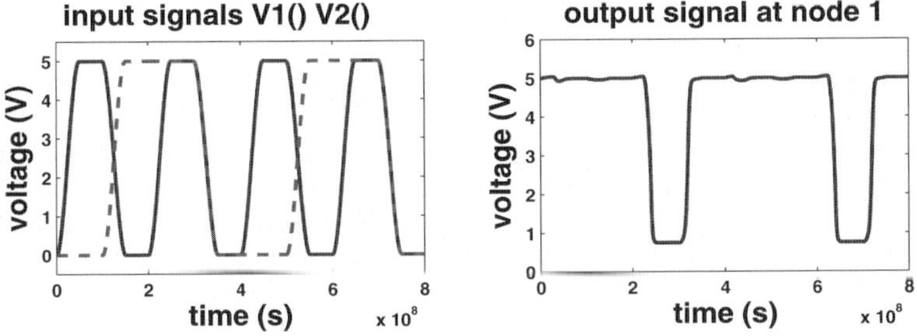

Fig. 2: Input signals (left) and reference solution for the output signal (right)

been computed with both RADAU5 and DASPK3.1, see [1], a variant of DASSL, cf. [5]. For the numerical integration, different tolerances from 10^{-3} to 10^{-12} have been used. Figure 3 depicts the obtained accuracies of the node potential at node 1, i.e. the output signal, and the computation time on a Pentium 4 desktop PC with 1.4GHz and 1GByte RAM. Here, 'error' means the maximum absolute difference of the computed solution and the reference solution at intervals of 0.1 ns.
It can be seen, that both methods manage to reproduce the reference solution. DASPK cannot achieve a greater accuracy than 10^{-5}, while RADAU gets to 10^{-8}. With DASPK, the index does not seem to have a great influence on efficiency. In the index 1 case however, RADAU5 gains gains one order of magnitude in accuracy, starting at tolerances of 10^{-7}, that is roughly in the region of an obtained accuracy of 10^{-3}.

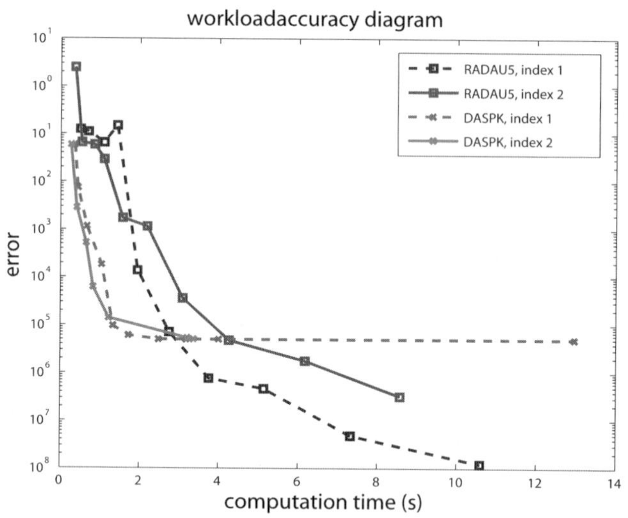

Fig. 3: workload vs. error for the NAND gate example

4 Conclusions

We have presented a method for reducing the index of circuit equations under some restrictions on the elements of the network, see Remarks 2 and 3. The method relies on circuit topology only and is, thus, applicable even for large scale systems - in contrast to known algebraic index reduction methods, cf. [13]. The equations arising with the use of DCS can be interpreted as the DAE after the minimal extension procedure with an extra elimination step. The emulated-sources approach differs from the minimal extension as it is specially conceived for the structure of circuit equations. The method is principally independent of the used method for equation setup and also of the solver used to actually solve these equations. It represents a one-time preprocessing step, so little to no extra work is required after the preprocessing. The procedure may increase the size of the system to solve, depending on the number of CV loops and involved capacitances. However, recent tests with a 4-Bit adding unit, composed of NAND gates as in the example, showed an increase of only about 15% in the number of variables. Tests in the previous section with a relatively small example show satisfactory results and confirm the claim that the extra work for index reduction pays off with respect to accuracy of the obtained solution, especially for higher accuracy requirements. Other examples that have been presented e.g. in [3] also show an improvement of the computation time but they are purely academic and have been omitted here.

References

1. DASPK3.1, a software program. www.engineering.ucsb.edu/~cse/software.html, January 5th, 2007.
2. S. Bächle. Index reduction for differential-algebraic equations in circuit simulation. Technical Report MATHEON 141, Technical University of Berlin, 2004.
3. S. Bächle. *Numerical Solution of Differential-Algebraic Systems Arising in Circuit Simulation*. PhD thesis, Technical University Berlin, 2007 (in preparation).
4. S. Bächle and F. Ebert. Graph theoretical algorithms for index reduction in circuit simulation. Technical Report MATHEON 245, Technical University of Berlin, 2005.
5. K.E. Brenan, S.L. Campbell, and L.R. Petzold. *Numerical Solution of Initial-Value Problems in Differential Algebraic Equations*. SIAM, Philadelphia, PA, 1996.
6. D. Estévez Schwarz. A step-by-step approach to compute a consistent initialization for the MNA. *Int. J. Circ. Theor. Appl*, 30:1–16, 2002.
7. D. Estévez Schwarz and R. Lamour. The computation of consistent initial values for nonlinear index-2 differential-algebraic equations. Technical report, Humboldt-University Berlin, 2001.
8. D. Estévez-Schwarz and C. Tischendorf. Structural analysis for electric circuits and consequences for the MNA. *Int. J. Circ. Theor. Appl.*, 28:139–159, 1998.
9. M. Günther and U. Feldmann. CAD-based electric-circuit modeling in industry, II. impact of circuit configuration and parameters. *Surveys on Mathematics for Industry*, 8:131–157, 1999.
10. E. Hairer and G. Wanner. *Solving Ordinary Differential Equations II - Stiff and Differential-Algebraic Problems*. Springer, Berlin Heidelberg, 2nd edition, 1996.
11. H. Khakzar, A. Mayer, and R. Oetinger. *Entwurf und Simulation von Halbleiterschaltungen mit SPICE*. expert Verlag, 1992.
12. P. Kunkel and V. Mehrmann. Index reduction for differential-algebraic equations by minimal extension. *ZAMM*, 84:579–597, 2004.
13. P. Kunkel and V. Mehrmann. *Differential-Algebraic Equations. Analysis and Numerical Solution*. EMS Publishing House, Zürich, Switzerland, 2006.
14. R. Melville. PSIM — pedagogical simulator. Modular circuit simulation software, Columbia University, May 2004.
15. C. Tischendorf. *Solution of index-2 differential-algebraic equations and its application in circuit simulation*. PhD thesis, Humboldt-Universität zu Berlin, 1996.

Application of 2D Nonuniform Fast Fourier Transforms Technique to Analysis of Shielded Microstrip Circuits

Raimond Grimberg, Adriana Savin and Sorin Leitoiu

National Institute of R&D for Technical Physics 47 D. Mangeron Blv., Iasi, 700050, Romania grimberg@phys-iasi.ro

1 Introduction

Characterization of microstrip discontinuities is an important task in computer aided design (CAD) of microstrip circuits. Many methods for modeling the discontinuities have been developed, such as integral equation [1],[2] spectral domain approach [3], finite difference time domain [4], and finite element method [5].
he method of moments (MoM) is a core engine for analysis of microstrip circuits [6]. Most of the Central Processing Unit (CPU) time is consumed in evaluation of the MoM matrix elements since Greens functions converge slowly and a large number of basis functions are required for expanding surface current densities on conductor. The MoM matrix elements are computed by two-dimensional discrete fast Fourier transforms (2D-FFT). In this method, however, the mesh scheme is restricted to be uniform. Obviously, uniform grids are very inefficient for analysis of a general microstrip circuit because electrical currents have fast variations along microstrip edges, thus, time local discretization becomes a need for accurate analysis of the whole circuit. In [7], in order to reduce the number of unknowns, the currents are expanded by a linear combination of the current distributions of the first few resonant modes of the circuit. However, there are yet required very fine discretization and a 2D FFT of large size to find solutions of the resonant modes.
In this paper, a two dimensional non uniform fast Fourier transform (2D NUFFT) [8] incorporated into the spectral domain approach (SDA) is developed for analysis of microstrip circuits. The mesh scheme for the microstrip circuit can be very flexible, although each subdivision must be rectangular.

2 NUFFT algorithm

The FFT is a fast algorithm for calculating discrete Fourier transforms, and has multiple applications in electrical and electromechanic engineering, physics, applied mathematics etc. It requires that the sampled data should be equally spaced.
The idea of the NUFFT is to approximate a non-uniform sample point in the space domain by interpolating over the sampled uniform Fourier basis using a FFT with finite nonzero coefficients.
Consider the following summation with unequally spaced output data d_p:

$$d_p = \sum_{k=-\frac{N_f}{2}}^{\frac{N_f}{2}-1} f_k e^{jks_p} \qquad p = 1, 2, ...N_a \tag{1}$$

where the sampled modes $s_p \in [\pi, \pi]$ can be unequally spaced. The input sequence f_{l_0} can be equally or unequally spaced, and N_f and N_a are the numbers of input and output data points, respectively.
To evaluate eq.1 with unequally spaced s_p, the key step of the NUFFT is to approximate each e^{jks_p} with a sum of weighted complex exponentials at $q+1$ equally spaced modes in the vicinity of s_p as following:

$$e^{jks_p} \approx \varphi_k^{-1} \sum_{l=-\frac{q}{2}}^{\frac{q}{2}} \gamma_l\left(s_p\right) e^{j\left(\nu_p + l - \frac{q}{2} - 1\right)2\pi \frac{k}{mN_f}} \tag{2}$$

The accuracy factors φ_k (normalization constant) in eq.2 are chosen to minimize the error of approximation in the square sense. Here q is an even positive integer and $m \geq 2$ is an index indicating the over sampling rate of the approximation. The sampling points of the complex exponentials on the right hand side of eq.2 collocate with those of a regular FFT with size $mN_f \geq 2N_f$. A larger value of m improves accuracy. In eq.2, $\nu_f = \left[s_p m \frac{N_f}{2\pi}\right]$ denotes the integer nearest to $s_p m \frac{N_f}{2\pi}$. For each s_p, the $q+1$ interpolation coefficients, $\gamma_l\left(s_p\right)$, $1 \leq l \leq q+1$ are given as [8], [9]

$$\bar{\gamma} = \bar{\bar{F}}^{-1}\bar{P} \tag{3}$$

where the entries of the $(q+1) \times (q+1)$ matrix $\bar{\bar{F}}$ and $(q+1) \times 1$ column vector $\bar{P}$ are

$$F_{ln} = \begin{cases} \frac{e^{j(l-n)/m\pi} - e^{-j(l-n)/m\pi}}{1 - e^{-j2l - n\pi/mN_f}} & l \neq n \\ N_f & l = n \end{cases} \tag{4}$$

$$P_l(s_p) = j\frac{sin\left(\frac{2l-q-2\delta_p-3}{2m}\pi\right)}{1 - e^{-j(2l-q-2\delta_p-3)/mN_f\pi}} + j\frac{sin\left(\frac{2l-q-2\delta_p-1}{2m}\pi\right)}{1 - e^{-j(2l-q-2\delta_p-1)/mN_f\pi}} \tag{5}$$

and

$$\delta_p = \frac{s_p m N_f}{2\pi} - \nu_p \tag{6}$$

3 2D NUFFT algorithm

In figure 1, the cross is a nonuniform sample point (x_t, y_s), $-\pi \leq x_t \leq \pi$ and $-\pi \leq y_s \leq \pi$, and the circles and large black dots are $(q+1) \times (q+1)$ uniformly oversampled grid points (X_i, Y_i), which are called the square neighborhood of (x_t, y_s) therein. We are going to evaluate the following 2D Fourier transform:

$$D_{st} = \sum_{m=-\frac{M}{2}}^{\frac{M}{2}-1} \sum_{n=-\frac{N}{2}}^{\frac{N}{2}-1} G_{mn} e^{jmx_t} e^{jny_s} \tag{7}$$

where D_{st} and G_{mn} are finite complex sequences and M and N are even integers.
In the similar manner as in the non-uniform fast Fourier transform (NUFFT) algorithm

$$e^{jmx_t} e^{jny_s} = \varphi_{mn}^{-1} \sum_{p=-\frac{q}{2}}^{\frac{q}{2}} \sum_{g=-\frac{q}{2}}^{\frac{q}{2}} r_{pg}(x_t, y_s) e^{j(\nu_t + p)2\pi \frac{m}{CM}} \cdot e^{j(u_s + g)2\pi \frac{n}{CN}} \tag{8}$$

where $\nu_t = \left[x_t \frac{CM}{2\pi}\right]$ and $u_s = \left[y_s \frac{CN}{2\pi}\right]$ denote the integers nearest to $x_t \frac{CM}{2\pi}$ and $y_s \frac{CN}{2\pi}$, respectively. The accuracy factors j_m are chosen to minimize the error of eq.8 in the least square sense [8], [9].

Substituting eq.8 into eq.7 yields

$$D_{st} = \sum_{p=-\frac{q}{2}}^{\frac{q}{2}} \sum_{g=-\frac{q}{2}}^{\frac{q}{2}} r_{pg}(x_t, y_s) H_{pg}(s,t) \quad (9)$$

where

$$H_{pg}(s,t) = \sum_{m=-\frac{M}{2}}^{\frac{M}{2}-1} \sum_{n=-\frac{N}{2}}^{\frac{N}{2}-1} G_{mn} \varphi_{mn}^{-1} e^{j(\nu_t+p)2\pi \frac{m}{CM}} \cdot e^{j(u_s+g)2\pi \frac{n}{CN}} \quad (10)$$

The calculation of eq.10 can be performed by regular 2D FFT of size CMxCN. In eq.9 the interpolated coefficients r_{pg} can be obtained by two sets of $(q+1)^2$ NUFFT coefficients, i.e. the square neighborhood in figure 1.

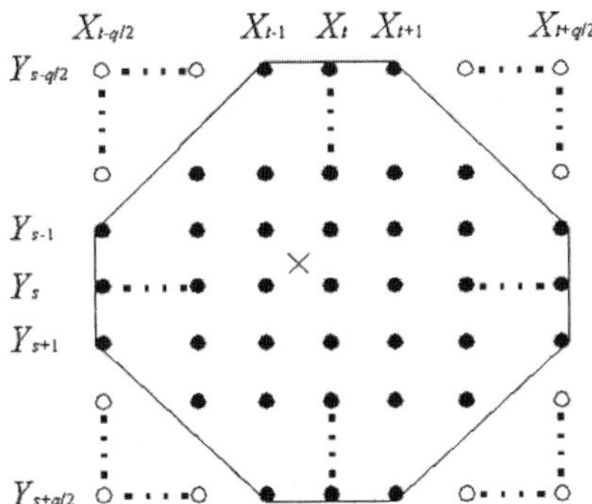

Fig. 1: 2 D NUFFT algorithm: the cross is a nonuniform sample point (x_t, y_s); the circles and large block dates are $(q+1) \times (q+1)$ uniformly oversampled grid points (x_i, y_i) - called the square neighbourhood of (x_t, y_s); the line delimits octagonal or non square neighbourhood (large block dots) from essential neighbouring squares, represented only by blank dots

4 Incorporating the 2D NUFFT into the SDA

For a microstrip circuit enclosed in a rectangular shielded box of dimensions axbxc, one of the spatial domain Green's functions can be written as

$$G_{xx}(x,x',y,y') = \sum_m \sum_n \tilde{G}_{xx} cos(k_{xm}x') sin(k_{yn}y') cos(k_{xm}x) sin(k_{yn}y) \quad (11)$$

where $k_{xm} = m\pi/a$, $K_{yn} = n\pi/b$ and G_{xx} is the Green's function in the spectral domain. Other Greens functions can be expressed in a similar manner.
In the solution procedure, asymmetric rooftop functions are used to expand current densities on conductors, and the half rooftop functions in [10] are used for modeling the external source and load terminals. Let the current densities be expressed as

$$J(x,y) = \sum a_{x\alpha} J_{x\alpha}(x,y) + \sum b_{y\beta} J_{y\beta}(x,y) \quad (12)$$

where $a_{x\alpha}$ and $b_{y\beta}$ are unknown constants to be determined. The Fourier transform of the basis functions can be easily derived. For example, let the αth basis function for the current in the x direction be

$$J_{x\alpha} = J_{xx}(x, x_\alpha)J_{xy}(y, y_\alpha) \tag{13}$$

where

$$J_{xx}(x, x_\alpha) = \begin{cases} \frac{x-x_\alpha}{\Delta x_{\alpha 1}} + 1, \; x_\alpha - \Delta x_{\alpha 1} \le x \le x_\alpha \\ \frac{x-x_\alpha}{\Delta x_{\alpha 2}} + 1, \; x_\alpha \le x \le x_\alpha + \Delta x_{\alpha 2} \\ 0 \qquad otherwise \end{cases} \tag{14}$$

$$J_{xy}(y, y_\alpha) = \begin{cases} 1 \;\; y_\alpha - \frac{\Delta y_\alpha}{2} \le y \le y_\alpha + \frac{\Delta y_\alpha}{2} \\ 0 \;\; otherwise \end{cases}$$

Their 2D Fourier transform can be derived as

$$\tilde{J}_{xx} = \frac{cos(k_{xm}x_\alpha) - cos(k_{xm}(x_\alpha - \Delta x_{\alpha 1}))}{\Delta x_{\alpha 1} k_{xm}^2} + \frac{cos(k_{xm}x_\alpha) - cos(k_{xm}(x_\alpha + \Delta x_{\alpha 1}))}{\Delta x_{\alpha 2} k_{xm}^2} \tag{15}$$

$$\tilde{J}_{xy} = \frac{cos\left(k_{yn}\left(y_\alpha - \frac{\Delta y_\alpha}{2}\right)\right)}{k_{yn}} - \frac{cos\left(k_{yn}\left(y_\alpha + \frac{\Delta y_\alpha}{2}\right)\right)}{k_{yn}}$$

The transformations (15) are trigonometric functions weighted by powers k_{xm} or k_{yn}. It can be validated that the transforms of basis functions for currents in the y direction can be expressed in a similar way.
The Galerkin's procedure is used to set up the final MOM matrix, whose (d, e) element can be expressed as

$$z_{xx}(d, e) = \sum_m \sum_n \tilde{G}_{xx}(m, n)\tilde{J}_{xd}(m, n)\tilde{J}_{xe}(m, n) \tag{16}$$

After some trigonometric calculus, evaluation of $z_{uv}(d, e)$ given by eq.16 can be reduced to

$$\sum_{m=-\frac{M}{2}}^{\frac{M}{2}-1} \sum_{n=-\frac{N}{2}}^{\frac{N}{2}-1} \frac{\tilde{G}_{uv}(m, n)}{k_{xm}^g k_{yn}^h} \times \begin{cases} sin\left(k_{xm}\left(x_d \pm x_e\right)\right) cos\left(k_{yn}\left(y_d \pm y_e\right)\right) \\ cos\left(k_{xm}\left(x_d \pm x_e\right)\right) sin\left(k_{yn}\left(y_d \pm y_e\right)\right) \end{cases} \tag{17}$$

where $u, v = x$ or y and g and $h = 2, 3$ or 4.
Figure 2 summarizes the procedure for establishing the final MOM matrix:

- first step partition the circuit and find the required 2D NUFFT interpolation coefficients for four sets of sampling points $(x_d \pm x_e, y_d \pm y_e)$
- second step evaluate the double summations of products in eq.17 by the 2D NUFFT . If the impressed and load currents are in the same directions, only five cells of 2D NUFFTs will be needed.
- last step recombine the five sum terms to set up the final MOM matrix.

When the currents on input and output feed lines are obtained, the complex amplitudes of the incident and reflected current waves can be extracted by using the generalized pencil of function (GPOF) method [11], and the scattering parameters can be obtained via standard circuit theory.

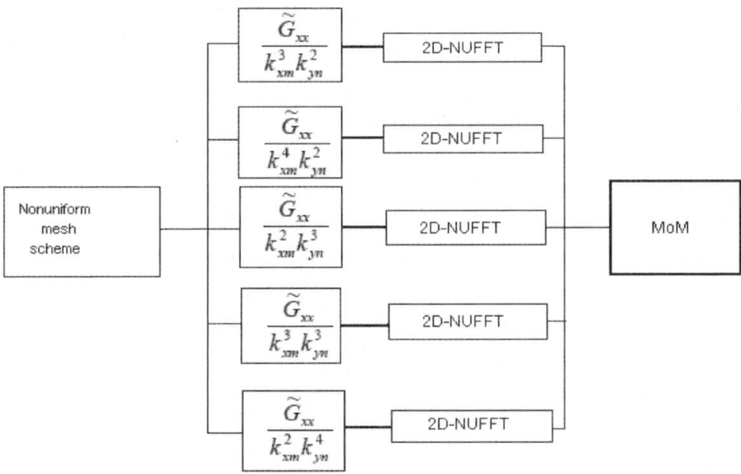

Fig. 2: The procedure for establishing the final MoM matrix.

5 Results

One example, a compact miniaturized hairpin resonator is used to demonstrate the proposed technique for the analysis of microstrip circuit. Consider a hairpin resonator with geometrical dimensions presented in figure 3. The thickness of the dielectric substrate is $f = 1.27mm$ and electrical permittivity is $\varepsilon_r = 10.2$. The hairpin resonator is embedded in a shielded box with dimension $23.6 \times 18.15 \times 16mm^3$. The sampling points (centers of mesh cells) are chosen according to $(S/2)\left[1 + cos\left(k\pi/T\right)\right]$ where $k = 1, 2, ..., T$. S can an be the length or width of the conductor. The sampling points are presented in figure 4.

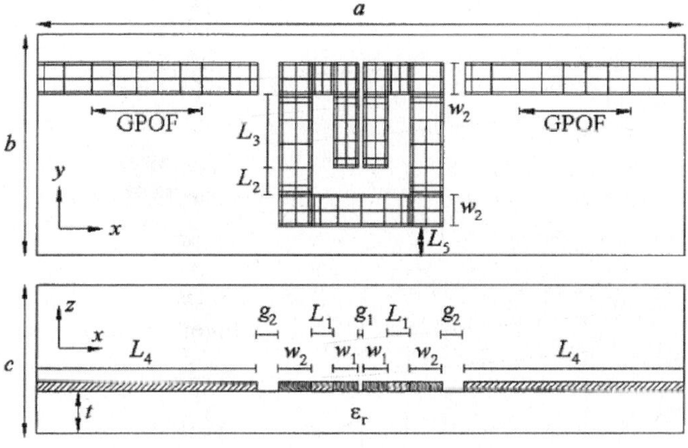

Fig. 3: Hairpin resonator.

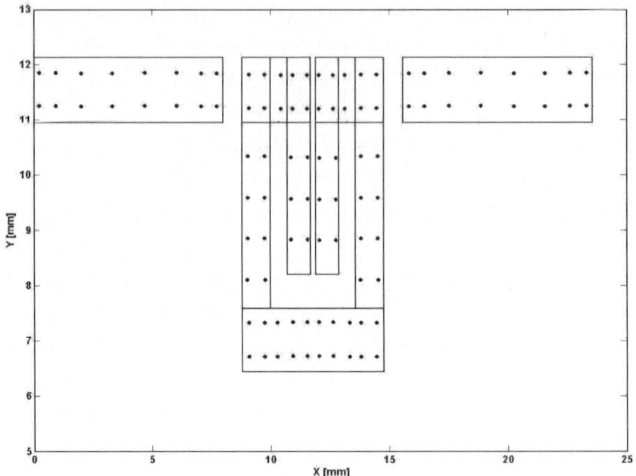

Fig. 4: Hairpin resonator with sampling points.

We select $q = 6$, that suppose a L_2 error (norm's error) of 5.51×10^{-6} and CPU time $128.84s$, which is sufficient for the precision of calculus. If $q = 8$, the L_2 diminishes with an approximate one order and CPU time will double.
We have written a numerical code in Matlab 7.0 which allows calculation of the S-parameter of this hairpin resonator. The matrix S, named scattering matrix, was calculated, because it can relatively easily be measured in the domain of microwaves [12], and thus experimental data exists, which allow the comparison between numerical calculus and experiment. The elements of S were calculated as function of the impedance matrix (Z parameters) from eqs.16 and 17 in the basis of usual relations [12].
The CPU time has been 257.01 seconds on a PC with Pentium IV processor at 3.2GHz and with 2Gb RAM.
The dependencies $|S_{11}|$ and $|S_{21}|$ vs frequency are presented in Figure 5.
To validate the proposed method and the numerical code, the dimensions of the hairpin resonator were identically taken as those from [13]. Comparing the results, the relative error of the calculated values reported to experimental ones presented in [13] was smaller than 1.2%.

6 Conclusions

A 2D NUFFT technique incorporated with SDA has been proposed for efficient analysis of microstrip circuits in a rectangular enclosure. In this method, the mesh scheme has good flexibility since conductors can be discretized into fine cells near the edges and relatively large cells in region with smooth current densities. The comparison between experimental data and calculated data via our model validates the method.

Acknowledgement
This paper is supported by the Romanian Ministry of Education and Research, Excellence Research Program, under Contract 9 CEEX I 03 / 06.10.2005, acronym nEDA and CNCSIS Grant 982.

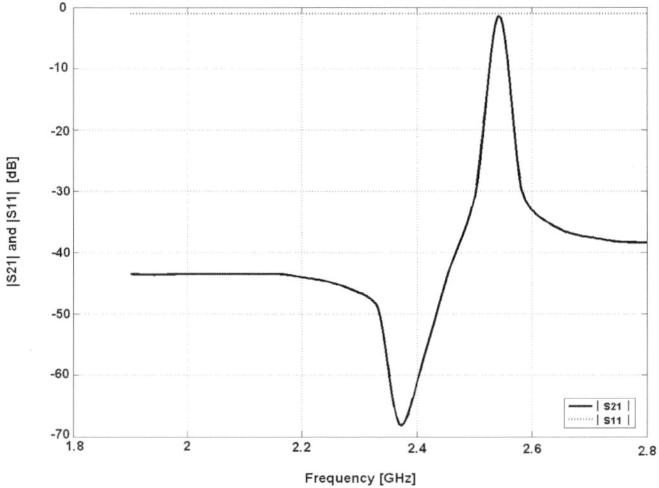

Fig. 5: The dependences $|S_{11}|$ and $|S_{12}|$ vs. frequency

References

[1] G. V. Eleftheriades, J. R. Mosig, and M. Guglielmi, "An efficient mixed potential integral equation technique for the analysis of shielded MMICs", in Proc. 25th Eur. Microwave Conf., Sep. 1995, pp. 825829

[2] E. Drake, R. R. Boix, M. Horno, and T. K. Sarkar, "Effect of substrate dielectric anisotropy on the frequency behavior of microstrip circuits", IEEE Trans. Microw. Theory Tech., vol. 48, no. 8, Aug.2000, pp. 13941403

[3] R. H. Jansen, "The spectral-domain approach for microwave integrated circuits", IEEE Trans. Microw. Theory Tech., vol. MTT-33, no. 10, Oct. 1985, pp. 10431056

[4] D. Bica and B. Beker, "Analysis of microstrip discontinuities using the spatial network method with absorbing boundary conditions", IEEE Trans. Microw. Theory Tech., vol. 44, no. 7, Jul. 1996, pp. 11571161

[5] R.W. Jackson, "Full wave, finite element analysis of irregular microstrip discontinuities", IEEE Trans. Microw. Theory Tech., vol. 37, no. 1, Jan. 1989, pp. 8189

[6] E. S. Tony and S. K. Chaudhuri, "Analysis of shielded lossy multilayered-substrate microstrip discontinuities", IEEE Trans. Microw. TheoryTech., vol. 49, no. 4, Apr. 2001, pp. 701711

[7] C. J. Railton and S. A. Meade, "Fast rigorous analysis of shielded planar filters", IEEE Trans. Microw. Theory Tech., vol. 40, no. 5 May 1992, pp. 978985

[8] Q. H. Liu and N. Nguyen, "An accurate algorithm for nonuniform fast Fourier transforms (NUFFT's)", IEEE Microw. Guided Wave Lett., vol. 8, no. 1, Jan. 1998, pp. 1820

[9] K. Y. Su and J. T. Kuo, "A two-dimensional nonuniform fast Fourier transform (2-D NUFFT) method and its applications to the characterization of microwave circuits", in AsiaPacific Microwave Conf., Seoul, Korea, Nov. 47, 2003, pp. 801804

[10] R. C. Hsieh and J. T. Kuo, "Fast full-wave analysis of planar microstrip circuit elements in stratified media", IEEE Trans. Microw. Theory Tech., vol. 46, no. 9, Sep. 1998, pp. 12911297

[11] Y. Hua and T. K. Sarkar, "Generalized pencil-of-function method for extracting poles of an EM system from its transient response", IEEE Trans. Antennas Propag., vol. 37, no. 2, Feb. 1989, pp. 229234

[12] Agilent Test & Measurement Application Note 95-1; Agilent Technologies. S-Parameter Techniques for Faster, More Accurate Network Design, www.agilent.com
[13] M. Sagawa, K. Takahashi, and M. Makimoto, "Miniaturized hairpin resonator filters and their application to receiver front-end MIC's", IEEE Trans. Microw. Theory Tech., vol. 37, no. 12, Dec. 1989, pp. 19911996

A Filter Design Framework with Multicriteria Optimization Based on a Genetic Algorithm

Neag Marius[1], Marina Topa[2], Liviu Nedelea[3], Lelia Festila[4], and Vasile Topa[4]

Technical University of Cluj-Napoca, Str Ctin Dacovicuiu,nr 15,400020 Cluj-Napoca,
Marius.Neag@bel.utcluj.ro

Abstract - The paper presents a method for generating filter transfer functions optimized in respect to a wide range of behavioral and implementation criteria. The optimization engine is the SQP algorithm, with the main parameters provided by a genetic algorithm tailored to this application.

Keywords - classical approximations, behavioral and implementation requirements, SQP algorithm, genetic algorithm.

1 Introduction

Continuous time filters are widely used functional blocks, from simple anti-aliasing filters preceeding ADCs to high-spec channel-select filters in integrated RF transceivers. Filter design is challenging, even more when the system has to meet a wide set of constraints. Numerous CAD tools for filter design are available [The04], [Fil], but most of them are based on the classical transfer functions, which meet only requirements related to the magnitude or phase responses.
This paper proposes a new method for deriving filter transfer functions which takes into consideration not only the gain and phase requirements, but also the peak overshoot, the rise and settling times and even the values of the quality factor of the biquads in a cascade implementation of the filter.
Section II provides a general mathematical description of the filter. In Section III the multicriteria optimization method used for deriving filter transfer functions for a wide range of requirements is presented. Next its enhancement by devising the genetic algorithm into the optimization procedure is described. In Section V a design example is given.

2 Mathematical description of a time-continuous filter

The paper focuses on the most challenging step in designing a filter: the so-called approximation, which is the derivation of the filter transfer function. Classical approximations, as Butterworth, Chebyshev, Cauer, available in usual filter design software are meeting only gain or phase requirements. Usually filters have to meet also additional practical specifications as peak overshoot, rise- and settling-time and/or implementation properties as spread of same-type components values, biquads quality factors, tunability etc. Our CAD tool solves this problem and takes several specifications into account.
The method starts from the framework proposed in [VB99]. A general conti- nuous-time filter is defined by its transfer function with n complex-conjugate pole pairs:

$$p_k = a_k + jb_k,\ \bar{p_k} = a_k - jb_k, \quad a_k, b_k < 0, k = 1, ..., n \tag{1}$$

and r complex-conjugate zero pairs:

$$z_l = c_l + jd_l,\ \bar{z_l} = c_l - jd_l, \quad c_l < 0,\ b_k \leq 0, l = 1, ..., r \tag{2}$$

where $r \leq n$. The symbolic expression of the transfer function is:

$$H(s) = \frac{\prod_{k=1}^{n} \left(a_k^2 + b_k^2\right) \cdot \prod_{l=1}^{r} (s - z_l)(s - \bar{z_l})}{\prod_{l=1}^{r} (c_l^2 + d_l^2) \cdot \prod_{k=1}^{n} (s - p_k)(s - \bar{p_k})} \tag{3}$$

Next, all behavioral properties - the magnitude, phase and step response - are expressed in symbolic form as well [NTN03].

3 Multicriteria optimization in filter transfer function design

An ideal normalized filter has unitary magnitude in the pass-band, zero magnitude in the stop-band, a zero transition band, a linear phase and fast step-response with low overshoot. A real analog filter approximates (some of) the above-mentioned behavioral properties with inevitable errors. As no classical approximation can cope with both magnitude and phase requirements, let alone additional constraints (as peak overshoot, value of quality factor, etc.), one needs to derive a multicriteria transfer function in order to meet the requirements mentioned above. The first steps towards such a transfer function were achieved in the framework presented in [VB99], [NTN03] and the solution was obtained by a SQP (sequential quadratic programming) optimization procedure. The objective function for weighted multicriteria optimization was:

$$F = W_{pb}\sigma_{pb} + W_{tb}\sigma_{tb} + W_{sb}\sigma_{sb} + W_{lp}\sigma_{lp} + W_{lp}\sigma_{Q} + W_{lp}\sigma_{t} \tag{4}$$

where W_{pb} is the weight on magnitude response in passband,W_{tb} the weight on magnitude response in transition band,W_{sb} the weight on magnitude response in stopband,W_{lp} the weight on deviation from linear phase,W_Q the weight on quality factor cost,W_t the weight on percent overshoot cost. The graphical response areas σ_{pb}, σ_{tb}, σ_{sb} measure the deviation from ideal magnitude response for a lowpass filter in the pass, transition and stop-band:

$$\begin{cases} \sigma_{pb} = \dfrac{1}{\omega_p} \int_0^{\omega_p} \left(|H(j\omega)| - 1\right)^2 d\omega \\ \sigma_{tb} = \dfrac{1}{\omega_s - \omega_p} \int_{\omega_p}^{\omega_s} \left(|H(j\omega)| - \dfrac{\omega - \omega_s}{\omega_p - \omega_s}\right)^2 d\omega \\ \sigma_{sb} = \dfrac{1}{10^d\omega_s - \omega_s} \int_{\omega_s}^{10^d\omega_s} \left(|H(j\omega)|\right)^2 d\omega \end{cases} \tag{5}$$

where $10^d\omega_s \approx \infty$. Similarly, one can derive the symbolic forms of the deviation from W_{lp} the weight on deviation from linear phase the ideal quality factor σ_Q and peak overshoot in the step response σ_t.[VB99]. A user-selected classical approximation is used as the initial guess in the multicriteria optimization. The input is a set of data derived from the filter specifications passband ripple D_a, passband corner frequency Ω_p, stopband attenuation *A* and stopband corner frequency Ω_s by using a correction factor for extra margin. During the multicriteria optimization process the parameters are computed by inspection of the gain plots. For example, in the pass-band, the ripple d_a and the corner frequency ω_p are obtained as follows:

$$\begin{aligned} \omega \leq |p_k|,\ d_a &= -20\log(max|H(j\omega)t| - min|H(j\omega)| \\ \omega_p \geq max|p_k|,\ H(j\omega_p) &= min|H(j\omega)| \end{aligned} \tag{6}$$

where p_k are the poles of the transfer function. In the stop-band, the parameters stopband attenuation a and stopband corner frequency ω_s are computed as follows:

$$\omega_1 \leq \omega_p,\ \omega_2 \geq \omega_s,\ \ d_a = -20\log(max|H(j\omega_1)| - min|H(j\omega_2)|) \qquad (7)$$
$$0 < \omega < \omega_p,\ \omega_s < min|z_k|,\ \ |H(j\omega_s)| = max|H(j\omega)| - 10^{-0.1a}$$

where z_k are the zeros of the transfer function. Other parameters as slope, overshoot may be computed also by general inspection of the corresponding plotted graphs. The SQP optimization algorithm is performed on a normalized low-pass filter. The variables are the locations of the poles pk and zeros zl, which define the transfer function (equations (1)(3)). The multicriteria optimization problem was first solved by a simple repetitive (iterative) running of the procedure, with different values for the weights or different types of the initial guesses [NTN03]. This method was strongly influenced by the designer experience and very accurate solutions with respect to the requirements were hard to be obtained. The new proposed extension of the method consists in devising a genetic algorithm able to provide the values of these weights [HH04]. The complete flowchart of the proposed method is presented in Fig. 1. The filter specifications vector is a complete set of the prescribed values for the parameters of interest:

$$\mathbf{S} = [S_1...S_N]^t. \qquad (8)$$

A subset, called $\mathbf{S}*$, comprising only D_a, Ω_p, A, Ω_s, is used to calculate the parameters of the initial classical approximation; its type (Butterworth, Chebyshev or Cauer) is chosen by the user. Note that in order to create the necessary headroom for the optimization in the next step, tighter specifications are considered for the classical approximation; these are derived by multiplying the subset $\mathbf{S}*$ with a set of pre-distorting factors, called here the pre-distorting vector $\mathbf{P}$, generated by the natural genetic algorithm. By comparing $\mathbf{S}$ with the corresponding parameters of the initial (classical approximation) filter, one can derive the deviation vector:

$$\triangle = [\sigma_{pb}\sigma_{tb}\sigma_{sp}...] \qquad (9)$$

The procedure continues as long as the deviations from the prescribed specifications are not in the imposed limits.

4 Genetic algorithm for multicriteria optimization

In our method a chromosome $\mathbf{C}_i$ of the population consists of a set of N pre-distorting coefficients and 5 weighting factors (see 5 weights in equation (4)). The chromosome has N_v elements, where $N_v = N + 5$.

$$\mathbf{C}_i = [\mathbf{P}_i\mathbf{W}_i]^t, \qquad (10)$$

For the genetically mutated $\mathbf{P}_i$ and $\mathbf{W}_i$, the SQP algorithm determines the locations of the new poles p_k and zeros z_k, such that the objective function (4) is minimized. The corresponding filter parameters are estimated numerically and compared with the specifications $\mathbf{S}$. The resulting differences are gathered in an error vector which determines the fitness function of the chromosome $\mathbf{C}_i$ (Fig. 1)

$$E_i = [\varepsilon_{1i}\varepsilon_{Ni}]^t \qquad (11)$$

The goal is to get at least one chromosome for which the filter parameters are close enough to the specifications; the values of the tolerances (limit error) set by the user determines the duration of the optimization procedure. The continuous genetic algorithm has six operation stages:

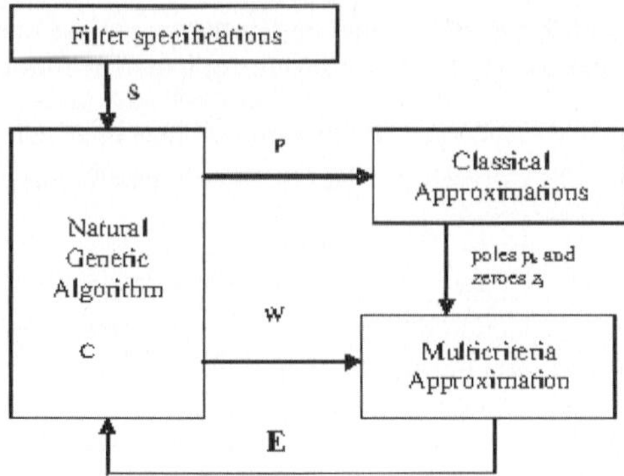

Fig. 1: The stucture of the proposed method

- Generate initial population - The population consists of a set of N_p chromosomes, therefore the program generates an $N_p x N_v$ array of continuous random values. The matrix is stored in a database. All variables are normalized and take continuous values between 0 and 1 with three digit precision (the range of a uniform random number generator).
- Natural Selection - The transfer function associated with each chromosome is generated and the error vector (11) is derived. They are ranked from the lowest to highest cost, considering a certain priority order among the $\mathbf{E}_i$ elements (11). Of the Np chromosomes in a given generation, only the top N_k are kept for mating; the value of N_k is determined by the selection rate, set by the user.
- Pairing - The N_k most-fit chromosomes form the mating pool, from each two mothers and fathers pair in some random fashion. Each pair produces two offsprings that contain traits from each parent. Both the parents and their offsprings survive to be part of the next generation.
- Mating - The offsprings are derived as combination of the parents using the Radcliff approach [HH04].
- Mutations - The number of variables that will be mutated is determined by the mutation rate, set by the user. The total number of mutations results as the product of the mutation rate, $(N_p - 1)$ and N_v. Next, the position (the row and column numbers in the population matrix) of the variables to be mutated are chosen in a random process. A mutated variable is replaced by a new value, also randomly generated.
- Next Generation - The next generation is obtained from all previews steps and it is stored in a new database. The algorithm procedure is repeated until one of the following three events happens: (a) a solution is found; (b) the population average remains constant for a number of generations (no evolution); (c) the algorithm reaches the maximum number of generations N_G, set by the user. The input data for the genetic algorithm procedure are: population size N_p, number of generations N_G, selection and mutation rates. The main parameters are stored at each generation: the errors average, population average and mean cost. The algorithm is written in C++, using Paradox database and SQL [Cel05].

5 A design example

Let us design a high-pass filter to the following specifications: $D_a = 0.5dB20\%$, $A = 60dB10\%$, $\Omega_p = 30MHz7\%$, $\Omega_s = 10MHz10\%$, D_t passband group delay deviation$= 30ns10\%$. For the genetic algorithm the following restrictions were set: $N_p = 300$, $N_G = 250$, selection rate$= 50\%$ and mutation rate$= 30\%$. Table 1 summarizes the specifications and the main parameters of three filters from the designing process. The Ist and IInd filter are Cauer ones and meet either predistorted requirements (Ist filter) or not distorted requirements (IInd filter). The IIIrd filter is obtained as a result of an optimization procedure. One can easily see that group delay specifications are not reached at all by the IInd filter, but gain requirements are in the imposed limits. The Ist filter is the initial guess for the optimization procedure; most of its parameters are far away from the imposed requirements. After running a multicriteria optimization combined with the genetic algorithm on the initial guess (Ist filter), the solution (IIIrd filter) was obtained. It meets all the specs (compare second and last columns in Table 1). Figure 2 shows the magnitude response and group delay of the initial guess (dashed line) and the multicriteria optimized filters (line) for comparison. There is an important improvement in the group delay characteristics without significant losses in the gain plot.

Table 1: Main parameters of the filters in the process

Parameter	Specs	Ist filter (initial guess)	IInd filter (Cuer)	IIIrd filter (optimized filter)
Ripple in PassBand	$0.4 \div 0.6\ dB$	1.2 dB	0.55 dB	0.55 dB
StopBand attenuation	$54 \div 66\ dB$	60 dB	56 dB	56 dB
PassBand frequency	$27.9 \div 32.1\ MHz$	30.062 MHz	28.079 MHz	28.079 MHz
StopBand Frequency	$9 \div 11\ MHz$	11.909 MHz	10.598 MHz	10.598 MHz
Group Delay Deviation in PassBand	$27 \div 33\ ns$	97.749 ns	72.557 ns	31.311 ns

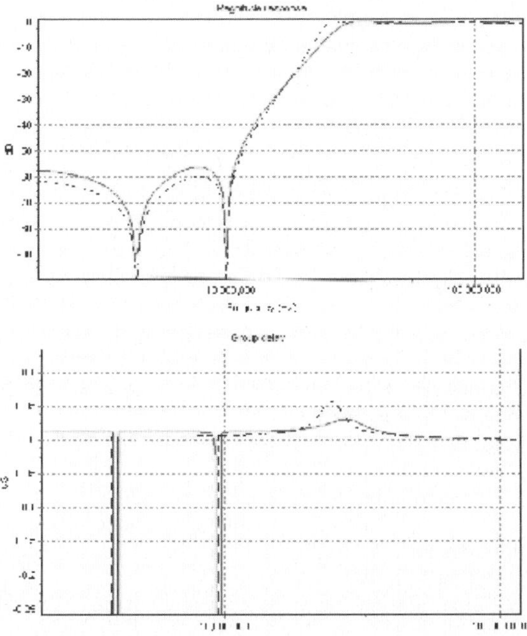

Fig. 2: **M**agnitude response and group delay of the multicriteria optimized filter

Figure 3 presents the average population during the operation of the genetic algorithm. There is no evolution after the 98th generation.

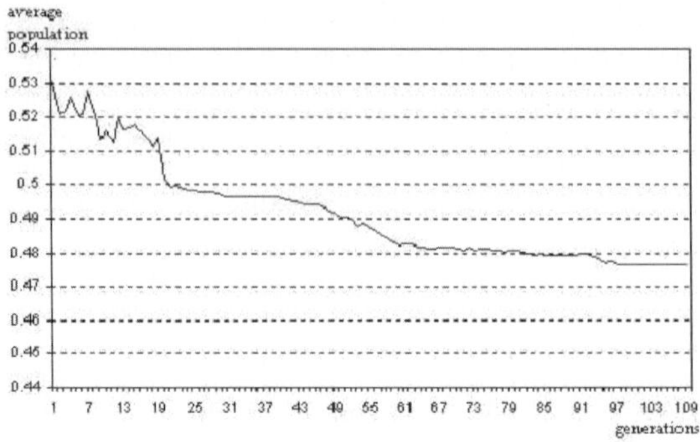

Fig. 3: The evolution of the average popultion in genetic algorithm

6 Conclusions

The paper presents a framework for designing filter transfer functions with a wide range of behavioural and implementation requirements, which is a necessity for electronic circuits designers. No available software is reported in the literature. The method starts with deriving an initial guess, obtained by classical approximations. Then this solution is refined using a multicriteria optimization procedure that combines numerical optimization with symbolic computation. The weighting factors in the objective function are controlled by a genetic algorithm. This algorithm ensures good convergence even for less experienced designers. The whole procedure is included in a software package, which designs cascaded active filter. Further work will include introduction of additional requirements and an increased range of classical approximations types handled by the design framework.

References

[The04] Thede, Les, Practical Analog and Digital Filter Design, Artech House, Inc., Norwood, MA, USA (2004)

[Fil]Filter Solutions http://www.filter-solutions.com; FilterLab http://www. microchio.com; Filter Wiz Pro http://www.schematica.com

[HH04] Haupt, R. L., Haupt, Sue Ellen, Practical Genetic Algorithms - Second Edition, John Wiley & Sons, Inc., New Jersey, USA (2004)

[Cel05] Celko, J., SQL Programming Style, Elsevier Inc, San Francisco, CA, USA (2005)

[VB99] Venkata, N. D., Evans, Brian L. An automated framework for multi- criteria optimization of analog filter designs, IEEE Trans on Circ. & Syst.- II: Analog and Digital Signal Processing, Vol. 46, No. 8,pp. 981-990 (1999)

[NTN03] Nedelea, L., Topa, Marina, Neag, M. Computer-Aided Network Function Approximation for Multicriteria Filter Design, SCS'2003 - Int. Symp.on Sign., Circ.& Syst., Iasi, Romania, pp. 81-84 (2003)

Thermal Network Method in the Design of Power Equipment

C. Gramsch[1], A. Blaszczyk[2], H. Löbl[3] and S. Grossmann[3]

[1] hagenuk KMT GmbH, Rderaue 41, 01471 Radeburg, Germany,
Gramsch.C@sebakmt.com
[2] ABB Corporate Research, 5405 Baden-Daettwil, Switzerland,
Andreas.Blaszczyk@ch.abb.com
[3] Technical University Dresden, Institute of Electrical Power Systems and High Voltage Engineering, 01062 Dresden, Germany,
[Loebl, Grossmann]@ieeh.et.tu-dresden.de

Abstract - The paper presents basic principles of the thermal networks method. Modeling of network elements has been shown for convection and radiation. A concept of hierarchical thermal network models for complex geometries has been explained. A formulation of a mass transfer model and a new iterative method of coupling ventilation and thermal networks have been introduced. The last section includes an example of the thermal network computation and a comparison with test results.

1 Introduction

The Thermal Network Method (TNM) is based on a substitution of an arbitrary 3D geometry by a circuit consisting of thermal resistances, capacitances and heat sources. For such a network the currents correspond to heat flow and the nodal potentials to temperatures. Due to similarity of mathematical formulations the electrical circuit programs can be used to obtain a solution. The basic advantage of the thermal network analysis is the fast computation time: steady state computations of large models can be performed within a few seconds. Therefore the TNM is very suitable for parameter studies and became popular as a tool supporting the industrial design. A drawback of TNM is the creation of the network, in particular transition from the real geometry to a network based model. This drawback can be mitigated by applying hierarchical modelling approach and reusable library elements with ready-to-use representations of the whole devices.
An important effect that can be modelled using TNM technology is ventilation. For complex models of power devices consisting of many compartments we introduced a concept of ventilation networks that allow a comprehensive analysis of mass transfer coupled with TNM. In this paper we describe the hierarchical TNM approach including ventilation networks and its application to real design cases.

2 Basic Concept of Thermal Networks

The basic concept of substituting the geometrical objects by a thermal network model is shown for an example of a coated conductor carrying electrical current I, Fig. 1. The current generates temperature dependent power losses that are conducted along the conductor (in metal) as well as through the insulation layer and dissipated via convection and radiation. A part of the generated heat can be stored in the conductor material represented by the capacitance C (used for the transient computations only). In this paper we present basic formulas for the calculation of the circuit elements corresponding to convection R_{conv} and radiation R_{rad}. More details and formulations for other network elements (R_{cond}, R_{coat}, C, P) are included in [Hol02, Boe05, Loe99].

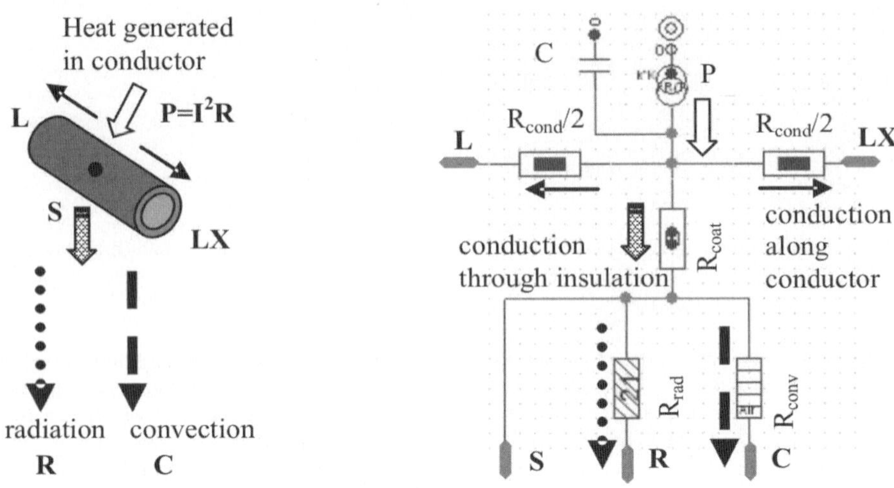

Fig. 1: Thermal network model of coated conductor

The thermal resistance of convection can be expressed as follows:

$$R_{conv} = \frac{1}{\alpha_{conv} A_{conv}} \tag{1}$$

with the convection coefficient α_{conv} and the surface area A_{conv}. The calculation of α_{conv} is based on the similarity theory [Hol02, Boe05, Loe99], which requires evaluation of the characteristic numbers of Nusselt (Nu), Grasshof (Gr), Prandtl (Pr) and Reynolds (Re). The following basic relationship can be used:

$$\alpha_{conv} = \frac{Nu \cdot \lambda_{fluid}}{l_{ch}} \tag{2}$$

where λ_{fluid} is the thermal conductivity of fluid and l_{ch} the characteristic length (e.g. the height of a vertical plate or the diameter of a horizontal cylinder). The Nusselt number is calculated for the natural convection with coefficients c_1 and n_1:

$$Nu = c_1(GrPr)^{n_1} \tag{3}$$

while for the forced convection with coefficients c_2 and n_2:

$$Nu = c_2(RePr)^{n_2}. \tag{4}$$

The values of coefficients c_1, n_1, c_2, n_2 are obtained experimentally for typical geometrical configurations [Boe05]. The products $GrPr$, $RePr$ and λ_{fluid} depend on temperature [Loe99]. The implemented convection resistances can be used for the laminar and turbulent flow models in different fluids (air, SF_6, oil, H_2O).
The radiation resistance is expressed in a similar way as (1):

$$R_{rad} = \frac{1}{\alpha_{rad} A_{rad}}. \quad (5)$$

The radiation coefficient α_{rad} is based on the Stefan-Boltzmann-constant $\sigma = 5.67 \cdot 10^{-8} W/(m^2 K^4)$, the emissivity ϵ_{12} between the radiating and absorbing surfaces and the absolute temperatures T_1 and T_2 of both surfaces:

$$\alpha_{rad} = \frac{\epsilon_{12} \cdot \sigma(T_1^4 - T_2^4)}{T_1 - T_2}. \quad (6)$$

The value of ϵ_{12} depends on emissivity of the emitting and absorbing surfaces, their surface area and the viewing factor between them. In case of a conductor located in a free space the emissivity of the conductor outer surface can be used for ϵ_{12}.
Due to the temperature dependency of convection and radiation resistances as well as power losses the network problem is nonlinear. This kind of problems can be efficiently solved using programs for analysis of electric circuits like Spice or its commercial derivates (www.pspice.com). The nonlinear resistors have been implemented as voltage controlled current sources.

3 Hierarchical Approach

The generation of a thermal network for a complex device is a time consuming process. To make it easier and faster we introduced a concept of hierarchical thermal networks. In hierarchical approach we create networks consisting not only of primitive elements representing physical phenomena like resistors or sources but also models of whole components. For example, the network scheme of a conductor shown in Fig. 1 consists of 5 thermal resistances, one capacitance and one source. These elements can be wrapped together into a new element representing the whole conductor with pins corresponding to the heat conduction (L,LX), convection (C), radiation (R) and the outer surface (S). The example in Fig. 2 shows application of the new "coated conductor" element (denoted here as CN_gICYL1) in a network model representing encapsulated conductor. In this model additionally radiation (RRA1), convection (RCO1/2) and eddy losses (PI2R1) related to enclosure walls as well as ventilation (RFR1) have been included. The network in Fig. 2 can be again wrapped into a new "encapsulated conductor" element and applied in higher hierarchical levels.
For complex devices we use up to 5 hierarchical network levels. The hierarchical approach allows a better management of large models and reusability of components.

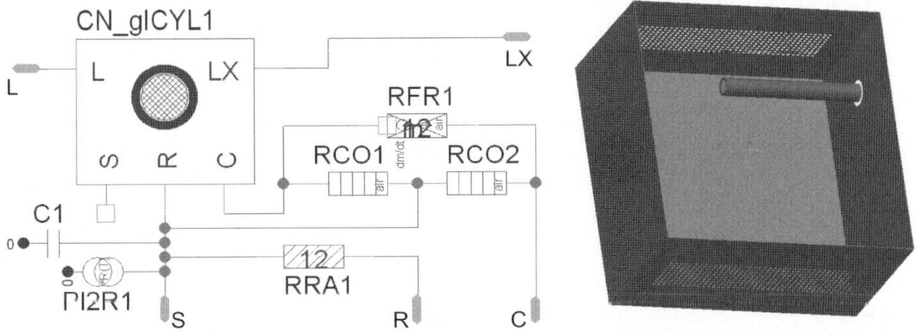

Fig. 2: Example of a hierarchical thermal network representing coated conductor in a ventilated enclosure

4 Modeling of ventilation

4.1 Formulation

Air insulated electrical power system devices are usually manufactured with openings in the outer walls in order to ensure better natural ventilation, e.g. compact substations, air insulated medium voltage switchgear and control panels. Typically, a compartment including heat sources has at least one inlet and one outlet that enable transfer of air as shown in Fig. 3.

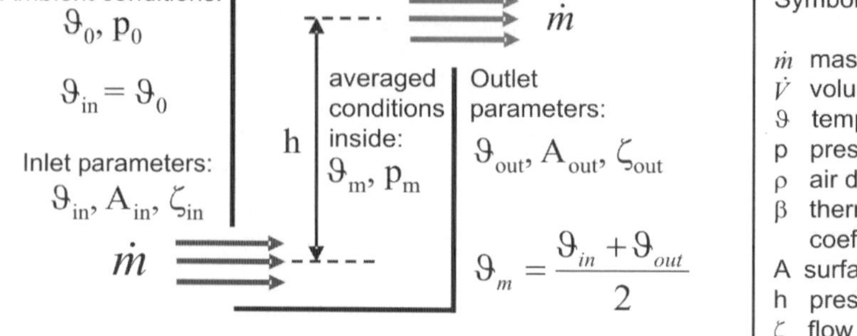

Fig. 3: Basic relationships for the natural ventilation of a compartment

The heat power P transferred via ventilation inside of a compartment can be calculated as:

$$P = \dot{m} \cdot c_p(\vartheta_{out} - \vartheta_{in}) = 2\dot{m} \cdot c_p(\vartheta_m - \vartheta_0) \tag{7}$$

and consequently the resistor representing ventilation heat flow through the outlet is expressed as follows (R_v is connected between the inner air node at temperature ϑ_m and the air node outside):

$$R_v = \frac{1}{2\dot{m} \cdot c_p} \tag{8}$$

where c_p is the specific heat of air.

The mass flow $\dot{m}$ can be obtained from the basic relationship between the velocity v and the pressure difference Δp for turbulent flows through barriers (like inlet and outlet) as follows [RSS95] (see explanation of symbols in Fig. 3):

$$\Delta p = \zeta \frac{1}{2} \rho \cdot v^2. \tag{9}$$

After substituting in (9) the velocity v by mass flow and introducing the temperature dependency of ρ:

$$v = \frac{\dot{V}}{A} = \frac{\dot{m}}{\rho \cdot A} \quad , \tag{10}$$

$$\rho = \frac{\rho_{0^\circ C}}{1 + \beta \cdot \vartheta} \tag{11}$$

the following basic formula can be obtained:

$$\Delta p = \dot{m}^2 \cdot S = \dot{m} \cdot S_m \tag{12}$$

where S denotes the flow resistance calculated as a function of flow parameters ϑ, A, ζ (see explanation of symbols in Fig. 3) while S_m is a nonlinear, mass flow dependent resistance ($S_m = \dot{m} \cdot S$) for the use in ventilation networks (as described in the next subsection). In the configuration shown in Fig. 3 the pressure difference occurs due to flow through both openings including inlet and outlet. Consequently the total pressure drop can be expressed as a sum:

$$\Delta p = \Delta p_{in} + \Delta p_{out} = \dot{m}^2 (S_{in} + S_{out}) \tag{13}$$

where the flow resistances S for inlet and outlet are calculated according to the following formula (see also Fig. 3):

$$S_{\{in,out\}} = \zeta_{\{in,out\}} \frac{1}{2A^2_{\{in,out\}}} \frac{1 + \beta \cdot \vartheta_{\{in,out\}}}{\rho_{0^\circ C}}. \tag{14}$$

The pressure difference Δp can be calculated based on density difference between the air outside and inside of the compartment:

$$\Delta p = g \cdot h \cdot \rho_{0^\circ C} \left(\frac{1}{1 + \beta \cdot \vartheta_0} - \frac{1}{1 + \beta \cdot \vartheta_m} \right). \tag{15}$$

The formulas (14)-(15) applied to (13) allow to calculate the mass flow and consequently the temperature dependent value of R_v in (8). The ventilation resistor is usually placed between the gas nodes representing air inside and outside of a compartment, see resistor RFR1 shown in Fig. 2.

4.2 Ventilation Networks

Based on formula (12) complex nonlinear networks representing ventilation mass flows can be created. Similarly to thermal networks we use in case of ventilation networks the analogy to electric circuits: the electric current is represented by the mass flow $\dot{m}$ while the voltage by the pressure difference Δp. An example of a ventilation problem that requires a network based analysis is shown in Fig. 4.

For the model in Fig. 4a we create two networks: a ventilation network that allows analysis of mass flow, Fig. 4b, as well as a more complex thermal network for the calculation of temperatures and heat flow, Fig 4c. The main purpose of the ventilation network is to obtain the mass flow values that can be applied in ventilation resistors of the thermal network. The iterative computation procedure includes the following steps:

1. Assume initial values of the nodal temperatures ϑ in the thermal network.
2. For known temperatures calculate the values of resistors S and the pressure differences Δp between all openings according to (14), (15) and create the nonlinear ventilation network, Fig. 4b.
3. Solve the ventilation network in Fig. 4b and apply the obtained mass flow values to calculate the values of R_v resistors (8).
4. Solve the thermal network in Fig. 4c and obtain the new values of nodal temperatures ϑ.
5. Check the differences between the new values of nodal temperatures and the corresponding values applied for computations in step 2. If the differences are small enough stop iterations, otherwise go to step 2.

Typically a few iteration steps are sufficient to achieve the convergence (changes of nodal temperatures below 0.1 K) in case of real ventilation problems.

5 Example

An example of a complex power device is shown in Fig. 5a. Based on hierarchical thermal network approach we computed steady state temperatures along the phase conductors. The deviations between computations and tests are for most measured points in the range of 3 K, see Fig. 5b. The created model allows simulation of mass transfer phenomena and has been used to study different ventilation solutions.

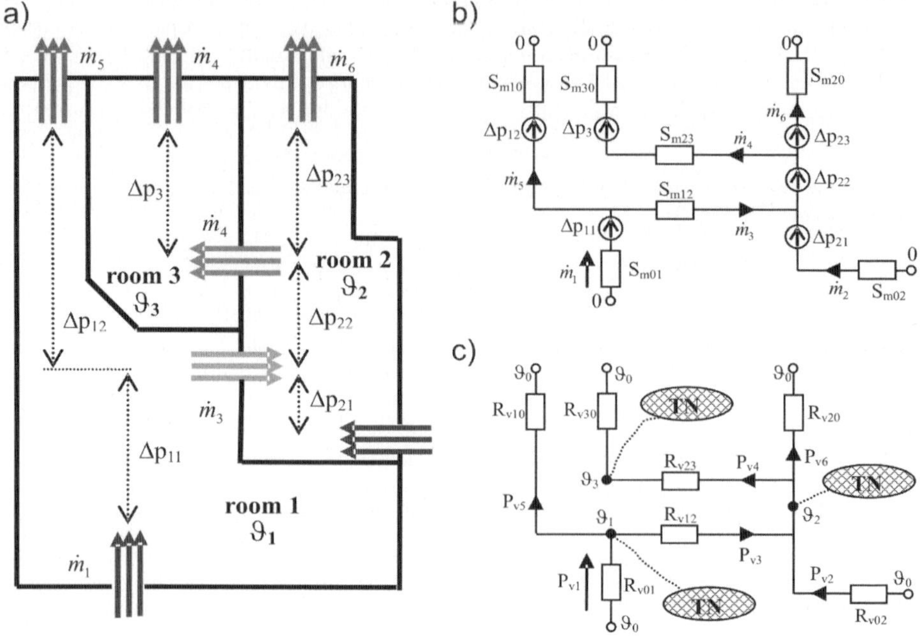

Fig. 4: Example of an arrangement with air flow between compartments (a) and the corresponding models of (b) ventilation network and (c) thermal network. (The configuration of compartments and the ventilation openings is the same as in the arrangement shown in Fig. 5.)

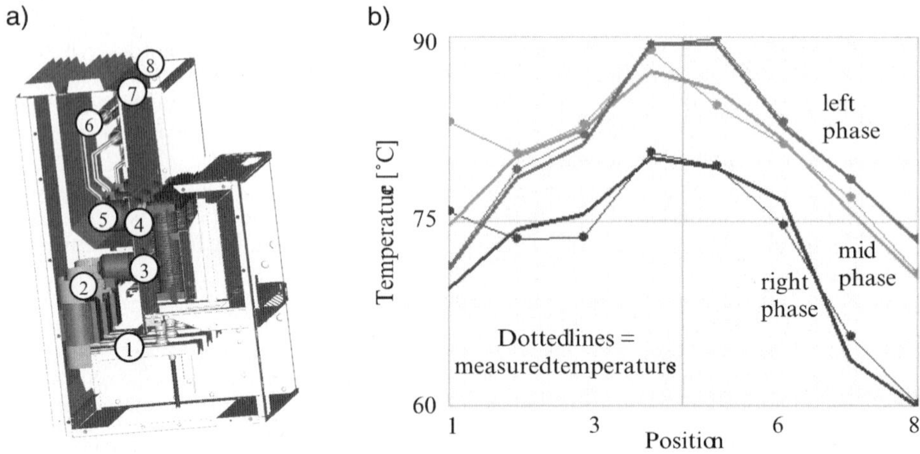

Fig. 5: Air insulated medium voltage switchgear arrangement. The numbers shown in circles on geometry view (a) denote positions of measurement/calculation points at the metal surface of the current carrying conductors. These positions correspond to the values shown on the X-axis of the temperature distribution (b)

6 Conclusion

Thermal network method has been effectively applied to thermal design of complex power devices like high and medium voltage switchgear, transformers, bushings and circuit-breakers.

The hierarchical modeling approach allows efficient handling of complex geometries. A good agreement with experimental results can be achieved. The fast computation times enable comprehensive parameter studies for modeled devices.

References

[Hol02] J. P. Holman, “Heat Transfer” , McGraw-Hill Higher Education, 9th edition, 2002

[Boe05] H. Bhme, “Mittelspannungstechnik”, Verlag Technik Berlin Mnchen, 2. Auflage, 2005

[Loe99] H. Lbl, “Basis of Thermal Networks”, (unpublished) Dresden University of Technology, 1999.

[RSS95] Recknagel, Sprenger, Schramek, “Taschenbuch fr Heizung und Klimatechnik”, R. Oldenbourg Verlag GmbH, Mnchen, 67. Auflage, 1995

Hierarchical Mixed Multirating in Circuit Simulation

Michael Striebel[1] and Michael Günther[2]

[1] Infineon Technologies Austria AG, Siemensstr. 2, A-9500 Villach, Austria
michael.striebel2@infineon.com
[2] Bergische Universität Wuppertal, Departement of Mathematics, Chair of Applied Mathematics/Numerical Analysis, D-42097 Wuppertal, Germany.
guenther@math.uni-wuppertal.de

Summary. In most applications large integrated circuits comprise subcircuits of different functionality causing heterogeneous transient behaviour. Multirate methods exploit local latency by using different stepsizes according to the subcircuits' activity levels at each time point. Following the idea of mixed multirate for ordinary differential equations a hierarchical ROW-based multirate method that can deal with an arbitrary number of subcircuits is developed.

1 Modular Modelling

Large electrical circuits are usually built up in a modular way. Several subcircuits of different functionality are developed separately, or taken from some library, and glued together.
To enable the embedding of a single subcircuit into a complex design, each module needs to have terminals at which it can connect to its surrounding. For this purpose a subset of its nodes is declared "pins". Only at these pins a current flow across the subcircuit's border is permitted. We identify each single module with an index $\lambda \in \mathbb{N}$ and introduce the vector $\jmath_{P_\lambda} \in \mathbb{R}^{n_{P_\lambda}}$ of the pin currents. Furthermore, we agree that they are leaving the unit at the terminals. Therefore, the mapping of the pin currents to the corresponding nodes of the subcircuit can be described by the additional incidence matrix $A_{P_\lambda} \in \{0,1\}^{n_{e_\lambda} \times n_{P_\lambda}}$, where n_{e_λ} and $n_{P_\lambda} \ll n_{e_\lambda}$ are the number of nodes and pins, respectively in the λth subcircuit.
Regarding fundamental units, i. e. modules that do not contain other subcircuits' instantiations, charge oriented modified nodal analyis (MNA) [3] yields network equations of the following form:

* The author is indebted to Infineon Technologies München, and especially to Drs. Feldmann and Schultz, for supporting his PhD project.

$$
\begin{aligned}
0 &= A_{C_\lambda}\dot{q}_\lambda + A_{R_\lambda} r(A_{R_\lambda}^t e_\lambda) + A_{L_\lambda} \jmath_{L_\lambda} + A_{V_\lambda} \jmath_{V_\lambda} + \\
&\quad + A_{I_\lambda} \imath_\lambda(t) + \boxed{A_{P_\lambda} \jmath_{P_\lambda}}, && \text{(1a)}\\
0 &= \dot{\Phi}_\lambda - A_{L_\lambda}^t e_\lambda, && \text{(1b)}\\
0 &= A_{V_\lambda}^t e_\lambda - v_\lambda(t), && \text{(1c)}\\
0 &= q_\lambda - q_{C_\lambda}(A_{C_\lambda}^t e_\lambda), && \text{(1d)}\\
0 &= \Phi_\lambda - \phi_{L_\lambda}(\jmath_{L_\lambda}). && \text{(1e)}
\end{aligned}
$$

From the outside each subcircuit can be regarded as a black box element with a specific number of terminals. Therefore, it can be instantiated in a more complex subcircuit. A hierarchical circuit design is generated in this way. However, in the following we consider circuits consisting of $r \in \mathbb{N}$ subcircuits that that can be described according to (1). Note that in a hierarchical design this representation can be obtained from fanning out the subcircuits' hierarchical structure. First, we describe some strategies to couple r subcircuits.
A *master circuit*, which acts as a carrier network presents the most flexible strategy. It consists of n_M master nodes and r subcircuit instantiations as the only elements. As the pin currents leave the subcircuits and flow into the master nodes, the master's topology can be described by the incidence matrix

$$A_M = (A_{Z_1}, \ldots, A_{Z_r}),$$

with $A_{Z_\lambda} \in \{0, -1\}^{n_M \times n_{P_\lambda}}$ assembling the pin currents of the λth subcircuit.
Kirchoff's current law, applied to the master nodes yields

$$A_{Z_1} \jmath_{P_1} + \cdots + A_{Z_r} \jmath_{P_r} = 0. \tag{2a}$$

As the subcircuits, or *slaves*, are attached to the master's nodes, the appropriate nodal voltages have to coincide, i. e.

$$A_{P_\lambda}^t e_\lambda + A_{Z_\lambda}^t e_M = 0, \quad \text{for each } \lambda = 1, \ldots, r. \tag{2b}$$

The system (1,2) with $\lambda = 1, \ldots, r$ determines the slaves' quantities e_λ, $\jmath_{L_\lambda}$, $\jmath_{V_\lambda}$, q_λ, Φ_λ, i. e. node potentials, currents through inductors and voltage sources, charges and fluxes, and the pin currents $\jmath_{P_\lambda}$ as well as the master's node potentials e_M.
A *peer-to-peer network* where interrelated subcircuits are connected directly by short-circuit is another strategy. The compound is described by r subcircuits and a set w of n_W coupling currents. A subcircuit's pin currents $A_{P_\lambda} \jmath_{P_\lambda}$ is replaced by a specific selection $A_{W_\lambda} w$ of the coupling currents, where also the coupling currents' direction is allowed for by the matrix $A_{W_\lambda} \in \{-1, 0, 1\}^{n_{e_\lambda} \times n_W}$. As no additional nodes are needed the interconnection is described by

$$A_{W_1}^t e_1 + \ldots + A_{W_r}^t e_r = 0, \tag{3}$$

reflecting that voltages of connected subcircuits have to coincide. Again, (1,3) serve as a model for a modular design.

Compatibility

The shorts to the master nodes (2b) and to other subcircuits' pins (3), respectively, can be understood as paths built by *virtual voltage sources* that provide a zero-voltage drop [1]. Therefore, there is an upward compatibility of the two approaches as each virtual voltage source can be added to the set of voltage sources of one subcircuit.
The model equations of the *master circuit* and the *peer-to-peer network* approach can be written as a differential-algebraic equation (DAE):

$$
\left.\begin{aligned}
0 &= \mathcal{A}_\lambda \dot{y}_\lambda + f_\lambda(x_\lambda) + \mathcal{A}_{\lambda,\text{ext}} x_\text{ext},\\
0 &= y_\lambda - g_\lambda(x_\lambda),
\end{aligned}\right\} \quad \text{for } \lambda = 1, \ldots, r \tag{4a}
$$

$$0 = \mathcal{A}_{\text{ext},1} x_1 + \cdots + \mathcal{A}_{\text{ext},r} x_r \tag{4b}$$

with the charges/fluxes in y_λ, the node potentials and currents through inductors and voltage sources (and the pin currents) in x_λ and the coupling currents (or the master nodes' potentials) in x_{ext}, respectively. On the basis of the well known argument we omit time dependency.
In the following we will stick to the *peer-to-peer network* for analysis and to the *master circuit* approach for the implementation. Actually, in circuit simulators that support a hierarchical circuit design the the pin currents are not needed explicitly as additional unknowns. There they arise naturally from the factorisations of the matrices that display the subcircuits' inner topology and their contribution to the master circuits (see [9]). However, we introduce them as additional unknowns for some reasons we will mention later on.

Index Properties

The network equations (4) consists of a set of r subcircuit models (4a) coupled by (4b). Restricting ourself to index-1 problems we demand:

(C1) The overall system (4a,4b) has index 1 (with respect to $x_1, \dots, x_r, x_{\text{ext}}$).
(C2) All systems (4a) define index-1 systems with respect to x_λ (with $x_{\text{ext}} = w$ given as input).

From [3] we know, that (C2) holds, if there are neither CV-loops nor LI-cutsets in the subcircuits. Note, that the virtual voltage sources are regarded as current sources that provide the current w in this context. Following these lines, we can show that (C1) holds if there are also no loops of capacitors, voltage sources and virtual voltage sources in the overall circuit [1, 7]. Under these conditions (4) is equivalent to the semi-explicit system:

$$\left.\begin{aligned} \dot{y}_\lambda &= f_\lambda(z_\lambda, w), \\ 0 &= h_\lambda(y_\lambda, z_\lambda, w), \end{aligned}\right\} \quad \text{for } \lambda = 1, \dots, r \tag{5a}$$

$$0 = g(z_1, \dots, z_r)\ , \tag{5b}$$

where f_λ, h_λ are linear in w and y_λ, w, respectively and g is linear in $z_1, \dots, z_r$.

2 Multirate Methods

The modular design of complex systems often implies heterogeneous transient behaviour. Different parts of the overall system make different demands on the stepsizes in order to guarantee given error tolerances. The basic idea of *multirate methods* is to integrate each subsystem with its appropriate stepsize and thus prevent parts to be integrated with smaller stepsizes than necessary.
We associate *activity levels* with the range of the stepsizes at each timepoint. Subsystems that propose a small stepsize are called *active* and those content with a large stepsize are called *latent*. Fig. 1 shows the typical multirate situation: One large *macrostep* can be applied to the latent part whereas the active part can only be integrated with small *microsteps* without leaving the prescribed accuracy range. In multirate schemes an overhead is caused by the necessity to balance approximations on different timegrids, i. e. each timestep for one part needs information about the other parts' state. To save expenses in comparison with classical schemes, this overhead has to be outbalanced by the reduction of computational costs for the less active parts' discretisation. Hence systems showing heterogeneous transient behaviour are said to have *multirate potential* if the different timescales are widely seperated, the latent parts are larger than the active ones and the coupling amongst subsystems representing different activity levels is weak.

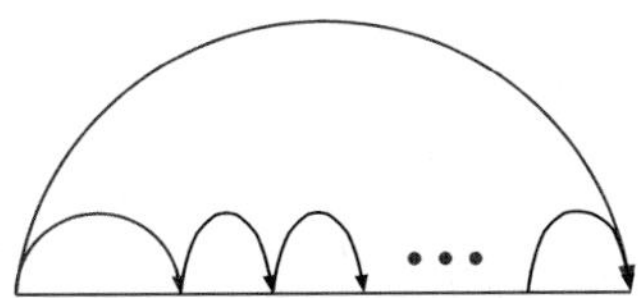

Fig. 1 Macro- and microsteps: stepsize selection.

Various multirate strategies emerge from addressing the aspects of *underlying method*, *mediation across the time grids* and *sequence of computation*. Onestep and multistep based approaches working with interpolation, extrapolation and Runge-Kutta (RK) like formulas to map the transition between the different timegrids are being analysed. For an overview see [4]. Furthermore, there is a refinement strategy being investigated for both BDF methods [8] and onestep methods [5].
With regards to circuit simulation we intend to develop a Rosenbrock-Wanner (ROW) based onestep multirate scheme because here stepsize selection and error control in terms of node potentials and currents is possible and no nonlinear equation has to be solved.

2.1 Mixed Multirate for ODEs

We start with a system of coupled ordinary differential equations

$$\dot{y}_L = f_L(y_L, y_A), \quad y_L(t_0) = y_{L,0}, \tag{6a}$$
$$\dot{y}_A = f_A(y_L, y_A), \quad y_A(t_0) = y_{A,0}, \tag{6b}$$

and suppose that, at the actual timepoint t_0, the latent part (subscript L) can be integrated with one macrostep $\mathcal{H}_L$ whereas a sequence of q microsteps $\mathcal{H}_{A,1}, \ldots, \mathcal{H}_{A,q}$ is needed for the active part (subscript A) to reach $t_0 + \mathcal{H}_L$.
In its most general way the onestep formalism of this procedure is

$$y_{L,1} = y_{L,0} + \sum_{i=1}^{s_L} b_i^L \cdot k_i^L, \tag{7a}$$
$$y_{A,\mu} = y_{A,\mu-1} + \sum_{i=1}^{s_A} b_i^A \cdot k_i^{A,\mu} \quad (\mu = 1, \ldots, q), \tag{7b}$$
$$k_i^L = \Phi_L(\mathcal{H}_L; y_{L,0}, Y_i^A, k_1^L, \ldots, k_{s_L}^L) \quad (i = 1, \ldots, s_L), \tag{7c}$$
$$k_i^{A,\mu} = \Phi_A(\mathcal{H}_{A,\mu}; y_{A,\mu-1}, Y_i^{L,\mu}, k_1^{A,\mu}, \ldots, k_{s_A}^{A,\mu}) \quad (i = 1, \ldots, s_A), \tag{7d}$$

where $\Phi_\mathcal{L}$ denotes an $s_\mathcal{L}$ stage RK or ROW scheme with coefficients $b^\mathcal{L}$, $A^\mathcal{L}$, $B^\mathcal{L}$, $\Gamma^\mathcal{L}$ $(\mathcal{L} = L, A)$. For ROW schemes the equations (7c,d) which determine the increments are linear in k_i^L, $k_i^{A,\mu}$, respectively.
As the subsystems are coupled, the computation of the increments $k^\mathcal{L}$ for each part depends on information on the other one at some supporting timepoints:

$$Y_i^A \approx y_A(t_0 + \alpha_i^L \mathcal{H}_L) \quad (i = 1, \ldots, s_L), \tag{8a}$$
$$Y_i^{L,\mu} \approx y_L(t_0 + \sum_{\nu=1}^{\mu-1} \mathcal{H}_{A,\nu} + \alpha_i^A \mathcal{H}_{A,\mu}) \quad (i = 1, \ldots, s_A; \mu = 1, \ldots, q). \tag{8b}$$

Mixed multirate [2] is characterised by a "compound step" and a series of "later microsteps". In the former the macrostep (7a) and the first microstep (7b) $(\mu = 1)$ are done at once. Y_i^A, $Y_i^{L,1}$ are determined in RK-like manner. For this, additional coefficient $\delta^{\mathcal{L}\bar{\mathcal{L}}}$, $\nu^{\mathcal{L}\bar{\mathcal{L}}}$ $(\{\mathcal{L}, \bar{\mathcal{L}}\} = \{L, A\})$ are introduced and the increments $k_i^{A,1}$ and k_i^L are scaled with the *stepsize ratio* $\mathfrak{m} = \frac{\mathcal{H}_L}{\mathcal{H}_{A,1}}$ and its inverse $\mathfrak{m}^{-1}$, respectively. Note, that $\mathfrak{m} = q$ if $\mathcal{H}_{A,1} \equiv \cdots \equiv \mathcal{H}_{A,q}$.
For the later microsteps *dense output* formulas are applied to get reasonable values $Y_i^{L,2}, \ldots, Y_i^{L,q}$. Dense output formulas a-posteriori provide cheap numerical approximations, e. g.

$$y_L(t_0 + \theta \cdot \mathcal{H}_L) \approx y_{L,0} + \sum_{i=1}^{s} b_i^{\text{ds}}(\theta) \cdot k_i^L \quad \text{with} \quad b_i^{\text{ds}}(\theta) = \sum_{k=1}^{s} b_{ik} \cdot \theta^k \tag{9}$$

on the interval $[t_0, t_0 + \mathcal{H}_L]$, i. e. with $\theta \in [0, 1]$.

2.2 Mixed multirate scheme for coupled index-1 DAE systems

The next step towards a multirate method for the system (5) that is capable of handling an arbitrary amount of activity levels, is to restrict to a two level scheme for the index-1 system

$$\begin{aligned} \dot{y}_L &= f_L(z_L, w) & \dot{y}_A &= f_A(z_A, w) \\ 0 &= h_L(y_L, z_L, w) & 0 &= h_A(y_A, z_A, w) \\ & & 0 = g(z_L, z_A). \end{aligned} \tag{10}$$

The coupling current w, affecting both subsystems, is assumed to be latent.
Using a ROW scheme with $s(= s_L = s_A)$ stages, the compound step for the index-1 DAE-system (10), with weights $b^{\mathcal{L}}$ and increments $l^{\mathcal{L}}$, $k^{\mathcal{L}}$, $p^{\mathcal{L}}$ reads:

$$\begin{pmatrix} y_{L,1} \\ z_{L,1} \\ w_1 \end{pmatrix} = \begin{pmatrix} y_{L,0} \\ z_{L,0} \\ w_0 \end{pmatrix} + (b^L)^t \begin{pmatrix} l^L \\ k^L \\ p \end{pmatrix}, \quad \begin{pmatrix} y_{A,1} \\ z_{A,1} \end{pmatrix} = \begin{pmatrix} y_{A,0} \\ z_{A,0} \end{pmatrix} + (b^A)^t \begin{pmatrix} l^A \\ k^A \end{pmatrix}. \tag{11a}$$

According to (7c,d) the stage increments are defined by the linear system

$$M^\star \cdot (l_{L,i}, k_{L,i} | l_{A,i}, k_{A,i} | p_i)^t = \text{RHS}_i, \quad \text{for } i = 1, \ldots, s \tag{11b}$$

with $M^\star =$

$$\left(\begin{array}{cc|cc|c} \mathrm{I}_{y_L} & -\mathcal{H}_L \gamma^L \frac{\partial f_L}{\partial z_L} & & & -\mathcal{H}_L \gamma^L \frac{\partial f_L}{\partial w} \\ -\gamma^L \frac{\partial h_L}{\partial y_L} & -\gamma^L \frac{\partial h_L}{\partial z_L} & & & -\gamma^L \frac{\partial h_L}{\partial w} \\ \hline & & \mathrm{I}_{y_A} & -\mathcal{H}_A \gamma^A \frac{\partial f_A}{\partial z_A} & -\frac{1}{\mathfrak{m}} \cdot \mathcal{H}_A \nu^{AL} \frac{\partial f_A}{\partial w} \\ & & -\gamma^A \frac{\partial h_A}{\partial y_a} & -\gamma^A \frac{\partial h_A}{\partial z_A} & -\frac{1}{\mathfrak{m}} \cdot \nu^{AL} \frac{\partial h_A}{\partial w} \\ \hline & -\gamma^L \frac{\partial g}{\partial z_L} & & -\mathfrak{m} \cdot \nu^{LA} \frac{\partial g}{\partial z_A} & \end{array}\right)$$

and a right-hand side RHS_i depending on both stepsizes $\mathcal{H}_L, \mathcal{H}_A$, the stepsizeratio $\mathfrak{m} := \frac{\mathcal{H}_L}{\mathcal{H}_A}$, the increments $l_j^{\mathcal{L}}, k_j^{\mathcal{L}}, p_j$ of the former steps $j = 1, \ldots, i-1$ and a set of coefficients $b^{\mathcal{L}}$, $A^{\mathcal{L}}$, $B^{\mathcal{L}}$, $\Gamma^{\mathcal{L}}$, $D^{\mathcal{L}\bar{\mathcal{L}}}$, $N^{\mathcal{L}\bar{\mathcal{L}}}$ ($\{\mathcal{L}, \bar{\mathcal{L}}\} = \{L, A\}$).
In the *later microsteps* it remains, to solve $[\dot{y}_A = f_A, 0 = h_A]$ with respect to y_A, z_A and $w(t)$ entering the right-hand-side. We want to avoid additional interpolation schemes for the coupling part. As we introduced the coupling currents (pin currents) as additional unknowns, we can obtain an approximation to $w(t_0 + \theta \cdot \mathcal{H}_L)$ in a cheap way via dense output formulas (see (9)).
For lack of space we refer to [7] for an accurate definition.

MA-Trees and MA-series

Order conditions for the method's coefficients have to be derived and fulfilled to get approximations of prescribed accuracy. Derivatives of numerical approximation (11) and exact solution (10) soon become hardly manageable, e. g. the first derivative of the coupling consists of two elementary differentials:

$$\dot{w} = S^{-1} \frac{\partial g}{\partial z_L} \left(-\frac{\partial h_L}{\partial z_L}\right)^{-1} \frac{\partial h_L}{\partial y_L} f_L + S^{-1} \frac{\partial g}{\partial z_A} \left(-\frac{\partial h_A}{\partial z_A}\right)^{-1} \frac{\partial h_A}{\partial y_A} f_A$$

with Schur complement $S := -\left[\frac{\partial g}{\partial z_L} \left(-\frac{\partial h_L}{\partial z_L}\right)^{-1} \frac{\partial h_L}{\partial w} + \frac{\partial g}{\partial z_A} \left(-\frac{\partial h_A}{\partial z_A}\right)^{-1} \frac{\partial h_A}{\partial w}\right]$

Therefore, we extend the *B-series'* concept to the theory of *MA-series*, where we have to deal with five types of nodes and the stepsize ratio $\mathfrak{m}$ entering the order conditions. To be able to do this, we have to demand

$$D^{\mathcal{L}\bar{\mathcal{L}}} + N^{\mathcal{L}\bar{\mathcal{L}}} = A^{\mathcal{L}} + \Gamma^{\mathcal{L}}.$$

Two conditions arise for order 1, eight more for order 2 and again 36 more conditions have to be fulfilled to get a method of order 3.
This applies to the compound step only. The later microsteps are a combination of a classical DAE-solver and a dense-output scheme that has provides approximations of appropriate accuracy. For a detailed discussion see [7].

3 Hierarchical mixed multirate

Aiming at a multirate method that can deal with an arbitrary amount of activity levels, *hierarchical mixed multirate* seems to be the most feasible approach. The main idea is to nest compound steps and later micro-steps in a way, that at each time merely a two-level multirate scheme is engaged. At each timepoint of integration each subsystem has either the status

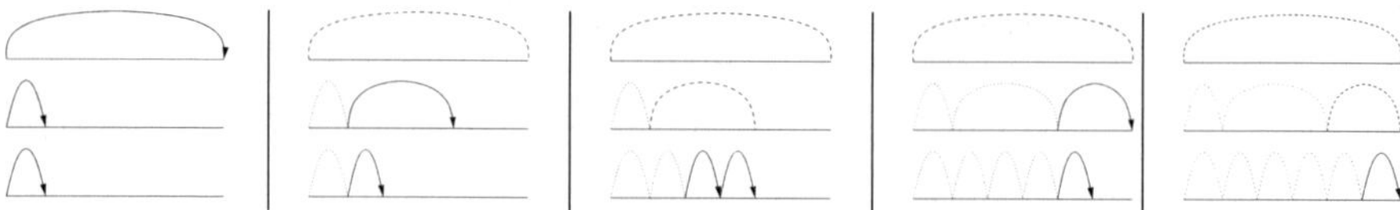

Fig. 2: Hierarchical mixed multirate for three blocks

asleep or *latent* or *active*.

A part is *asleep* if the last timepoint at which an approximation is available is beyond the current one. The set of not sleeping subsystems is split into a subset of systems that propose a large individual stepsize and one containing those, demanding a small step. The former ones are called *latent*, the latter ones *active*.
While the set of not sleeping systems can be decomposed in this way, a *compound step* is applied. Otherwise *later microsteps* are executed. The sleeping subsystems contribute to the current step via dense-output (Fig. 2).

Numerical Tests

A hierarchical mixed multirate method of order 2 with and embedded scheme of order 1 for stepsize control for network equations (4) has been embedded into Qimonda's in-house simulator `titan`.
Great importance is attached to the problem of traversing signals, forcing sleeping subsystems to "wake up", i. e. causing an a-posteriori rejection of a macrostep t_{n-1} to $t_n = t_{n-1} + \mathcal{H}_L$. In this implementation the detection of such situations is based on comparing pin voltages of connected subsystems: each timepoint $t_{\text{wup}} \in (t_{n-1}, t_n)$ where the deviation between the voltages computed with the not sleeping part and their equivalent that is recalculated by dense-output formulas applied to the sleeping part becomes too large, is considered a wake up point. However, not the whole macrostep is restored, but there is a re-initialisation at t_{wup}, again using dense-output formulas to get appropriate initial values. Furthermore, starting at t_{wup} a fixed number $n_{\text{frc}} \in \mathbb{N}$ of singlerate steps is done to stabilise integration.
It has been tested, amongst others, with an inverter chain that comprises 500 inverter stages. These are arranged in five blocks, each containing 100 inverters, see Fig. 3. At node "in$_M$" we apply a single pulse of width 17 ns, see Fig. 4 (left). Moreover Fig. 4 (right) shows how this signal passes each blocks' output node. The timepoints (60.14, 113.7, 166.8 and 220.2 ns) at which the blocks 2-5 awake are detected automatically (with $n_{\text{frc}} = 100$). Furthermore, each single block is touched only about one fourth as often as the singlerate integration (ROW-scheme of order 2 with the same set of coefficients) implies. For a detailed report on the numerical test we refer to [7].

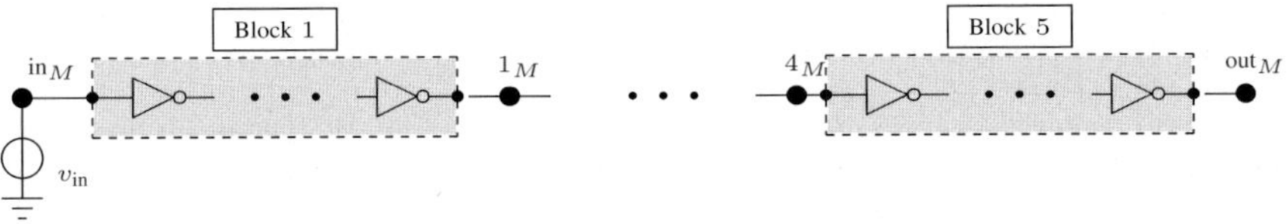

Fig. 3: Distributed inverter chain

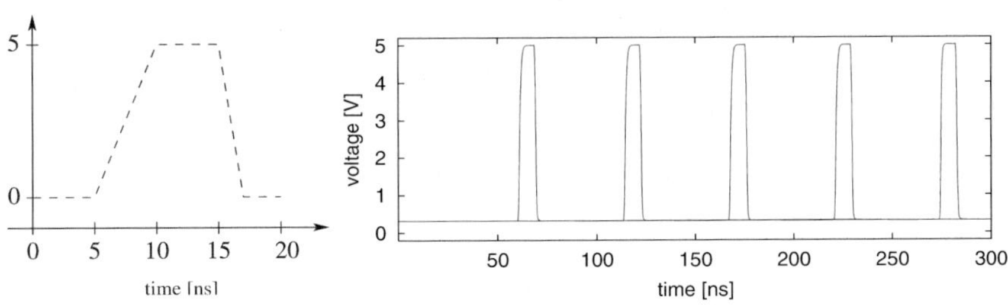

Fig. 4: Inverter chain: input signal (left), passing signal (right)

4 Conclusion

A multirate scheme for circuit simulation that can deal with an arbitrary amount of subsystems has been derived. Domain decomposition of large electrical circuits has been reached by introducing extra variables. The hierarchical multirate method has been embedded in a sophisticated industrial simulators.
Future tasks are a partitioning strategy and stepsize control that are tailored to multirate needs. Higher order schemes and extensions to higher index problems are desirable. Concerning wake up situation, also the currents at the pins must be taken into account. Furthermore the stepsize control, especially after synchronisation points or wake up situation are encountered, should be improved.

References

1. M. Arnold and M. Günther. Preconditioned dynamic iteration for coupled differential-algebraic systems. *BIT*, 41(1):1–25, 2001.
2. A. Bartel, M. Günther, and Kværnø. Multirate methods in electrical circuit simulation. In A. M. Anile, V. Capasso, and A. Greco, editors, *Progress in Industrial Mathematics at ECMI 2000*, volume 1 of *Mathematics in Industry*, pages 258–265. Springer, 2002.
3. D. Estévez Schwarz and C. Tischendorf. Structural analysis for electric circuits and consequences for MNA. *Int. J. Circ. Theor. Appl.*, 28:131–162, 2000.
4. M. Günther, U. Feldmann, and E. J. W. ter Maten. Modelling and discretization of circuit problems. In *[6]*, chapter 6, pages 523–650. Elsevier B.V., 2005.
5. V. Savcenco, W. H. Hundsdorfer, and J. G. Verwer. A multirate time stepping strategy for stiff odes. *To appear in BIT*.
6. W. H. A. Schilders and E. J. W. ter Maten, editors. *Numerical Methods in Electromagnetics*, volume XIII of *Handbook of Numerical Analysis*. Elsevier, North Holland, 2005.
7. M. Striebel. *Hierarchical Mixed Multirating for Distributed Integration of DAE Network Equations in Chip Design*. Number 404 in Fortschritt-Berichte VDI Reihe 20. VDI-Verlag Düsseldorf, 2006.

8. A. Verhoeven, A. El Guennouni, E. J. W. ter Maten, and R. Mattheij. Multirate methods for the transient analysis of electrical circuits. *PAMM, Proceedings to GAMM 2005*, 5:821 – 822, 2005.
9. U. Wever and Q. Zheng. Parallel Circuit Simulation on Workstation Clusters. In H. Neunzert, editor, *Progress in Industrial Mathematics at ECMI 94*, pages 274–284. John Wiley & Sons and B.G. Teubner, 1996.

Automatic Partitioning for Multirate Methods

A. Verhoeven[1], B. Tasić[2], T.G.J. Beelen[2], E.J.W. ter Maten[1,2], and R.M.M. Mattheij[1]

[1] Technische Universiteit Eindhoven, Den Dolech 2, 5600 MB, The Netherlands
averhoev@win.tue.nl

[2] DMS, NXP Semiconductors B.V., High Tech Campus 48, 5656 AE Eindhoven, The Netherlands bratislav.tasic@philips.com

Summary. The (nonlinear) transient analysis of electrical circuit models plays an important role in circuit design. Multirate time integration can be able to achieve the same accuracy for much lower costs. An essential assumption is the existence of a good partition of the circuit in a slow and fast part. This paper describes how this can be done automatically.

1 Introduction

Analogue electrical circuits are usually modeled by differential-algebraic equations (DAEs) of the following type:

$$\frac{d}{dt}\left[\mathbf{q}(t,\mathbf{x})\right]+\mathbf{j}(t,\mathbf{x})=\mathbf{0}. \tag{1}$$

The vector-valued functions $\mathbf{q},\mathbf{j}$ are constructed by Modified Nodal Analysis and represent the charges and currents in the network model. The state vector $\mathbf{x}(t)\in\mathbf{R}^d$ represents the nodal voltages and the currents through the voltage-defined elements like voltage sources and inductors and depends on the time variable t. A common analysis is the transient analysis, which computes the solution $\mathbf{x}(t)$ of this nonlinear DAE along the time interval $[0,T]$ for a given initial state. Often, parts of electrical circuits have multirate behaviour, which means that some variables are slowly varying, compared to other variables.

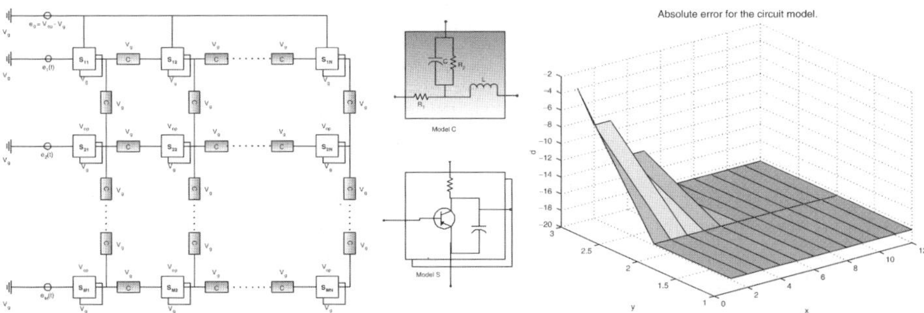

Fig. 1: At the left the circuit diagram of test example and at the right the typical shape of the corresponding error vector for $M=3, N=6$.

Figure 1 shows an example, which is a two-dimensional scalable circuit with $M\times N$ inverters, which has $M\times 2N$ nodal voltages. The subcircuits are connected with linear filters which were chosen such that only 3 subcircuits are active and nearly decoupled from the other subcircuits.

This property implies that the shape of the error vector looks like an iceberg, which is also visible in the right picture in Figure 1.
For those circuits it is very attractive to partition the model and use multirate methods [1, 3, 5, 6, 8]. First we will give a brief introduction to multirate in section 2. Partitioning is an important attribute for multirate methods. Together with the time-steps it can be used to control the local discretization error. Because still little attention has been payed to this topic, the subsequent sections deal with this. Finally also some results are shown from the application to multirate simulations.

2 Multirate transient algorithm

2.1 The DAE system

For a multirate method it is necessary to partition the variables and equations into an active (A) and a latent (L) part. This can be done by the user or automatically. Let $\mathbf{B}_A \in \mathbf{R}^{d_A \times d}$ and $\mathbf{B}_L \in \mathbf{R}^{d_L \times d}$ with $d_A + d_L = d$ be selection matrices, which satisfy $\mathbf{B}_A \mathbf{B}_A^T = \mathbf{I}_{d_A}, \mathbf{B}_L \mathbf{B}_L^T = \mathbf{I}_{d_L}, \mathbf{B}_A \mathbf{B}_L^T = \mathbf{O}, \mathbf{B}_L \mathbf{B}_A^T = \mathbf{O}$ where $\mathbf{I}_d \in \mathbf{R}^{d^2}$ is the identity matrix and $\mathbf{O}$ the zero matrix. Then the variables and functions can be split in active (A) and latent (L) parts: $\mathbf{x} = \mathbf{B}_A^T \mathbf{x}_A + \mathbf{B}_L^T \mathbf{x}_L$, $\mathbf{q} = \mathbf{B}_A^T \mathbf{q}_A + \mathbf{B}_L^T \mathbf{q}_L$ and $\mathbf{j} = \mathbf{B}_A^T \mathbf{j}_A + \mathbf{B}_L^T \mathbf{j}_L$. Because of the properties of $\mathbf{B}_A, \mathbf{B}_L$ we have $\mathbf{x}_A = \mathbf{B}_A \mathbf{x}$, $\mathbf{x}_L = \mathbf{B}_L \mathbf{x}$, $\mathbf{q}_A = \mathbf{B}_A \mathbf{q}$, etc. Now equation (1) is equivalent to the following partitioned system:

$$\frac{d}{dt}\left[\mathbf{q}_A(t, \mathbf{x}_A, \mathbf{x}_L)\right] + \mathbf{j}_A(t, \mathbf{x}_A, \mathbf{x}_L) = \mathbf{0}, \tag{2}$$

$$\frac{d}{dt}\left[\mathbf{q}_L(t, \mathbf{x}_A, \mathbf{x}_L)\right] + \mathbf{j}_L(t, \mathbf{x}_A, \mathbf{x}_L) = \mathbf{0}. \tag{3}$$

For DAEs the partition should be properly chosen such that the parts (2) and (3) are also solvable. Furthermore it is a nice property if also the stability and index are preserved. In practice this is not always the case.

2.2 Compound-Fast BDF algorithm

Multirate methods integrate both parts using different time-steps H and h. For circuits with multirate behaviour, like the previous example, the multirate factor $m = \frac{H}{h}$ should be a large number.
We use the BDF Compound-Fast algorithm on both the coarse and refined grid. This method is described in [8, 7]. It splits the circuit model in a slow and fast part. Each iteration first the complete system is integrated by one large compound step. The active part is relaxed during the Newton process. Afterwards only the fast part is integrated (or "refined"), while the connected slow interface variables are replaced by interpolated values. In [8] it is proved that this is a numerically stable scheme if the both parts are weakly coupled and the active part is asymptotically stable.
The discretization error $\widehat{\mathbf{e}} \in \mathbf{R}^d$ is controlled by independent stepsize control of the large compound step and of the much smaller refinement steps [7]. It has been shown that the active local error $\mathbf{B}_A \widehat{\mathbf{e}}$ consists of a discretization part and an interpolation part, which depend on H and h, respectively. If these parts are equal to $(1 - w)$TOL and wTOL, respectively, where $w \in [0, 1]$ is a balance number, the active local error is equal to the tolerance level TOL. The proposed compound steps are equal to $\min\{H_C, H_I\}$, where H_C keeps the latent discretization error at TOL and H_I keeps the interpolation error at wTOL.

2.3 Efficiency analysis of multirate methods

Although we introduced the Compound-Fast BDF algorithm, the next analysis is valid for a much larger family of multirate methods. Let W_C, W_R be the computational work per timestep for the compound phase and the refinement phase and define the workload ratio by $E = \frac{W_R}{W_C}$. Let W_S be the computational work per step for the standard singlerate version, which satisfies $\frac{W_S}{W_C} = F \approx 1$. If H, h are the average compound and refinement steps and $m = \frac{H}{h}$ is the average multirate factor, then a multirate method on $[0, T]$ will need the following computational workload:

$$W_{\text{mult}} = W_R \frac{T}{h} + W_C \frac{T}{H} = W_C T (E \frac{1}{h} + \frac{1}{H}) = W_C \frac{T}{h} (E + \frac{1}{m}), \tag{4}$$

while a singlerate method with step h_s would need $W_{\text{sing}} = W_S \frac{T}{h_s}$. Thus we have the following speed-up factor for the multirate method

$$S = \frac{W_{\text{sing}}}{W_{\text{mult}}} = \frac{W_S \frac{1}{h_s}}{W_C \frac{1}{h}(E + \frac{1}{m})} = F \frac{h}{h_s} \frac{1}{\frac{1}{m} + E} \approx \frac{h}{h_s} \frac{1}{\frac{1}{m} + E}. \tag{5}$$

Here m is the multirate factor which is large if the dynamics of the refined part are more active than the other slow part. The ratio E is determined by the partition and describes the relative costs of a refinement step which depends on the size d_A of the refinement part. It applies that $S \to \frac{F}{E} \frac{h}{h_s}$ for $m \to \infty$ and $S \to Fm \frac{h}{h_s}$ for $E \to 0$. Clearly, we get a large speed-up factor if m is large and E is small. Only if $S > 1$ it could be attractive to use for instance the multirate version of a certain integration scheme.
The multirate factor m can be approximated by the ratio between the proposed steps for the next step of a multirate and singlerate method $\hat{m} = \frac{H_{\text{new}}}{h_{\text{new}}}$. Here $H_{\text{new}}, h_{\text{new}}$ are the proposed stepsizes for the next compound step or singlerate step, respectively. Any integration method has an algorithm for $H_{\text{new}}, h_{\text{new}}$, which always depend on the integration order p. In fact $\hat{m}$ depends on the estimated local error vector $\widehat{\mathbf{e}}$ and approximately behaves like

$$\hat{m} = \left(\frac{\|\mathbf{B}_A \widehat{\mathbf{e}}\|}{\|\mathbf{B}_L \widehat{\mathbf{e}}\|} \right)^{\frac{1}{p+1}}. \tag{6}$$

The workload ratio E is approximated by

$$\hat{E} = \left(\frac{d_A}{d} \right)^{\alpha}, \tag{7}$$

where $\alpha \in (1, 3)$ depends on the application. By default we use $\alpha = 2$. Note that it is also possible to model E by a parameterized rational function of d_A and d, where the parameters can be identified by using experimental data.

3 Automatic partitioning

This section gives first some algorithms for partitioning. Afterwards it is shown how these algorithms can be applied dynamically. Finally some implementation issues are discussed.

3.1 The optimal partition

Although h can be smaller than h_s, we assume that

$$F \frac{h}{h_s} \approx 1 \text{ is independent of the partition.}$$

Thus the partition is optimal if $\hat{S} = \frac{1}{\frac{1}{m} + E}$ achieves its maximal value. Let the index sets of the active and latent parts be $\text{ind}_A, \text{ind}_L$ of lengths d_A, d_L such that $d = d_A + d_L$. Then both m and E are functions of $\text{ind}_\text{A}, \text{ind}_\text{L}$. Thus the optimal partition satisfies:

$$\max_{\text{ind}_A,\text{ind}_L} S \approx \max_{\text{ind}_A,\text{ind}_L} \frac{1}{\frac{1}{m} + E}. \tag{8}$$

The exact solution of optimization problem (8) is an 0-1 program which has exponential complexity. It is simply not possible to consider all possible transitions of the index sets. Therefore, we should use approximations which e.g. only determine the optimum for all transitions in subsets of size k. Then the algorithm has a polynomial complexity $O(d^k)$.

3.2 Partitioning algorithms

We considered the following types of algorithms:

1. The first type tries to find a solution of (8) by only considering the multirate factor m, which is approximated by $\hat{m}$ from (6). It is assumed that H_{new} only depends of $\mathbf{B}_L\hat{\mathbf{e}}$, which is the latent discretization error estimate of the compound step. Then the particular latent element is determined which has the largest local error element $\hat{e}_i$ and should be refined such that $\hat{m}$ becomes larger. Also the active element which has the smallest $\hat{e}_i$ is determined. Thus the most active latent element and the most latent active element are selected. Then all four possible transitions of these two found elements are compared with respect to their corresponding estimated speed-up factors and the optimal transition is performed. Iteratively the transition with maximum estimated speed-up factor is performed until convergence. This algorithm needs an initial guess to start with. It could be the previous partition, a partition computed by another algorithm or it is given by the user.
2. All nodes become refined when the estimated local error satisfies

$$\hat{e}_i > \epsilon_{\text{rel}} \|\hat{\mathbf{e}}\|_{\max}, \tag{9}$$

   where $\epsilon_{\text{rel}} < 1$.
3. A different approach considers the absolute criterion

$$\hat{e}_i > \epsilon_{\text{abs}}, \tag{10}$$

   where $\epsilon_{\text{abs}} <$TOL. The tolerance level TOL is given by the user such that the local error satisfies $\|\hat{\mathbf{e}}\| \leq$TOL.
4. A fourth approach computes the needed stepsizes per element from the local error vector $\hat{\mathbf{e}}$ and detects the largest gap between the stepsizes. Then this gap is used to separate the system in a fast and a slow part.

All four algorithms only need the information contained in the local error vector $\hat{\mathbf{e}}$. The first algorithm really tries to optimize the speed-up factor, while the second algorithm just uses a tolerance level for the relative activity. For algorithm 2 and 3 the values of ϵ_{abs} or ϵ_{rel} are free and have to be chosen. The optimal values depend on the properties of the vector $\hat{\mathbf{e}}$. In practice algorithm 2 works better than 3 because then the partition only depends on the ratio between the errors of the active and latent parts. The approach (10) depends much more on the quality of the error estimation itself. In this paper we restricted our attention to algorithm 1 and 2.

There are also some add-ons for the algorithms which could improve the performance. One could always refine the nodes which are connected to sources if they can suddenly become active. Furthermore, a safety region of distance γ around the active part can be used. This means that all nodes whose distance to the active part is at most γ are also refined. This makes it possible to predict sudden wake-ups and reduces the number of repartitionings.

For real-world applications it might be necessary to take care with the already existing structure, e.g. a hierarchical structure for circuit models. Instead of element-wise partitioning the submodels are treated like connected blocks which can be active or latent.

3.3 Dynamically changing partitioning

For e.g. digital circuits it is impossible to apply multirate with a static partition. Then dynamical partitioning techniques [2, 4] are needed which are able to follow the moving active part. This means that the partition should be updated during the multirate time-integration. Because repartitionings are not cheap, it is not allowed to change the partition during the refinement. Thus repartitionings only can occur just after the compound step or just after the refinement phase. There exist the following three alternatives:

A The partition is updated just after the compound step by use of the computed error vector at the coarse grid.
B The partition is updated just after the refinement phase where also the active errors of the refined time-grid are used.
C First a singlerate step is done and then its error vector is used to repartition.

Methods B and C are better suited for the stepsize control and the relaxation of the Newton process, because the active part of the coarse error vector in method A is not accurate. Nevertheless, method A has been used for the numerical experiments in the next section because it is better able to detect sudden wake-ups of latent variables. Figure 2 shows a combination of these methods. Note that method A is applied with the restriction that only latent elements can be transfered to the refinement part. By keeping the old partition for an acceptable speed-up factor $\hat{S}$, the number of repartitionings is reduced.

3.4 Consequences for multirate algorithm

Dynamical partitioning also has some consequences for the existing multirate algorithm. Firstly, there is the storage problem. Now, each node can have its own time-points, theoretically. Because the lengths of these time-grids will differ for each unit, it is impossible to store the solutions and time-grids in a normal array. Furthermore, for multistep methods there is the initialization problem for the waked up fast nodes which were slow during the previous compound step. Restarting with onestep methods, like Euler Backward, can reduce the gained efficiency. We use the previous coarse-grid polynomial also as a predictor polynomial for the new refinement phase using upgraded results. A nice property of dynamical partitioning is that the estimated speed-up factor also can be used to decide whether the next step will be a compound step or a singlerate step. This is a benefit of the Compound-Fast method actually.

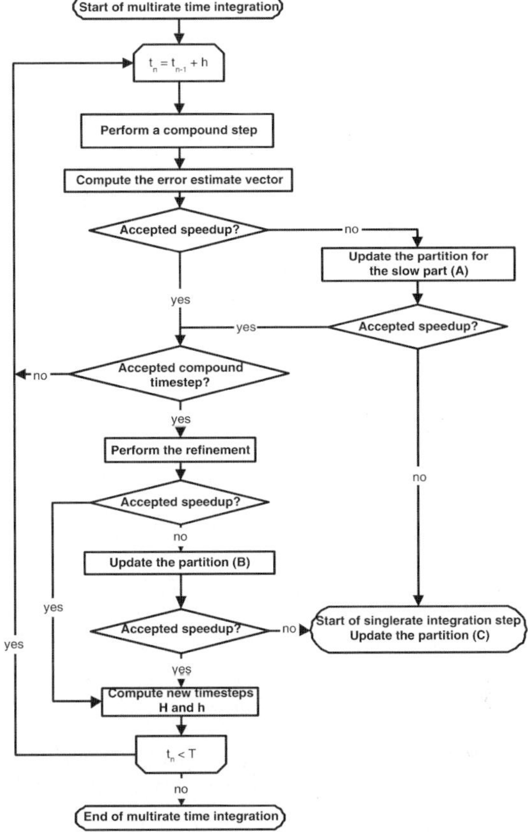

Fig. 2: The multirate algorithm with dynamically changing partitions

4 Numerical experiments

4.1 Model problems

In this section we show how dynamical partitioning works in practice for some model problems in MATLAB. Firstly, we look at the circuit model (ODE) of an inverter chain, which is described in more detail in [2]. It is a chain consisting of 500 inverters. If we excite the first node by a short pulse, a voltage wave is traveling through the chain from left to right. Secondly, we look at the circuit model which is shown in Figure 1 for $M = 5, N = 10$. The subcircuits are connected by C-elements that can filter the voltages and currents. The circuit is driven by M voltage sources which can have different frequencies. The location of the active part is controlled by the C-elements and the voltage sources. We used the voltage sources $e_i = \frac{5}{2}(1 - \cos(\omega_i t))$, where $\omega_1 = 100 \cdot 10^9$, and for $i > 1$, $\omega_i = 10^9$.

4.2 Results for inverter chain

For TOL $= 10^{-2}$, $N = 100$ we did an experiment on $[0, 75 \text{ ns}]$ by several dynamical partitioning algorithms of type A. Algorithm 1 is used with the workload model (7) but for different values of α. In all cases at most 4 iterations are performed during a compound step. Algorithm 2 is used for different values of ϵ_{rel}. All algorithms use $\gamma = 3$ as overlap value. Table 1 shows the results. Note that n_i and k_i are the numbers of timesteps and Newton iterations, respectively. Clearly, for each case the number of refinement steps, n_R, is much larger than the number of compound steps, n_C. It is larger than the number of steps for the singlerate version because of error control reasons. In the column below av($\frac{d_A}{d}$) the average relative size of the active part is shown. The required CPU time also includes the repartitioning time effort.

Method	α	ϵ_{rel}	n_C	n_R	k_C	k_R	av($\frac{d_A}{d}$)(%)	time (s)	S
Singlerate			1340	0	5440	0	0	266	
1	2		82	1651	1008	3415	16	87	3.1
1	$\frac{3}{2}$		94	1663	996	3429	15	86	3.1
2		10^{-1}	166	1953	1313	4034	9	100	2.7
2		10^{-2}	97	2001	1225	4105	16	105	2.5
2		10^{-3}	94	1992	1637	4093	22	133	2.0

Table 1: Statistics of singlerate and multirate method using algorithms 1 and 2 of type A for the inverter chain model.

If we compare the algorithms for lower accuracy TOL $= 10^{-1}$ it appears that the method 2 does not converge in contrast to 1. Also for other experiments it appears that the method 1 implies better convergence than 2. All methods are able to follow the active wave front. Figure 3 shows the timepoints per element for case (1) with $\alpha = \frac{3}{2}$. The two flanks of the traveling wave are easy to observe in the picture.
For small N the speed-up factor is relatively small because of the overhead. But for large N it is indeed possible to get a large speed-up. This is also the case for increasing accuracy.

4.3 Results for Matrix circuit

If the Matrix circuit has a fixed latent part it is possible to get high speed-up factors while keeping the accuracy [7]. In this case also the previous techniques can be used to find the static partition in an automatic way. For really dynamically changing partitions problems can occur because of ill-conditioned systems during the refinement.

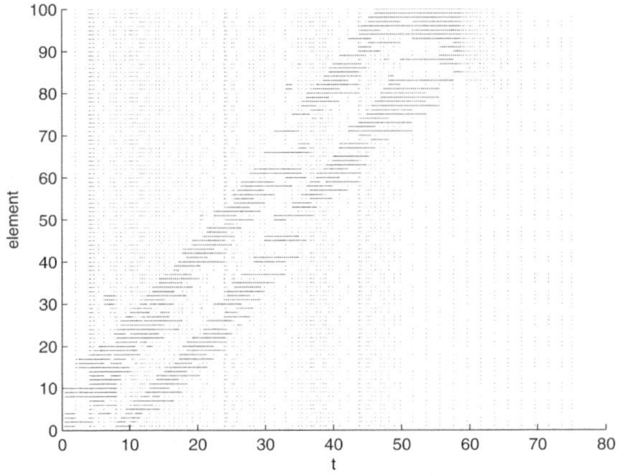

Fig. 3: Timepoints per element for the inverter chain (case 1 with $\alpha = \frac{3}{2}$).

5 Conclusion

We studied two partitioning algorithms. It appears that algorithm 1 is much more robust than algorithm 2, which strongly depends on the relative tolerance. We showed how these techniques can be applied in a dynamical way. For the ODE model of an inverter chain it already appeared that dynamical partitioning indeed can be attractive to use. For more general DAE models still research is necessary. We also intend to apply these partitiong algorithms to industrial circuit models in Pstar[3].

References

1. C.W. Gear, D.R. Wells: *Multirate linear multistep methods*, BIT, 24, pp. 484-502, 1984.
2. M. Günther, P. Rentrop. *Partitioning and multirate strategies in latent electrical cicuits* International Series of Numerical Mathematics, Vol.117, Birkhäuser Verlag Basel, pp. 33-60, 1994.
3. M. Günther, A. Kværnø, P. Rentrop. *Multirate partitioned Runge-Kutta methods* BIT, Vol.41, pp. 504-514, 2001.
4. V. Savcenco, W.H. Hundsdorfer, J.G. Verwer. *A multirate time stepping strategy for stiff ODEs*, To appear in BIT, Preprint,[4] 2006.
5. S. Skelboe. *Adaptive partitioning techniques for ordinary differential equations*, BIT, Numerical Mathematics issue in memory of G. Dahlquist, 2006.
6. M. Striebel. *Hierarchical Mixed Multirating for Distributed Integration of DAE Network Equations in Chip Design* PhD Thesis, Bergischen Universität Wuppertal, 2006.

[3] This is the inhouse analog circuit simulator provided by NXP Semiconductors and also used at Philips.

[4] `http://www.cwi.nl/~savcenco/MR_BIT.pdf`

7. A. Verhoeven, B.Tasić, T.G.J. Beelen, E.J.W. ter Maten, R.M.M. Mattheij. *Error analysis of BDF Compound-Fast multirate method for differential-algebraic equations*, Submitted for publ., Preprint,[5] 2007.
8. A. Verhoeven, A. El Guennouni, E.J.W. ter Maten, R.M.M. Mattheij: *A general compound multirate method for circuit simulation problems*, In: A.M. Anile, G. Ali, G. Mascali: *Scientific Computing in Electrical Engineering*, Series Mathematics in Industry, ECMI, Vol. 9, pp. 143-150, 2006.

[5] http://www.win.tue.nl/~averhoev/JNAIAM_errorcontrol.pdf

Simulation of Quasiperiodic Signals via Warped MPDAEs Using Houben's Approach

Julia Greb and Roland Pulch

Bergische Universität Wuppertal, Fachbereich Mathematik und Naturwissenschaften, Lehrstuhl für Angewandte Mathematik und Numerische Analysis, Gaußstr. 20, D-42119 Wuppertal (Germany). pulch@math.uni-wuppertal.de

Summary. A multidimensional model yields an alternative strategy for the numerical simulation of frequency-modulated signals. Thus the differential algebraic equations (DAEs), which describe an electric circuit, change into warped multirate partial differential algebraic equations (MPDAEs). Houben [6] introduced an approach for solving efficiently initial-boundary value problems of such MPDAE systems. Thereby, envelope-modulated solutions of the DAEs are reproduced. In this paper, the technique is analysed for obtaining quasiperiodic solutions of the DAEs. The crucial question is if biperiodic solutions of the MPDAEs are generated automatically by Houben's approach provided that the initial values of a biperiodic solution are applied.

1 Introduction

In radio-frequency applications, electric circuits often produce oscillatory signals with widely-separated time scales. For example, the amplitude as well as the frequency of a high-frequency oscillation may change relatively slowly. A numerical simulation of the circuit demands to solve the corresponding time-dependent system of differential algebraic equations (DAEs), see [4]. Thus the simulation becomes inefficient, since fast oscillations limit the step size in time, whereas the slow time scale determines the total time interval.
A multivariate signal model yields an alternative strategy, where each separate time scale is given an own variable. Brachtendorf et al. [1] introduced the corresponding system of multirate partial differential algebraic equations (MPDAEs), which yields an efficient simulation of purely amplitude-modulated signals. Narayan and Roychowdhury [7] generalised the approach for signals, which are amplitude-modulated (AM) as well as frequency-modulated (FM). Accordingly, a system of warped MPDAEs arises, where the determination of an appropriate local frequency function is crucial for the efficiency of the multidimensional model. Rough choices produce unnecessary oscillations in the multivariate solutions of the system.
Houben [5, 6] introduced a minimisation criterion with respect to partial derivatives, which shall reduce oscillatory behaviour in the solutions of the MPDAE system. This strategy yields a formula for the unknown local frequency function depending on the multivariate solution. The approach can be used to solve initial-boundary value problems of MPDAEs, which reproduce envelope-modulated solutions of the DAEs.
The direct determination of quasiperiodic solutions of the DAEs demands to solve biperiodic boundary value problems of corresponding MPDAEs, see [10]. A method of characteristics

can be used to compute biperiodic solutions of an MPDAE system efficiently, see [8]. This technique becomes inappropriate in case of initial-boundary value problems.
We investigate the performance of Houben's strategy when initial values from a biperiodic solution are given. If the resulting solution is biperiodic, too, then a method for biperiodic boundary value problems can be constructed based on the original strategy. Although the method of characteristics still seems to be superior for biperiodic problems, the results give more insight in the properties of Houben's method.

2 Multidimensional Model

The mathematical model of electric circuits yields a system of DAEs, see [4]. We consider a system of the form

$$\frac{\mathrm{d}\mathbf{q}(\mathbf{x})}{\mathrm{d}t} = \mathbf{f}(\mathbf{b}(t), \mathbf{x}(t)), \qquad \begin{array}{ll} \mathbf{x} : \mathbb{R} \to \mathbb{R}^k, & \mathbf{q} : \mathbb{R}^k \to \mathbb{R}^k, \\ \mathbf{b} : \mathbb{R} \to \mathbb{R}^l, & \mathbf{f} : \mathbb{R}^l \times \mathbb{R}^k \to \mathbb{R}^k, \end{array} \tag{1}$$

where $\mathbf{x}$ denotes unknown node voltages and branch currents. We assume that the predetermined input signals $\mathbf{b}$ vary relatively slowly. In contrast, the solution $\mathbf{x}$ shall include high-frequency oscillations, whose amplitude as well as frequency are changed slowly by the input signals. Thus the signals $\mathbf{x}$ include widely-separated time scales. Hence solving the DAEs (1) demands a huge number of time steps and a transient analysis becomes inefficient.
Brachtendorf et al. [1] introduced a multivariate signal model for purely AM signals, where each time scale is assigned an own variable. Narayan and Roychowdhury [7] generalised this model for signals including AM as well as FM. In case of two time scales, a multivariate function (MVF) $\hat{\mathbf{x}} : \mathbb{R}^2 \to \mathbb{R}^k$ and a local frequency amplification function $\nu : \mathbb{R} \to \mathbb{R}$ of the signal $\mathbf{x}$ are introduced. Thus an efficient model is achieved by decoupling the time scales. Consequently, the system of DAEs (1) is transformed into the system of warped MPDAEs

$$\frac{\partial \mathbf{q}(\hat{\mathbf{x}})}{\partial t_1} + \nu(t_1)\frac{\partial \mathbf{q}(\hat{\mathbf{x}})}{\partial t_2} = \mathbf{f}(\mathbf{b}(t_1), \hat{\mathbf{x}}(t_1, t_2)), \qquad \begin{array}{l} \hat{\mathbf{x}} : \mathbb{R}^2 \to \mathbb{R}^k, \\ \nu : \mathbb{R} \to \mathbb{R}. \end{array} \tag{2}$$

We assume $\mathbf{q}, \hat{\mathbf{x}} \in C^1$ and $\mathbf{b}, \mathbf{f}, \nu \in C^0$. The local frequency amplification ν is a priori unknown, too. The input signals vary just slowly and thus do not require a multivariate description. An arbitrary solution of the MPDAEs (2) yields a solution of the DAEs (1) using the reconstruction

$$\mathbf{x}(t) = \hat{\mathbf{x}}\left(t, \int_0^t \nu(\sigma)\mathrm{d}\sigma\right), \tag{3}$$

i.e., the MVF includes the original signal. In this general case, ν represents a local frequency amplification and thus ν is physically dimensionless.
This model is suitable only if the fast time scale is periodic, since we want to resolve many oscillations in a bounded and relatively small multidimensional domain. Hence two types of problems arise. Firstly, initial-boundary value problems of the system (2) read

$$\hat{\mathbf{x}}(0, t_2) = \mathbf{h}(t_2), \quad \hat{\mathbf{x}}(t_1, t_2) = \hat{\mathbf{x}}(t_1, t_2 + 1) \quad \text{for all } t_1 \geq 0,\ t_2 \in \mathbb{R} \tag{4}$$

with a predetermined periodic function $\mathbf{h} : \mathbb{R} \to \mathbb{R}^k$. The period is standardised to 1 and thus the second argument t_2 of the MVF becomes dimensionless. Hence ν in (3) includes the magnitude of the fast time scale and exhibits the physical dimension of a frequency now. The problems (4) are solved in a domain $[0, T] \times [0, 1]$ for some $T > 0$. Secondly, biperiodic boundary value problems exhibit the conditions

$$\hat{\mathbf{x}}(t_1, t_2) = \hat{\mathbf{x}}(t_1 + T_1, t_2) = \hat{\mathbf{x}}(t_1, t_2 + 1) \quad \text{for all } t_1, t_2 \in \mathbb{R}, \tag{5}$$

which correspond to the domain $[0, T_1] \times [0, 1]$. In this case, the input signals as well as the local frequency function have to be T_1-periodic. Note that we require a smooth solution of (2) to fulfil the biperiodicity condition (5).

Applying (3), solutions satisfying (4) reproduce envelope-modulated signals, whereas solutions fulfilling (5) yield quasiperiodic signals. Note that both envelope-modulated and quasiperiodic signals include FM here.
For appropriate choices of the local frequency functions, the corresponding MVF exhibits a simple structure in $[0,T] \times [0,1]$. Thus we can compute the solution of the MPDAEs (2) using a relatively low number of grid points and achieve an efficient numerical simulation. The desired solution of the DAEs (1) is reconstructed via (3).
Solutions of the MPDAEs (2) corresponding to different local frequency functions are interconnected by a transformation, see [9]. If $\hat{\mathbf{x}}$ is a MVF satisfying the system for the local frequency ν, then the transformed MVF

$$\hat{\mathbf{y}}(t_1,t_2) := \hat{\mathbf{x}}\left(t_1, t_2 + \int_0^{t_1} \nu(\sigma) - \mu(\sigma)\;\mathrm{d}\sigma\right) \tag{6}$$

represents a solution of the system with local frequency μ. The initial values at $t_1 = 0$ are invariant in this transformation. Thus, for solving initial-boundary value problems (4), the local frequencies are completely free parameters, which can be used to achieve an efficient representation. In case of biperiodic problems (5), an additional requirement is necessary to preserve the periodicity in the slow time scale, namely

$$\int_0^{T_1} \mu(\sigma)\;\mathrm{d}\sigma = \int_0^{T_1} \nu(\sigma)\;\mathrm{d}\sigma, \tag{7}$$

which means that the average frequency coincides.

3 Houben's Method

A suitable local frequency function for representing the signals efficiently is unknown a priori. Inappropriate selections cause undesired oscillations in the MVFs, see [9]. Houben [5, 6] formulated the minimisation problem

$$s(t_1) := \int_0^1 \|\tfrac{\partial \mathbf{q}(\hat{\mathbf{x}})}{\partial t_1}(t_1,u)\|^2\;\mathrm{d}u \quad \longrightarrow \quad \text{min.} \qquad \text{for each } \; t_1 \geq 0 \tag{8}$$

using the Euclidean norm $\|\cdot\|$. Thus oscillatory behaviour is reduced via minimising the impact of the partial derivative with respect to the slow time scale. For example, a method of lines can be employed to solve the initial-boundary value problem (2),(4). Hence a corresponding optimal solution allows for using relatively large step sizes in the numerical simulation.
The demand (8) implies a necessary condition for an optimal solution:

$$\nu(t_1) = \frac{\int_0^1 \langle \mathbf{f}(\mathbf{b}(t_1), \hat{\mathbf{x}}(t_1,u)), \tfrac{\partial \mathbf{q}(\hat{\mathbf{x}})}{\partial t_2}(t_1,u)\rangle\;\mathrm{d}u}{\int_0^1 \|\tfrac{\partial \mathbf{q}(\hat{\mathbf{x}})}{\partial t_2}(t_1,u)\|^2\;\mathrm{d}u} \qquad \text{for all } \; t_1 \geq 0 \tag{9}$$

with the Euclidean inner product $\langle\cdot,\cdot\rangle$. This formula can be used to eliminate the unknown local frequency function. Thus initial-boundary value problems can be solved by proceeding in the slow time scale. Furthermore, the condition (9) is equivalent to the orthogonality relation

$$\int_0^1 \langle \tfrac{\partial \mathbf{q}(\hat{\mathbf{x}})}{\partial t_1}(t_1,u), \tfrac{\partial \mathbf{q}(\hat{\mathbf{x}})}{\partial t_2}(t_1,u)\rangle\;\mathrm{d}u = 0 \quad \text{for all } \; t_1 \geq 0. \tag{10}$$

In the following, we assume the existence of a smooth biperiodic solution $\hat{\mathbf{z}}$ corresponding to the periodic local frequency κ. Let $\mathbf{h} := \hat{\mathbf{z}}(0,\cdot)$ be its initial values. We investigate the results from the initial-boundary value problem (4) applying Houben's technique. In [3], the case of ordinary differential equations ($\mathbf{q}(\mathbf{x}) \equiv \mathbf{x}$) has already been considered.
An arbitrary solution $\hat{\mathbf{x}}, \nu$ of the MPDAEs (2) with the same initial values $\mathbf{h}$ can be obtained from the biperiodic solution $\hat{\mathbf{z}}$ via the transformation (6). We define the quantity

$$c := \int_0^{T_1} \nu(\sigma) - \kappa(\sigma)\;\mathrm{d}\sigma. \tag{11}$$

Using the transformation (6), it follows

$$\hat{\mathbf{x}}(T_1, t_2) = \hat{\mathbf{z}}(T_1, t_2 + c) = \hat{\mathbf{z}}(0, t_2 + c) = \hat{\mathbf{x}}(0, t_2 + c) \quad \text{for all } t_2 \in \mathbb{R}. \tag{12}$$

We have achieved the following result.

Theorem 1. *If $\hat{\mathbf{x}}, \nu$ is an arbitrary solution of the system (2) with initial values from a biperiodic solution, then it holds $\hat{\mathbf{x}}(T_1, t_2) = \hat{\mathbf{x}}(0, t_2 + c)$ for all t_2, i.e., the end values represent a time shift of the initial values.*

In Houben's approach, the question is if this time shift is equal to zero or not. In the formula (9) for the corresponding local frequency function, the arising integrals are invariant with respect to a shift in the fast time scale t_2. Thus the following theorem holds.

Theorem 2. *If $\hat{\mathbf{x}}, \nu$ is a solution of the system (2) with initial values from a biperiodic solution and satisfying (9), then it holds $\nu(0) = \nu(T_1)$ and thus the local frequency function is periodic.*

Note that this theorem does not imply that the corresponding MVF is biperiodic. Nevertheless, the local frequency becomes periodic and thus a biperiodic solution may result from Houben's approach. However, a proof is still missing.
On the other hand, a minimisation demand for biperiodic solutions has been introduced in [9]. Similar to this approach, we consider the formulation

$$\gamma := \int_0^{T_1} \int_0^1 \left\| \frac{\partial \mathbf{q}(\hat{\mathbf{x}})}{\partial t_1}(v, u) \right\|^2 \, du \, dv \quad \longrightarrow \quad \text{min.} \tag{13}$$

here. A variational calculus yields the necessary condition

$$\int_0^1 \left\langle \frac{\partial^2 \mathbf{q}(\hat{\mathbf{x}})}{\partial {t_1}^2}(t_1, u), \frac{\partial \mathbf{q}(\hat{\mathbf{x}})}{\partial t_2}(t_1, u) \right\rangle du = 0 \quad \text{for all } t_1 \geq 0, \tag{14}$$

which an optimal solution has to satisfy. Thereby, the periodicity of the solution in t_1 is crucial to obtain this requirement. A biperiodic solution, which minimises (8), also represents a minimum of (13). This fact yields the following statement.

Theorem 3. *Given a biperiodic solution $\hat{\mathbf{x}}, \nu$ of the system (2), which is minimal with respect to Houben's criterion (8), then the MVF $\hat{\mathbf{x}}$ satisfies the orthogonality property (10) as well as (14).*

This result indicates that a solution obtained by Houben's approach is not biperiodic in general. If it is biperiodic, then two orthogonality properties are satisfied, which do not seem to be equivalent. Likewise, we consider an optimal biperiodic solution with respect to (13). This solution may become better by a transformation (6) to the optimal local frequency (9). Consequently, we may loose the periodicity as the price to be paid for the further reduction of the impact of partial derivatives.

4 Illustrative Example

As test example, we consider a voltage controlled oscillator, which is illustrated in Fig. 1 (left). The mathematical model of this circuit can be written as a system of ordinary differential equations (ODEs). We apply this formulation, since the examinations on the periodicity do not differ significantly if ODEs instead of DAEs are considered. The system reads

$$\dot{u} = (-\imath_R(u) - \imath) / (Cb(t)), \qquad \dot{\imath} = u/L \tag{15}$$

with the node voltage u and the branch current $\imath$. For the input signal, we choose the slowly varying oscillation

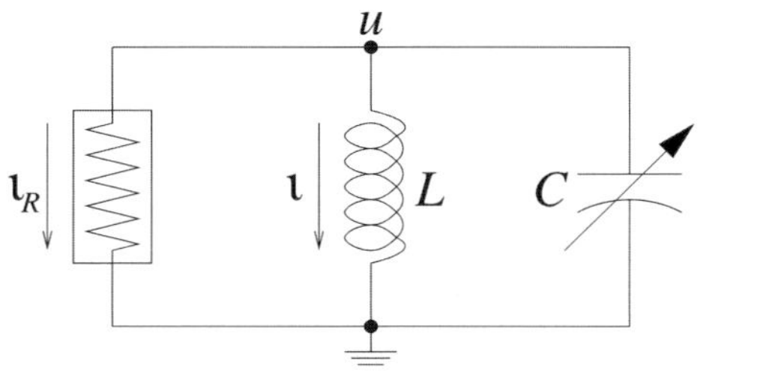

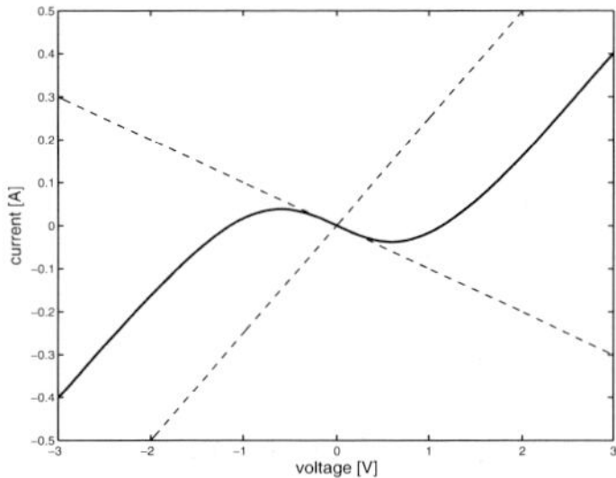

Fig. 1: Circuit diagram of voltage controlled oscillator (left) and current-voltage relation $\imath = \imath_R(u)$ of nonlinear resistor (right).

$$b(t) = 1 + 0.8 \cos\left(\tfrac{2\pi}{T_1} t\right) \quad \text{with } T_1 = 1 \text{ ms} \quad (f := T_1^{-1} = 1 \text{ kHz}). \tag{16}$$

The current-voltage relation of the nonlinear resistor is given by

$$\imath_R(u) = (G_0 - G_\infty)U_k \tanh(u/U_k) + G_\infty u. \tag{17}$$

The used parameters are $C = 1$ nF, $L = 1$ μH, $U_k = 1$ V, $G_0 = -0.1$ A/V, $G_\infty = 0.25$ A/V. Fig. 1 (right) shows the corresponding relation (17).
For constant input $b \equiv 1$, the system (15) exhibits a periodic limit cycle with a frequency of about 4 MHz. The input signal (16) changes the capacitance and thus introduces a frequency modulation. Since the input is periodic, a quasiperiodic signal arises. Consequently, we transform the ODEs (15) into a system (2) of partial differential equations (PDEs). We compute a biperiodic solution of the system via the method presented in [9]. Its initial values are used to apply Houben's strategy now.
To solve the initial-boundary value problem (4), we use a method of lines. The integrals in (9) are replaced by finite sums evaluated on the lines. The derivatives with respect to t_2 are substituted by BDF2-formulae, see [2], which are applied in the PDEs (2) as well as in the local frequencies (9). The arising system of ODEs is solved by trapezoidal rule in the interval $[0, T_1]$, where a relatively high accuracy is demanded in the step size control.
Firstly, we apply $m = 100$ lines in the semidiscretisation to demonstrate the optimal solution. Fig. 2 shows the resulting optimal local frequency, which is periodic in view of our discussions. The local frequency is physically reasonable, since it becomes low for high capacitances and vice versa. The corresponding optimal MVFs are illustrated in Fig. 3. On the one hand, we recognise that $\hat{u}$ is nearly constant in the slow time scale, which is caused by the minimisation. On the other hand, $\hat{\imath}$ exhibits a slight change in the slow time scale, which describes an AM signal and thus can not be reduced further.
Secondly, we compare the initial values at $t_1 = 0$ with the end values at $t_1 = T_1$ for several numerical simulations using different numbers of lines, namely $m = 25, 50, 100$. Table 1 demonstrates the maximum of the differences obtained from the discrete values on the lines. We recognise that the differences become smaller for an increasing accuracy in the method. This behaviour indicates that the exact solution is biperiodic or nearly (except for small differences) biperiodic. In [3], other numerical simulations, where a Van-der-Pol-oscillator is used, indicate that it can not be excluded that the resulting solution is biperiodic, too.

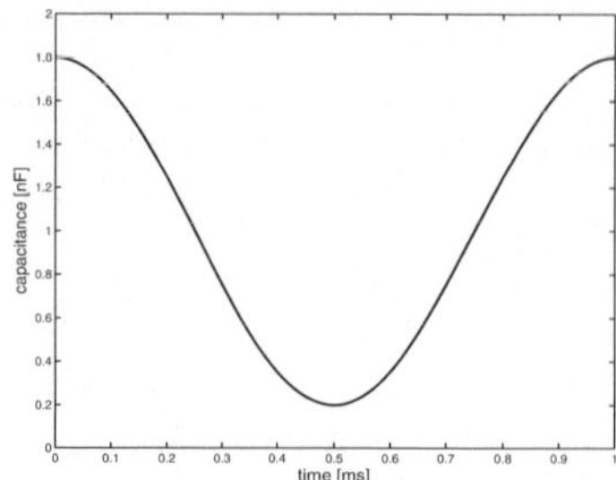

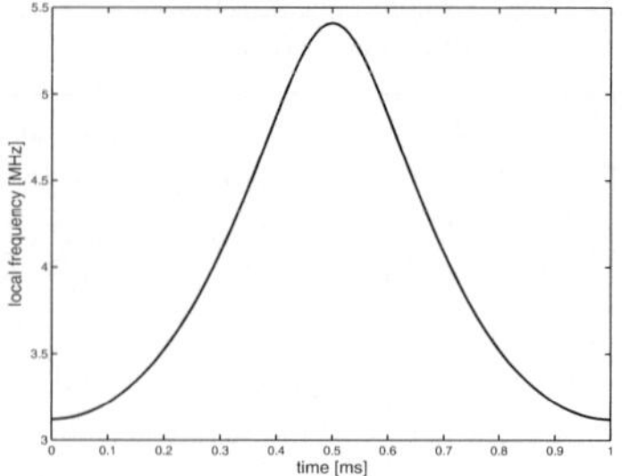

Fig. 2: Capacitance Cb [nF] (left) and optimal local frequency ν [MHz] (right).

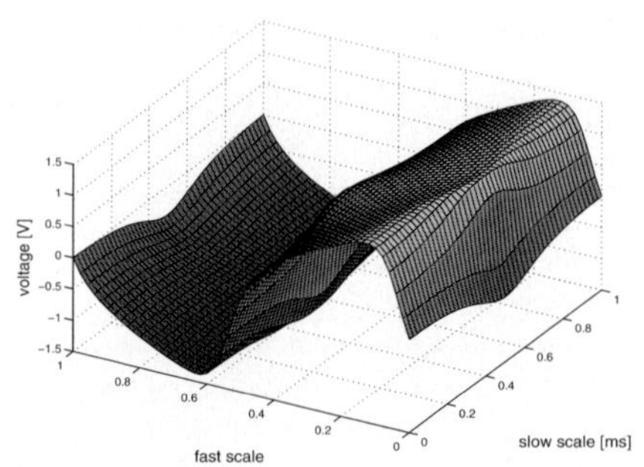

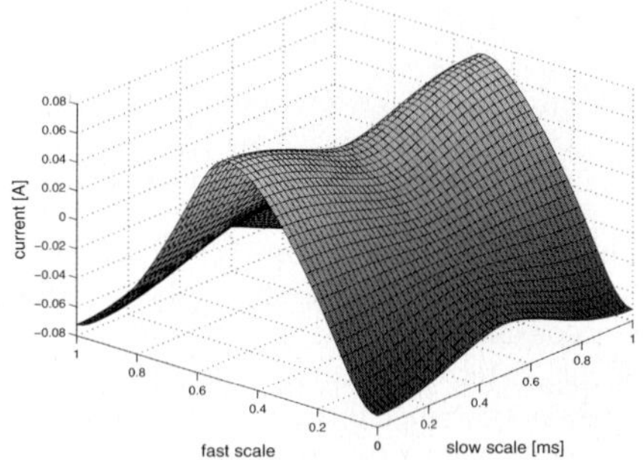

Fig. 3: Optimal MVFs $\hat{u}$ [V] (left) and $\hat{\imath}$ [A] (right).

Table 1: Maximum differences between initial and end values.

number of lines	$m = 25$	$m = 50$	$m = 100$
$\max \lvert\hat{u}(0,\cdot) - \hat{u}(T_1,\cdot)\rvert$	$8 \cdot 10^{-2}$	$2 \cdot 10^{-2}$	$4 \cdot 10^{-5}$
$\max \lvert\hat{\imath}(0,\cdot) - \hat{\imath}(T_1,\cdot)\rvert$	$3 \cdot 10^{-3}$	$8 \cdot 10^{-4}$	$1 \cdot 10^{-6}$

5 Conclusions

The approach of Houben permits to solve initial-boundary value problems of warped MPDAEs efficiently, which yields envelope-modulated signals. We consider initial values of a biperiodic solution to investigate the determination of quasiperiodic signals. It follows that the resulting local frequency function becomes periodic in this case. However, it is still an open question if the corresponding MVF is always biperiodic. We performed numerical simulations with Houben's strategy via a method of lines. The results illustrate that it can not be excluded that the arising solution is automatically biperiodic. In practice, the resulting MVFs seem to be biperiodic or at least nearly (except for a small difference) biperiodic. If the solution is exactly biperiodic, then a method for computing biperiodic solutions of the warped MPDAEs can be constructed based on Houben's technique. For example, the initial conditions in the method of lines are just replaced by periodic boundary conditions.

References

1. H. G. Brachtendorf, G. Welsch, R. Laur, and A. Bunse-Gerstner. Numerical steady state analysis of electronic circuits driven by multi-tone signals. *Electrical Engineering*, 79:103–112, 1996.
2. C.W. Gear. Simultaneous numerical solution of differential-algebraic equations. *IEEE Trans. on Circuit Theory*, 18:89–95, 1971.
3. J. Greb. Optimale Simulation von frequenzmodulierten Signalen. Master's thesis, University of Wuppertal, 2006.
4. M. Günther and U. Feldmann. CAD based electric circuit modeling in industry I: mathematical structure and index of network equations. *Surv. Math. Ind.*, 8:97–129, 1999.
5. S.H.M.J. Houben. *Circuits in Motion: The numerical simulation of electrical oscillators*. PhD thesis, Technical University Eindhoven, 2003.
6. S.H.M.J. Houben. Simulating multi-tone free-running oscillators with optimal sweep following. In W.H.A. Schilders, E.J.W. ter Maten, and S.H.M.J. Houben, editors, *Scientific Computing in Electrical Engineering*, volume 4 of *Mathematics in Industry*, pages 240–247. Springer, 2004.
7. O. Narayan and J. Roychowdhury. Analyzing oscillators using multitime PDEs. *IEEE Trans. CAS I*, 50(7):894–903, 2003.
8. R. Pulch. Multi time scale differential equations for simulating frequency modulated signals. *Appl. Numer. Math.*, 53(2-4):421–436, 2005.
9. R. Pulch. Variational methods for solving warped multirate PDAEs. Preprint BUW-AMNA 05/01 (University of Wuppertal), April 2005.
10. J. Roychowdhury. Analysing circuits with widely-separated time scales using numerical PDE methods. *IEEE Trans. CAS I*, 48(5):578–594, 2001.

Part III

Computational Electromagnetics

Part III

Computational Electromagnetics

RF & Microwave Simulation with the Finite Integration Technique – From Component to System Design*

I. Munteanu[1] and T. Weiland[2]

[1] Computer Simulation Technology, Bad Nauheimer Straße 19, D-64289 Darmstadt, Germany munteanu@cst.com
[2] Technische Universität Darmstadt, Institut für Theorie Elektromagnetischer Felder, Schloßgartenstraße 8, D-64289 Darmstadt, Germany
weiland@temf.tu-darmstadt.de

Summary. The paper presents a historical review and the current state-of-the-art of the Finite Integration Technique (FIT), method which has been successfully used for almost 30 years for the solution of electromagnetic field problems. The presented applications are in the range of high-end RF and microwave technologies.

Keywords—Finite Integration Technique, FEM, 3D field simulation, microwaves, high-frequency, numerical techniques

1 A Short Historical Review

The Finite Integration Technique, for short FIT [1] was first proposed almost 30 years ago, as a method for the simulation of electromagnetic fields and of various coupled problems. The key idea was to use in the discretization the integral, rather than the differential form of Maxwells equations. This early intuition proved to be correct and to have numerous theoretical, algorithmic and numerical advantages. Moreover, recently the same viewpoint seems to become predominant also in a historically completely different method, the finite element method [2]. FIT was first proposed with application to the solution of Maxwells equations in frequency domain (Fig. 1). It was the first eigenmode algorithm able to reliably eliminate all spurious modes [1] whereas for other methods, such as the FEM, a solution to this issue could be found only 10 years later.
The first application of FIT to eddy current problems was presented one year after the first paper (1978) [3], followed by (to mention just a few) the extension of FIT to FDTD-like schemes, including the extension to r-f-z coordinate system (1980) [4], application to triangular meshes (1987) [5], waveguide boundary conditions to allow accurate S-parameter extraction from time domain simulations (1988) [6], stable subgridding algorithm (1995) [7], application to non-orthogonal grids, including triangular fillings (1998, 1999) [8], model order reduction in conjunction with FIT (2000) [9].
Around 1980, the FIT gained instantaneous fame in the international accelerator physics community, as the first code ever which was able to calculate transient field of charged particles at ultra-relativistic energies. This was the starting point of the MAFIA Collaboration (an acronym for "solving MAxwell's equations with the Finite Integration Algorithm"), a consortium of universities, research institutes, accelerator laboratories with the goal to develop an

* Invited Paper at SCEE-2006

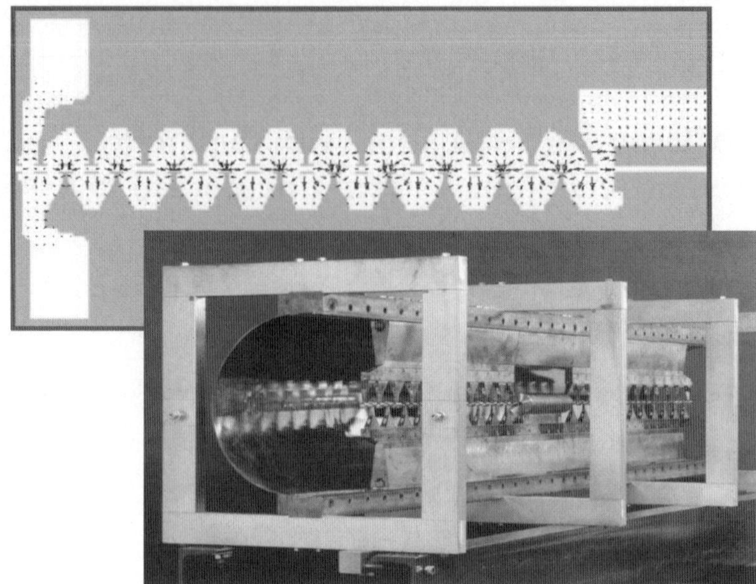

Fig. 1: High accuracy mode computation for a 22 Gap IH Ion Accelerator designed entirely based on FIT simulations; the mesh of the 20m long structure contained 3 million cells (simulation and experimental model)

FIT-based, general software package, for the solution of electromagnetic problems. The result of the 10 years of existence of this consortium was a full electromagnetic, thermal and particle tracking software, distributed to (and widely used in) research facilities from 26 countries.
The success of the Finite Integration Technique is probably mainly due to three factors. First, it is an algorithm with a sound theoretical foundation (among others, stability, orthogonality of numerically computed modes, energy and charge conservation were demonstrated in a very early stage). Second, it is applicable not only in frequency, but also in time domain, allowing thus the simulation of very large or very complex structures. Last but not least, it is applicable to a variety of mesh types.

2 The Finite Integration Technique

FIT generates exact algebraic analogues to Maxwells equations, which guarantee that the physical properties of fields are maintained in the discrete space, and lead to a unique solution. Maxwell's equations and the related material equations are transformed from the continuous to the discrete space by allocating electric voltages on the edges and electric fluxes on the faces of a grid ("primary grid") and magnetic voltages on the edges and magnetic fluxes on the faces of a second grid ("dual grid").
The use of integral degrees of freedom, i.e. voltages and fluxes, instead of field components (such as used in FDTD) allows not only a very elegant way of writing the matrix form of Maxwells equations, but also has important algorithmic-theoretical and numerical consequences [10]. In fact, measurable quantities are also of integral type: for instance, the electric field strength cannot be measured directly, but through the intermediary of the electric voltage along a very short path.
In the very first step, the problem has to be discretized in either coordinate meshes or tetrahedral meshes, as shown in Fig. 2.

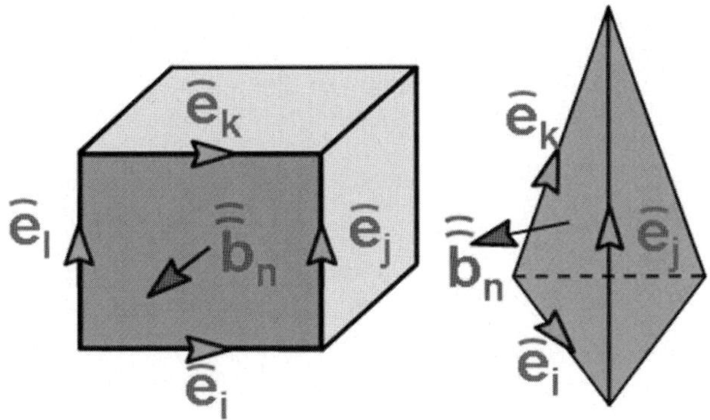

Fig. 2: FIT allocation of voltages and fluxes along edges and through surfaces, respectively

The next step is to write the first Maxwells equation on all surfaces of the elementary cells:

$$\int_{\partial A} \mathbf{E} \cdot \mathrm{d}\mathbf{s} = -\int_A \frac{\partial}{\partial t} \mathbf{B} \cdot \mathrm{d}\mathbf{A} \,, \tag{1}$$

where A denotes any open surface, ∂A is its boundary (a closed curve), $\mathrm{d}\mathbf{A}$ and $\mathrm{d}\mathbf{s}$ are the vectorial area and line element, respectively.
The result of the integration of (1) over the cell of the cube using the notations of Fig. 2 reads

$$\widehat{e}_\ell + \widehat{e}_k - \widehat{e}_j - \widehat{e}_i = -d/dt\, \widehat{\widehat{b}}_n \,. \tag{2}$$

In case of the tetrahedral cell the summation on the left hand side has one component less than in (2):

$$\widehat{e}_k - \widehat{e}_j - \widehat{e}_i = -d/dt\, \widehat{\widehat{b}}_n \,. \tag{3}$$

After collecting all electric and magnetic unknowns in vectors, one may write the discrete analogue of the Maxwell equation (1) in a simple matrix form as:

$$\mathbf{C}\widehat{\mathbf{e}} = -\frac{d}{dt}\widehat{\widehat{\mathbf{b}}} \,. \tag{4}$$

The "curl"-matrix **C** is a topological matrix (has only elements 0, +1 or -1) and represents the edges-to-faces incidence matrix on the primary grid. In an almost identical way one may write down the second Maxwell equation on a dual grid for the quantities $\widehat{\mathbf{h}}$, $\widehat{\widehat{\mathbf{d}}}$ and $\widehat{\widehat{\mathbf{j}}}$ respectively. Similarly, the remaining Maxwells equations may be transformed into a discrete set, so that one finally obtains the so-called **M**axwells **G**rid **E**quations produced by the FIT approach as:

$$\int_{\partial A} \mathbf{E} \cdot \mathrm{d}\mathbf{s} = -\int_A \frac{\partial}{\partial t} \mathbf{B} \cdot \mathrm{d}\mathbf{A} \quad \leftrightarrow \quad \mathbf{C}\widehat{\mathbf{e}} = -\frac{d}{dt}\widehat{\widehat{\mathbf{b}}} \,; \tag{5}$$

$$\int_{\partial A} \mathbf{H} \cdot \mathrm{d}\mathbf{s} = -\int_A \left(\frac{\partial \mathbf{D}}{\partial t} + \mathbf{J}\right) \cdot \mathrm{d}\mathbf{A} \quad \leftrightarrow \quad \widetilde{\mathbf{C}}\widehat{\mathbf{h}} = \widehat{\widehat{\mathbf{j}}} + \frac{d}{dt}\widehat{\widehat{\mathbf{d}}} \,; \tag{6}$$

$$\int_{\partial V} \mathbf{B} \cdot \mathrm{d}\mathbf{A} = 0 \quad \leftrightarrow \quad \mathbf{S}\widehat{\widehat{\mathbf{b}}} = 0 \,; \tag{7}$$

$$\int_{\partial V} \mathbf{D} \cdot \mathrm{d}\mathbf{A} = \int_V \rho \, \mathrm{d}v \quad \leftrightarrow \quad \widetilde{\mathbf{S}}\widehat{\widehat{\mathbf{d}}} = \mathbf{q} \,. \tag{8}$$

It is well known that there are two kinds of errors which affect the numerical solution. The space discretization error is inherent to *any* numerical method and results from the fact that the computational domain is discretized into a *finite* number of mesh cells. The "method discretization error" on the other hand results from the numerical discretization of the

continuous operators. For any given set of open surfaces (e.g. mesh faces), the right side of the equations (5-8) is *exact* in the sense that it has no method discretization error error involved (unlike the case in which the differential form of Maxwell's equations would be discretized by finite differences). The methods approximation only intervenes when the material relations are discretized.
The algebraic structure of the discrete material relations is independent of the method for the local field approximation and results in

$$\mathbf{D} = \varepsilon\mathbf{E} + \mathbf{P} \leftrightarrow \widehat{\widehat{\mathbf{d}}} = \mathbf{M}_\varepsilon \widehat{\mathbf{e}} + \varphi ; \tag{9}$$

$$\mathbf{B} = \mu\mathbf{H} + \mathbf{M} \leftrightarrow \widehat{\widehat{\mathbf{b}}} = \mathbf{M}_\mu \widehat{\mathbf{h}} + \widehat{\widehat{\mathbf{m}}} ; \tag{10}$$

$$\mathbf{J} = \sigma\mathbf{E} \leftrightarrow \widehat{\widehat{\mathbf{j}}} = \mathbf{M}_\sigma \widehat{\mathbf{e}} . \tag{11}$$

The matrices $\mathbf{M}_\varepsilon$, $\mathbf{M}_\mu$ are positive definite, while $\mathbf{M}_\sigma$ is in general semi-positive definite. On Cartesian meshes these matrices have a diagonal form. It must be noted that a similar idea for discretizing Maxwell's equations is used in the cell method by Tonti [11], although the material matrix discretization in this method generally leads to unsymmetric matrices.
The method not only works for virtually any kind of mesh, be it tetrahedral, non-orthogonal hexahedral or any other coordinate grid, but it may also employ various discretization methods in the classical language when modeling the material relations locally, including edge elements as used in many Finite Element approaches. On Cartesian meshes, equations (5-11) can be rewritten to yield the classical FDTD method [12].
The beauty of equations (5-8) is that they represent a one-to-one discrete counterpart of the corresponding continuous relations (just try it: read **C** as "curl", **S** as "divergence" !). This way, deriving any second-order equation in the discrete space can be done just as straightforwardly as in the continuous case.
Take for example the harmonic case (where the time derivatives are replaced by $i\omega$), in lossfree materials with no source currents. The use of the discrete material property relations (9-10) in equations (5-6) leads to

$$\mathbf{M}_{\mu^{-1}} \mathbf{C} \widehat{\mathbf{e}} = -i\omega \widehat{\mathbf{h}} ; \tag{12}$$

$$\widetilde{\mathbf{C}} \widehat{\mathbf{h}} = i\omega \mathbf{M}_\varepsilon \widehat{\mathbf{e}} . \tag{13}$$

The application of the dual curl operator $\widetilde{\mathbf{C}}$ to the relation (12), and the use of (13) in the right-hand side yields the discrete form of the well-known wave equation:

$$\widetilde{\mathbf{C}} \mathbf{M}_{\mu^{-1}} \mathbf{C} \widehat{\mathbf{e}} = \omega^2 \mathbf{M}_\varepsilon \widehat{\mathbf{e}} . \tag{14}$$

The generality of the FIT approach, applicable to different frequency ranges, from DC to THz, yields an ideal base for implementations in computer codes. Starting almost 30 years ago, a long series of codes has been written by several authors [13] [5], [14], [15] for many different applications ranging from statics over quasistatics to high frequency problems. Computer codes based on FIT have become a routine tool in industry and research.

3 State of the Art

The domain of RF and microwave simulation is characterized by a tremendous variety: many different types of devices (filters, connectors, antennas, cavities, ...), each requiring specific postprocessing capabilites; large variety in the geometric complexity and level of detail (e.g. rounded parts, thin layers, small details in otherwise large structures, complicated shapes), which imposes tough requirements for the generation of a discretization mesh; various types of materials (lossy, anisotropic, dispersive, ...); both narrow-band and wide-band applications, ranging from the MHz to multi-GHz.

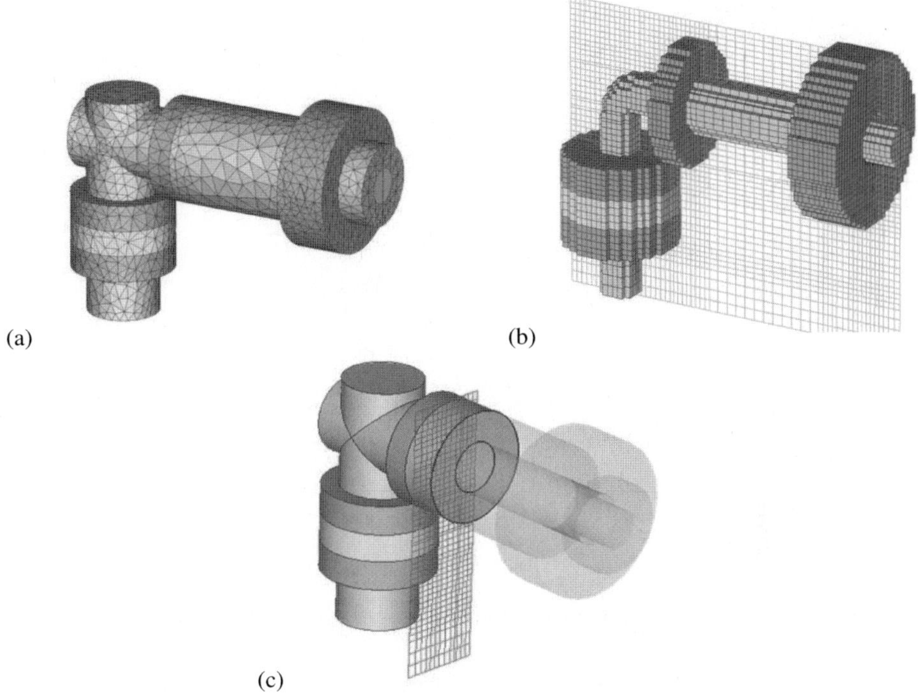

Fig. 3: Mesh types for a coaxial connector: **(a)** Tetrahedral; **(b)** hexahedral-staircase (some dielectric parts hidden for better visibility); **(c)** hexahedral-PBA mesh

While there is no single method which can solve every problem, due to its versatility FIT is probably the 3D numerical method which covers the largest possible spectrum of simulation needs. Some of them will be discussed in the present section.

3.1 Meshtypes

Before solving an electromagnetic problem, the structure needs to be spatially discretized, i.e. a discretization mesh needs to be mapped onto the structures geometry. One of the main strengths of FIT is the "translation" of Maxwells equations onto any (2D or 3D) given mesh. The most often employed discretization meshes are the tetrahedral, the staircase-hexahedral and the conformal-hexahedral meshes (Fig. 3).
The tetrahedral meshes (Fig. 3 a) have the advantage of allowing a good approximation of curved surfaces. Their main disadvantage is that such a mesh is not appropriate for time-domain algorithms: the resulting matrices (for any numerical method) can be efficiently solved in frequency-domain but due to their nondiagonal character, they are inefficient in time-domain algorithms. Last, but not least, it should be pointed out that the generation of the tetrahedral mesh is not a trivial task.
The classical hexahedral meshes (Fig. 3 b) have the advantage that they can be easily applied in both time- and frequency-domain algorithms. In time domain they lead to very memory- and computing-time-efficient algorithms. The mesh generation is quite straightforward, even for very complicated geometries. The main disadvantages of the classical hexahedral mesh are the staircase approximation of curved surfaces, sometimes with severe consequences on the solutions accuracy [16], and the fact that if a fine mesh is needed in a small zone of the structure, it will be extended through the entire computational domain.

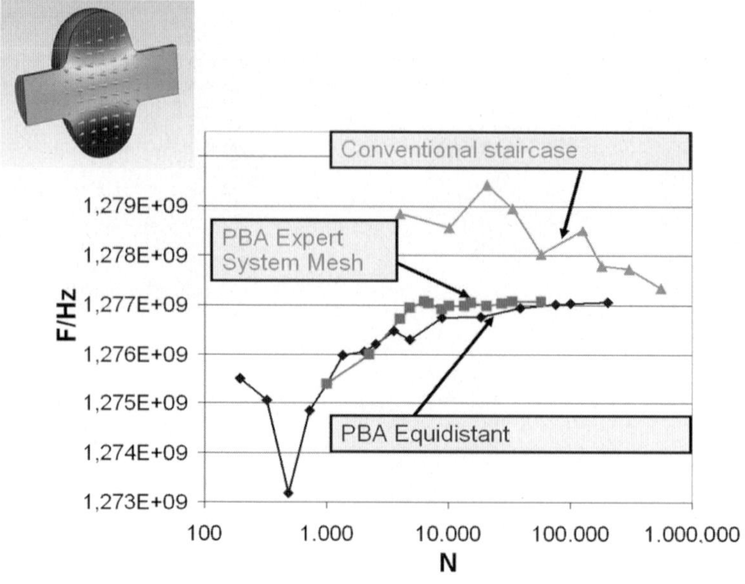

Fig. 4: Eigenfrequency variation for a simple cavity with the number of cells in the discretization mesh

Fortunately, there are solutions for both these disadvantages [17] [18] [19] [20], the most widely used being the PERFECT BOUNDARY APPROXIMATION (PBA)® (Fig. 3 c), which maintains all the advantages of the structured Cartesian grids, while allowing an accurate modeling of curved boundaries.
The PBA geometry improvement allows to use a much coarser mesh than in the staircase approximation, for the same required accuracy. This effect is illustrated in Fig. 4, which shows the evolution of a cavity eigenfrequency with the number of meshcells. To reach an accurate ($< 0.1\%$) value of 1.277 GHz for the eigenfrequency, 20000 meshcells are sufficient in the PBA mesh, whereas with a staircase mesh more than 500,000 meshcells are needed.
The PBA-extension Thin Sheet Technique™ (TST) allows the accurate geometric modeling of structures with thin layers, such as the curved patch antenna array of Fig. 5 (which is difficult to mesh with any numerical method).
A new development is the subgridding technique which allows the local refinement of the Cartesian grid, in regions with high field variation or with fine geometrical details. The total number of meshcells and the overall computing time are thus reduced considerably (Fig. 6). Note that in each refined cell the PBA® and TST™ algorithms allow for even more accurate description of the geometry. Unlike most known subgridding techniques, the one implemented in CST MWS has **guaranteed stability**, thus ensuring that the accuracy of the results is not negatively affected.
For every problem depending on its size and on its geometrical characteristics, there is a type of mesh which is optimal from the point of view of memory requirements and of the accuracy of geometry approximation. Disposing of several types of mesh generators within a single simulation environment can increase tremendously the efficiency of modeling RF components.

3.2 Time and Frequency Domain Simulations

Contrary to the common belief, FIT is not just a time-domain method: it offers, to mention just a few solver types, frequency domain (general-purpose, modal analysis and model order reduction), as well as time domain explicit and implicit.

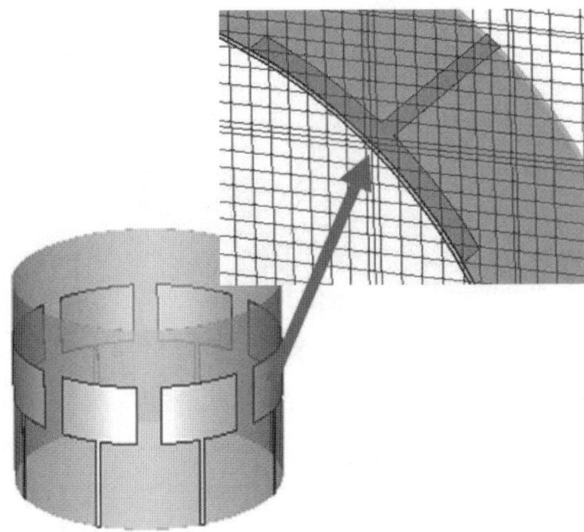

Fig. 5: Thin Sheet Technique applied to a curved patch antenna array. Staircase approximation, as well as too small mesh steps are avoided

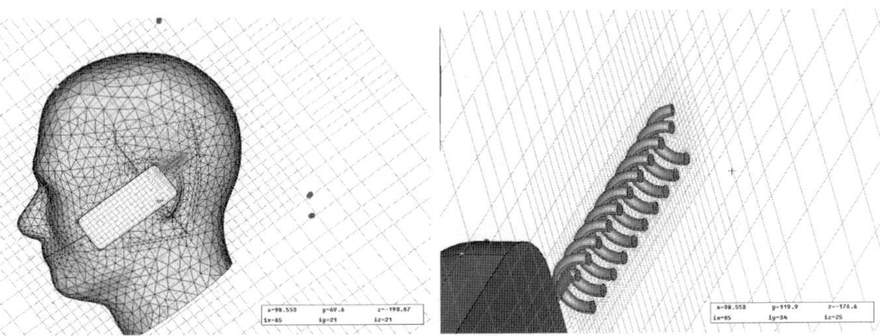

Fig. 6: Multilevel Subgridding Scheme showing the flexibility in this hierarchical type of mesh. The small cell area follows details in a similar way to unstructured meshes leading to a 10-fold reduction of the number of meshcells for the same solution accuracy

This way, one can often solve one and the same problem with completely different algorithms. This gives not only the possibility to choose the most efficient algorithm (in terms of computing time and memory requirements) d uring the time-consuming phase of the structure design, but it also offers independent results for cross checking in the verification and prototyping phase.

The time-domain (TD) solvers are the preferred ones when broadband results are required, since they deliver broadband results with a single run. A typical time-domain application is shown in Fig. 7: the full 3D simulation of a 30 meter-long airplane illuminated by a plane wave at 500 MHz. Although quite large (9 million cells), with the efficient FIT/PBA time domain algorithm it takes under two hours to simulate on a common PC. This problem would be simply too big for a volume-based frequency domain solution.

TD solvers are also a must if predefined or complicated time signals are needed in the simulation. Ultrawideband (UWB) antenna applications, or Time Domain Reflectometry (TDR) are just two examples.

Fig. 7: Surface currents (at 500 MHz) on an airplane illuminated by a plane wave

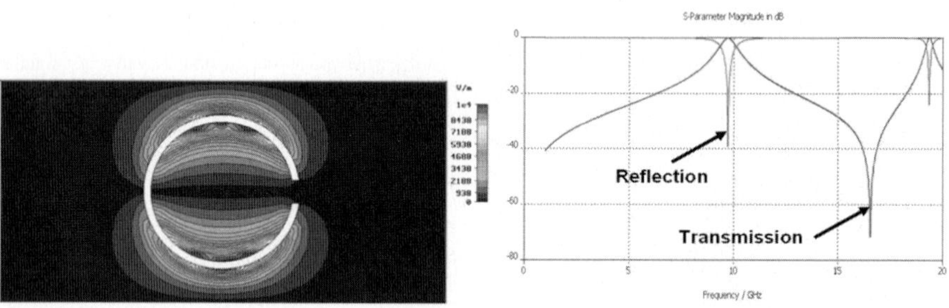

Fig. 8: **Left:** Electric field at 10 GHz in a Split Ring Resonator unit cell, **Right:** reflection and transmission coefficients

Frequency domain (FD) solvers on the other side may be the most efficient when narrow-band or single-frequency results are required, for small structures in terms of both number of cells and electrical size, or when periodic boundary conditions with nonzero phase shifts are needed.

An example of the latter are the Frequency Selective Surfaces (FSS). They are made up of a large (theoretically infinite) number of identical cells, so that in a first approximation just one unit cell (with periodic boundary conditions) can be simulated. Fig. 8 shows such a surface: a metallic surface with regularly-placed slots in the shape of split rings (only one repeating element is simulated). The surface is illuminated by a plane wave with a specific incidence. The frequency selective character is evident in reflection and transmission coefficient plot of Fig. 8 right: transmission is permitted only for some specific frequencies (here, around 9.8 GHz and around 19 Ghz).

It is the common belief that strongly resonant structures are also a case for FD-only, since the time signals within a resonant structure tend to oscillate for a very long time, making the TD simulation time also very long. However, the use of signal processing techniques for the time signals, such as the autoRegressive filtering, can change the situation again in favor of TD techniques. Moreover, if (relatively) broad-band results are required in a FD solution, the frequency range needs to be "intelligently" sampled, otherwise key-frequency points, such as resonances could be simply missed (this cannot happen with a TD technique). Advanced Multipoint frequency interpolation techniques can be used to interpolate between

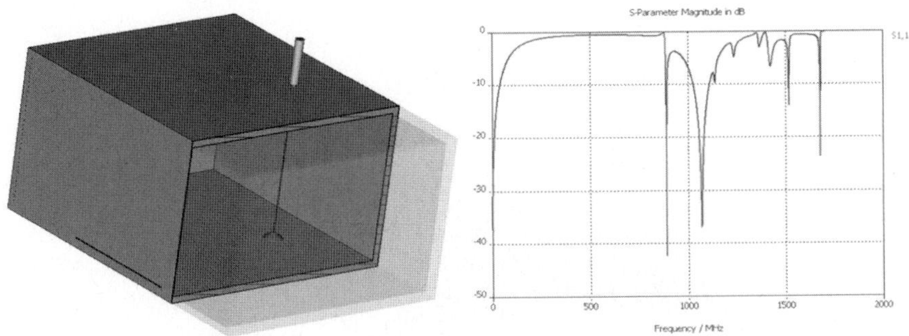

Fig. 9: **Left:** resonant cavity with excitation antenna and slot; **Right:** reflection coefficient S11

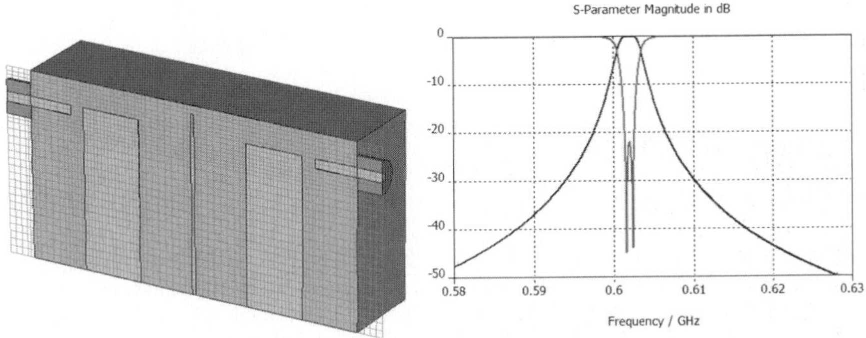

Fig. 10: **Left:** Crossection of an RF filter structure (Hexahedral PBA mesh is also shown). **Right:** S-parameters

calculated frequency points, in order to speed up the broadband calculation (the quite well-known Asymptotic Waveform Evaluation AWE method performs an extrapolation around every single calculated frequency point and may fail in case of complicated frequency behaviour).

For example, Fig. 9 shows the geometry of an EMC problem: a resonant cavity with a thin slot. From the S-parameter plot it is obvious that the structure is strongly resonant. The time domain solver (combined with an autoregressive filtering technique) and the frequency domain solver (combined with a powerful multi-point fast frequency sweep) were used. Interestingly enough, for this problem the time-domain solver is the quickest: To reach the same accuracy, it needed 8 minutes on a standard PC, while the frequency domain solver took more than 1 hour.

For special applications, yet another method might be the most efficient: the Model Order Reduction (MOR). This is the case of the classical RF filter shown in crossection in Fig. 10: due to the fine mesh, the initial system has a very high order of 130,000. This is reduced by MOR to a system with only about 1,000 unknowns, which can be then solved in roughly one minute on a standard PC.

Meshing alone is only half of the story: full efficiency is reached in combination with automatic optimization and full parameterization. New research has been done on automatic mesh generation, based on an expert system which takes into account not only the geometry, but also the physical properties of the device. The most recent trend is to combine the expert system with the classical automatic mesh adaptation.

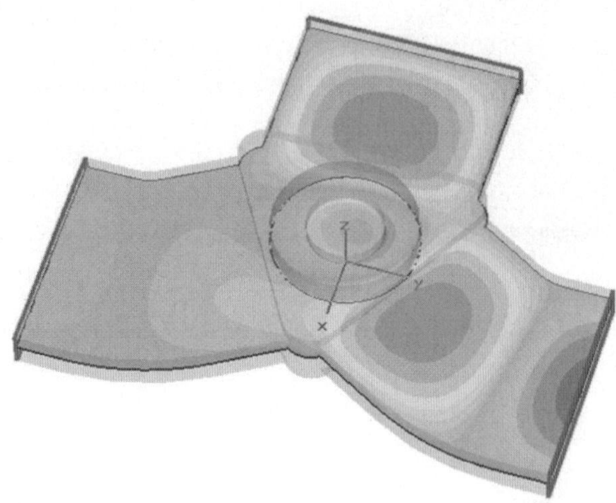

Fig. 11: Electric field inside a circulator structure, containing in the middle a cylinder made of gyrotropic ferrite material

3.3 Material Models

Material models are clearly one of the main "factors" for the accuracy of the electromagnetic simulation. Real materials most often obey to complicated relationships, are anisotropic or have properties which vary with frequency. In FIT, anisotropic, lossy, as well as dispersive materials with various dispersion behavior types (Debye first and second order, Lorentz, gyrotropic), are available. They allow the simulation of most complicated devices, such as plasma devices, or circulators. In Fig. 11, the electric field inside a circulator structure (with a gyrotropic-material cylinder placed in the middle) is depicted. It was obtained through an FIT-based magnetostatic simulation to determine the magnetization of the ferrite part, followed by high-frequency time domain simulation. The circulator effect is evident.
Recently, EMC/EMI concerns, as well as new medical investigation techniques require more and more simulations involving human body models. These are strongly inhomogeneous, contain many different dispersive materials, and require typically a large and fine mesh. Fig. 12 shows a human head model exposed to the radiation of a mobile phone. Due to the relatively large size of the model, the a time-domain simulation was used (FD would have required too much memory). The Specific Absorbtion Rate (SAR) within the head is shown in Fig. 12 (right).
Whereas 3D field simulation is perfectly suited for simulating components and devices of high complexity, it would be inefficient to apply to the simulation of entire systems: the numerical effort for obtaining a desired accuracy would simply be too large. For such applications, a hybrid approach is needed. FIT has been successfully integrated in such a design environment, in which numerical methods may be arbitrarily mixed: one may e.g. easily combine fully three dimensional blocks with planar solution tools, analytical solution or mode matching techniques, or virtually with any other technique that is capable of describing an element by some port behavior [21]. Most importantly, this open architecture approach allows to interface with other specialized software and thus eliminates the dependence on single proprietary software.
As an example, Fig. 13 shows a block schematic containing a patch antenna array block and its feeding network. The feeding network contains both lumped circuit elements (amplifier, resistors, capacitors, etc.), as well as 2D microstrip blocks (simulated with a planar field simulator).

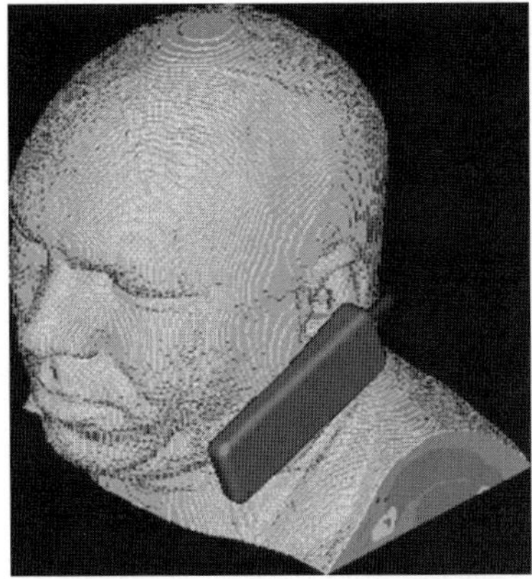

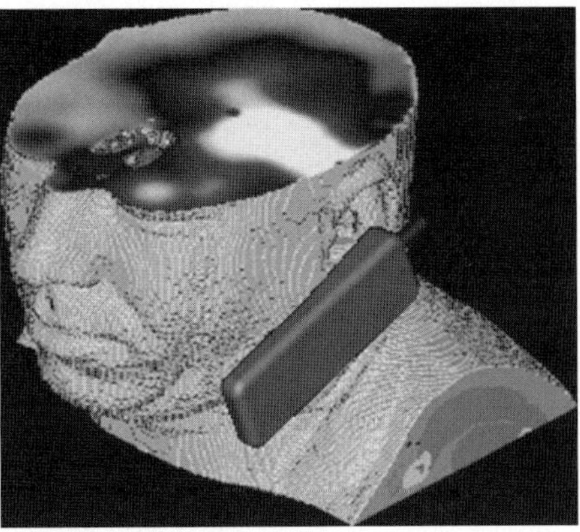

Fig. 12: EMI Study on the human head: **Left:** Human head with mobile phone. **Right:** Specific Absorbtion Rate (SAR) within the head, at eye-level. Light-grey colours indicate zones of high SAR

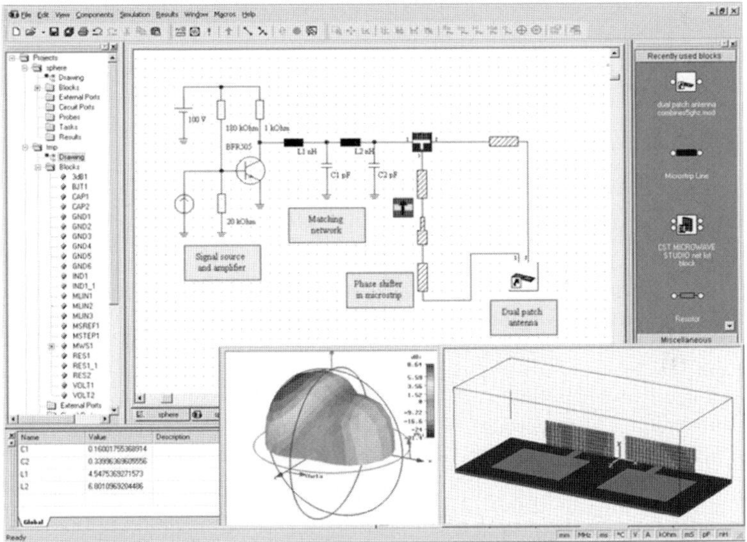

Fig. 13: Co-Simulation and co-optimization of a circuit-3D EM problem: patch antenna array with feeding network

The antenna array itself is simulated with a full 3D solver. After optimizing the circuit parameters to ensure a good matching at the frequency of interest, the combined farfield of the two patches can be obtained.

4 New Domains of Application and Future Trends

The needs of todays industry go more and more in the direction of computer simulation: operating frequencies grow and make existing design techniques difficult to apply; device complexity increases every year; prototyping becomes more expensive and time-consuming.

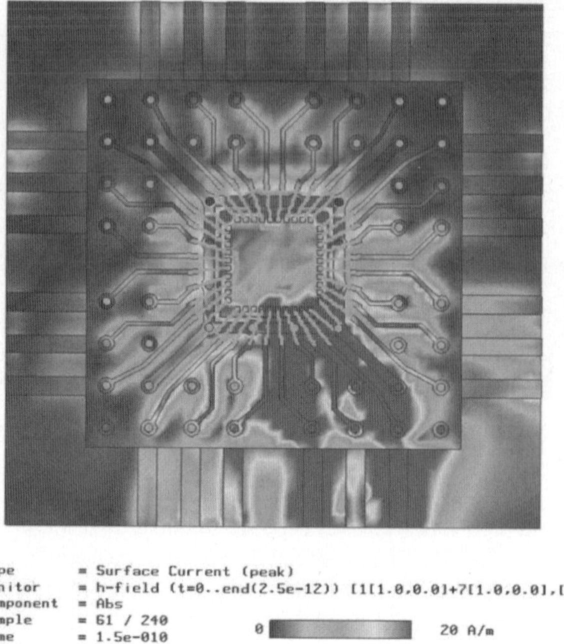

Fig. 14: Surface currents on an IC package at 10 GHz

High-frequency electromagnetic simulation today continues of course to be needed in the classical areas of microwave applications: filters, connectors, waveguide structures, antennas. The characteristic of this area of applications is that the models become very large, requiring very efficient algorithms in terms of computational complexity and memory requirements.
A new tendency is to apply field simulation to domains in which until recently only circuit-simulation techniques were needed: PCBs, integrated circuit components, etc. This is due to the partial failure of the circuit design techniques, when the operating frequencies of integrated circuits grow. Figure 14 presents such an example: the surface currents on a Ball Grid Array Package, at a quite high frequency, 10 GHz. The field effects are clearly visible, and would not be captured by any circuit simulation.
Another example of a challenging problem is shown in Fig. 15: the full layout and 3D view of a small portion of a PCB [22]. The geometrical complexity of the 8-layer structure requires a very fine discretization mesh of up to 10^9 mesh cells. The structure was simulated on a 24-CPU Intel-based cluster in order to obtain crosstalk between the lines at very high frequencies [23].
Last but not least, there are two other trends which became evident in the last few years. The first one is the increasing need for coupled problems (electromagnetic coupled with thermal, mechanical, or fluid dynamic systems). The second is the integration of electromagnetic simulation tools into major design flows (Cadence, Mentor Graphics, etc.), triggered by the increasing operating frequencies of todays integrated circuits, leading to field effects which cannot be taken into account by the existing circuit models.

Fig. 15: PCB Structure from IBM. **Left:** Layout view; **Right:** 3D view. The left picture also shows a possible partitioning of the structure, each portion being simulated on a different processor

5 Conclusion

The Finite Integration Technique, now 30 years young, is probably the numerical method for electromagnetic field simulation with the most dynamic development.
Due to its capability to solve electromagnetic problems in both time- and frequency-domain, to the variety of material properties, and to its exceptional numerical efficiency and accuracy, FIT was used worldwide in the simulation of a wide range of devices, from DC to THz. In the time-domain, the major break through was the introduction of the Perfect Boundary Approximation, which allows accurate modelling of curved surfaces while maintaining all advantages of time-domain algorithms.
Moreover, the Finite Integration Techniques theoretical background contributed, in the last decade, to fundamental changes of viewpoint for other numerical methods, such as the Finite Element Method.

6 Acknowledgment

The authors are grateful to the anonymous reviewers for the valuable suggestions for improvement.

References

1. Weiland, T.: A discretisation method for the solution of Maxwell's equations for six-component fields. International Journal of Electronics and Communication AEÜ **31** (1977) 116–120
2. Bossavit, A.: 'Generalized finite differences' in computational electromagnetics. In: Progress In Electromagnetics Research. Volume PIER32. (2001) 45–64
3. Weiland, T.: On the calculation of eddy currents in arbitrarily shaped, three dimensional, laminated iron cores. Part I: The method. Archiv für Elektrotechnik (AfE) **60**(6) (11 1978) 345–351
4. Weiland, T.: Transient electromagnetic fields excited by bunches of charged particles in cavities of arbitrary shape. In: Proc. of the XI-th International Conference on High Energy Accelerators, Geneva, Switzerland (1980) 570–575

5. van Rienen, U., Weiland, T.: Triangular discretization method for the evaluation of RF-fields in cylindrically symmetric cavities. IEEE Trans. on Magnetics **21**(6) (11 1985) 2317–2320
6. Klatt, R., Weiland, T.: A three dimensional code BCI that solves Maxwell's equations in the time domain. In: Proc. 3rd International IGTE Symposium, Graz, Austria (1988) 1–8
7. Thoma, P., Weiland, T.: A subgridding method in combination with the Finite Integration Technique. In: Proc. of the 25th European Microwave Conference. Volume 2. (1995) 770–774
8. Schuhmann, R., Weiland, T.: FDTD on nonorthogonal grids with triangular fillings. IEEE Trans. on Magnetics **35** (1999) 1470–1473
9. Munteanu, I., Wittig, T., Weiland, T., Ioan, D.: FIT/PVL circuit parameter extraction for general electromagnetic devices. IEEE Trans. on Magnetics **36** (2000) 1421–1425
10. Weiland, T.: Time domain electromagnetic field computation with finite difference methods. International Journ. Num. Modell. **9** (1996) 295–319
11. Tonti, E.: Finite formulation of the electromagnetic field. In: Progress In Electromagnetics Research. Volume PIER32. (2001) 317,356
12. Yee, K.S.: Numerical solution of initial boundary value problems involving Maxwell's equations in isotropic media. IEEE Trans. Antennas Propagat. **AP-14**(4) (1966) 302–307
13. Weiland, T.: TBCI and URMEL - new computer codes for wake field and cavity mode calculations. IEEE Transactions on Nucl. Sci. **NS 30** (1983) 2489–2491
14. Barts, T., al.: MAFIA - a three dimensional electromagnetic CAD system for magnets, RF structures and transient wake field calculations. In: Proc. of the Intern. Linear Accelerator Conference. Volume SLAC-303., Stanford University (1986) 276–278
15. CST GmbH: CST DESIGN ENVIRONMENT™ with its modules CST MICROWAVE STUDIO®, CST EM STUDIO™, CST PARTICLE STUDIO™ and CST DESIGN STUDIO™. www.cst.com (2006)
16. Munteanu, I., Hirtenfelder, F.: Convergence of the Finite Integration Technique on various mesh types. In: Proc. Of Gemic 05, Ulm, Germany (2005)
17. Krietenstein, B., Schuhmann, R., Thoma, P., Weiland, T.: The Perfect Boundary Approximation technique facing the big challenge of high precision field computation. In: Proc. of the XIX Intern. Linear Accelerator Conference. (1998) 860–862
18. Yu, W., Liu, Y., Su, T., Hunag, N., Mittra, R.: A robust parallel conformal finite-difference time-domain processing package using the MPI library. Ant. Prop. Magazine **47** (2005) 39–59
19. Schuhmann, R., Weiland, T.: Stability of the FDTD algorithm on nonorthogonal grids related to the spatial interpolation scheme. IEEE Trans. on Magnetics **34** (1998) 2751–2754
20. Schuhmann, R., Schmidt, P., Weiland, T.: A new Whitney-based material operator for the Finite-Integration Technique on triangular grids. In: Proc. of the COMPUMAG 2001. Volume 3. (2001) 102–103
21. Schuhmann, R., Weiland, T.: Open architecture solves large 3D puzzles. Microwaves & RF **41** (2002) 144–155
22. Deutsch, A.: IBM, private communications (2006)
23. Gjonaj, E., Perotoni, M., Weiland, T.: Large scale simulation of an integrated circuit package. In: 16th Conference on Electrical Performance of Electronic Packaging (EPEP). (2006)

The Energy Viewpoint in Computational Electromagnetics*

Francois Henrotte[1] and Kay Hameyer[1]

Institute of Electrical Machines, RWTH Aachen University
Schinkelstrae 4, D-52056 Aachen, Germany
fh@iem.rwth-aachen.de

1 Introduction

Opening a textbook on electromagnetism, it is likely that the first set of equations presented will be Maxwell's equations

$$\operatorname{curl} \mathbf{h} - \partial_t \mathbf{d} = \mathbf{j} \tag{1}$$

$$\operatorname{curl} \mathbf{e} + \partial_t \mathbf{b} = 0 \tag{2}$$

$$\operatorname{div} \mathbf{b} = 0 \tag{3}$$

$$\operatorname{div} \mathbf{d} = \rho^Q \tag{4}$$

complemented by a set of constitutive relations of the form

$$\mathbf{b} = \mu\, \mathbf{h} \quad , \quad \mathbf{d} = \varepsilon\, \mathbf{e} \quad , \quad \mathbf{j} = \sigma\, \mathbf{e} \tag{5}$$

with the mention that the first set are universal (always valid) and the second one contains any relation one would need to 'close the system' and be able to solve it. Electromagnetism is in this way seen as a set of fields whose evolution in time and distribution in space are ruled by partial differential equations (PDE) and constitutive relations. There is no place in this setting for any energy considerations.
Further in the same book however, some energy related notions are likely to be introduced. The magnetic energy, for instance, is usually defined as a functional of $\mathbf{b}$ or $\mathbf{h}$ (or even both). Different materials will be considered, starting with the simplest medium (vacuum) and proceeding in a bottom-up fashion towards more complex materials : linear, anisotropic, nonlinear, etc. Not for long however, because the definitions become quickly rather technical and fall outside the scope of a general monograph.
Classical presentations of the theory of electromagnetism leave thus the impression that energy aspects are by-products of the field theory, somehow accessory and difficult to exploit. The principles of Thermodynamics however are universal and they must apply to electromagnetic phenomena also. Maxwell's equations say actually something yet about electromagnetic energy conservation, but they do so in a way that makes it impossible to disentangle the different energy flows in presence. Moreover, classical presentations of the theory leave unanswered fundamental questions like

- What are the state variables in an electromagnetic system ?
- How are magnetic and electric energy defined in the general case ?
- What are the possible dissipation mechanisms ?
- How is magnetic energy converted into electric energy ?

* Invited Paper at SCEE-2006

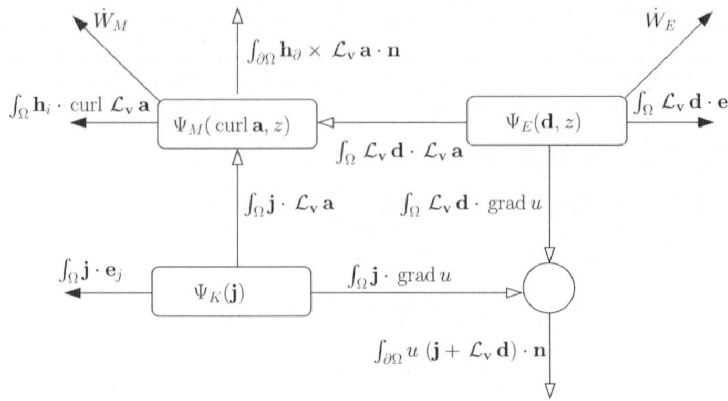

Fig. 1: EM energy flow diagram in the Euclidean space E.

- How is electromagnetic energy converted into other forms of energy ?
- etc.

Those shortcomings are particularly hampering when one deals with problems like the computation of local electromagnetic forces (energy conversion), magnetic hysteresis (energy dissipation) or magnetostriction (both) or multiphysics problems in general. For such problems, it really makes sense to dispose of a theory of electromagnetism where energy aspects are considered from the beginning and throughout.
After pursuing theoretical investigations in those domains, and accumulating along the way pieces of knowledge about how energy behaves in electromagnetic systems, a big picture has eventually, and somewhat unexpectedly, formed that gives rise to an energy-based theory of electromagnetism [1]. This representation of Electromagnetism takes the form of a flow diagram. It provides more information than the classical theory and gives answers to the questions listed above. Being expressed in integral form instead of by a set of PDE's, the governing equations can be established straightforwardly in arbitrary coordinate systems. Finally, the energy-based theory provides operative concepts, which clarify issues like hysteresis modelling and give many clues how to deal in a consistent way with coupling terms in multiphysics problems and parameters in reduced order models.

2 Energy flow diagram

The energy-based theory of electromagnetism is now briefly presented. More details can be found in [1]. The representation in an Euclidean space of the electromagnetic energy flow diagram is depicted in Fig. 1. The diagram consists of four interconnected energy reservoirs, each one associated with a state variable. The state variables are the two electromagnetic potentials, i.e. the magnetic vector potential $\mathbf{a}$ and the electric scalar potential u, and the two fields associated with electric charges, i.e. the electric displacement $\mathbf{d}$ and the current density $\mathbf{j}$.
The $\mathbf{a}-$reservoir contains the magnetic energy

$$\Psi_M(\text{curl}\,\mathbf{a}, z) \equiv \int_\Omega \rho_M^\Psi(\text{curl}\,\mathbf{a}, z), \tag{6}$$

which is the integral over the domain Ω under consideration of the magnetic energy density ρ_M^Ψ, a function of the induction $\text{curl}\,\mathbf{a}$ and possibly also of one or several additional non-electromagnetic quantities (e.g. the strain) represented in a generic way by the unspecified

variable z. Similarly, the $\mathbf{d}-$reservoir contains the electric energy $\Psi_E(\mathbf{d}, z)$. The $u-$reservoir is always empty. The $\mathbf{j}-$reservoir finally, contains the kinetic energy of the charge carriers, $\Psi_K(\mathbf{j}) = \alpha|\mathbf{j}|^2/2$, where α is a constant depending on the mass of the charge carriers. Except for superconductors, the inertia of charge carriers is negligible and the $\mathbf{j}-$reservoir can therefore be considered empty as well.
The co-moving time derivative $\mathcal{L}_\mathbf{v}$ is a time derivative that accounts for a possible motion or deformation of the domain Ω under consideration. It is also called material derivative but, as electromagnetic fields do not need material support, the name co-moving time derivative is preferred in this context. One has

$$\partial_t \Psi = \partial_t \int_\Omega \rho^\psi = \int_\Omega \mathcal{L}_\mathbf{v} \rho^\Psi, \tag{7}$$

where $\mathbf{v}$ is the velocity field. The co-moving time derivative allows obtaining the local form (partial differential equations) of global energy balances on moving domains. In the absence of motion, $\mathbf{v} \equiv 0$ and $\mathcal{L}_\mathbf{v} \equiv \partial_t$. See also [2, 3].
Whereas the internal flows (the flows connecting two reservoirs of the diagram) depend on the state variables only, the external flows depend on four generalised forces acting on the system. The dissipative forces $\mathbf{h}_i$, $\mathbf{e}_i$ and $\mathbf{e}_j$ are associated respectively with magnetic hysteresis, dielectric hysteresis and Joule losses. The surface generalised force $\mathbf{h}_\partial$ is associated with the magnetic energy crossing the surface of the system, by means of an electromagnetic wave or a boundary condition. The second surface flow represents the energy entering the system through the conductors crossing its surface. Finally, the flows $\dot{W}_M$ and $\dot{W}_E$ account respectively for the electric or magnetic energy converted into non-electromagnetic forms of energy (e.g. mechanical, chemical, etc.)
The structure of the diagram constitutes the basis of the theory. It tells something fundamental about how electromagnetic fields interact with matter and spacetime. It makes up a framework wherein any electromagnetic system, including dissipative and coupled ones, should inscribe.

3 Conservation equations

As the state variables are independent variables describing the system, they can be varied freely in order to obtain, following a variational line of argument, the conservation equations implied by the structure of the diagram. By expressing on the one hand energy conservation in integral form (the variation of energy in the reservoir is equal to the sum of all incoming fluxes minus the sum of all outgoing fluxes) at all nodes of the diagram and applying on the other hand the chain rule of derivatives to the algebraic expression of the energy functionals, two expressions are obtained for the variation of the energy in each reservoir that can be identified with each other. Conservation equations in local form are then derived by applying the fundamental lemma of Calculus of variations, with the arbitrary co-moving time derivatives of the state variables $\mathcal{L}_\mathbf{v}\, x$, $x = \mathbf{a}, \mathbf{d}, \mathbf{j}, u$, playing the role of the variations δx. The Euler-Lagrange equations obtained this way are

$$\operatorname{curl} \bar{\mathbf{h}} = \mathbf{j} + \mathcal{L}_\mathbf{v}\, \mathbf{d} \tag{8}$$

$$\bar{\mathbf{e}} = -\mathcal{L}_\mathbf{v}\, \mathbf{a} - \operatorname{grad} u \tag{9}$$

$$\mathbf{e}_j + \alpha \mathcal{L}_\mathbf{v} \mathbf{j} = -\mathcal{L}_\mathbf{v}\, \mathbf{a} - \operatorname{grad} u \tag{10}$$

$$0 = \operatorname{div} (\mathbf{j} + \mathcal{L}_\mathbf{v}\, \mathbf{d}) \tag{11}$$

on Ω and $\mathbf{h}_\partial = \bar{\mathbf{h}}$ on $\partial\Omega$, with the shorthand notations

$$\bar{\mathbf{h}} = \left(\partial_\mathbf{b}\, \rho_M^\Psi\right) (\operatorname{curl} \mathbf{a}, z) + \mathbf{h}_i \tag{12}$$

$$\bar{\mathbf{e}} = \left(\partial_\mathbf{d}\, \rho_E^\Psi\right) (\mathbf{d}, z) + \mathbf{e}_i \tag{13}$$

involving the Frchet derivatives $\partial_{\mathfrak{b}}$ and $\partial_{\mathfrak{d}}$.
On can recognize in (8) Ampere's law. Faraday's law is obtained by applying curl to (9). Equation (10) is a generalisation of Ohm's law, since the classical material law for conductors assumes $\alpha \equiv 0$ and $\mathbf{e}_j \equiv \sigma^{-1}\mathbf{j}$. Finally, (11) is redundant with (8), as a consequence of the fact that the $u-$reservoir is always empty.
Equation (12) shows that the magnetic field is composed of a reversible part $\mathbf{h}_r \equiv \partial_{\mathfrak{b}}\rho_M^{\Psi}$ that accounts for the magnetization phenomenon (alignment of microscopic magnetic moments), and an irreversible part $\mathbf{h}_i$ that accounts for the local dissipation process. The magnetic field $\bar{\mathbf{h}}$ is thus not a fundamental quantity but a composite one representing at the same time two different phenomena. A similar remarks holds for $\bar{\mathbf{e}}$.

4 Convex analysis

In order to draw all the benefit from the diagram presented in the previous section, some concepts from Convex analysis are useful. See e.g. [4] for a sufficient introduction to the subject.
Let X be a set. A **function** $f\ :\ \mathrm{dom}\, f \subset X \mapsto \mathbb{R}$ is defined by fixing a **domain** $\mathrm{dom}\, f \subset X$ and a **rule** $x \rightarrow f(x)$ that makes sense $\forall x \in \mathrm{dom}\, f$ with $f(x) \in \mathbb{R}$.[2] The **epigraph** of f is the subset of $X \times \mathbb{R}$ defined by epi $f = \{(x,z)\ :\ x \in \mathrm{dom}\, f, z \geq f(x)\}$. The function f is upper-bounded iff $\forall x \in \mathrm{dom}\, f$, $\exists\, \alpha \in \mathbb{R}\ :\ f(x) \leq \alpha$. The smallest upper bound for f is denoted by $\sup f$.
Let us suppose now that X is vector space. A subset $K \subset X$ is **convex** iff $\forall x, y \in K, ax + (1-a)y \in K$ with real $0 \leq a \leq 1$. A function $f\ :\ \mathrm{dom}\, f \subset X \mapsto \mathbb{R}$ is convex if its epigraph is convex.
Let us now additionally assume a **norm** $|x|$ is defined on the vector space X. This notion allows for consideration of convergence. The set $K \subset X$ is **closed** if it contains the limits of all its convergent sequences. The function f is **lower semi-continuous** if its epigraph is closed.
Let finally X and Y be two Hilbert spaces with the **scalar product** $\langle y, x\rangle, x \in X, y \in Y$. The **Legendre transform** of a function $\Psi\ :\ \mathrm{dom}\,\Psi \subset X \mapsto \mathbb{R}$ is the function $\Psi^*\ :Q \subset Y \mapsto \mathbb{R}$ defined by the rule

$$y \rightarrow \sup_{x \in \mathrm{dom}\,\Psi} \{x \mapsto \langle y, x\rangle - \Psi(x)\}. \tag{14}$$

and the domain Q that is the set of the points $y \in Y$ for which the function $x \rightarrow \langle y, x\rangle - \Psi(x)$ is upper bounded. It can be shown that the functions Ψ^* defined this way is convex and lower semi-continuous (clsc) and that $\Psi^{**} = \Psi$ if Φ is cslc itself.
The functions $\Psi\ :\ \mathrm{dom}\,\Psi \subset X \mapsto \mathbb{R}$ and $\Phi\ :\ \mathrm{dom}\,\Phi \subset Y \mapsto \mathbb{R}$, are said to be **dual** iff both

$$\begin{aligned} \Phi(y) &= \sup_{x \in \mathrm{dom}\,\Psi} \{x \mapsto \langle y, x\rangle - \Psi(x)\} \\ \Psi(x) &= \sup_{y \in \mathrm{dom}\,\Phi} \{x \mapsto \langle y, x\rangle - \Phi(y)\} \end{aligned} \tag{15}$$

are true. Dual functions are automatically clsc. Note that a pair of functions Φ and Ψ that are the Legendre transform of each other (i.e. $\Psi^* = \Phi and\ \Phi^* = \Psi$) are dual by definition but, as Q might be different from a prescribed domain $\mathrm{dom}\,\Phi$, Φ and Ψ might be dual without having $\Phi^* = \Psi$.
It is obvious from the definitions of Ψ and Φ that the inequality

$$\Lambda(x,y) = \Psi(x) + \Phi(y) - \langle y, x\rangle \geq 0 \tag{16}$$

holds $\forall x \in X$ and $\forall y \in Y$.

---

[2] Note that the domain $\mathrm{dom}\, f$ might be prescribed as being a subset only of the domain on which the rule $x \rightarrow f(x)$ is actually defined.

The **subdifferential** $\partial_x \Psi$ of the function $\Psi(x)$ is the set $\partial_x \Psi = \{y \in Y : \Psi(x') - \Psi(x) \geq \langle y, x' - x\rangle, \forall x' \in \operatorname{dom} \Psi\}$. The elements of that set are called **subgradients**. If the function $\Psi(x)$ happens to be differentiable at x, its **gradient** is the only element of $\partial_x \Psi$ and $y = \partial_x \Psi$ can be written instead of $y \in \partial_x \Psi$. An important result is that the inequality (16) becomes an equality if either $y \in \partial_x \Psi$ or $x \in \partial_y \Phi$. Finally, the applications $x \mapsto \partial_x \Psi$ and $y \mapsto \partial_y \Phi$ are **monotonous** in the sense that $\langle y_2 - y_1, x_2 - x_1\rangle \geq 0$ for any given $x_1, x_2 \in \operatorname{dom} \Psi$ and $\forall y_1 \in \partial_x \Phi(x_1), \forall y_2 \in \partial_x \Phi(x_2)$.

5 Applications

5.1 Formulations

In many problems encountered in electromagnetism, it is not necessary to solve the complete set of Maxwell equations. According to the dimensions and the time scale under consideration, the materials in presence and the configuration of the system, it happens often that simplifications are possible. Those simplifications consist generally in dropping terms in the full Maxwell's equations and weak formulations are then obtained by applying Galerkin's method to the simplified equations.
The alternative top-down approach, which consists in deriving weak formulations directly from the energy diagram, is not necessarily more straightforward but has nevertheless several advantages. Firstly, the assumptions done take on a physical justification this way, instead of a mathematical one. The different terms in the weak formulation also maintain their interpretation in terms of energy, so that they can be exploited to establish the global energy balance of the device or to express coupling terms in multi-physics problems.
Spelling out the wide variety of weak formulations encountered in computational electromagnetism would be fastidious. We are going to consider only electrostatics and magnetodynamics.

Electrostatics

The electrostatic regime is obtained by setting to zero the state variables $\mathbf{a}$ and $\mathbf{j}$ and preventing the system from any energy conversion, i.e. $\partial_t z \equiv 0$, $\dot{W}_M \equiv 0$, and assuming no motion, $\mathbf{v} \equiv 0 \Rightarrow \mathcal{L}_{\mathbf{v}} \equiv \partial_t$. Since dissipative forces act over time, it is also natural to assume $\mathbf{e}_i \equiv 0$ in a static problem. Two conservation equations then remain.
At node u, (11) becomes

$$\operatorname{div} \partial_t \mathbf{d} = \partial_t \operatorname{div} \mathbf{d} = 0, \tag{17}$$

which shows that the quantity $\operatorname{div} \mathbf{d}$ is conserved. The state variable $\mathbf{d}$ is therefore constrained. The vector potential $\mathbf{c}$ is then defined as a new unconstrained state variable, such that $\mathbf{d} = \mathbf{d}_0 + \operatorname{curl} \mathbf{c}$ with $\partial_t \mathbf{d}_0 = 0$, $\operatorname{div} \mathbf{d}_0 = \operatorname{div} \mathbf{d}$.
The conservation equation at node $\mathbf{d}$ in integral form,

$$\partial_t \Psi_E + \int_\Omega \operatorname{grad} u \cdot \partial_t \mathbf{d} = 0 \quad \forall\, \mathbf{d}(t), \tag{18}$$

becomes then

$$\int_\Omega \left\{\partial_{\mathbf{d}} \rho_E^\Psi(\mathbf{d}_0 + \operatorname{curl} \mathbf{c}) + \operatorname{grad} u\right\} \cdot \partial_t \operatorname{curl} \mathbf{c} = 0 \quad \forall\, \mathbf{c}(t), \tag{19}$$

and after an integration by part

$$\int_\Omega \partial_{\mathbf{d}} \rho_E^\Psi(\mathbf{d}_0 + \operatorname{curl} \mathbf{c}) \cdot \operatorname{curl} \partial_t \mathbf{c} + \int_{\partial\Omega} \operatorname{grad} u \times \partial_t \mathbf{c} \cdot \mathbf{n} = 0 \quad \forall\, \mathbf{c}(t). \tag{20}$$

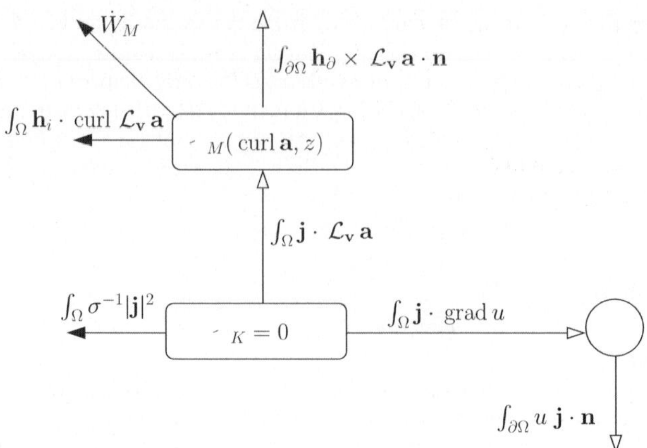

Fig. 2: EM energy flow diagram for the magnetodynamics regime.

This is the vector potential weak formulation for electrostatics. The arbitrary $\partial_t \mathbf{c}$ can be chosen equal to the shape functions of the field $\mathbf{c}$. At the boundary, either $\mathbf{c}$ (Dirichlet boundary condition) or $-\mathbf{n} \times \mathrm{grad}\, u$ (Neumann boundary condition) must be specified.
The formulation in terms of the scalar potential u, is obtained thanks to the concept of duality introduced above. The dual variables are in this case $x = \mathbf{d} \equiv \mathbf{d}_0 + \mathrm{curl}\, \mathbf{c}$ and $y = -\mathrm{grad}\, u$. Since (19) is a condition stronger than $\partial_\mathbf{d} \rho_E^\Psi \ni -\mathrm{grad}\, u$, the coenergy Φ_E defined by (15) as the dual of the energy Ψ_E satisfies the equality

$$\Phi_E = -\int_\Omega \mathrm{grad}\, u \cdot \mathbf{d} - \Psi_E, \tag{21}$$

so that

$$\begin{aligned} \partial_t \Phi_E &= -\int_\Omega \partial_t \mathrm{grad}\, u \cdot \mathbf{d} - \int_\Omega \mathrm{grad}\, u \cdot \partial_t \mathbf{d} - \partial_t \Psi_E \\ &= -\int_\Omega \partial_t \mathrm{grad}\, u \cdot \mathbf{d} - \int_\Omega \left\{ \mathrm{grad}\, u + \partial_\mathbf{d} \rho_E^\Psi \right\} \cdot \partial_t \mathrm{curl}\, \mathbf{c} \\ &= -\int_\Omega \partial_t \mathrm{grad}\, u \cdot \mathbf{d} \quad \forall u(t) \end{aligned}$$

by (19). Making now an integration by part, one has

$$\int_\Omega \partial_{\mathrm{grad}\, u} \Phi_E \cdot \mathrm{grad}\, \partial_t u = \int_\Omega \partial_t u\, \mathrm{div}\, \mathbf{d} - \int_{\partial\Omega} \partial_t u\, \mathbf{d} \cdot \mathbf{n} \quad \forall u(t)$$

with $\mathrm{div}\, \mathbf{d} = \mathrm{div}\, \mathbf{d}_0$ the charge density. This is the scalar potential formulation for electrostatics. At the boundary, either u (Dirichlet boundary condition) or $\mathbf{d} \cdot \mathbf{n}$ (Neumann boundary condition) must be specified.

Magnetodynamics

The magnetodynamics regime is obtained by setting $\mathbf{d} \equiv 0$. The corresponding energy diagram is depicted in Fig. 2. Dissipation (Joule and hysteresis) and electromechanical coupling ($\mathbf{v} \neq 0$) are going to be considered in this dynamical formulation, but additional dependencies represented by z are disregarded, as well as the kinetic energy of charge carrier, i.e. $\mathcal{L}_\mathbf{v}\, z = 0$,

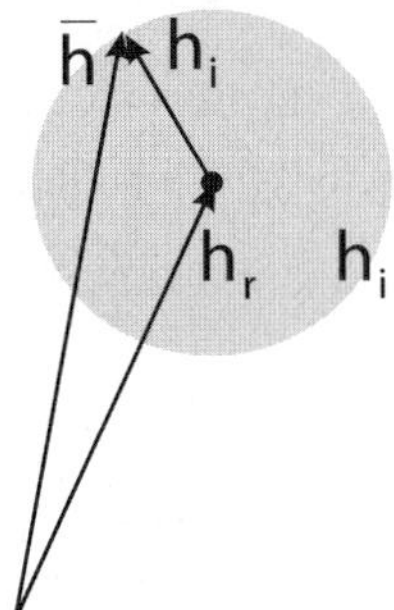

Fig. 3: Equilibrium equation (29). The grey circle represents the subgradient G.

$\Psi_K \equiv 0$. Energy conservation in integral form at node $\mathbf{a}$ and the application of the chain rule of derivatives to the magnetic energy Ψ_M write respectively

$$\partial_t \Psi_M = \int_\Omega \mathbf{j} \cdot \mathcal{L}_\mathbf{v}\, \mathbf{a} - \int_\Omega \mathbf{h}_i \cdot \operatorname{curl} \mathcal{L}_\mathbf{v}\, \mathbf{a} - \int_{\partial\Omega} \mathbf{h}_\partial \times \mathcal{L}_\mathbf{v}\, \mathbf{a} \cdot \mathbf{n} - \dot{W}_M$$

$$\partial_t \Psi_M = \int_\Omega \left\{\partial_\mathbf{b} \rho_M^\Psi\right\} (\operatorname{curl} \mathbf{a}) \cdot \operatorname{curl} \mathcal{L}_\mathbf{v}\, \mathbf{a} + \int_\Omega \left\{ \mathcal{L}_\mathbf{v}\, \rho_M^\Psi \right\} (\operatorname{curl} \mathbf{a})$$

and, after identification of both right hand sides,

$$0 = \int_\Omega \left\{\partial_\mathbf{b} \rho_M^\Psi (\operatorname{curl} \mathbf{a}) + \mathbf{h}_i \right\} \cdot \operatorname{curl} \mathcal{L}_\mathbf{v}\, \mathbf{a} - \int_\Omega \mathbf{j} \cdot \mathcal{L}_\mathbf{v}\, \mathbf{a} + \int_{\partial\Omega} \mathbf{h}_\partial \times \mathcal{L}_\mathbf{v}\, \mathbf{a} \cdot \mathbf{n}$$
$$+ \int_\Omega \left\{ \mathcal{L}_\mathbf{v}\, \rho_M^\Psi \right\} (\operatorname{curl} \mathbf{a}) + \dot{W}_M \qquad \forall\, \mathbf{a}(t). \tag{22}$$

Being independent of $\mathcal{L}_\mathbf{v}\, \mathbf{a}$, the last two terms must sum up to zero separately, which defines the power delivered by magnetic forces (See also [3]).

$$\dot{W}_M = - \int_\Omega \left\{ \mathcal{L}_\mathbf{v}\, \rho_M^\Psi \right\} (\operatorname{curl} \mathbf{a}). \tag{23}$$

The other terms make up the vector potential weak formulation of Magnetodynamics, with an imposed current density.
If on the other hand the dissipation force $\mathbf{e}_j$ is assumed to be an invertible function of $\mathbf{j}$ (one has for instance $\mathbf{e}_j = \sigma^{-1}\mathbf{j}$ for normal conductors), one can with (10) express $\mathbf{j} = f(\mathcal{L}_\mathbf{v}\, \mathbf{a} - \operatorname{grad} u)$ and substitute this in the weak formulation above in order to obtain the weak formulation of Magnetodynamics with imposed voltages. In practice, the voltage sources is not modelled explicitly and represented by a discontinuity of u over a given cross section of the conductor.

5.2 Magnetic hysteresis

The energy diagram indicates that the natural variable to represent the magnetic state of a material is the induction $\mathbf{b} \equiv \operatorname{curl} \mathbf{a}$. In the presence of hysteresis, this variable is subjected to a force $\mathbf{h}_r = \partial_\mathbf{b} \rho_M^\Psi$ deriving from a potential (the magnetic energy ρ_M^Ψ) and to a dissipative force $\mathbf{h}_i$. It is now shown how complying with this decomposition yields naturally a vector hysteresis model, in contrast to Preisach and Jiles-Atherton, which are basically scalar models. Starting from the vector potential formulation (22), using $\operatorname{curl} \bar{\mathbf{h}} = \mathbf{j}$ and making an integration by part, the conservation equation at node $\mathbf{a}$ in integral form can be put into the form of the First Principle of Thermodynamics $\partial_t \Psi_M = \dot{W} + \dot{Q}$ with

$$\dot{Q} = -\int_{\Omega} \mathbf{h}_i \cdot \dot{\mathbf{b}} \quad , \quad \dot{W} = \int_{\Omega} \bar{\mathbf{h}} \cdot \dot{\mathbf{b}} \tag{24}$$

and where $\dot{\mathbf{b}}$ is shorthand for curl $\mathcal{L}_{\mathbf{v}}\,\mathbf{a}$. It follows directly that

$$\int_{\Omega} \{\mathbf{h}_r - \bar{\mathbf{h}} + \mathbf{h}_i\} \cdot \dot{\mathbf{b}} = 0 \qquad \forall\, \mathbf{b}(t) \tag{25}$$

so that the conservation equation is $\bar{\mathbf{h}} = \mathbf{h}_r + \mathbf{h}_i$.
The principle of the dynamic hysteresis model is introduced by making a mechanical analogy. The dissipative phenomenon can be accurately represented by the friction force $\mathbf{h}_i = \mathbf{h}_i^{\kappa} + \mathbf{h}_i^{\lambda}$ obtained from the non-smooth non-negative convex potential

$$\dot{Q}(\dot{\mathbf{b}}) = -\int_{\Omega} \left\{ \kappa\, |\dot{\mathbf{b}}| + \lambda \dot{\mathbf{b}}^2 \right\} \leq 0. \tag{26}$$

Since the dissipation functional $\dot{Q}$ is a function of $\dot{\mathbf{b}}$, and *not* of $\mathbf{b}$ like Ψ_M is, the relation between $\dot{Q}$ and $\mathbf{h}_i$ is not a differential one (subgradient) but an algebraic one (a kind of division of $\dot{\rho}^Q$ by $\dot{\mathbf{b}}$). However, for a large class of dissipation functionals, this division can be expressed easily in terms subgradients of convex functionals thanks to the notion of homogeneous function. A homogenous function of order n is a function such that $f(\xi x) = \xi^n f(x)$. It has the property $x\partial_x f = nf$. This can be written $f/x = (\partial_x f)/n$, which is precisely the sought relation.
The quadratic term in (26) represents a viscous friction force. It stands for microscopic eddy currents induced in the material by the variation with time of induction. Since this term is a homogenous function of order 2 of $\dot{\mathbf{b}}$, one has

$$\mathbf{h}_i^{\lambda} = \frac{1}{2}\partial_{\dot{\mathbf{b}}}(\lambda \dot{\mathbf{b}}^2) = \lambda \dot{\mathbf{b}}. \tag{27}$$

The pinning phenomenon, which is at the origin of magnetic hysteresis, is on the other hand represented by the dry friction force associated with the term $\kappa\, |\dot{\mathbf{b}}|$. This term is not differentiable at $\dot{\mathbf{b}} = 0$, but, as it is a convex function, it has a subgradient G defined by

$$G = \{\mathbf{h}_i^{\kappa}, |\mathbf{h}_i^{\kappa}| \leq \kappa \text{ if } \dot{\mathbf{b}} = 0, \mathbf{h}_i^{\kappa} = \kappa\, \mathbf{e}_{\dot{\mathbf{b}}} \text{ if } \dot{\mathbf{b}} \neq 0\} \tag{28}$$

where $\mathbf{e}_{\mathbf{x}} \equiv \mathbf{x}/|\mathbf{x}|$. Since it is a homogeneous function of degree 1, one has $\mathbf{h}_i^{\kappa} = \partial_{\dot{\mathbf{b}}}\kappa|\dot{\mathbf{b}}|$, i.e. one can identify $\mathbf{h}_i^{\kappa}$ with the subgradient G.
The equilibrium equation writes finally

$$\bar{\mathbf{h}} - \mathbf{h}_r - \mathbf{h}_i^{\lambda} = \mathbf{h}_i^{\kappa} \in G. \tag{29}$$

The memory effect originates from the non-univocity of the friction force $\mathbf{h}_i^{\kappa}$ at $\dot{\mathbf{b}} = 0$. The subgradient, i.e. the set of possible forces $\mathbf{h}_i^{\kappa}$, is represented by the grey circle of radius κ in Fig. 3. If the tip of $\bar{\mathbf{h}}$ is inside the circle, one has $\dot{\mathbf{b}} = 0$ by (28), which implies $\dot{\mathbf{h}}_r = 0$. A given induction can thus persist although the applied magnetic field $\bar{\mathbf{h}}$ has decreased, whence the memory effect. If on the contrary the tip of $\bar{\mathbf{h}}$ tends to get out of the circle, $\mathbf{h}_r$ is updated according to the differential equation in time

$$\bar{\mathbf{h}} - \mathbf{h}_r - \mathbf{h}_i^{\lambda} = \kappa \mathbf{e}_{\dot{\mathbf{h}}_r}, \tag{30}$$

where we have noted that $\mathbf{e}_{\dot{\mathbf{b}}} = \mathbf{e}_{\dot{\mathbf{h}}_r}$. Details on the implementation can be found in [5]. The use of non-smooth functionals is essentially a theoretical issue. In the implementation, it amounts to a simple *if* statement.
This model is able to represent minor loops, Fig. 4. By combination of several submodels with different values of κ, the number of parameters of the model can be increased for a better accuracy. Fig. 5 shows the agreement obtained with 5 submodels. As this hysteresis model is based on a real physical description of the phenomenon, it makes sense to use it in a 3D model, even when the parameter identification has been done on basis of uniaxial quasi-static measurements.

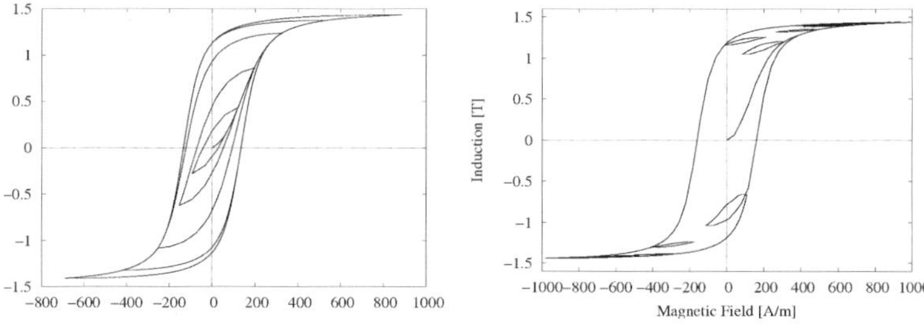

Fig. 4: Internal loops (left) and minor loops (right) are represented by the model.

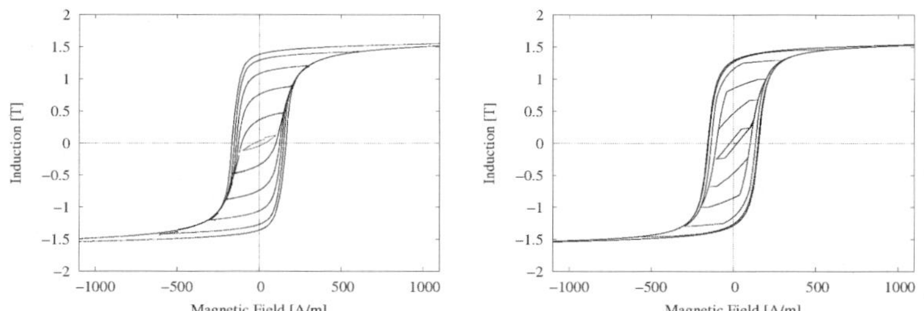

Fig. 5: Measurements (left) and model (right) obtained with 5 cells for electrical steel.

5.3 Model reduction

It is getting more and more important in modern computations to dispose of a concise, computationally tractable, but nevertheless accurate representation of a given large system, in order to allow real time computation, coupling with other parts of a larger system, etc. There are essentially two ways to create such simplified representations.

The first approach consists in truncating an asymptotical (in some sense) representation of the initial system. These are e.g. the Model Order Reduction (MOR) techniques, which are mostly applicable to linear problems [6]. In this case the initial and simplified representations are of the same nature. The approximation error is measured by the mathematical norm in terms of which the convergence of the asymptotical representation is expressed. As this norm has however scarcely a physical meaning, the neglected terms turn out often to have a significant impact on the physical properties of the reduced model. Therefore, special actions need be taken in order to preserve physical properties like passivity, stability, etc.

The second category gathers Parameter Identification methods, which are often based on energy criteria. When it comes to construct detailed models, energy turns out indeed very often to be the fundamental quantity to preserve. This holds also for Information Technology (IT) devices for instance, where information is carried over under the form of a propagating electromagnetic energy pulses, and distortion is a manifestation of energy diffusion. A good model is therefore a model able to account accurately for the energy stored in the system, and for the main energy flows entering the system, being converted inside it, or leaving it.

Various applications of reduction methods implicitly based on energy criteria can be found in the literature, see e.g. [7, 8, 9]. This approach can be seen also as the one that leads to the definition of RLC lumped parameters in electrical circuits. Lacking a unifying theoretical background, it has however not been identified yet as a specific method but the energy diagram introduced in this paper contributes to providing such a theoretical framework.

The identification method proceeds as follows. The first step consists in identifying subsystems that interact through controllable and monitorable channels. The variables that represent those interaction channels are usually averaged or global quantities in terms of which engi-

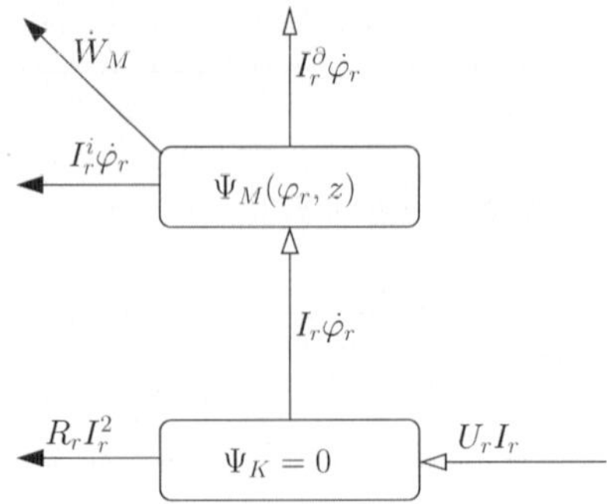

Fig. 6: EM energy flow diagram in scalar representation.

neers think to their system or measure it. They are thus natural variables for a reduced model and their number is usually limited. Now, the initial system and the reduced model have their own energy diagram. Since they both represent the same physical system, corresponding terms can be identified with each other. This gives the relations necessary to determine the parameters of the reduced model. It is enough to achieve this identification, to construct a map from the state variables of the large system onto the state variables of the reduced model. The definition of this map is based on the exploitation of existing regularities or simplifying features of the system.

Application to a synchronous electrical machine

As an example, a synchronous electrical machine is considered, for which one disposes of a detailed representation (e.g. a finite element model) in terms of the field state variables $\mathbf{a}, \mathbf{j}$ and u, and for which one wishes to extract a reduced model in terms of the corresponding scalar state variables φ, I_r and U_r, $r = 0, \ldots, N$, where N is the number of phases of the motor. The energy diagrams of the field representation and the scalar representation are depicted at Fig. 2 and 6 respectively.
The simplifying feature that allows reducing the model is the banal observation that the current density j can be written

$$\mathbf{j} = \sum_r I_r \mathbf{w}_r, \tag{31}$$

where the current shape functions $\mathbf{w}_r$ have support in the conducting regions $C \subset \Omega$. Note that (31) entails no approximation if the $\mathbf{w}_r$'s are allowed to depend on time.
Requiring now that the magnetic work is exactly represented, i.e. the corresponding energy flows in the field and scalar energy diagrams are equal,

$$\int_\Omega \mathbf{j} \cdot \mathcal{L}_\mathbf{v} \mathbf{a} \equiv \sum_r I_r \dot{\varphi}_r \quad \Rightarrow \quad \dot{\varphi}_r = \int_\Omega \mathbf{w}_r \cdot \mathcal{L}_\mathbf{v} \mathbf{a} \tag{32}$$

a mapping between $\dot{\varphi}$ and $\mathcal{L}_\mathbf{v} \mathbf{a}$ is obtained, whereas one needs a mapping between the state variables φ and $\mathbf{a}$. One makes therefore the *assumption* that the $\mathbf{w}_r$'s do not depend on time, so that one obtains the sought mapping

$$\varphi_r : \mathbf{a} \mapsto \mathbb{R} \quad , \quad \varphi_r = \int_\Omega \mathbf{w}_r \cdot \mathbf{a}. \tag{33}$$

This single approximation allows identifying all lumped parameters of the reduced model. It determines therefore also the domain of validity of the reduced model. The reduction is accurate if the actual $\mathbf{w}_r$'s do not vary *too much* in time. This assumption is true for ideal

coils, but it can also be fulfilled in a more restrictive way, e.g. for a given frequency in time-harmonic problems, or on a limited time interval for a linearised model.
Phase resistances are determined by identification of the dissipation functionals

$$R_r I_r^2 = \int_\Omega \sigma^{-1} |\mathbf{j}|^2 \quad \Rightarrow \quad R_r = \int_\Omega \sigma^{-1} |\mathbf{w}_r|^2. \tag{34}$$

There are two different ways to identify the magnetic energy, i.e. the inductance matrix, of the reduced model. Either one makes a global identification or a linearisation around a given working point. For a global identification, the inductance is defined as the matrix of multiplicative factors such that

$$\Psi_M(\mathrm{curl}\,\mathbf{a}) = \int_0^{\varphi_r(\mathbf{a})} L_{rs}^{-1} \varphi_s \,\mathrm{d}x \quad \Rightarrow \quad L_{rs}^{-1} \varphi_s = I_r. \tag{35}$$

The inductance is in this case a non-linear function of all state variables, and of the φ_r's in particular. In practice, the magnetic energy of the system or the fluxes are pre-computed by static finite element computations over the state space of the system (i.e for all rotor positions, $I_r, \dots$) and the computed values are stored in look-up tables.
This approach has two drawbacks. Firstly, the size of the look-up tables grows exponentially when the number of parameters increases. Secondly, differentiation of the stored values must be approximated numerically by finite differences. The discretisation of the state space must therefore be fine enough, yielding again an increase of the look-up table dimensions. One avoids one differentiation by storing directly the fluxes φ_r's, instead of the energy, but one numerical differentiation is still required to evaluate $U_r = R_r I_r + \dot{\varphi}_r$.
The second approach consists in linearising the magnetic behaviour of the system around a given working point. This approach is very useful when one wants to couple the reduced model of the motor with a high dynamic model of the supplying inverter. The state variables of the linearised model are denoted by $\delta\varphi_r$, and the governing equations are

$$L_{rs}^{-\partial} \delta\varphi_s + R_r^{-1} (\partial_t \delta\varphi_r - \delta U_r) = 0 \tag{36}$$

where $L_{rs}^{-\partial} \equiv (L_{rs}^{\partial})^{-1}$ denotes the inverse of the tangent inductance matrix of the reduced system. It is defined by

$$L_{rs}^{-\partial} = \partial_{\varphi_r} \partial_{\varphi_s} \Psi_M(\varphi^*) \quad \Rightarrow \quad L_{rs}^{-\partial} \dot{\varphi}_s = \dot{I}_r. \tag{37}$$

It can be shown it can be evaluated as follows

$$L_{rs}^{\partial} = W_{ri} J_{ij}^{-1} W_{sj} \quad , \quad W_{ri} = \int_\Omega \mathbf{w}_r \cdot \alpha_i \tag{38}$$

where J_{ij} is the Jacobian matrix of the non-linear system and α_i denotes the ith edge shape function.

5.4 Equivalent time-harmonic reluctivity

Another interesting application of energy-based parameter identification is the definition of equivalent material characteristics for time-harmonic models. Periodic phenomena are ubiquitous in electromagnetic applications but, due to magnetic saturation or the presence of non-linear electronic components, actual wave shapes are scarcely sinusoidal, which invalidates the phasor representation. Still, the complex formalism is so practical and offers so many useful mathematical properties that it is often worth in practice to seek for approximative phasor representations. In this case again, it is meaningful to adopt energy as identification criterion. In frequency domain, vector fields are represented by two vectors, e.g. for an harmonic induction field, one has $\mathbf{b}_\omega = \mathbf{b}_r + j\mathbf{b}_i$. The associated time domain vector field

$$\mathbf{b}_\omega(t) = \mathbf{b}_r \cos\omega t - \mathbf{b}_i \sin\omega t \tag{39}$$

describes an ellipsis in the three-dimensional geometrical space, of which the two axis are given by

$$\left.\begin{matrix} b_{max} \\ b_{min} \end{matrix}\right\} = \sqrt{\frac{|\mathbf{b}_r|^2 + |\mathbf{b}_i|^2}{2} \pm \Delta} \quad , \quad \Delta^2 = \left(\frac{|\mathbf{b}_r|^2 - |\mathbf{b}_i|^2}{2}\right)^2 + (\mathbf{b}_r \cdot \mathbf{b}_i)^2 \,.$$

The general relation between an induction phasor and a magnetic field phasor, when anisotropy is disregarded, is represented by a complex reluctivity $\nu = \nu_r + j\nu_i$, where ν_r and ν_i are real constants. The corresponding representation in time-domain is the operator

$$\nu = \nu_r + \frac{\nu_i}{\omega}\partial_t. \tag{40}$$

In time domain, considering magnetic hysteresis but disregarding anisotropy, the local relation between the induction vector $\mathbf{b}$ and the magnetic field vector $\mathbf{h}$ can be written formally

$$\mathbf{b}(t) = \mathcal{H}[\mathbf{h}, t], \tag{41}$$

where $\mathcal{H}$ denotes an hysteresis operator. Numerous theoretical and phenomenological representations of hysteresis operators can be found in literature. We use here the one presented in Sect. 5.2. The principle of the identification is now to determine ν_r and ν_i so that the energy balance of the equivalent material represented by the complex ν matches as closely as possible the energy balance of the hysteretic material represented by $\mathcal{H}$. Since we have two parameters to identify, we may impose two conditions.
Let us first assume that one has been able, for a given $\mathbf{b}_\omega(t)$, to determine a field $\mathbf{h}^\star(t)$ such that $\mathcal{H}[\mathbf{h}^\star, t] = \mathbf{b}_\omega(t)$. From this particular hysteresis curve, the model described in Sect. 5.2 can provide the value of the amount of energy dissipated over one period

$$Q^\star \equiv \int_0^T \mathbf{h}^\star \cdot \partial_t \mathbf{b}_\omega \tag{42}$$

and the amplitude of the fluctuation of the magnetic energy density ρ_M^Ψ

$$(\Delta\rho_M^\Psi)^\star \equiv \left\{ \max_{[0,T]} - \min_{[0,T]} \right\} \rho_M^\Psi(\mathbf{b}_\omega) = \rho_M^\Psi(b_{max}) - \rho_M^\Psi(b_{min}). \tag{43}$$

On the other hand, the magnetic field in the material represented by the complex ν

$$\mathbf{h}_\omega(t) \equiv \nu \mathbf{b}_\omega(t) = (\nu_r + \frac{\nu_i}{\omega}\partial_t)\, \mathbf{b}_\omega(t) \tag{44}$$

allows to write the energy balance

$$\mathbf{h}_\omega \cdot \partial_t \mathbf{b}_\omega = \partial_t \left\{ \frac{\nu_r}{2} |\mathbf{b}_\omega|^2 \right\} + \frac{\nu_i}{\omega} |\partial_t \mathbf{b}_\omega|^2, \tag{45}$$

where the bracketed term represents the magnetic energy density. One has therefore the two relations

$$\int_0^T \mathbf{h}_\omega \cdot \partial_t \mathbf{b}_\omega = \frac{\omega T}{2} \nu_i \left(|\mathbf{b}_r|^2 + |\mathbf{b}_i|^2\right) = \pi \nu_i \left(b_{min}^2 + b_{max}^2\right) \equiv Q^\star \tag{46}$$

$$\left\{ \max_{[0,T]} - \min_{[0,T]} \right\} \nu_r \frac{|\mathbf{b}_\omega|^2}{2} = \nu_r \Delta \equiv (\Delta\rho_M^\Psi)^\star \tag{47}$$

that allow identifying ν_r and ν_i.

6 Conclusion

The energy-based formulation of Electromagnetism is not just a re-formulation. It offers substantial improvements with regard to the classical theory, and in particular a stronger link with the universal principles of Thermodynamics. The purpose of this paper was to review the benefits of the energy-based formulation from the point of view of numerical simulations. We have shown that governing equations are obtained in a form that is directly usable by the finite element method and convex analysis. Moreover, all terms retain a clear physical understanding. This helps in the definition of coupling terms in multiphysics modelling and provides meaningful criteria for parameter identification.

References

1. Henrotte F., Hameyer K.: The structure of em energy flows in continuous media. *IEEE Transactions on Magnetics*, **42**(4):903–906, April 2006.
2. Schutz B.: *Geometrical methods of mathematical Physics*. Cambridge University Press, 1980.
3. Henrotte F., Hameyer K.: A theory of electromagnetic force formulae in continuous media. To be published in *IEEE Transactions on Magnetics*, **43**(4), April 2007.
4. Bossavit A.: Superconductivity: Homogenization of Bean's model in three dimensions and the problem of transverse conductivity. *IEEE Transactions on Magnetics*, **31**(3):1769–1774, May 1995.
5. Henrotte F., Hameyer K.: A dynamical vector hysteresis model based on an energy approach. *IEEE Transactions on Magnetics*, **42**(4):899–902, April 2006.
6. Antoulas A.C., Sorensen D.C., Gugercin S.: A survey of model reduction methods for large-scale systems. *Structured Matrices in Operator Theory, Numerical Analysis, Control, Signal and Image Processing*, Contemporary Mathematics, AMS publications, 2001.
7. Mohammed O.A., Liu S., Liu Z.: Physical modeling of PM synchronous motors for integrated coupling with Machine drives. *IEEE Transactions on Magnetics*, **41**(5):1628–1631, May 2005.
8. Mohammed O.A., Liu Z., Liu S., Abed N.Y.: Finite-element-based nonlinear physical model of iron-core transformers for dynamic simulations. *IEEE Transactions on Magnetics*, **42**(4):1027–1030, April 2006.
9. Henrotte F., Podoleanu I., Hameyer K.: Staged modelling: a methodology for developing real-life electrical systems applied to the transient behaviour of a permanent magnet servo motor. COMPEL **22**(4):1066–1076, 2003.

Newton and Approximate Newton Methods in Combination with the Orthogonal Finite Integration Technique

H. De Gersem[1], I. Munteanu[2] and T. Weiland[1]

[1] Technische Universität Darmstadt, Institut für Theorie Elektromagnetischer Felder, Schloßgartenstraße 8, D-64289 Darmstadt, Germany
degersem@temf.tu-darmstadt.de

[2] Computer Simulation Technology, Bad Nauheimer Straße 19, D-64289 Darmstadt, Germany munteanu@cst.com

Summary. The paper describes the application of the Newton method in conjunction with the Finite Integration Technique. Even on orthogonal grid pairs, the material matrices become nondiagonal and lead to higher algorithmic complexity. A uni-directional version of these matrices provides a computationally inexpensive alternative. The paper compares and discusses the two algorithms' order of convergence and their computational complexity.

Keywords—electromagnetic field simulation, Finite Integration Technique, Newton method, nonlinear constitutive equations

1 Introduction

The Finite Integration Technique (FIT) [WEI96] is often applied in combination with an orthogonal, staggered grid pair [CW02], leading to sparser algebraic matrices than in the case of unstructured grids, and to a higher efficiency of the numerical simulation scheme.
In this paper, we show that these beneficial algebraic properties do not apply when using the Newton method for linearizing a nonlinear field problem. We analyse the reasons for this and formulate an approximate Newton method that overcomes this. For selected test models, we show in which cases the approximate Newton method should be preferred over the exact Newton method and in which cases such approximation is not recommended. The classical successive approximation method is used for comparison.

2 Finite Integration Technique

2.1 Discretization of the Maxwell equations

The Finite Integration Technique (FIT) [WEI96] is a discretization method for vectorial partial differential equations, first proposed in 1977 for discretizing the Maxwell equations. The discretization process is carried out on a primary-dual, staggered grid complex $(G, \widetilde{G})$ (Fig. 1a). The considered degrees of freedom are *global* ones, such as electric voltages $\widehat{\mathbf{e}}_p$ along primary edges L_p, magnetic fluxes $\widehat{\widehat{\mathbf{b}}}_p$ through primary faces A_p, or magnetic voltages $\widehat{\mathbf{h}}_p$ along dual

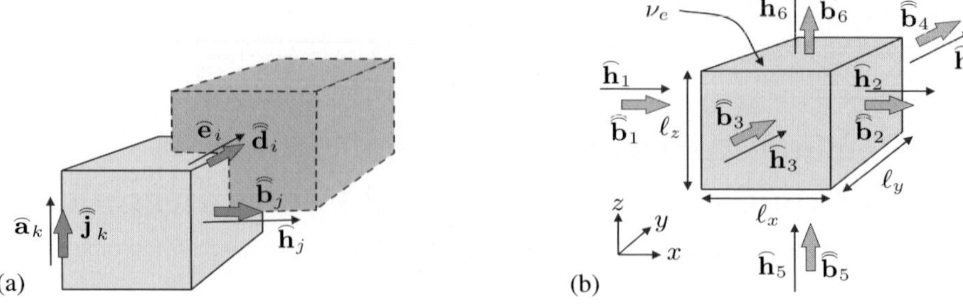

Fig. 1: (a) Primary-dual grid pair. (b) Local numbering of primary faces and dual edges associated at a primary grid cell

edges $\tilde{L}_p$. The discrete divergence operators $\mathbf{S}$ and $\widetilde{\mathbf{S}}$ and the discrete curl operators $\mathbf{C}$ and $\widetilde{\mathbf{C}}$ on the primary grid G and the dual grid $\tilde{G}$, respectively, are represented by incidence matrices containing 0, 1 and -1.
The discretized Maxwell equations become [WEI96]:

$$\mathbf{S}\widehat{\widehat{\mathbf{b}}} = 0 \tag{1}$$

$$\mathbf{C}\widehat{\mathbf{e}} = -\frac{d}{dt}\widehat{\widehat{\mathbf{b}}} \tag{2}$$

$$\widetilde{\mathbf{S}}\widehat{\widehat{\mathbf{d}}} = \mathbf{q} \tag{3}$$

$$\widetilde{\mathbf{C}}\widehat{\mathbf{h}} = \widehat{\widehat{\mathbf{j}}} + \frac{d}{dt}\widehat{\widehat{\mathbf{d}}}\,. \tag{4}$$

The metric and material properties are described by the discretized material relations $\widehat{\widehat{\mathbf{d}}} = \mathbf{M}_\varepsilon\widehat{\mathbf{e}}$, $\widehat{\mathbf{h}} = \mathbf{M}_\nu\widehat{\widehat{\mathbf{b}}}$, $\widehat{\widehat{\mathbf{j}}} = \mathbf{M}_\sigma\widehat{\mathbf{e}}$.
In this paper, we analyze nonlinear magnetostatic formulations $\nabla \times (\nu\nabla \times \mathbf{A}) = \mathbf{J}_{\mathrm{s}}$, whose FIT-discretized counterpart reads $\widetilde{\mathbf{C}}\mathbf{M}_\nu\mathbf{C}\widehat{\mathbf{a}} = \widehat{\widehat{\mathbf{j}}}_{\mathrm{s}}$, where $\mathbf{M}_\nu = \mathbf{M}_\nu(\widehat{\mathbf{a}})$ nonlinearly depends on $\widehat{\mathbf{a}}$. This dependence is given by the $B-H$ curve.
The solution of the discretized magnetostatic system of equations corresponds to finding the root of the nonlinear matrix function

$$\mathbf{F}(\widehat{\mathbf{a}}) = \widetilde{\mathbf{C}}\mathbf{M}_\nu\mathbf{C}\widehat{\mathbf{a}} - \widehat{\widehat{\mathbf{j}}}_{\mathrm{s}}\,. \tag{5}$$

2.2 Assembly of linear material matrices

In FIT, the assembly of linear material matrices is efficiently organized as a loop over the primary faces. In more complicated cases, however, it is more convenient to assemble material matrices element by element by introducing a local numbering at the level of a single material cell V_e. Let the number of primary faces and the number of primary edges be n_{pf}, and the matrix $\mathbf{Q}_e$ be a $6 \times n_{\mathrm{pf}}$-matrix for selecting a local vector associated with cell V_e from the corresponding global vector, i.e., $\widehat{\widehat{\mathbf{b}}}_e = \mathbf{Q}_e\widehat{\widehat{\mathbf{b}}}$ and $\widehat{\mathbf{h}}_e = \mathbf{Q}_e\widehat{\mathbf{h}}$. The primary face areas and the lengths of the parts of the dual edges inside the primary cell are collected in the 6×6 diagonal matrices $\mathbf{S}_e$ and $\tilde{\mathbf{L}}_e$, respectively.
With these notations, the *local reluctivity matrix* is computed by

$$\mathbf{M}_{\nu,e} = \tilde{\mathbf{L}}_e\nu_e\mathbf{S}_e^{-1}\,, \tag{6}$$

with ν_e the reluctivity of the material in the cell e, and the global reluctivity matrix is assembled by $\mathbf{M}_\nu = \sum_e \mathbf{Q}_e^T\mathbf{M}_{\nu,e}\mathbf{Q}_e$.

3 Newton Method

3.1 Iteration scheme

The Newton method applied to (5) reads

$$\mathbf{F}'\left(\widehat{\mathbf{a}}^{(k)}\right)\delta\widehat{\mathbf{a}}^{(k+1)} = -\mathbf{F}\left(\widehat{\mathbf{a}}^{(k)}\right). \tag{7}$$

The Jacobian $\mathbf{F}'(\widehat{\mathbf{a}})$ is derived from the algebraic formulation (5) as follows

$$\mathbf{F}'(\widehat{\mathbf{a}}) = \widetilde{\mathbf{C}}\frac{\mathrm{d}}{\mathrm{d}\widehat{\widehat{\mathbf{b}}}}\left(\mathbf{M}_\nu\widehat{\widehat{\mathbf{b}}}\right)\mathbf{C} = \widetilde{\mathbf{C}}\mathbf{M}_{\nu_\mathrm{d}}\mathbf{C},$$

where $\mathbf{M}_{\nu_\mathrm{d}}$ is the *differential reluctivity matrix*.
Similarly as for the chord reluctivity matrix $\mathbf{M}_\nu$, the differential reluctivity matrix can be assembled cell per cell: $\mathbf{M}_{\nu_\mathrm{d}} = \sum_e \mathbf{Q}_e^T\mathbf{M}_{\nu_\mathrm{d},e}\mathbf{Q}_e$.
The cell differential reluctivity matrices $\mathbf{M}_{\nu_\mathrm{d},e}$ are computed by differentiating (6) with respect to the fluxes at the local faces:

$$\mathbf{M}_{\nu_\mathrm{d},e} = \frac{\mathrm{d}}{\mathrm{d}\widehat{\widehat{\mathbf{b}}}_e}\left(\mathbf{M}_{\nu,e}\widehat{\widehat{\mathbf{b}}}_e\right) \tag{8}$$

$$= \mathbf{M}_{\nu,e} + \tilde{\mathbf{L}}_e\,\widehat{\widehat{\mathbf{b}}}_e\frac{\mathrm{d}\nu_e}{\mathrm{d}B^2}\frac{\mathrm{d}B^2_{\mathrm{cell},e}}{\mathrm{d}\widehat{\widehat{\mathbf{b}}}_e}\,\tilde{\mathbf{S}}_e^{-1}, \tag{9}$$

where $\frac{d\nu_e}{dB^2}$ follows from evaluating the material characteristic. According to the local numbering depicted in Fig. 1b, the square of the magnitude of the magnetic flux density in cell e can be determined by averaging the x, y and z components:

$$B^2_{\mathrm{cell},e} = \left(\frac{\widehat{\widehat{\mathbf{b}}}_1 + \widehat{\widehat{\mathbf{b}}}_2}{2S_{e,x}}\right)^2 + \left(\frac{\widehat{\widehat{\mathbf{b}}}_3 + \widehat{\widehat{\mathbf{b}}}_4}{2S_{e,y}}\right)^2 + \left(\frac{\widehat{\widehat{\mathbf{b}}}_5 + \widehat{\widehat{\mathbf{b}}}_6}{2S_{e,z}}\right)^2. \tag{10}$$

The square root of this value is used to determine the working point on the nonlinear characteristic. The differentiation of (10) leads to the 1-by-6 vector

$$\frac{\mathrm{d}B^2_{\mathrm{cell},e}}{\mathrm{d}\widehat{\widehat{\mathbf{b}}}_e} = \left[\frac{\widehat{\widehat{\mathbf{b}}}_1+\widehat{\widehat{\mathbf{b}}}_2}{2S^2_{e,x}}\;\frac{\widehat{\widehat{\mathbf{b}}}_1+\widehat{\widehat{\mathbf{b}}}_2}{2S^2_{e,x}}\;\frac{\widehat{\widehat{\mathbf{b}}}_3+\widehat{\widehat{\mathbf{b}}}_4}{2S^2_{e,y}}\;\frac{\widehat{\widehat{\mathbf{b}}}_3+\widehat{\widehat{\mathbf{b}}}_4}{2S^2_{e,y}}\;\frac{\widehat{\widehat{\mathbf{b}}}_5+\widehat{\widehat{\mathbf{b}}}_6}{2S^2_{e,z}}\;\frac{\widehat{\widehat{\mathbf{b}}}_5+\widehat{\widehat{\mathbf{b}}}_6}{2S^2_{e,z}}\right]. \tag{11}$$

The second term of (9) is not symmetric, therefore the overall system matrix in (7) is also nondiagonal and nonsymmetric. The reason is that different shape functions are combined in the material matrices, which catalogues the FIT in the family of the Petrov-Galerkin techniques: The reluctivity matrix considers fluxes discretized by some kind of primary facet functions and voltages discretized by functions defined along dual edges [SW00]. Despite the nonsymmetry, when the applied material characteristics are monotonically increasing, one can prove that the differential reluctivity matrix is positive definite. For solving the resulting magnetostatic system (7), a Krylov subspace solver for nonsymmetric systems such as the BiCG Stabilized (BiCGStab) method, has to be applied. Instead of algebraic multigrid techniques, only single-level algebraic preconditioners such as e.g. Incomplete LU-factorization (ILU) are directly applicable. The nondiagonal character of the *differential* reluctivity matrix reflects a coupling between spatial directions. As will be shown in the numerical examples, this is an essential property for taking *cross-magnetization* into account and for ensuring second order convergence of the Newton method.

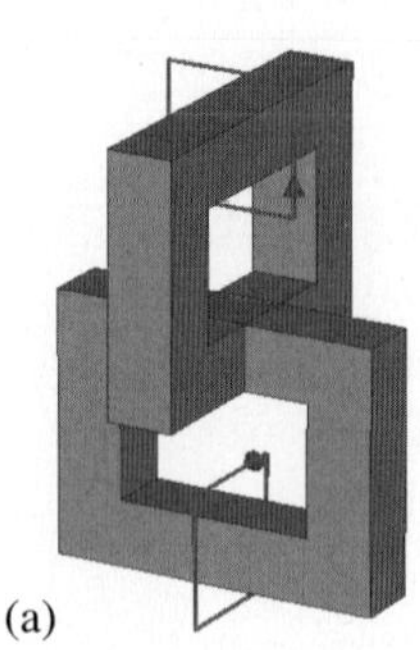
(a)

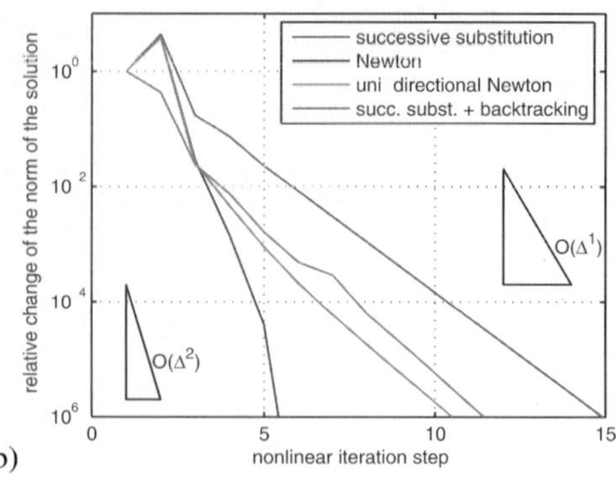

(b)

Fig. 2: Double C-core configuration: (a) Geometry, (b) Convergence of different linearization techniques

3.2 Uni-directional differential reluctivity matrix

To overcome the drawbacks of the denser reluctivity matrix, the dense differential reluctivity tensor $\overline{\overline{\nu}}_{\mathrm{d},e} = \nu_e \overline{\overline{\mathbf{1}}} + \widehat{\mathbf{b}}_e \frac{\mathrm{d}\nu_e}{\mathrm{d}B^2} \frac{\mathrm{d}B^2_{\mathrm{cell},e}}{\mathrm{d}\widehat{\mathbf{b}}_e}$ can be replaced by a diagonal approximation $\overline{\overline{\nu}}^{(e)}_{\mathrm{d,uni}} = \overline{\overline{\mathbf{1}}}\, \nu^{(e)}_{\mathrm{d,uni}}$ on the basis of the scalar differential reluctivity $\nu^{(e)}_{\mathrm{d,uni}} = \partial H_e / \partial B_e$ [DW00]. This *uni-directional* differential reluctivity matrix $\mathbf{M}_{\nu_{\mathrm{d,uni}}}$ is diagonal and therefore does not include cross-magnetization effects.

4 Numerical Examples

4.1 C-magnet

To compare the Newton and uni-directional Newton methods, a simple test model was used. It consists of a C-core driven by a single-wire coil generating a magnetic flux through an iron cube, and an additional core and winding arranged around the nonlinear piece of iron in a perpendicular direction (Fig. 2). The relative permeability of the C-core is taken as high as 10^6 in order to enforce an almost straight flux through the nonlinear piece of material. The currents in the coils have different magnitudes and are chosen such that the nonlinear part is highly saturated.
The convergence behaviour for the various methods is plotted in Fig. 2b. The convergence criterion is the relative correction of the solution in the Euclidian norm $||\widehat{\mathbf{a}}^{k+1} - \widehat{\mathbf{a}}^k||/||\widehat{\mathbf{a}}^k||$. It is seen that the Newton approach features second-order convergence while the successive substitution with and without backtracking does not even guarantee first-order convergence. Although originally also of second order, the convergence of the uni-directional Newton approach breaks down at a relatively high error, i.e., at 10^{-2}. The explanation is that in this device the magnetic flux in one direction drives the nonlinear iron piece into saturation, which also influences the magnetic flux in the perpendicular direction. In the uni-directional Newton method, the true differential reluctivity tensor is replaced by a diagonal approximation. Thus, the coupling between space directions is only weakly taken into account, namely when determining the new working point between two nonlinear steps iterations.

4.2 Nuclotron-magnet

The behaviour of the successive substitution, Newton and uni-directional Newton methods has been tested on a real-life application: a superconductive dipole magnet (Fig. 3) designed for the

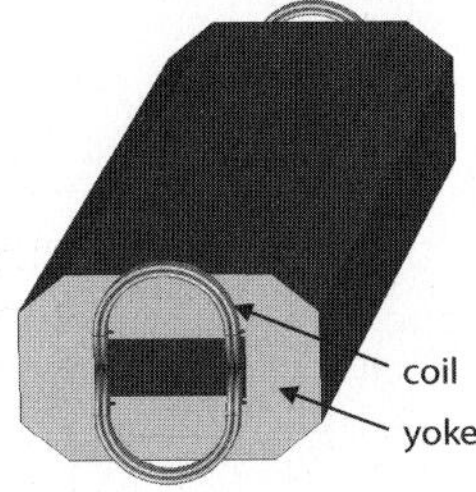

Fig. 3: Geometry of the Nuclotron magnet device.

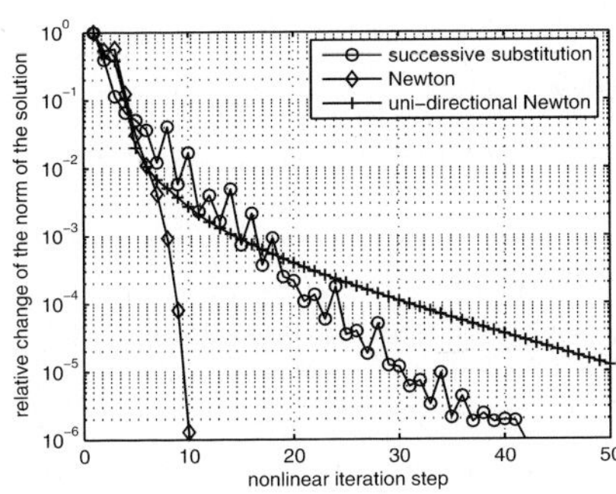

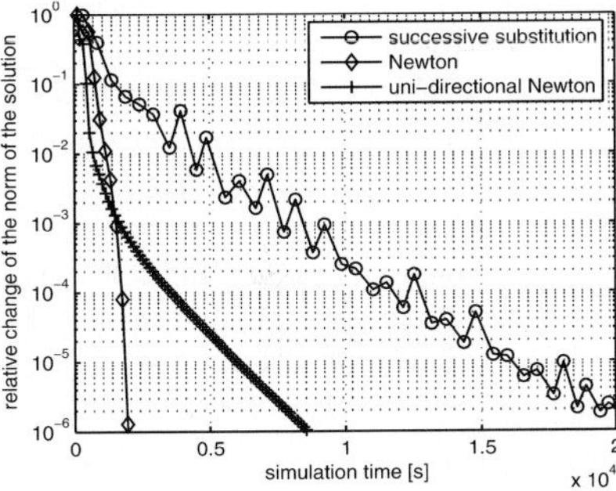

Fig. 4: Comparison of convergence for the nuclotron magnet example. Left: number of nonlinear iteration steps. Right: simulation time.

new Facility for Antiproton and Ion Research (FAIR) [KMZ04] to be built at the Gesellschaft für Schwerionenforschung (GSI) in Darmstadt, Germany. The magnet consists of a superconductive coil wound inside a ferromagnetic yoke with additional shimming and air slits to guarantee a homogeneous magnetic field in the magnet aperture. The geometry was meshed with 14616 cells corresponding to approximately 40000 degrees of freedom for the magnetic vector potential. The comparison of the convergence of the nonlinear iteration according to the number of nonlinear iteration steps indicates the beneficial properties of the Newton method (Fig. 4a). However, when comparing the three linearization approaches according to the simulation time (Fig. 4b), the advantage of the Newton method is less pronounced, mainly because of the significantly higher computational cost of assembling the denser material matrix.

5 Conclusions

In contrast to the chord reluctivity matrix, the differential reluctivity matrix arising in the Newton-FIT discretization of a magnetostatic formulation has off-diagonal entries even on an orthogonal grid pair. These entries reflect the connection between the spatial directions through the nonlinearity of the material characteristic. The attempt to construct a diagonal differential reluctivity matrix leads to the uni-directional Newton method which, for the numerical examples in the paper, outperformed the exact Newton method when a tolerance of only 10^{-2} was required for the nonlinear loop.

Acknowledgment

This paper was sponsored by the Gesellschaft für Schwerionenforschung (GSI), Darmstadt, Germany, under the grant DA/WE2.

References

[CW02] Clemens M., Weiland T.: Magnetic field simulation using conformal FIT formulations. IEEE Transactions on Magnetics, **38**, 389–392 (2002)

[DW00] Drobny S, Weiland T.: Numerical calculation of nonlinear transient field problems with the Newton-Raphson method. IEEE Transactions on Magnetics, **36**, 809–812 (2000)

[KMZ04] Kalimov A., Moritz G., and Zeller A.: Design of a superferric dipole magnet with high field quality in the aperture. IEEE Transactions on Applied Superconductivity, **14**, 271–274 (2004)

[WEI96] Weiland, T.: Time domain electromagnetic field computation with finite difference methods. International Journ. Num. Modell., **9**, 295–319 (1996)

[SW00] Schuhmann, R., Weiland, T.: The nonorthogonal finite integration technique applied to 2D- and 3D-eigenvalue problems. IEEE Transactions on Magnetics, **36**, 897–901 (2000)

Transient Simulation of a Linear Actuator Discretized by the Finite Integration Technique

Mariana Funieru, Herbert De Gersem and Thomas Weiland

Technical Universität Darmstadt, Institut für Theorie Elektromagnetische Felder,
Schloßgartenstraße 8, D-64289 Darmstadt, Germany
funieru/degersem/thomas.weiland@temf.tu-darmstadt.de

Abstract: - Two armatures of a linear actuator are discretized by the Finite Integration Technique at two independent three-dimensional Cartesian grids. The fixed and the moving armature are coupled at a sliding surface, situated in the middle of the air-gap. The moving armature is displaced in x-direction, on a plane parallel to the coupling plane and the generated forces are calculated. The mobile armature is made from laminated iron, in which the eddy current losses are negligible, but the fixed armature is made from massive iron in which the eddy currents have a significant influence. A coupled transient simulation is carried out, considering both the magnetic and the mechanic behavior of the actuator.

1 Introduction

Many electromagnetic devices have 2 parts that are moving relatively to each other. Usually, the relative position can be simulated by remeshing for each displacement position, either the whole geometry or just a part of it. In this paper, however, two different independent meshes are coupled at a common interface, which eliminates the need for remeshing and decreases the computational time.
In order to model the dynamic behavior of the actuator, a transient model combining the equation of motion with an electromagnetic field is simulated.
In this paper, the Finite Integration Technique is accomplished by a sliding-surface technique in order to account for nonlinear motion.

2 Electromagnetic Model

The studied linear actuator is a linear, hybrid stepper motor that employs the principle of reluctance forces. The actuator is developed and traded by PASIM Direktantriebe [PASIM] and is commonly applied in industrial applications where accurate linear positioning is required.
The fixed and the moving armature of the 3-D linear actuator (Fig. 1) are modeled independently, with the help of the commercial package EM Studio [CST]. The Finite Integration Technique [Weil96] is used for discretizing the electromagnetic field formulation.
The 2 meshes are connected at a sliding surface situated in the middle of the air gap. The fixed armature (rail) is made from massive iron, whereas the modules of the mobile armature are made from laminated iron. A pole pitch consists of a single tooth and a single slot of the rail.
The model in the z-direction (perpendicular to the cross-sectional plane shown in Fig. 1) is discretized by only one mesh cell. Consequently end-effects occurring within the real machine are not considered within the FIT model. The solver is programmed in C++ and allows the coupling of the two meshes and the coupling of the magnetic model with a mechanical

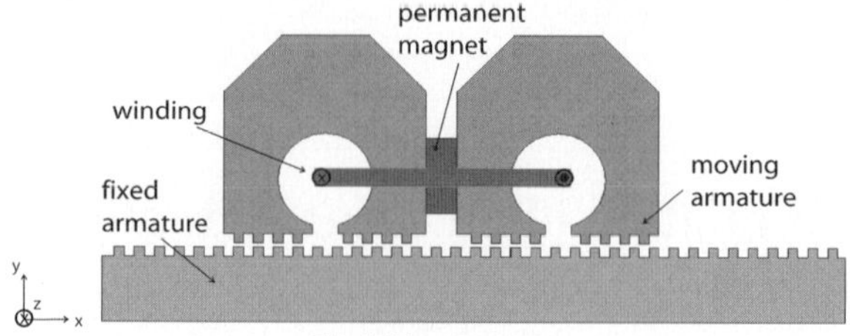

Fig. 1: Linear actuator

model. It is in this way possible to completely model the actuator's behavior by calculating the generated force and how the force is influenced by the relative position of the armatures. The magnetoquasistatic formulation discretized by the FIT in terms of the magnetic vector potential integrated along primary edges, $\widehat{\mathbf{a}}$, reads:

$$\widetilde{\mathbf{C}}\mathbf{M}_\nu\mathbf{C}\widehat{\mathbf{a}} + \mathbf{M}_\sigma\frac{d\widehat{\mathbf{a}}}{dt} = \widehat{\widehat{\mathbf{j}}}_s - \widetilde{\mathbf{C}}\mathbf{M}_\nu\widehat{\widehat{\mathbf{b}}}_\mathrm{r}\ , \tag{1}$$

where $\mathbf{C}$ and $\widetilde{\mathbf{C}}$ represent the discrete curl matrices at the primary and dual grid respectively, $\widehat{\widehat{\mathbf{j}}}_s$ is the current applied in the coils, $\widehat{\widehat{\mathbf{b}}}_\mathrm{r}$ is the remanence of the permanent magnets integrated over primary facets, $\mathbf{M}_\nu$ is the reluctivity matrix and $\mathbf{M}_\sigma$ is the conductivity matrix. More details about the spatial discretization can be found in [Weil96], [Ion05].
The mobile armature is displaced in the x-direction, Dirichlet boundaries condition are applied at the exterior boundary of the model. The formulation couples the $\widehat{\mathbf{a}}_x$ and $\widehat{\mathbf{a}}_z$ components of the magnetic vector potential allocated at the set of edges tangential to the common interface [DeGe05]. The moving and the fixed meshes do not match at the common interface and therefore, a 2D interpolation has to be applied. The degrees of freedom (dofs) $\widehat{\mathbf{a}}_\mathrm{mv}$ allocated at the moving side of the interface have to be coupled to the dofs $\widehat{\mathbf{a}}_\mathrm{fx}$ allocated at the fixed side of the coupling interface, here represented by a sliding operator $\mathbf{k}_\mathrm{slid} = \mathbf{k}_\epsilon\mathbf{k}^p_\mathrm{shift}$, i.e., $\widehat{\mathbf{a}}_\mathrm{mv} = \mathbf{k}_\mathrm{slid}\widehat{\mathbf{a}}_\mathrm{fx}$, where $\mathbf{k}_\mathrm{shift}$ is the *shift operator*, responsible for the displacement by an integer number $p = [\alpha/\Delta\theta]$ of grid lines, α is the displacement and $\Delta\theta$ is the mean value of the distances between two adjacent grid lines. $\mathbf{k}_\epsilon$ is the *interpolation operator*, responsible for the displacement by a fraction of a grid cell.
For example, the displacement of a slightly non-equidistant grid by 1 grid line corresponds to the operator: $\mathbf{k}_\mathrm{slid} = \mathbf{k}_\mathrm{shift} = \begin{bmatrix} 0 & 1 & 0 & \dots & 0 \\ 0 & 0 & 1 & \dots & 0 \\ \dots & \dots & \dots & \dots & \dots \\ 0 & \dots & 0 & 0 & 0 \end{bmatrix}$. The operator: $\mathbf{k}_\mathrm{slid} = \mathbf{k}_\epsilon = \begin{bmatrix} 1-\epsilon_1 & \epsilon_1 & 0 & \dots & 0 \\ 0 & 1-\epsilon_2 & \epsilon_2 & \dots & 0 \\ \dots & \dots & \dots & \dots & \dots \\ 0 & \dots & 0 & 0 & 1-\epsilon_n \end{bmatrix}$ corresponds to a displacement by the fractions $\epsilon_i = \frac{\theta_{\mathrm{mv,i}} - \theta_{\mathrm{fx,i-p}} + p\Delta\theta}{\theta_{\mathrm{fx,i-p+1}} - \theta_{\mathrm{fx,i-p}}}$, $i = 1...n$.
This 2D interpolation technique assumes that the meshes at both sides of the interface are staggered, i.e., each dof from the slave side of the interface is connected with 2 dofs from the master side of the interface. In case of a highly non-uniform grid, this assumption is no longer valid and dangling edges in the z-direction can occur on the master mesh. These master edges are not connected to edges at the slave side of the interface. This problem is alleviated by linearly interpolating the dofs of the neighboring coupled edges at the master side onto the dangling edges at the master side.
The decoupled system of equations obtained by independently discretizing both FIT models reads:

$$\begin{bmatrix} \mathbf{K}_\mathrm{fx} & 0 \\ 0 & \mathbf{K}_\mathrm{mv} \end{bmatrix} \begin{bmatrix} \widehat{\mathbf{a}}^{n+1}_\mathrm{fx} \\ \widehat{\mathbf{a}}^{n+1}_\mathrm{mv} \end{bmatrix} = \begin{bmatrix} 0 \\ \widehat{\widehat{\mathbf{j}}}_\mathrm{mv} \end{bmatrix} - \begin{bmatrix} 0 \\ \mathbf{B}_{\mathbf{r},\mathrm{mv}} \end{bmatrix} + \begin{bmatrix} \frac{\mathbf{M}_{\kappa,\mathrm{fx}}}{\Delta t} & 0 \\ 0 & \frac{\mathbf{M}_{\kappa,\mathrm{mv}}}{\Delta t} \end{bmatrix} \begin{bmatrix} \widehat{\mathbf{a}}^{n}_\mathrm{fx} \\ \widehat{\mathbf{a}}^{n}_\mathrm{mv} \end{bmatrix} \tag{2}$$

where $\mathbf{K}_{\rm fx} = \widetilde{\mathbf{C}}_{\rm fx}\mathbf{M}_{\nu,\rm fx}\mathbf{C}_{\rm fx} + \frac{\mathbf{M}_{\kappa,\rm fx}}{\mathbf{\Delta t}}$ and $\mathbf{K}_{\rm mv} = \widetilde{\mathbf{C}}_{\rm mv}\mathbf{M}_{\nu,\rm mv}\mathbf{C}_{\rm mv} + \frac{\mathbf{M}_{\kappa,\rm mv}}{\mathbf{\Delta t}}$ are the stiffness matrices of the fixed and moving armatures, respectively. The integrated magnetic vector potentials $\widehat{\mathbf{a}}_{\rm fx}$ and $\widehat{\mathbf{a}}_{\rm mv}$ are the unknowns in the models, $\mathbf{M}_{\kappa,\rm fx}$ and $\mathbf{M}_{\kappa,\rm mv}$ are the conductivity matrices, $\widehat{\widehat{\mathbf{j}}}_{\rm mv}$ is the source current and $\mathbf{B}_{\mathbf{r},\rm mv} = \widetilde{\mathbf{C}}_{\rm mv}\mathbf{M}_{\nu,\rm mv}\widehat{\widehat{\mathbf{b}}}_{\rm r}$ is the contribution of the permanent magnet.
The coupling is done by interpolating the dofs from the fixed to the moving part. The interpolation procedure is represented by a projection $\mathbf{P}$ mapping an arbitrary vector of dofs upon a vector satisfying the interface conditions, i.e., $\widehat{\mathbf{a}}_{\rm mv} = \mathbf{P}\widehat{\mathbf{a}}_{\rm fx}$ [DeGe04]. The coupled system (2) to be solved is:

$$\mathbf{P}^{\mathbf{H}} \begin{bmatrix} \mathbf{K}_{\rm fx} & 0 \\ 0 & \mathbf{K}_{\rm mv} \end{bmatrix} \mathbf{P} \begin{bmatrix} \widehat{\mathbf{a}}_{\rm fx}^{n+1} \\ \widehat{\mathbf{a}}_{\rm mv}^{n+1} \end{bmatrix} =$$

$$= \mathbf{P}^{\mathbf{H}} \left(\begin{bmatrix} 0 \\ \widehat{\widehat{\mathbf{j}}}_{\rm mv} \end{bmatrix} - \begin{bmatrix} 0 \\ \mathbf{B}_{\mathbf{r},\rm mv} \end{bmatrix} + \begin{bmatrix} \frac{\mathbf{M}_{\kappa,\rm fx}}{\Delta t} & 0 \\ 0 & \frac{\mathbf{M}_{\kappa,\rm mv}}{\Delta t} \end{bmatrix} \begin{bmatrix} \widehat{\mathbf{a}}_{\rm fx}^{n} \\ \widehat{\mathbf{a}}_{\rm mv}^{n} \end{bmatrix} \right) . \tag{3}$$

The reluctivity matrix $\mathbf{M}_\nu$ is constructed by FIT [Weil96] by:
$\mathbf{M}_\nu = \mathrm{diag}(\tilde{l})\ \nu\ \mathrm{diag}(S)^{-1}$, where $\mathrm{diag}(\tilde{l})$ is the diagonal matrix of dual-lengths, $\mathrm{diag}(S)$ is the diagonal matrix of primary-surfaces and ν is a diagonal matrix of reluctivities [Weil96]. The uni-directional Newton method [DeGe06] is used to ensure a rapid convergence of the nonlinear iteration. The method uses a differential reluctivity:

$$\nu_d = \nu + \frac{\partial \nu}{\partial \mid B \mid} \mid B \mid \ , \tag{4}$$

where the material matrix is constructed on the basis of ν_d, i.e.,
$\mathbf{M}_{\nu,\rm d} = \mathrm{diag}(\tilde{l})\ \nu_d\ \mathrm{diag}(S)^{-1}$. The magnetostatic system equation changes to:

$$\mathbf{P}^{\mathbf{H}} \begin{bmatrix} \mathbf{K}_{\rm fx,d} & 0 \\ 0 & \mathbf{K}_{\rm mv,d} \end{bmatrix} \mathbf{P} \begin{bmatrix} \mathbf{p}_{\rm fx}^{n+1,k+1} \\ \mathbf{p}_{\rm mv}^{n+1,k+1} \end{bmatrix} = \mathbf{P}^{\mathbf{H}} \begin{bmatrix} \mathbf{K}_{\rm fx} & 0 \\ 0 & \mathbf{K}_{\rm mv} \end{bmatrix} \mathbf{P} \begin{bmatrix} \widehat{\mathbf{a}}_{\rm fx}^{n+1,k} \\ \widehat{\mathbf{a}}_{\rm mv}^{n+1,k} \end{bmatrix} -$$

$$- \mathbf{P}^{\mathbf{H}} \begin{bmatrix} 0 \\ \widehat{\widehat{\mathbf{j}}}_{\rm mv} \end{bmatrix} - \mathbf{P}^{\mathbf{H}} \begin{bmatrix} 0 \\ \mathbf{B}_{\mathbf{r},\rm mv} \end{bmatrix} + \mathbf{P}^{\mathbf{H}} \begin{bmatrix} \frac{\mathbf{M}_{\kappa,\rm fx}}{\Delta t} & 0 \\ 0 & \frac{\mathbf{M}_{\kappa,\rm mv}}{\Delta t} \end{bmatrix} \begin{bmatrix} \widehat{\mathbf{a}}_{\rm fx}^{n} \\ \widehat{\mathbf{a}}_{\rm mv}^{n} \end{bmatrix} \tag{5}$$

where the superscript n stands for the transient iteration, k stands for the nonlinear iteration and the term $\mathbf{p}$ is the increment to the solution. The solution $\widehat{\mathbf{a}}$ reads: $\widehat{\mathbf{a}}_{\rm fx}^{n+1,k+1} = \widehat{\mathbf{a}}_{\rm fx}^{n+1,k} - \mathbf{p}_{\rm fx}^{n+1,k+1}$. The system matrices read: $\mathbf{K}_{\rm fx,d} = \widetilde{\mathbf{C}}_{\rm fx}\mathbf{M}_{\nu,\rm fx,d}\mathbf{C}_{\rm fx} + \frac{\mathbf{M}_{\kappa,\rm fx}}{\mathbf{\Delta t}}$ and $\mathbf{K}_{\rm mv,d} = \widetilde{\mathbf{C}}_{\rm mv}\mathbf{M}_{\nu,\rm mv,d}\mathbf{C}_{\rm mv} + \frac{\mathbf{M}_{\kappa,\rm mv}}{\mathbf{\Delta t}}$.

3 Mechanical Model

The displacement of the moving part obeys:

$$m\frac{d^2x}{dt^2} + c\frac{dx}{dt} + kx = F_x \ , \tag{6}$$

where the first term is related to the inertia of the moving armature, the second term represents the friction force and the third is the spring force. Here, m is the mass of the moving armature, c is the mechanical damping coefficient and k is the spring coefficient. F_x is the magnetic force in the x-direction which is obtained by integrating the Maxwell Stress Tensor on a surface that encloses the mobile armature [Ion05].
The mechanical equation can be written as a system of first order differential equation:

$$\frac{d}{dt}\begin{bmatrix} x(t) \\ v_x(t) \end{bmatrix} = \begin{bmatrix} 0 & 1 \\ \frac{-k}{m} & \frac{-c}{m} \end{bmatrix} \begin{bmatrix} x(t) \\ v_x(t) \end{bmatrix} + \begin{bmatrix} 0 \\ \frac{F_x(t)}{m} \end{bmatrix} , \tag{7}$$

and is resolved numerically by an explicit Euler scheme, yielding:

$$\begin{bmatrix} x^{k+1} \\ v_x^{k+1} \end{bmatrix} = \begin{bmatrix} x^k \\ v_x^k \end{bmatrix} + \begin{bmatrix} 0 & 1 \\ \frac{-k}{m} & \frac{-c}{m} \end{bmatrix} \begin{bmatrix} x^k \\ v_x^k \end{bmatrix} \Delta t + \begin{bmatrix} 0 \\ \frac{F_x^k}{m} \Delta t \end{bmatrix} . \tag{8}$$

4 Examples

The moving armature is constructed from laminated iron with a negligible conductivity, whereas the rail is made from massive iron, with a conductivity of $5e6$ S/m. Eddy currents are induced due to the movement.

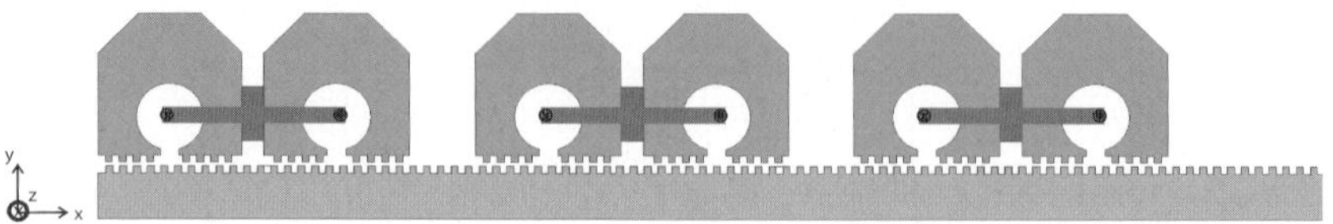

Fig. 2: A 3-phased linear actuator.

Single-Phase Machine. A first field simulation is carried out for a single module of the actuator (Fig. 1). A stable and an instable equilibrium position is encountered for every excitation current. As can be seen in Fig. 5 a), the applied force vanishes at a displacement of half of a pole pitch. Hence, a single module does not feature a self-starting capability.

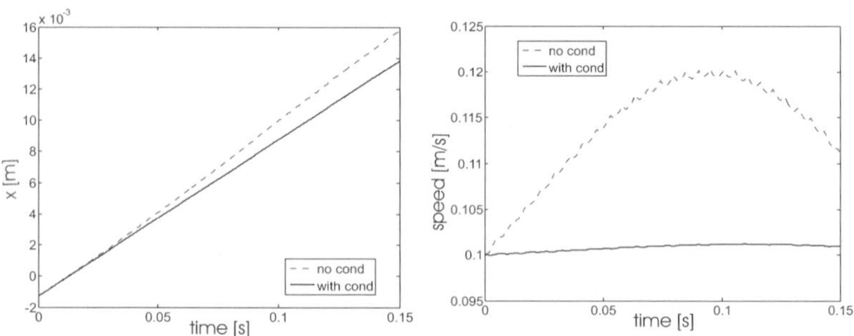

Fig. 3: Position and speed for a block current.

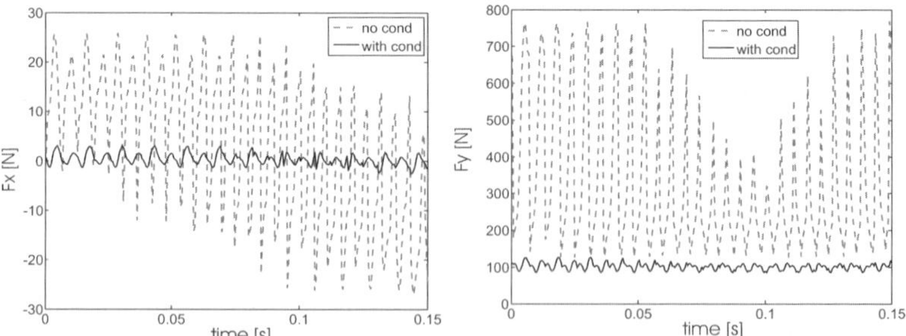

Fig. 4: Displacement force F_x and vertical attraction force F_y for a block current.

In Fig. 3, the position versus time and the speed versus time in the case of a linear actuator with and without conductivity, launched with an initial speed of 0.1 m/s is shown. In the presence

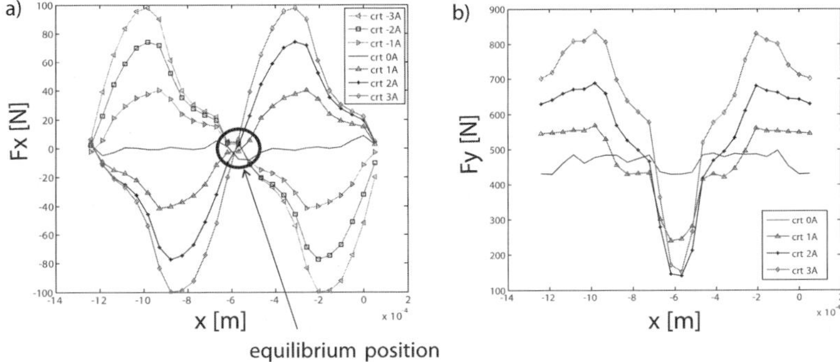

Fig. 5: Forces versus position: F_x a) and F_y b).

of eddy current effects, a substantially smaller acceleration is found. Also, the vertical attraction force F_y in the case of a block current with the amplitude of 3 A and the displacement force F_x have smaller values in the case of a conductive rail material (Fig. 4).

Three-Phase Machine. Secondly, a simulation of three-phase configuration is carried out. The simulation makes use of the static force-position characteristic which is calculated for a single pole pitch τ_p and for different excitations currents. The permanent magnets magnetically prestress the magnetic circuit in a certain direction. Hence, except for the situation of the equilibrium positions (Fig.5 a), applying a positive or a negative current to the actuator's winding, will displace the module to the left or to the right. The 3-phase machine (Fig. 2) consists of three modules where each neighboring module is displaced by an integer number of pole pitches plus a third part of a pole pitch. The performance of the three-phase actuator is computed by a semi-analytical model using the static force-position characteristic. The excitation currents of the different modules are dephased in time. In the first simulation, sinusoidal currents with an amplitude of $\hat{I} = 3$ A are applied (Fig. 6):

$$i_1 = \hat{I} sin(\frac{x}{\tau_p} 2\pi) \; ; \; i_2 = \hat{I} sin(\frac{x + \frac{\tau_p}{3}}{\tau_p} 2\pi) \; ; \; i_3 = \hat{I} sin(\frac{x + \frac{2\tau_p}{3}}{\tau_p} 2\pi) \, . \tag{9}$$

In Eq. (9), the currents are directly expressed in terms of the relative positions of the armatures. This reflects the synchronous nature of the current excitation where the time variation of the currents is linked to the actuator's speed, i.e., $x = v_x \, t$.

The forces for all three modules are derived from the static force-position characteristic by:

$$F_{x1} = F_x(x, i_1) \; ; \; F_{x2} = F_x(x + \frac{\tau_p}{3}, i_2) \; ; \; F_{x3} = F_x(x + \frac{2\tau_p}{3}, i_3) \, . \tag{10}$$

The total force is the sum of the three forces. The 3-phase actuator is simulated by combining the mechanical equation (8), the expressions (9) - (10) and the static force-position characteristic, organized as a look-up table. The cascaded simulation approach applied to the 3-phase actuator neglects eddy-current effects in the rail.

In a second simulation, a controlled block current of $I = 3$ A is applied (Fig. 7):

$$i_1 = \begin{cases} -I \text{ if } x < \frac{\tau_p}{2} \\ I \quad \text{else} \end{cases} ; \; i_2 = \begin{cases} -I \text{ if } x + \frac{\tau_p}{3} < \frac{\tau_p}{2} \\ I \quad \text{else} \end{cases} ; \; i_3 = \begin{cases} -I \text{ if } x + \frac{2\tau_p}{3} < \frac{\tau_p}{2} \\ I \quad \text{else} \end{cases} .$$

This control scheme is relatively primitive and leads to a highly oscillating force.The sinusoidal current excitations achieves a smaller force, but also exhibits smaller oscillations. It is therefore possible to reduce the oscillations in the thrust force by an adequate control.

5 Conclusions

A linear hybrid stepper motor is simulated. The fixed and the moving part are independently discretized by the FIT and coupled at a sliding surface. The 3-phase actuator was calculated

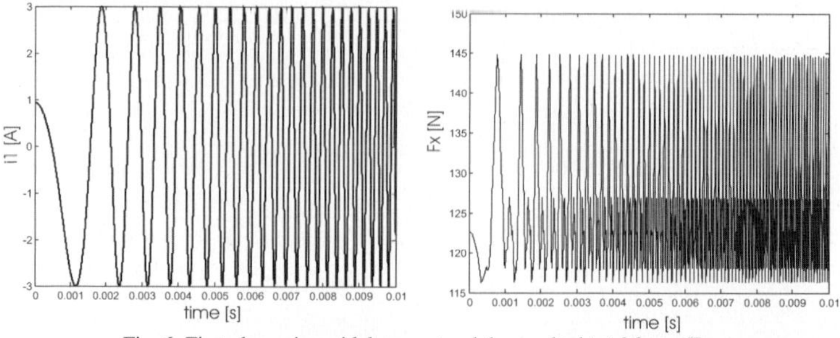

Fig. 6: First phase sinusoidal current and the resulted total force F_x.

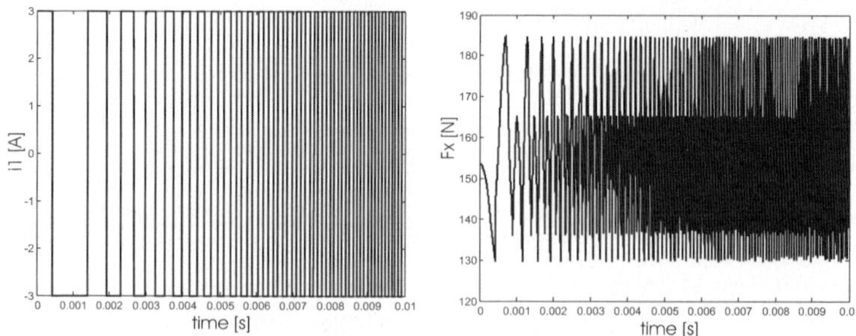

Fig. 7: First phase block current and the resulted total force F_x.

semi-analytically using the numerically obtained forces. A block current produces a bigger thrust than a sinusoidal current, but causes large oscillations.
A single-phase actuator constructed from a single module has stable and instable equilibrium positions. Due to the conductivity of the standstill armature, eddy current losses reduce the forces.

Acknowledgement
The authors thank PASIM Direktantriebe for providing the data of the actuator used as an example in this paper.

References

[CST] http://www.cst.com/
[DeGe04] H. De Gersem, M. Wilke and T. Weiland: Efficient modeling techniques for complicated boundary conditions applied to structured grids. In: COMPEL, Vol. **23**, 904–912 (2004)
[DeGe05] H. De Gersem, M. Ion and T. Weiland: Trigonometric Interpolation at Sliding Surfaces and in Moving Bands of Electrical Machine Models. In: COMPEL, Vol. **25**, No. 1, 31–42 (2006)
[DeGe06] H. De Gersem, I. Munteanu and T. Weiland: Comparison of Newton and Approximate Newton Methods for the Orthogonal Finite Integration Technique. Presented at Scientific Computing in Electrical Engineering (SCEE06), Sinaia, Romania (2006)
[Ion05] M. Ion, H. De Gersem and T. Weiland: Sliding-Surface Interface Conditions and Force Calculation for a Linear Actuator Discretized by the Finite Integration Technique. In: XVII International Symposium on Electromagnetic Fields (ISEF2005), Baiona, Spain, 15-17 Sept (2005).
[PASIM] http://www.direktantriebe.com/
[Weil96] T. Weiland: Time Domain Electromagnetic Field Computation with Finite Difference Methods. In: International Journal of Numerical Modelling, Vol. **9**, 295–319 (1996).

Reduced Order Electromagnetic Models for On-Chip Passives Based on Dual Finite Integrals Technique

Gabriela Ciuprina, Daniel Ioan and Diana Mihalache

Politehnica University of Bucharest, Electrical Engineering Department, Numerical Methods Laboratory, Spl. Independentei 313, 060042, Bucharest, Romania lmn@lmn.pub.ro

Abstract – An efficient methodology to extract reduced order models for electromagnetic devices is presented. To solve field-circuits coupled problems, the electromagnetic field equations are discretized by the dual Finite Integration Technique (dFIT), a numerical method which allows the accuracy control of the extracted parameters. Several techniques are used to accelerate the extraction process, such as minimal virtual boundary, minimal mesh and minimal frequency samples set. The frequency characteristic of the device is then approximated by a rational function of appropriate degree in order to extract the reduced order model and its SPICE equivalent circuit. The behavior of the synthesized model extracted with proposed algorithm, in the case of passive on-chip devices placed on silicon substrate shows good agreement with respect to the measurements.

1 Introduction

With the shrinking of on-chip devices, according to the Moore law, the operating frequency is increased. In this context, the literature reflects much interest in the computation of frequency-dependent characteristics of on-chip interconnects and passive components [1]. Generally, the electromagnetic simulation by numerical methods is still considered too time consuming to be a viable solution for computer aided design of integrated systems. However, the validity of any other new approach is checked by comparison with them. This is why the International Technology Roadmap for Semiconductors (ITRS, www.public.itrs.net) declared the high-frequency modelling ($>$ 5 GHz) of interconnect and on-chip passives as a grand challenge that should be solved in order to continue the pace of progress that was witnessed in the last three decades.
The goal of the present paper is to present techniques that speed-up the numerical electromagnetic simulation of on-chip passive components, in order to make it suitable for the CAD environments and current Electronic Design Automation (EDA) frameworks. These techniques resulted from the research carried on within the European projects FP5/IST/Codestar (www.imec.be/codestar) and FP6/IST/ STREP/Chameleon RF (www.chame
leon-rf.org). In these projects were designed, fabricated and characterized by experimental measurements a series of test structures comprising among others on-chip passive components. The spiral inductor presented in Fig.1 is a typical example of such structures, used as benchmarks to validate several modeling and simulation methodologies developed during the projects duration.

Fig. 1: A typical on-chip component (www.imec.be/codestar)

2 The Numerical Method

The combination of ALLROM technique [2] with the Very Fast Simulation (VFS) [3] strategy leads to the following algorithm for modeling of the passive components:

A) *Grid calibration* – a minimal orthogonal grid is successively refined, until the extracted resistances and capacitances are accurate enough;
B) *Virtual boundary calibration* – the computational domain is successively extended, until the extracted inductance is accurate enough;
C) *Frequency analysis* – using the grid resulted after refining and extension process, the frequency dependent matrix of the circuit functions $\mathbf{Z}(\omega) = \mathbf{R}(\omega) + j\omega\mathbf{L}(\omega)$ and $\mathbf{Y}(\omega) = \mathbf{Z}(\omega)^{-1} = \mathbf{G}(\omega) + j\omega\mathbf{C}(\omega)$ are computed in a minimal set of frequency samples, solving the Maxwell equations, by FIT;
D) *Optimal order of the compact model* – compact models of increasing order and their SPICE equivalent circuits are extracted and simulated in the frequency domain, until the result is close to previous computed $\mathbf{Y}(\omega)$;
E) *Validation* – based on the results of the SPICE simulation in frequency domain, the scattering parameters $\mathbf{S}(\omega)$ are computed and compared with the measurements, for a series of test structures, of practical interest - the Codestar benchmarks.

Below is detailed how this algorithm achieves the optimal tradeoff between solution error and the required computational resources.

A) Grid calibration by dFIT

The matrices of resistance **R** and capacitance **C** for the multi-polar passive components are extracted from the solution of the 3D field-problems of the static current distribution in conductive domains combined electrostatic field distribution in insulators. Apparently, two simple 3D Laplace problems with Dirichlet boundary conditions for the scalar potential have to be solved. However in these problems, it is not an easy task to handle the singularities of charge distribution at edges and corners of conductors. It has been well noticed that the convergence of capacitance versus discretisation is rather slow due to these singularities, especially in the most commonly used first-order FDM, FIT, FEM, or BEM based on collocation or Galerkin method [4]. The high order method we propose is suitable for non-homogeneous dielectrics being based on Finite Integrals Technique - FIT, and thus not requiring Green function. The improvement of FIT, able to handle the singularities of charge distribution is called *dual Finite Integration Technique (dFIT)* [5]. The main idea of dFIT is to solve the field problem two times, the first time using the primary grid and the second time using the secondary, dual grid. The numerical solution of dFIT

$$\mathbf{R_d} = \frac{\mathbf{R_p} + \mathbf{R_s}}{2}, \tag{1}$$

is the average of primary and secondary solutions of the problem of the static current-distribution.

Theorem 1. *Theorem on dFIT scissors relationship:*
The primary $\mathbf{R_p}$ *and the secondary* $\mathbf{R_s}$ *solutions are the lower and upper bounds respectively of the exact solution* $\mathbf{R}$*:*

$$\mathbf{R_p} \leq \mathbf{R} \leq \mathbf{R_s} \tag{2}$$

and therefore

$$\varepsilon = \frac{||\mathbf{R_d} - \mathbf{R}||}{||\mathbf{R_d}||} \leq \frac{||\mathbf{R_p} - \mathbf{R_s}||}{||\mathbf{R_d}||} \tag{3}$$

Here $\mathbf{A} \leq \mathbf{B}$ *means* $\mathbf{x^T}(\mathbf{A} - \mathbf{B})\mathbf{x} \leq \mathbf{0}$ *for any vector* $\mathbf{x}$*. A similar statement states for the capacitance matrix, extracted from the solution of the electrostatic problem.*
The proof of this theorem [5] is based on following remarks: the exact value of the electric power $\mathbf{i^T R i}$ is higher than the power $\mathbf{i^T R_p i}$ of primary FIT solution, which is curl-conform, while the power $\mathbf{i^T R_s i}$ of the secondary FIT solution, which is div-conform is lower than the exact energy.
This surprising behavior may be explained considered a (curl-conform) scalar potential which interpolate the voltages in the nodes of the primary grid and a (div-conform) vector potential which interpolate the edge-values of the dual grid. The curl-conform and div-conform solutions provide lower and upper bounds of the exact solution, regardless the used interpolation. P.w.l. interpolation on the grid cells generate diagonal discrete Hodge operators with simplest expressions, which provide the complementary numerical solutions have as graphs the dual FIT grids.
The accuracy provided by dFIT is similar to that of the second order FEM, but dFIT has additional advantages, requiring less CPU time to compute the solution, providing a reliable method to evaluate the numerical error of the solution and first of all simplicity. In order to understand the outstanding power of the proposed method, let consider a simple example of a 2D field problem: the resistance of a L-shaped conducting sheet. The coarsest grid in this case contains three rectangular cells and the FIT solution on this grid has a relative error of 17.2%, while the solution provided by dFIT has an error of only 3.3%. If the harmonic average (the conductance average) is used then the numerical error became 1.4%, while the computational effort is reduced to only 15 arithmetic operations. In fig. 2 are presented the resistance stamps applied in each of the three primary and secondary cells, and how they are assembled in the L-shaped computational domain.
The initial coarsest mesh is successively refined, dividing each cell in 8 sub-cells and the refined process continues until the difference between the two bounds became low enough. This difference is an excellent error estimator for the FIT solution, but it is too pessimistic for dFIT (which provides in practical cases a 10 times better accuracy). What is more important is the higher order of the convergence rate for dFIT, as compared with first order traditional FIT [5].

B) Virtual boundary calibration

The inductance matrix $\mathbf{L}$ is extracted from the solution of the 3D magneto-static field problem. The relative important magnetic energy of the field outside the truncated computational domain makes the dc inductance very sensitive to the position of the virtual boundary

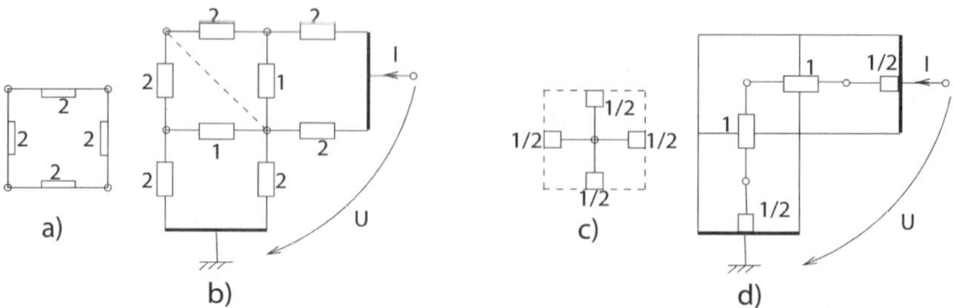

Fig. 2: Dual circuit for the L-shape resistor: a) Primary stamp b) Primary dFIT circuit ($R_p = 16/7R_{sq}$) c) Secondary stamp d) Secondary dFIT circuit ($R_s = 3R_{sq}$)

and the conditions on it. A large variety of open boundary techniques were developed for static and quasi-static electromagnetic field problems [6]. Each of them has advantages and shortcomings. While there is not an all-purpose powerful technique, the selection of the best technique, suitable for our case, is not trivial. The authors experimented with two approaches: Equivalent Layer of Open Boundary condition – ELOB and the Strategic Dual Image – SDI technique [7]. The first approach uses a layer of one cell thickness, with artificial material properties, mapped on the boundary. Optimal choice of layer parameters yields the first-order Asymptotic Boundary Condition (ABC) for the magnetic scalar potential φ:

$$\frac{\partial \varphi}{\partial r} + \frac{\varphi}{r} = 0, \tag{4}$$

which is actually a homogeneous Robin boundary condition. In the second approach the problem is solved two times with homogeneous Dirichlet (DBC) and Neumann (NBC) boundary conditions, and the arithmetic or the harmonic average of extracted parameters,

$$\mathbf{L_a} = \frac{\mathbf{L_{NBC}} + \mathbf{L_{DBC}}}{2}, \quad or \quad \mathbf{L_h} = 2((\mathbf{L_{NBC}})^{-1} + (\mathbf{L_{DBC}})^{-1})^{-1}, \tag{5}$$

respectively is adopted as the numerical solution. The two dual solutions allow the error estimation for the extracted parameters.

Theorem 2. *Theorem on scissors relationship for boundary conditions: Inductance matrix* $\mathbf{L_{NBC}}$ *computed with NBC and inductance matrix* $\mathbf{L_{DBC}}$ *computed with DBC are lower and upper bounds respectively of the exact solution* $\mathbf{L}$*, as well as of* $\mathbf{L_{ELOB}}$*, for any non-negative values of ELOB material parameters:*

$$\mathbf{L_{NBC}} \leq \mathbf{L} \leq \mathbf{L_{DBC}}, \quad \mathbf{L_{NBC}} \leq \mathbf{L_{ELOB}} \leq \mathbf{L_{DBC}}, \tag{6}$$

and therefore

$$\varepsilon_a = \frac{||\mathbf{L_a} - \mathbf{L}||}{||\mathbf{L_a}||} \leq \frac{||\mathbf{L_{NBC}} - \mathbf{L_{DBC}}||}{||\mathbf{L_a}||}, \tag{7}$$

$$\varepsilon_h = \frac{||\mathbf{L_h} - \mathbf{L}||}{||\mathbf{L_h}||} \leq \frac{||\mathbf{L_{NBC}} - \mathbf{L_{DBC}}||}{||\mathbf{L_h}||}. \tag{8}$$

The Euclidian norm is natural in this case, but the result is valid for other equivalent norms. The proof of this theorem is based on the following lemma: a given current distribution "stores" higher magnetic energy $\mathbf{i}^T\mathbf{L}'\frac{\mathbf{i}}{2} \leq \mathbf{i}^T\mathbf{L}''\frac{\mathbf{i}}{2}$ in materials with higher permeability $\mu'(r) \leq \mu''(r)$. Neumann and Dirichlet boundary conditions are degenerate cases of the ELOB condition, with $\mu \to 0$ and $\mu \to \infty$, respectively.

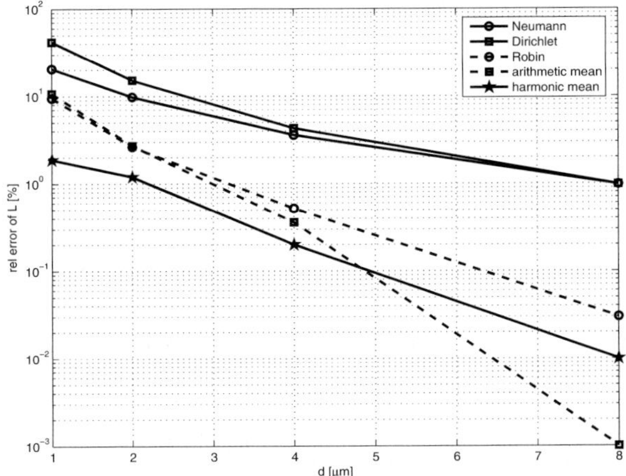

Fig. 3: Numerical error of inductance vs. the boundary position.

To improve the accuracy of the brute force truncation of the outer boundary, an iterative scheme is proposed; virtual boundary shifted (inflated) until the threshold of estimated error is reached. The boundary of the computational domain is pushed outward recursively with a fixed geometric progression.
Apparently, ELOB is better then SDI approach, because it does not requires solving the field problem two times. However, the two dual solutions provide a robust error estimator, and, based on it, the extension process can be stopped near the optimal position of the virtual boundary.
Fig.3 shows the relative error of the inductance extracted from the magnetic field distribution with several boundary conditions for the magnetic scalar potential, such Neumann, Dirichlet, Robin, as well as the arithmetic and harmonic average between inductances computed as function of boundary position. As we expected all errors tend to zero when the virtual boundary goes to infinity. However the convergence of the average values is more accelerated than the classical Neumann and Dirichlet boundary condition. We notice again the increasing of the order of convergence in the case of the dual approach.

C) Frequency analysis

The frequency dependent circuit-function $\mathbf{Y}(\omega)$ is extracted from the solution of the electromagnetic field problem with appropriate boundary conditions [8]. The full-wave approach provides the best approximation. However in the practical case of the today on-chip components, the electro-magnetic-quasi-static (EMQS) approach is sufficient. That means to accept magneto-quasi-static approximations inside metallic conductors and electro-quasi-static approximations inside insulator dielectrics and in the semiconductor substrate. In this way, the propagation occurs only along the interface surfaces between conductors and insulators, along interconnection wires.
The final grid obtained after the successive refinement in stage A and successive extension in stage B is used in the frequency analysis. Since the numerical solution of Maxwell equations is computed by dFIT, which requires two numerical solutions on two dual grids, the combination with SDI approach to model open boundary is a natural choice. This combination is proven to be in practical cases about 300 times more efficient than traditional FIT+NBC approach with same accuracy. To build numerical scissors for the exact solution, a practical approach we propose is to use the dual (complementary) solutions, solving the Maxwell Grid Equations

two times, and computing the admittance matrix using the two dual-staggered grids and two kind of boundary conditions, for a sequence of frequency samples ω:

- $\mathbf{Y_p}(\omega)$ is computed by FIT on the primary grid with ELOB parameters: $\varepsilon_r = M >> 1, \mu_r = 1$;
- $\mathbf{Y_s}(\omega)$ is computed by FIT on the secondary grid with ELOB parameters: $\varepsilon_r = 1, \mu_r = M >> 1$.

According to the theorems above, the exact matrices of capacity $\mathbf{C}$ and inductance $\mathbf{L}$, extracted from the static field solutions, and the average matrices $\mathbf{C_a} = (\mathbf{C_p} + \mathbf{C_s})/2$, $\mathbf{L_a} = (\mathbf{L_p} + \mathbf{L_s})/2$ comply with the inequalities: $\mathbf{C_s} \leq \mathbf{C}$ (and $\mathbf{C_a}$) $\leq \mathbf{C_p}$; $\mathbf{Lp} \leq \mathbf{L}$ (and $\mathbf{L_a}$) $\leq \mathbf{L_s}$. By averaging the two admittance, a numerical solution $\mathbf{Y_a}(\omega) = (\mathbf{Y_p}(\omega) + \mathbf{Y_s}(\omega))/2$ is generated, which provides a better accuracy than any of two direct extracted admittance $\mathbf{Y_p}(\omega)$ or $\mathbf{Y_s}(\omega)$, at least at low frequencies. The dynamic dFIT+ELOB model is a parallel connection of two dual semi-models with different ELOB conditions. The equivalent extracted capacitance and conductance correspond to the arithmetic average, while the equivalent extracted inductance and resistance correspond to the harmonic average.
The scissor theorems are results of the energy minimization in the case of the solutions of the elliptic PDE, therefore they are strictly valid only for Electrostatic and Magnetostatic fields. The general Maxwell equations are hyperbolic PDE, and then there is not expected to be a similar scissor theorem valid for full-wave fields.
An *Adaptive Frequency Sampling (AFS)* algorithm automatically selects the interpolation points, reducing their number and the global computational effort, because the frequency analysis is by far the most time consuming step.

D) Extraction of the compact model

In this algorithm step, the frequency characteristic $\mathbf{Y}(\omega)$ of the analyzed component is approximated by rational functions using the *Vector Fitting* procedure [9] and then a SPICE equivalent circuit is synthesized by the *Differential Equation Macromodel* [10]. The couple of procedures is iterated, successively increasing the order of the extracted model. Compact models of increasing order and their equivalent circuits are extracted and simulated in the frequency domain with SPICE, until the result is close to $\mathbf{Y}(\omega)$ previously computed, on the frequency range of interest. In this way, the compact model and its SPICE equivalent circuit for the given components having an optimal order are generated.

3 Validation of results - CODESTAR benchmarks

The described procedure was validated by comparing simulation results with measured data for a series of 16 benchmarks developed within FP5/IST/ CODESTAR European project. Test structures relevant for several technologies Al-SiO2, Cu-Lowk, with different geometries (meander resistor, MIM capacitors, spiral inductors, several transmission lines and other more complex configurations) were designed, fabricated and characterized. On wafer measurement up to 40 GHz were carried out and specific de-embedding procedures were applied, in order to eliminate the parasitic effects inherent at such frequencies.
Fig.4 shows the good agreement between measurement and simulation of scattering parameters for the benchmark no. 3 (a spiral inductor SP_SMALL over Silicon substrate). Errors less than 5% between simulation of extracted models (of order up to 10) and measurements are obtained for all the standard test structures.
More data related to the benchmark results are included in the Table 1. Files corresponding to the standard benchmarks described in the first four columns of this table (macromodels, reduced order models and measured frequency characteristics) may be found at www.lmn.pub.ro/codestar.

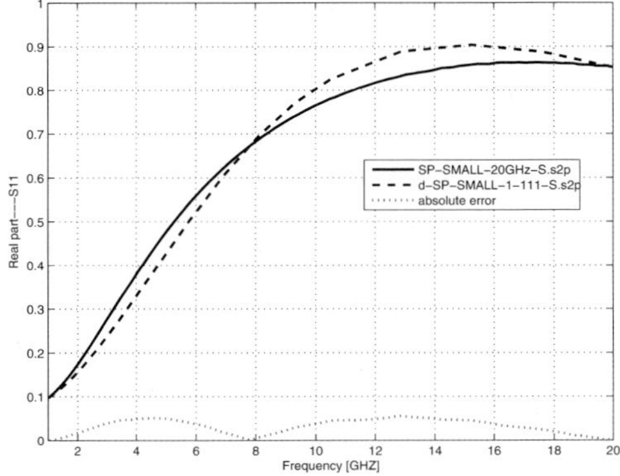

Fig. 4: Comparison between measurements and simulation.

The proposed strategy has proven to be an efficient methodology for modeling and simulation of advanced on-chip passive components. It consists of a series of techniques, each having its characteristic efficiency, measured by the degree-of-freedom (dof) reduction rate. Due to its higher order, dFIT requires a lower number of grid nodes compared with classical FIT, in order to obtain the same accuracy of the numerical solution. Therefore, due to necessary finer grid FIT requires more DOFs to describe the same phenomena. For instance in the case of the Codestar benchmarks FIT with 330 000 DOFs has the same accuracy as dFIT with 10 000 DOFs. In this sense, the efficiency of the order reduction of dFIT related to FIT is about 33 higher. Despite of the necessity to solve the problem two times, as dFIT requires, the global computational effort is drastically reduced, due to the smaller size of the linear systems which are solved. If the systems are solved with direct methods for full matrices the necessary CPU times is therefore more than 15 000 higher in the FIT case than in the dFIT case. Of course this impressive speed-up is lower when sparse matrix and/or iterative methods are used.

Table 1: Benchmarks results

Benchmark	Meander resistor	MIM capacitor	Spiral inductor	Transmission line	Coupled inductors	SP-CMIM LC cell
Nodes of initial mesh	368,200	833,280	596,068	2,860,441	681,876	458,304
Macromodel DOFs	19,510	29,925	39,920	19,972	43,138	29,862
No. of AFS in 0-20GHz	11	15	17	12	15	35
ALLROM CPU time [s]	145	3326	4278	161	2969	9467
Rel. error mes-sim [%]	1.4	2.5	13.6	5.0	15	20
Reduced order	4	4	4	10	4	8
Rel. error red-sim [%]	0.16	0.02	0.05	1.3	0.	0.5

On the other hand, the open boundary conditions we propose proves to be in the case of Codestar test structures about 10 times more efficient than the classical Neumann boundary conditions. It means that the computational domain extension with ELOB can be reduced so that the number of DOFs is ten times lower than in the case of the domain with classical Neumann conditions. This boundary shrinking is made keeping the numerical error at an acceptable level. Comparing the Vector Fitting with other aposteriori Krylov-type methods for order reduction, such as PRIMA, it was noticed a better efficiency in the order reduction with the former approach, which provided models having for the Codestar test structures ten times less state variables than PRIMA models. An explanation of the Vector Fitting better behavior could be its capability to consider only the system response in given frequencies, while the Kylov-type methods take care of the system behavior in all frequency space, including very high frequency range not-relevant for our case.
Combining these rates of the order-reduction efficiency, the ALPROM approach we propose became an effective methodology for the model extraction. Taking care of the order reduction in each modelling stage, the final result is better than the model reduced only with aposteriori methods. Moreover, working with macro-models reduced at previous stages, each modelling step requires lower computational resources, such as CPU time and memory.

References

[1] J. Zheng, et. al.: IEEE Trans. Microwave Theory Tech, **Vol. 48**, 1443–1451 (2000)

[2] D. Ioan, G.Ciuprina, M.Radulescu, M.Piper: IEEE CEFC 2004, Digest Book, Korea (June 6-9)

[3] D. Ioan, G. Ciuprina, M. Radulescu, and E. Seebacher: Compact Modeling and Fast Simulation of On-Chip Interconnect Lines, IEEE Transactions of Magnetics. **Vol. 42 No. 4**, 547–550 (April 2006)

[4] J. Zhao: Proc. of DAC 2000, Los Angeles, California, ACM, 536–539

[5] D. Ioan, M.Radulescu, G.Ciuprina: Scientific Computing in Electrical Engineering,**Vol. 4**, 248–256, Springer (2004)

[6] Q.Chen, A. Konrad, IEEE Trans. Magnetics, **Vol 33, No.1**, 663–676

[7] Daniel Ioan, Gabriela Ciuprina, Marius Radulescu: Absorbing Boundary Conditions for Compact Modeling of on-chip Passive Structures, COMPEL: The International Journal for Computation and Mathematics in Electrical and Electronic Engineering, **Vol. 25, No.3**, 625–659 (2006)

[8] D. Ioan, I. Munteanu: Proc. of Japanese Romanian Joint Seminar on Applied Electromagnetics and Mechanics JRJSAEM'98, p. 1. Kiryu, Gunma, Japan, (1998)

[9] B.Gustavsen, A.Semlyen: IEEE Trans. Power Delivery,**vol. 14**, 1052–1061

[10] T. Palenius, J. Roos: IEEE Trans. Microwave Theory and Techniques, **vol. 52, no.9**, 22

Techniques to Reduce the Equivalent Parallel Capacitance for EMI Filters Integration

Adina Racasan, Calin Munteanu, Vasile Topa, and Claudia Racasan

Department of Electrotechnics, Technical University of Cluj-Napoca, C. Daicoviciu 15, 400020 Cluj-Napoca, Romania, Adina.Racasan@et.utcluj.ro

1 Introduction

One of the major goals in designing of the integrated EMI filters is to improve their high-frequency characteristics. To achieve this, special technologies need to be developed, including the mechanisms for suppression of the equivalent parallel capacitance (EPC) and of the equivalent series inductance (ESL), in spite of increasing the high-frequency losses. In this light, the main goal of the paper is to develop and analyse the effectiveness of several EPC-reducing technologies. The study is performed using the numerical analysis software Maxwell Q2D Extractor that is able to give at the end of the numerical analysis process the values of the lumped per-unit-length capacitance or inductance of the geometrical structure proposed. There are calculated the EPC of the four single winding structures and of the coupled windings.
The main component of an EMI filter is the low pass filter; therefore, in order to develop the integrated low pass filter, the integrated L-C structure must be carefully studied and modelled. The planar integrated L-C structure consists of alternating layers of conductors, dielectrics, insulation and ferrite materials that produce an integrated structure with similar terminal characteristics as the lumped components. The exploded view of an integrated L-C structure was shown in Figure 1 [WLD03]. The integrated L-C winding consists of a dielectric substrate with conductor windings directly deposited on both sides, thus resulting in a structure having both sufficient inductance and capacitance. This realizes the equivalent integrated capacitance as well as the inductance. By appropriately terminating the four terminals A, B, C and D of the integrated L-C winding, the same structure could be configured as equivalent L-C series resonator, parallel resonator or low pass filter. To integrate the EMI filter, the L-C low pass filter configuration is used, where AD is the input port and CD is the output.

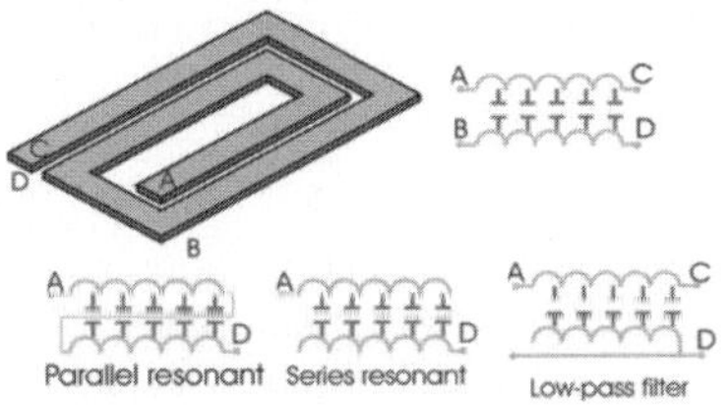

Fig. 1: The integrated LC structure

The schematic to illustrate the integrate EMI filter composition is shown in Figure 2. The exploded view of the physical structure is shown in Figure 3 [SW03].

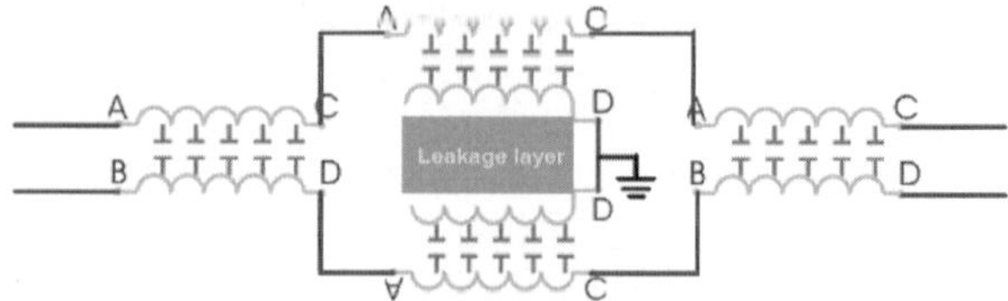

Fig. 2: Integrated EMI filter composition

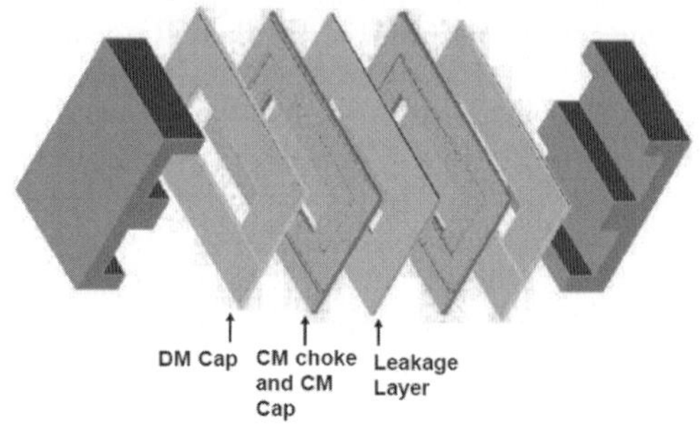

Fig. 3: Physical structure of integrated EMI filter

The existing integrated L-C technologies and design methodologies were mostly developed for high-frequency power passive components integration in order to achieve high efficiency and high power density. Since functions and requirements are different for passive components in EMI filters, special technologies need to be developed for EMI filter integration.

2 Numerical Examples

To evaluate the effectiveness of the EPC-reducing technologies, the EPC of the four single winding structures shown in Figure 4 (a-d) are calculated by using the Ansoft Maxwell Field Solver - 2D Extractor modules. Each figure in Figure 4 (a-d) is the cross-section view of a half winding window of the ferrite cores. The blue rectangles are the cross-sections of spiral winding conductor. All the conductors have the same dimensions, which are 1.2 x 0.3 mm. The relative permittivity of the materials used in the simulation is given in Table 1. In order to perform the numerical simulations a 2D working space has been considered. The discretization mesh consists of about 20000 linear triangular elements for each structure.

Materials	Ferrite	Air	Copper	Kapton	Ceramic
ε_r	12	1	1	3.6	84

Table 1: Material properties used in the simulation

The structure shown in Figure 4a is the original structure, which has two winding layers and six turns per layer. The first winding layer is an integrated L-C winding, consisting of a thin copper winding, a ceramic layer and a thick copper winding. The second winding layer is a normal copper-foil winding. The thickness of the insulation kapton between winding layers is 0.1 mm. The structure shown in Figure 4b is similar to that of Figure 4a, except the insulation kapton thickness is increased to 0.5 mm. The structure shown in Figure 4c replaces kapton in Figure 4b with air. The structure shown in Figure 4d is the staggered winding structure. To achieve non-overlapping windings, the total number of winding layers is increased to four and

Structure	(A)	(B)	(C)	(D)
EPC (pF)	93.6	23.81	10.3	10.7

Table 2: Calculated EPCs of four structures

the number of turns per layer is reduced to three, accordingly. Assuming a linear voltage distribution along the winding length, the equivalent capacitance is calculated based on equation $C_e = 2W_E V^{-2}$, where W_E is the stored electric field energy and V is the winding terminal voltage. The calculation results are given in Table 2. It is evident that the EPC of the proposed staggered winding structure in Figure 4d is more than 9 times smaller than that of the original structure shown in Figure 4a.

3 EPC of coupled windings

The calculated capacitance in Table 2 is the EPC of a single winding. For Common Mode (CM) chokes, there are two magnetically-coupled windings; hence the total equivalent structural winding capacitance will be increased. The equivalent circuit of two coupled windings with winding capacitance is shown in Figure 5.

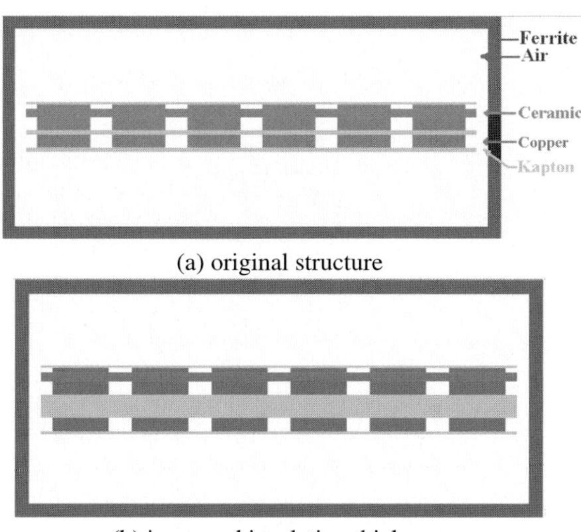

(a) original structure

(b) increased insulation thickness

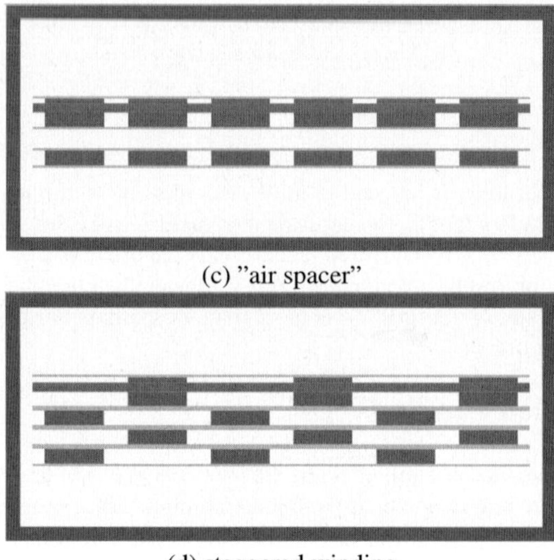

(c) "air spacer"

(d) staggered winding

Fig. 4: FEA Simulation models of different winding structures

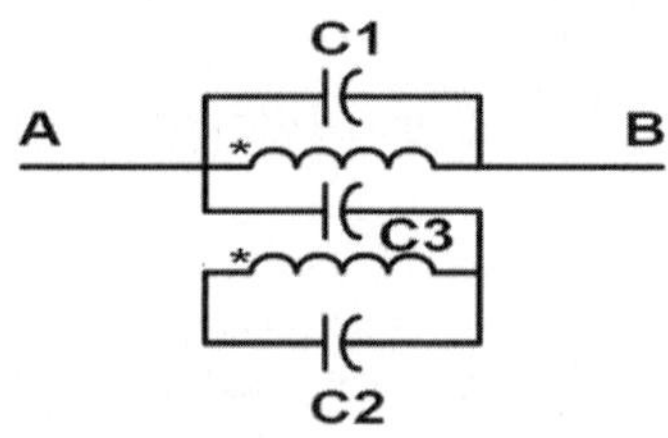

Fig. 5: EPC of two coupled windings

Under common mode excitation [WCWLO05], the equivalent circuit can be simplified to Figure 6.

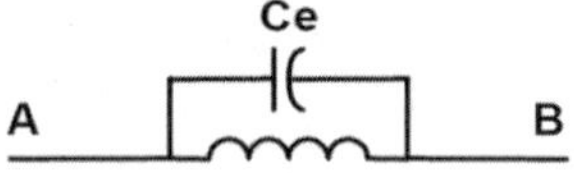

Fig. 6: Simplified circuit under CM excitation

The equivalent winding capacitance of the coupled windings is $C_e = C_1 + C_2 + C_3$, where C_1 and C_2 are the winding capacitances of each winding, and C_3 is the structural capacitance between windings. So the EPC of the CM choke is at least the sum of the EPCs of each winding. The FEA simulation model of a planar CM choke is shown in Figure 7, with a staggered winding structure for each winding. The calculated EPC under CM excitation is 28.237 pF.
To reduce the increased EPC caused by magnetic coupling, the two windings of CM chokes can be interleaved. Under common mode excitation, the two interleaved windings can be regarded as a single winding from an electrostatic point of view. Hence the total equivalent

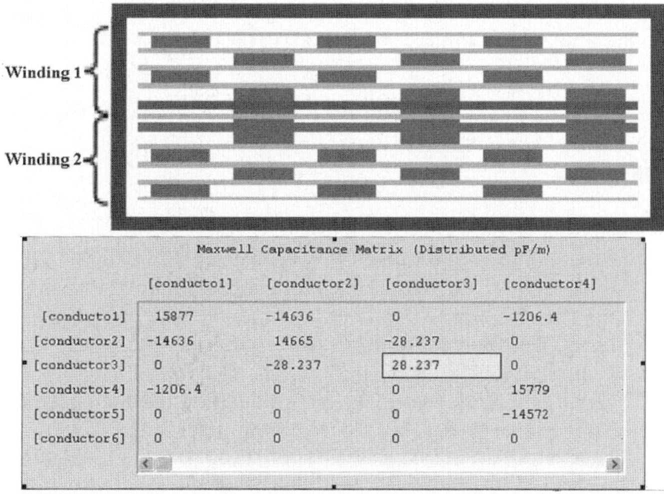

Maxwell Capacitance Matrix (Distributed pF/m)

	[conductol]	[conductor2]	[conductor3]	[conductor4]
[conductol]	15877	-14636	0	-1206.4
[conductor2]	-14636	14665	-28.237	0
[conductor3]	0	-28.237	28.237	0
[conductor4]	-1206.4	0	0	15779
[conductor5]	0	0	0	-14572
[conductor6]	0	0	0	0

Fig. 7: Two staggered windings not interleaved

winding capacitance will be equal to the structural capacitance of a single winding. Figure 8 shows a structure in which the staggered and interleaved winding techniques are combined. With the same material and geometry parameters, the calculated equivalent winding capacitance is only 8 pF.

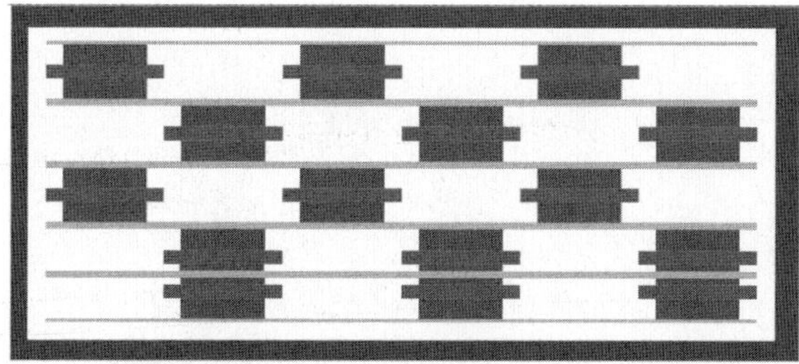

Fig. 8: Staggered and interleaved windings

4 Conclusions

The paper proposes several techniques for reducing the EPC of the integrated EMI filters, such as: "air spacer", increased insulation thickness, staggered winding, for the case of the single winding structure and two staggered windings not interleaved, respectively the staggered and interleaved winding for the case of the two coupled windings. By applying these technologies, the EPC of a constructed prototype is reduced from originally around 100 pF to less than 10 pF.

Acknowledgement. The authors are grateful to the Ministry of Education and Research for the support within the frame of the Project 9 CEEX I 03 / 06.10.2005.

References

[WLD03] Van Wyk, J. D., Lee, F. C., Dushan, B., Deb, K.: A future Approach to Integration in Power Electronics, Proceedings of IEEE IECON (2003)

[SW03] Strydom, J. T., Van Wyk, J. D.: Volumetric Limits of Planar Integrated Resonant Transformers: A 1 mHz Case Study, IEEE Transactions on Power Electronics, Vol. 18, No. 1, pp. 236 247 (2003)

[WLOW05] Wang, S., Lee, F. C., Odendaal, G. W., Van Wyk, J. D.: Improvement of EMI Filter Performance With Parasitic Coupling Cancellation, IEEE Transactions on Power Electronics, Vol. 20, No. 5, pp. 1221 1228 (2005)

[WCWLO05] Wang, S., Chen, R., Van Wyk, J. D., Lee, F. C., Odendaal, G. W.: developing Parasitic Cancellation Technologies To Improve EMI Filter Performance for Switching Mode Power Supplies, IEEE Transactions on Electromagnetic Compatibility, Vol. 47, No. 4, pp. 921 929 (2005)

Buffered Block Forward Backward (BBFB) Method Applied to EM Wave Scattering from Homogeneous Dielectric Bodies

Conor Brennan[1] and Diana Bogusevschi[2]

[1] Dublin City University brennanc@eeng.dcu.ie
[2] Dublin City University diana@eeng.dcu.ie

Abstract - The buffered block forward backward method is described and applied to the problem of scattering from dielectric cylinders. Numerical results suggest that the method converges in the case of scattering from bodies closed at infinity while it produces divergent results in the case of scattering from closed bodies.

1 Introduction

The problem of wave scattering computation is of central importance in the domain of computational electromagnetics. There are different techniques for solving such problems, ranging from asymptotic methods which are efficient but less accurate to full wave methods which are highly accurate but computationally intensive. In this paper we describe the Buffered Block Forward Backward Method (BFBB) which has been used in the past to efficiently compute scattering from perfectly conducting bodies. We formulate the problem of scattering from homogeneous dielectric bodies using the coupled Electric Field Integral Equation (EFIE) formulation. While scattering from a perfectly conducting object can be described in terms of a single integral equation describing fields external to the scatterer, scattering from a homogeneous dielectric body is described in terms of coupled electric field integral equations involving expressions for both interior and exterior fields. The integral equations are discretized by applying the method of moments with N suitable basis and testing functions. This results in a matrix equation that has to be solved for the unknown basis function amplitudes.

$$\mathbf{ZJ} = \mathbf{V} \tag{1}$$

$\mathbf{Z}$ is a $N \times N$ complex valued dense matrix. $\mathbf{J}$ and $\mathbf{V}$ are column vectors of length N. For objects small in terms of the wavelength this equation can be solved by direct matrix inversion. For larger objects we are forced to use iterative methods.

Iterative Solvers

The matrix equation obtained after applying the Method of Moments can not be solved using direct inversion because of its prohibitive size. Instead iterative solvers are used so that the solution $\mathbf{J}$ is sequentially build up. Golub and van Loan present a very good overview of iterative methods in [1]. There are two main categories of iterative solvers, *stationary* and *non stationary*. Non stationary methods are usually based on the development of a Krylov sub space for the $\mathbf{Z}$ matrix. The most common non stationary solvers are the Conjugate Gradient (CG) method and its varations such as Generalised Minimum Residual (GMRES). The

practical use of these solvers for electromagnetic wave scattering applications is thoroughly described by Peterson, Ray, and Mittra in [2]. Examples of stationary methods include Gauss-Seidel, Jacobi, Successive Overrelaxation and Symmetric Successive Overrelaxation The last two are variations on the basic Gauss-Seidel Method.
The block versions of these algorithms are derived in a straightforward fashion from the classical stationary algorithms. We note that stationary solvers were historically not favored in solving electromagnetic wave scattering problems due to their questionable convergence properties. Nevertheless in [3] West and Sturm showed that in the situation when these algorithms converge they can outperform the non stationary solvers. However they noted that the classical stationary methods can diverge when applied to bodies which exhibit the possibility of much multiple scattering. The advantage of the non stationary algorithms is that they are much less affected by the geometry of the scattering body. The non stationary methods will converge in most cases, even though at a much slower rate as the amount of multiple scattering increases.

2 Coupled Electric Field Integral Equations for homogeneous dielectric bodies

In this work the problem of scattering from a dielectric body is formulated using the the Coupled Electric Field Integral Equations. The derivation of these equations can be found in [1]:

$$E_z^{inc}(t) = K_t(t) + jk_0\eta_0 A_z^{(0)} + \{\frac{\partial F_y^{(0)}}{\partial x} - \frac{\partial F_x^{(0)}}{\partial y}\}_{S+} \tag{2}$$

$$0 = -K_t(t) + jk_d\eta_d A_z^d + \{\frac{\partial F_y^{(d)}}{\partial x} - \frac{\partial F_x^{(d)}}{\partial y}\}_{S-} \tag{3}$$

The formulation invokes the surface equivalence principle and expresses the fields interior and exterior to the scatterer in terms of vector potentials $\mathbf{A}$ and $\mathbf{F}$. These potentials are written in terms of the tangential magnetic field (electric current) and the tangential electric field (the so called magnetic current $\mathbf{K}$). Applying the method of moments with N pulse basis functions and Dirac Delta testing functions yields [1]

$$\begin{bmatrix} \mathbf{A} & \mathbf{B} \\ \mathbf{C} & \mathbf{D} \end{bmatrix} \begin{bmatrix} \mathbf{j} \\ \mathbf{k} \end{bmatrix} = \begin{bmatrix} \mathbf{E} \\ \mathbf{0} \end{bmatrix} \tag{4}$$

Each of $\mathbf{A}, \mathbf{B}, \mathbf{C}, \mathbf{D}$ is a $N \times N$ matrix. In order to best describe the forward backward method we explicitly rewrite equation (4) illustrating the matrix entries

$$\begin{bmatrix} A_{11} & A_{12} & A_{13} & \dots & A_{1N} & B_{11} & B_{12} & B_{13} & \dots & B_{1N} \\ A_{21} & A_{22} & A_{23} & \dots & A_{2N} & B_{21} & B_{22} & B_{23} & \dots & B_{2N} \\ \vdots & \vdots & \vdots & \vdots & \vdots & \vdots & \vdots & \vdots & \vdots & \vdots \\ A_{N1} & A_{N2} & A_{N3} & \dots & A_{NN} & B_{N1} & B_{N2} & B_{N3} & \dots & B_{NN} \\ C_{11} & C_{12} & C_{13} & \dots & C_{1N} & D_{11} & D_{12} & D_{13} & \dots & D_{1N} \\ C_{21} & C_{22} & C_{23} & \dots & C_{2N} & D_{21} & D_{22} & D_{23} & \dots & D_{2N} \\ \vdots & \vdots & \vdots & \vdots & \vdots & \vdots & \vdots & \vdots & \vdots & \vdots \\ C_{N1} & C_{N2} & C_{N3} & \dots & C_{NN} & D_{N1} & D_{N2} & D_{N3} & \dots & D_{NN} \end{bmatrix} \begin{bmatrix} j_1 \\ j_2 \\ \vdots \\ j_N \\ k_1 \\ k_2 \\ \vdots \\ k_N \end{bmatrix} = \begin{bmatrix} E_1 \\ E_2 \\ \vdots \\ E_N \\ 0 \\ 0 \\ \vdots \\ 0 \end{bmatrix} \tag{5}$$

The unknowns can be re-arranged to sequentially run through the unknowns in each domain $j_1, k_1, j_2, k_2, \cdots j_N, k_N$ rather than first running through the unknown electric current amplitudes $j_1, j_2, \cdots j_N$ and then the magnetic current amplitudes $k_1, k_2, \cdots k_N$. This trivial re-arrangement yields

$$\begin{bmatrix} A_{11} & B_{11} & A_{12} & B_{12} & A_{13} & B_{13} & \dots & A_{1N} & B_{1N} \\ C_{11} & D_{11} & C_{12} & D_{12} & C_{13} & D_{13} & \dots & C_{1N} & D_{1N} \\ A_{21} & B_{21} & A_{22} & B_{22} & A_{23} & B_{23} & \dots & A_{2N} & B_{2N} \\ C_{21} & D_{21} & C_{22} & D_{22} & C_{23} & D_{23} & \dots & C_{2N} & B_{2N} \\ \vdots & \vdots & \vdots & \vdots & \vdots & \vdots & \vdots & \vdots & \vdots \\ A_{N1} & B_{N1} & A_{N2} & B_{N2} & A_{N3} & B_{N3} & \dots & A_{NN} & B_{NN} \\ C_{N1} & D_{N1} & C_{N2} & B_{N2} & C_{N3} & D_{N3} & \dots & C_{NN} & D_{NN} \end{bmatrix} \begin{bmatrix} j_1 \\ k_1 \\ j_2 \\ k_2 \\ \vdots \\ j_N \\ k_N \end{bmatrix} = \begin{bmatrix} E_1 \\ 0 \\ E_2 \\ 0 \\ \vdots \\ E_N \\ 0 \end{bmatrix} \tag{6}$$

This can be re-written more compactly as

$$\begin{bmatrix} \mathbf{Z}_{11} & \mathbf{Z}_{12} & \dots & \mathbf{Z}_{1N} \\ \mathbf{Z}_{21} & \mathbf{Z}_{22} & \dots & \mathbf{Z}_{2N} \\ \vdots & \vdots & \vdots & \vdots \\ \mathbf{Z}_{N1} & \mathbf{Z}_{N2} & \dots & \mathbf{Z}_{NN} \end{bmatrix} \begin{bmatrix} \mathbf{J}_1 \\ \mathbf{J}_2 \\ \vdots \\ \mathbf{J}_N \end{bmatrix} = \begin{bmatrix} \mathbf{V}_1 \\ \mathbf{V}_2 \\ \vdots \\ \mathbf{V}_N \end{bmatrix} \tag{7}$$

where $\mathbf{Z}_{mn}$ is a 2×2 matrix containing interactions between the unknowns j_m, k_m and j_n, k_n.

$$\mathbf{Z}_{mn} = \begin{bmatrix} A_{mn} & B_{mn} \\ C_{mn} & D_{mn} \end{bmatrix} \tag{8}$$

If we group basis function domains together into M groupings each containing $\frac{N}{M}$ basis functions we can write a block version of equation (8) as

$$\begin{bmatrix} \tilde{\mathbf{Z}}_{11} & \tilde{\mathbf{Z}}_{12} & \dots & \tilde{\mathbf{Z}}_{1M} \\ \tilde{\mathbf{Z}}_{21} & \tilde{\mathbf{Z}}_{22} & \dots & \tilde{\mathbf{Z}}_{2M} \\ \vdots & \vdots & \vdots & \vdots \\ \tilde{\mathbf{Z}}_{M1} & \tilde{\mathbf{Z}}_{M2} & \dots & \tilde{\mathbf{Z}}_{MM} \end{bmatrix} \begin{bmatrix} \tilde{\mathbf{J}}_1 \\ \tilde{\mathbf{J}}_2 \\ \vdots \\ \tilde{\mathbf{J}}_M \end{bmatrix} = \begin{bmatrix} \tilde{\mathbf{V}}_1 \\ \tilde{\mathbf{V}}_2 \\ \vdots \\ \tilde{\mathbf{V}}_M \end{bmatrix} \tag{9}$$

where $\tilde{\mathbf{Z}}_{mn}$ contains the interactions between all basis functions in groups m and n. A forward-backward solver finds a global solution by solving a sequence of problems, each one describing the surface current in one grouping. By 'marching' the currents forward and backward from group to group a solution can be found in manner that can be more efficient than using other iterative solvers. A block forward backward proceeds by

$$\tilde{\mathbf{Z}}_{mm}\tilde{\mathbf{J}}_m^{(k+\frac{1}{2})} = \tilde{\mathbf{V}}_m - \sum_{n<m} \tilde{\mathbf{Z}}_{mn}\tilde{\mathbf{J}}_n^{(k+\frac{1}{2})} - \sum_{n>m} \tilde{\mathbf{Z}}_{mn}\tilde{\mathbf{J}}_n^{(k)} \tag{10}$$

$$\tilde{\mathbf{Z}}_{mm}\tilde{\mathbf{J}}_m^{(k+1)} = \tilde{\mathbf{V}}_m - \sum_{n<m} \tilde{\mathbf{Z}}_{mn}\tilde{\mathbf{J}}_n^{(k+\frac{1}{2})} - \sum_{n>m} \tilde{\mathbf{Z}}_{mn}\tilde{\mathbf{J}}_n^{(k+1)} \tag{11}$$

In [4] and [5] a variation on the block successive over relaxation method is introduced and applied to the case of scattering from perfectly conducting objects. Rather than solving for the unknowns in each group individually the interactions with neighbouring groups (referred to as buffer regions) are included in order to suppress spurious diffraction effects that would otherwise arise and cause the solution to diverge. Specifically the following equation describes the forward sweep of the BBFB scheme [4]:

$$\begin{bmatrix} \tilde{\mathbf{Z}}_m & \tilde{\mathbf{Z}}_{m(m+1)} \\ \tilde{\mathbf{Z}}_{(m+1)m} & \tilde{\mathbf{Z}}_{(m+1)(m+1)} \end{bmatrix} \begin{bmatrix} \tilde{\mathbf{J}}_m^{(k+\frac{1}{2})} \\ \tilde{\mathbf{B}}_{m+1} \end{bmatrix} = \begin{bmatrix} \tilde{\mathbf{V}}_m \\ \tilde{\mathbf{V}}_{m+1} \end{bmatrix} - \begin{bmatrix} \tilde{\mathbf{L}}_m \\ \tilde{\mathbf{L}}_{m+1} \end{bmatrix} - \begin{bmatrix} \tilde{\mathbf{U}}_m \\ \tilde{\mathbf{U}}_{m+1} \end{bmatrix} \tag{12}$$

In the forward sweep group $m+1$ acts as a buffer zone for group m. $\tilde{\mathbf{B}}_{m+1}$ is a dummy unknown used to temporarily compute the unknowns in group $m+1$ in order to allow their accurate interaction with the unknowns in group m. The last two quantities on the right incorporate scattering from other groups and are given by

$$\begin{bmatrix} \tilde{\mathbf{L}}_m \\ \tilde{\mathbf{L}}_{m+1} \end{bmatrix} = \begin{bmatrix} \sum_{n<m} \tilde{\mathbf{Z}}_{mn} \tilde{\mathbf{J}}_n^{(k+\frac{1}{2})} \\ \sum_{n<m} \tilde{\mathbf{Z}}_{(m+1)n} \tilde{\mathbf{J}}_n^{(k+\frac{1}{2})} \end{bmatrix} \tag{13}$$

and

$$\begin{bmatrix} \tilde{\mathbf{U}}_m \\ \tilde{\mathbf{U}}_{m+1} \end{bmatrix} = \begin{bmatrix} \sum_{n>m+1} \tilde{\mathbf{Z}}_{mn} \tilde{\mathbf{J}}_n^{(k+\frac{1}{2})} \\ \sum_{n>m+1} \tilde{\mathbf{Z}}_{(m+1)n} \tilde{\mathbf{J}}_n^{(k+\frac{1}{2})} \end{bmatrix} \tag{14}$$

In the backward sweep the group $m-1$ acts as a buffer for group m and the process is updated as

$$\begin{bmatrix} \tilde{\mathbf{Z}}_{(m-1)(m-1)} & \tilde{\mathbf{Z}}_{(m-1)m} \\ \tilde{\mathbf{Z}}_{m(m-1)} & \tilde{\mathbf{Z}}_{mm} \end{bmatrix} \begin{bmatrix} \tilde{\mathbf{B}}_{m-1} \\ \tilde{\mathbf{J}}_m^{(k+1)} \end{bmatrix} = \begin{bmatrix} \tilde{\mathbf{V}}_{m-1} \\ \tilde{\mathbf{V}}_m \end{bmatrix} - \begin{bmatrix} \tilde{\mathbf{L}}_{m-1} \\ \tilde{\mathbf{L}}_m \end{bmatrix} - \begin{bmatrix} \tilde{\mathbf{U}}_{m-1} \\ \tilde{\mathbf{U}}_m \end{bmatrix} \tag{15}$$

where the lower sum in $\tilde{\mathbf{L}}$ is now over $n < m - 1$ and the upper sum in $\tilde{\mathbf{U}}$ is over $n > m$.

3 Results

The first example presented involves the computation of scattering from an infinite surface separating a homogeneous dielectric from free-space. The surface is illuminated by fields from a line-source radiating at $300MHz$ as depicted in figure (1). To solve the problem numerically we must truncate the surface which will introduce non-physical diffraction effects. Obviously, the longer the surface considered the lesser these effects and the more accurate the fields will be in the central $18m$ section under the source. In order to ascertain how far away the truncation must be to achieve acceptable accuracy in the central region we solved a sequence of problems, where the straight sections of the surface leading to $\pm\infty$ were truncated at different points. It was found that truncating the straight segments at $(-6, 0)$ and $(6, 0)$ produced fields in the central region which differed to those obtained by truncating at $(-8, 0)$ and $(8, 0)$ by on average 0.13%. We concluded that truncating the surface at $(-6, 0)$ and $(6, 0)$ gives results in the central section which are essentially identical to those which would be obtained by considering an infinite surface. We then applied the BBFB to solve this truncated problem.
For this example there were 1320 unknowns which, for the purposes of applying the block forward-backward method were grouped together into groups of 10. No buffer zones were used. Figure (2) shows the electric current after only 3 iterations compared to that obtained using direct matrix inversion. Figure (3) shows the magnetic current after 3 iterations compared to that obtained by direct matrix inversion. We see that excellent agreement is achieved in both cases. To see how the algorithm performs as the number of iterations is increased consider figure (4) which shows the convergence properties of the BBFB. Specifically it plots the normalised error in the solution of the matrix equation

$$\log_{10} \frac{||\mathbf{V} - \mathbf{Z}\mathbf{J}^{BBFB}||}{||\mathbf{V}||}$$

against iteration number where $\mathbf{J}^{BBFB}$ denotes the surface fields computed using the BBFB. The norm used is the L_2 norm. The results confirm that the BBFB produces convergent results as the number of iterations is increased.
The second example applies the BBFB to a closed dielectric cylinder schematically shown in figure (5). The line-source is positioned at $(-5, 0)$ and is radiating at $300MHz$. The radius of the cylinder is $2m$ and it is centred at the origin. The constitutive parameters of the cylinder

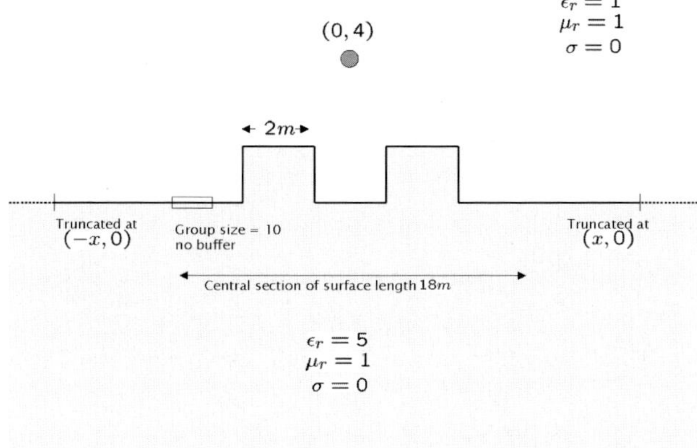

Fig. 1: Dielectric Homogeneous Surface Closed at infinity

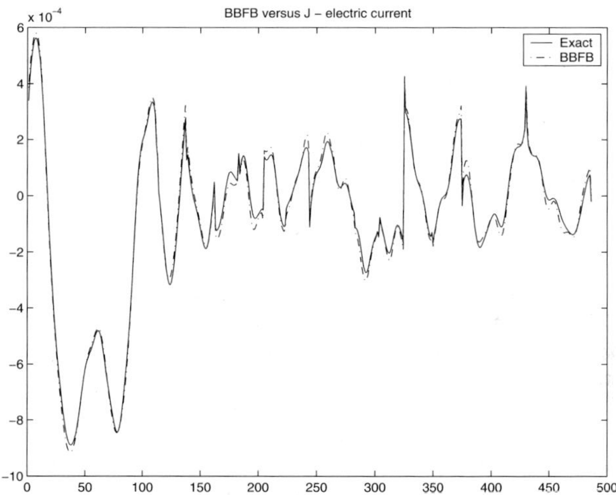

Fig. 2: Results for Electric Currents for the homogeneous dielectric surface closed at infinity after 3 iterations. The index on the x-axis refers to the index number of the basis functions used.

are $\epsilon_r = 15 - 0.01\jmath$, $\mu_r = 1$. Discretising at a rate of 10 unknowns per wavelength leads to 972 unknowns. The groups were identified by imposing a rectangular grid over the cylinder as depicted in figure (5). Each strip in the rectangular grid was $\frac{\lambda_0}{2} = 0.5m$ in width and the groups consisted of all basis functions residing in the various strips as indicated. In practice this produced 12 groups, containing up to 115 discretisation points each. The BBFB was then applied and the fields were marched forwards and backwards through the structure.
The convergence of the BBFB when applied to this problem is confirmed by examining figure (6). The algorithm produces convergent results as the number of iterations increases and reasonable results can be obtained with only a few iterations. Figure(7) shows the electric current after 3 iterations compared to the exact value obtained using direct matrix inversion. Figure (8) shows the magnetic current after 3 iterations compared to the exact value. Both display satisfactory agreement despite only requiring a small amount of computational effort.

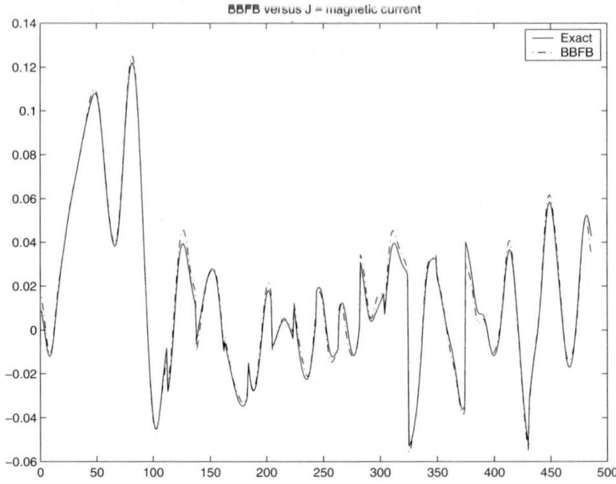

Fig. 3: Results for Magnetic Currents for the homogeneous dielectric surface closed at infinity after 3 iterations. The index on the x-axis refers to the index number of the basis functions used.

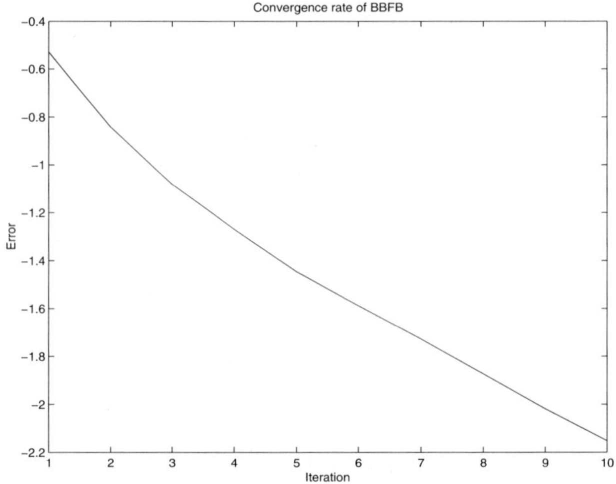

Fig. 4: Normalised error $\log_{10} \frac{\|\mathbf{V}-\mathbf{ZJ}\|}{\|\mathbf{V}\|}$ of BBFB algorithm applied to truncated surface.

4 Conclusions

A buffered block forward backward method has been presented for solving 2D scattering problems involving homogeneous dielectric bodies. Numerical results suggest that the method produces quickly convergent results when applied both to scattering from a body closed at infinity (such as a surface) and to scattering from closed bodies.

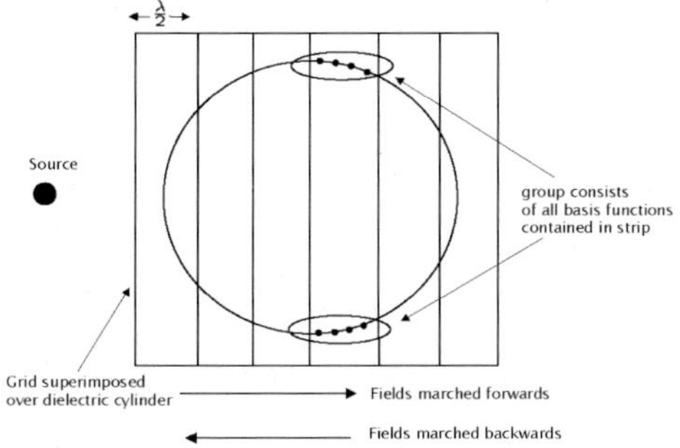

Fig. 5: Closed Homogeneous Dielectric body

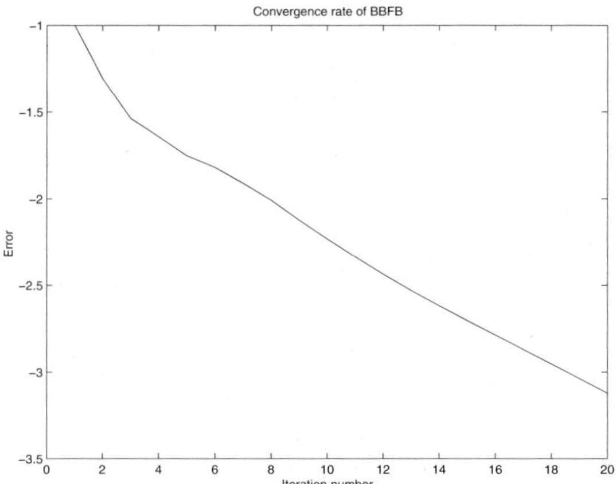

Fig. 6: Normalised error $\log_{10} \frac{\|\mathbf{V}-\mathbf{ZJ}\|}{\|\mathbf{V}\|}$ of BBFB algorithm applied to cylinder.

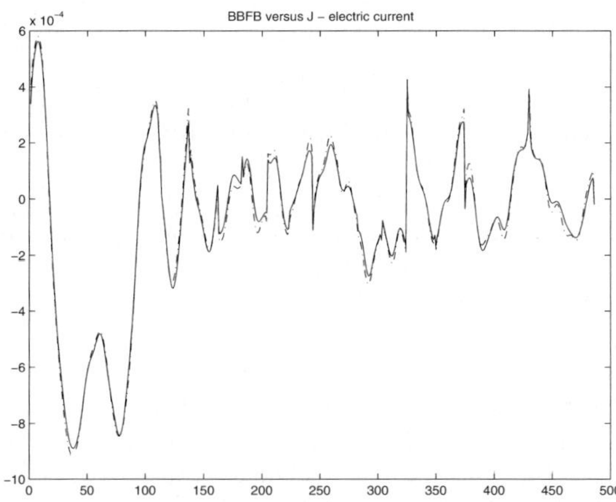

Fig. 7: Results for Electric Currents for the closed homogeneous dielectric cylinder after 3 iterations. The index on the x-axis refers to the index number of the basis functions used.

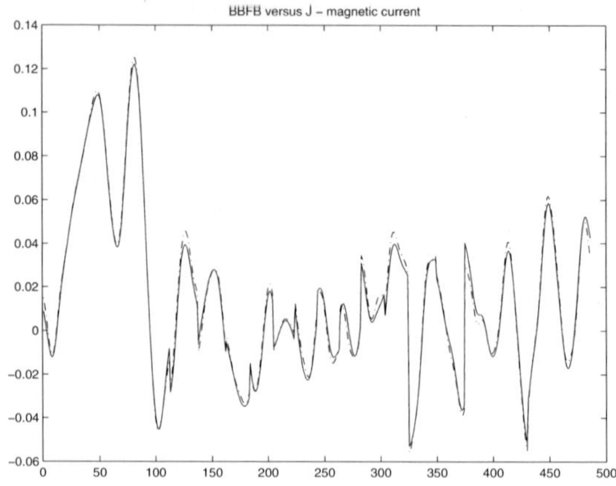

Fig. 8: Results for Magnetic Currents for the closed homogeneous dielectric cylinder after 3 iterations. The index on the x-axis refers to the index number of the basis functions used.

5 Acknowledgments

This publication has emanated from research conducted with financial aid from Science Foundation Ireland.

References

1. Matrix Computations G. H. Golub, C. F. van Loan, The Johns Hopkins University Press, Baltimore and London, 1996
2. Computational Methods for Electromagnetics A. F. Peterson, S. L. Ray, R. Mittra, IEEE Press, New York, 1998
3. On iterative Approaches for Electromagnetic Rough-Surface Scattering Problems, J. C. West, J. M. Sturm, IEEE Transactions on Antennas and propagation, Vol. 47, No. 8 , August 1999
4. A Novel Iterative Solution of the Three Dimensional Electric Field Integral Equation C. Brennan, P. J. Cullen, M. Condon, IEEE transactions on Antennas and Propagation, Vol.52, No. 10, October 2004
5. Convergence analysis for buffered block forward-backward (BBFB) method applied to EFIE, C. Brennan, D. Bogusevschi, IEEE Antenna and Propagation Symposium 2006, New Mexico USA.

Symmetric Coupling of the Finite-Element and the Boundary-Element Method for Electro-Quasistatic Field Simulations

T. Steinmetz[1], N. Gödel[1], G. Wimmer[1], M. Clemens[1], S. Kurz[1], M. Bebendorf[2], and S. Rjasanow[3]

[1] Department of Electrical Engineering, Helmut-Schmidt-University, D-22043 Hamburg, Germany t.steinmetz@hsu-hh.de
[2] Mathematical Institute, University of Leipzig, D-04109 Leipzig, Germany bebendorf@math.uni-leipzig.de
[3] Mathematical Institute, Saarland University, D-66041 Saarbrcken, Germany rjasanow@num.uni-sb.de

Summary. The electrodynamic simulation of 3D high-voltage technical devices can be performed under the electro-quasistatic assumption. In order to avoid large spatial discretization domains, a Finite-Element-Method (FEM) is coupled to a Boundary-Element-Method (BEM) which implicitly asserts the electrophysical asymptotic attenuation condition. A symmetric FEM-BEM coupled formulation in time domain is presented. Numerical results are shown for the simulation of a three dimensional high-voltage application.

1 Introduction

In this paper, transient simulations under the electro-quasistatic (EQS) assumption are presented, where the time derivative of the magnetic flux density in the induction law is omitted. These simulations can be performed for the analysis of technical devices for which electromagnetic wave propagation effects are negligible and where the electric energy density of the problem is much greater than the magnetic energy density. Typically, these conditions are valid for applications from high-voltage technology or microelectronics. Electro-quasistatic simulations in the time domain have already been presented using volume-based discretization schemes e.g. in [2, 9, 12]. The methods provide the simulation of inhomogenous and nonlinear material behavior. However, the volume-based methods have in common that the simulation domain has to be bounded, implicating that one has to provide boundary conditions on the boundary of the domain. This stands for a specific disadvantage of these methods, especially in the simulation of free-standing high-voltage devices for which usual boundary conditions cannot model the electrical far field accurately. In the frame of this paper, this problem is addressed by a symmetric coupling of the FEM and the BEM. Symmetric BEM or FEM-BEM formulations, respectively, have been proposed e.g. in [10] for the solution of stationary flow field problems, in [11] for the solution of magnetic field problems or in [6] for the computation of eddy current problems. Here, a symmetric FEM-BEM coupled formulation for electro-quasistatic field computations in the time domain is presented which takes the electrophysical asymptotic attenuation of the electric scalar potential, $|\varphi(\mathbf{r})| = O(|\mathbf{r}|^{-1})$ for $|\mathbf{r}| \to \infty$, into account.
The paper is organized as follows: In section 2, the governing differential equation for EQS fields is stated and the domain decomposition in use is described. The symmetric coupled

formulation is derived in section 3 while section 4 is devoted to the time discretization and solution of the resulting linear systems of equations. Numerical results are presented in section 5.

2 Domain Decomposition for Electro-Quasistatic Fields

2.1 Transient Electro-Quasistatic Fields

Introducing the electro-quasistatic assumption $\partial_t \mathbf{B}(\mathbf{r},t) = 0$ into Maxwell's equations, a scalar potential function $\varphi(\mathbf{r},t)$ exists which allows to compute the resulting irrotational electric field via $\mathbf{E}(\mathbf{r},t) = -\operatorname{grad}\varphi(\mathbf{r},t)$. As a consequence, together with Faraday's law, the governing differential equation describing electro-quasistatic fields reads

$$-\operatorname{div}\left(\left(\kappa(\varphi,\mathbf{r}) + \varepsilon(\mathbf{r})\,\partial_t\right)\operatorname{grad}\varphi(\mathbf{r},t)\right) = 0.$$

Here, the electric conductivity is denoted by κ, while the electric permittivity is denoted by ε. Because of the relevance in technical applications, the electric conductivity is assumed to depend on the governing electric field, thus on the governing electric potential φ.

2.2 Model Problem

As inhomogenous electric permittivities and field-strength dependent electric conductivities can easily be taken into account by the finite-element method while the physical attenuation condition can be incorporated by the boundary-element method, the computational domain is decomposed accordingly, see Fig. 1. In Ω_{FEM}, $\varepsilon = \varepsilon(\mathbf{r})$ as well as $\kappa = \kappa(\varphi,\mathbf{r})$ are valid,

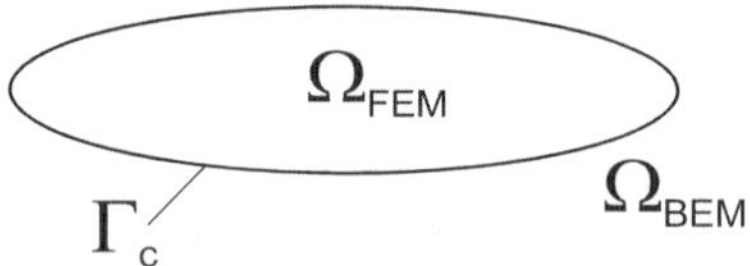

Fig. 1: Geometry of the model problem

whereas in Ω_{BEM} $\varepsilon = \varepsilon_0$ as well as $\kappa = 0$ holds. In the following, the dependence of the position vector $\mathbf{r}$ as well as the time t are omitted. With these conditions, the model problem is defined as follows:

$$-\operatorname{div}\left(\kappa(\varphi) + \varepsilon\partial_t\right)\operatorname{grad}\varphi = 0 \text{ in } \Omega_{\mathrm{FEM}}, \tag{1}$$

$$-\operatorname{div}\left(\varepsilon_0\partial_t\right)\operatorname{grad}\varphi = 0 \text{ in } \Omega_{\mathrm{BEM}}, \tag{2}$$

in the unbounded domain $\Omega = \Omega_{\mathrm{FEM}} \cup \Gamma_{\mathrm{c}} \cup \Omega_{\mathrm{BEM}}$ with the interface boundary Γ_{c}. Furthermore, $\overline{\Omega}_{\mathrm{FEM}} = \Omega_{\mathrm{FEM}} \cup \Gamma_{\mathrm{c}}$ and $\overline{\Omega}_{\mathrm{BEM}} = \Omega_{\mathrm{BEM}} \cup \Gamma_{\mathrm{c}}$ holds. The normal vector on the boundary Γ_{c} is assumed to be directed from the domain Ω_{FEM} to the domain Ω_{BEM}.

2.3 Finite-Element Formulation

For the closure of Ω_{FEM}, the standard variational formulation of (1) can be achieved by multiplication with a test function v and application of Green's first integral theorem:

$$\int\limits_{\Omega_{\mathrm{FEM}}} (\operatorname{grad}\varphi)\,(\kappa + \varepsilon\partial_t)\,(\operatorname{grad} v)\,\mathrm{d}\Omega - \int\limits_{\Gamma_{\mathrm{c}}} (\kappa + \varepsilon\partial_t)\,\gamma_1^{\mathrm{int}}\varphi\,\gamma_0^{\mathrm{int}} v\,\mathrm{d}\Gamma = 0, \tag{3}$$

with the interior trace operator γ_0^{int} and the operator of the interior co-normal derivative γ_1^{int}. The second integral term allows for the coupling to the boundary element formulation.

2.4 Boundary-Element Formulation

The solution of (2) in Ω_{BEM} can be formulated based on Kirchhoff's representation formula,

$$\partial_t \varphi\left(\mathbf{r}\right) = \int_{\Gamma_c} \gamma_0^{\prime\,\mathrm{ext}} \frac{1}{4\pi|\mathbf{r}-\mathbf{r}'|} \gamma_1^{\prime\,\mathrm{ext}} \partial_t \varphi\left(\mathbf{r}'\right) \mathrm{d}\Gamma' - \int_{\Gamma_c} \gamma_1^{\prime\,\mathrm{ext}} \frac{1}{4\pi|\mathbf{r}-\mathbf{r}'|} \gamma_0^{\prime\,\mathrm{ext}} \partial_t \varphi\left(\mathbf{r}'\right) \mathrm{d}\Gamma', \quad (4)$$

considering the time derivative which occurs in (2). Application of the exterior trace operator γ_0^{ext} and the operator of the exterior co-normal derivative γ_1^{ext}, respectively, on eqn. (4) results in a system of boundary integral equations

$$\begin{pmatrix} \gamma_0^{\mathrm{ext}} \partial_t \varphi \\ \gamma_1^{\mathrm{ext}} \partial_t \varphi \end{pmatrix} = \begin{pmatrix} \frac{1}{2}\mathcal{I} + \mathcal{K} & -\mathcal{V} \\ -\mathcal{D} & \frac{1}{2}\mathcal{I} - \mathcal{K}' \end{pmatrix} \begin{pmatrix} \gamma_0^{\mathrm{ext}} \partial_t \varphi \\ \gamma_1^{\mathrm{ext}} \partial_t \varphi \end{pmatrix} \quad (5)$$

with the factor $1/2$ for points on a smooth boundary [4]. In (5), the single-layer potential operator $\mathcal{V}$, the hypersingular integral operator $\mathcal{D}$ and the double-layer potential operator $\mathcal{K}$ and its adjoint $\mathcal{K}'$, respectively, are used. The identity operator is denoted by $\mathcal{I}$. Furthermore,

$$\mathcal{C}^{\mathrm{ext}} = \begin{pmatrix} \frac{1}{2}\mathcal{I} + \mathcal{K} & -\mathcal{V} \\ -\mathcal{D} & \frac{1}{2}\mathcal{I} - \mathcal{K}' \end{pmatrix}$$

is usually denoted as Caldern-projector.

3 Symmetric FEM-BEM Coupling

3.1 Continuous Formulation

The BEM formulation (5) can be coupled to the finite-element formulation (3) by expressing $\gamma_1^{\mathrm{int}} \varphi$ by the second boundary integral equation of the system (5), [4, 3], and by applying the interface conditions:

$$\gamma_0^{\mathrm{int}} \varphi = \gamma_0^{\mathrm{ext}} \varphi, \quad (6)$$

$$(\kappa + \varepsilon \partial_t)\, \gamma_1^{\mathrm{int}} \varphi = (\varepsilon_0 \partial_t)\, \gamma_1^{\mathrm{ext}} \varphi. \quad (7)$$

Eqn. (7) expresses the normal continuity of the total (conduction + displacement) current density. According to (7), substituting $(\varepsilon_0 \partial_t)\, \gamma_1^{\mathrm{ext}} \varphi$ for $(\kappa + \varepsilon \partial_t)\, \gamma_1^{\mathrm{int}} \varphi$ in (3), and inserting $\varepsilon_0 \gamma_1^{\mathrm{ext}} \partial_t \varphi$ from (5) yields the variational equation

$$\int_{\Omega_{\mathrm{FEM}}} (\operatorname{grad} \varphi)\, (\kappa + \varepsilon \partial_t)\, (\operatorname{grad} v)\, \mathrm{d}\Omega -$$

$$\int_{\Gamma_c} \varepsilon_0 \left(-\mathcal{D} \gamma_0^{\mathrm{ext}} \partial_t \varphi + \left(\frac{1}{2}\mathcal{I} - \mathcal{K}' \right) \gamma_1^{\mathrm{ext}} \partial_t \varphi \right) \gamma_0^{\mathrm{int}} v \, \mathrm{d}\Gamma = 0. \quad (8)$$

Another variational equation is obtained from the first equation of (5), by multiplication with another test function τ:

$$\int_{\Gamma_c} \varepsilon_0 \left(\left(-\frac{1}{2}\mathcal{I} + \mathcal{K} \right) \gamma_0^{\mathrm{ext}} \partial_t \varphi - \mathcal{V} \gamma_1^{\mathrm{ext}} \partial_t \varphi \right) \tau \, \mathrm{d}\Gamma = 0. \quad (9)$$

By (8) and (9), the coupled scalar potential problem is described in the unbounded domain $\Omega = \Omega_{\mathrm{FEM}} \cup \Gamma_c \cup \Omega_{\mathrm{BEM}}$.

3.2 Discrete Formulation

The variational problem given with eqns. (8) and (9) can be discretized using the Galerkin scheme with piecewise linear basis functions for φ and piecewise constant basis functions for $\gamma_1^{\mathrm{ext}}\varphi$. This proceeding results in finite-element stiffness matrices $\mathbf{C}$ for the electrical conductivity and $\mathbf{P}$ for the electrical permittivity. The potential operators of the boundary integral equations are discretized using the Galerkin-scheme, too, resulting in the single-layer potential matrix $\mathbf{V}$, the hypersingular potential matrix $\mathbf{D}$ and the double-layer potential matrix $\mathbf{K}$. Additionally, a mass matrix $\mathbf{M}$ is introduced. This leads to the following matrix equation, which is a system of ordinary differential equations (ODE) in the time domain:

$$\begin{pmatrix} \mathbf{C}_{\mathrm{ii}} & \mathbf{C}_{\mathrm{ic}} & 0 \\ \mathbf{C}_{\mathrm{ci}} & \mathbf{C}_{\mathrm{cc}} & 0 \\ 0 & 0 & 0 \end{pmatrix} \begin{pmatrix} \mathbf{\Phi}_{\mathrm{i}} \\ \mathbf{\Phi}_{\mathrm{c}} \\ \mathbf{t} \end{pmatrix} + \begin{pmatrix} \mathbf{P}_{\mathrm{ii}} & \mathbf{P}_{\mathrm{ic}} & 0 \\ \mathbf{P}_{\mathrm{ci}} & \mathbf{P}_{\mathrm{cc}} + \mathbf{D} & \left(-\frac{1}{2}\mathbf{M}^T + \mathbf{K}^T\right) \\ 0 & \left(-\frac{1}{2}\mathbf{M} + \mathbf{K}\right) & -\mathbf{V} \end{pmatrix} \frac{\mathrm{d}}{\mathrm{d}t} \begin{pmatrix} \mathbf{\Phi}_{\mathrm{i}} \\ \mathbf{\Phi}_{\mathrm{c}} \\ \mathbf{t} \end{pmatrix} = 0. \tag{10}$$

Herein, the vector of the degrees of freedom (DoF) is divided into three partitions. The partitions $\mathbf{\Phi}_{\mathrm{i}}$ and $\mathbf{\Phi}_{\mathrm{c}}$ represent the DoF of the scalar potential inside the domain Ω_{FEM} and on the interface boundary Γ_{c}, respectively. The third partition contains the DoF of the co-normal derivative values $\mathbf{t}$ of the scalar potential on the interface boundary Γ_{c}. The latter two partitions are needed to evaluate the scalar potential in Ω_{BEM} by Kirchhoff's representation formula.
Both matrices in eqn. (10) are symmetric for two reasons: On the one hand, the finite-element stiffness matrices $\mathbf{C}$, $\mathbf{P}$ as well as the boundary-element single-layer potential matrix $\mathbf{V}$ and the hypersingular matrix $\mathbf{D}$, respectively, are symmetric. On the other hand, the non-vanishing secondary-diagonal matrix blocks related to $-\frac{1}{2}\mathbf{M} + \mathbf{K}$ are transposed to each other.
Because the Galerkin discretization of the BEM operators in (10) results in dense matrix blocks, low-rank approximations of $\mathbf{V}$, $\mathbf{D}$ and $\mathbf{K}$, generated by the adaptive-cross-approximation (ACA, see [1, 8]), are used here. These approximated matrix blocks are stored in the $\mathcal{H}$-matrix format [5]. Hence, matrix assembly as well as matrix-vector multiplications of BEM matrix blocks are enabled with almost linear complexity.

4 Solution of the ODE System

4.1 Time Integration Scheme

The system (10) is of the form $\mathbf{H}_\kappa \mathbf{\Phi} + \mathbf{N}_\varepsilon \frac{\mathrm{d}}{\mathrm{d}t}\mathbf{\Phi} = 0$. In order to perform the time integration, a singly-diagonal-implicit-Runge-Kutta-method (SDIRK) with four internal stages is applied to the system, [2]. In each stage of the time integrator, a nonlinear system of equations,

$$\left(\mathbf{N}_\varepsilon + a_{ii}\Delta t \mathbf{H}_\kappa\left(\mathbf{\Phi}_i^n\right)\right)\mathbf{\Phi}_i^n = \mathbf{N}_\varepsilon \left(\mathbf{\Phi}^{(n)} + \sum_{j=1}^{i-1} a_{ij}\mathbf{\Phi}'^{(n)}_j\right), \tag{11}$$

has to be solved. Here, $\mathbf{\Phi}'^{(n)}_i = \left(\mathbf{\Phi}_i^{(n)} - \mathbf{\Phi}^{(n)} - \sum_{j=1}^{i-1} a_{ij}\mathbf{\Phi}'^{(n)}_j\right)/a_{ii}$ and $\mathbf{\Phi}_4^{(n)} = \mathbf{\Phi}^{(n+1)}$ hold. In (11), the time step length is denoted by Δt while a_{ij} denotes coefficients of the specific SDIRK3(2) method in use.

4.2 Solution of the Linear Systems

The resulting systems of linear equations (11) are, as a rule, symmetric but indefinite (the matrix blocks $\mathbf{C}$, $\mathbf{P}$ and $\mathbf{D}$ are positive definite whereas the matrix block $-\mathbf{V}$ is negative definite). Therefore, a biconjugate-gradient-stabilized (BiCGStab) method is used for their numerical

solution. An algebraic multigrid method (AMG) is used as a preconditioner for the FEM degrees of freedom solely, whereas the BEM degrees of freedom are currently not preconditioned.
The application of an efficient preconditioner of the whole system matrix will improve the solution process further on. Its construction is subject of our ongoing research, but is not in the focus of the paper at hand.

5 Numerical Results

5.1 Charged Sphere

The first numerical example shows a sphere with radius r_0. Inside the sphere, the scalar potential is set to $1\,\mathrm{V}\cdot\sin(\omega t)$. A high electric permittivity results in a nearly homogenous distribution of the scalar potential inside the sphere and on its surface. The scalar potential is computed by the proposed FEM-BEM coupling with the surface of the ball as coupling boundary. In the frame of this example, the electric conductivity is set to zero in order to compare the simulation result with an analytical solution. The discrete model consists of 4.623 nodes in the domain Ω_{FEM} and 1.369 nodes and 2.734 triangles on Γ_{c}, respectively. Fig. 2 shows the scalar potential distribution at a time instant of maximum voltage at Γ_{c}. Inside the sphere, the scalar

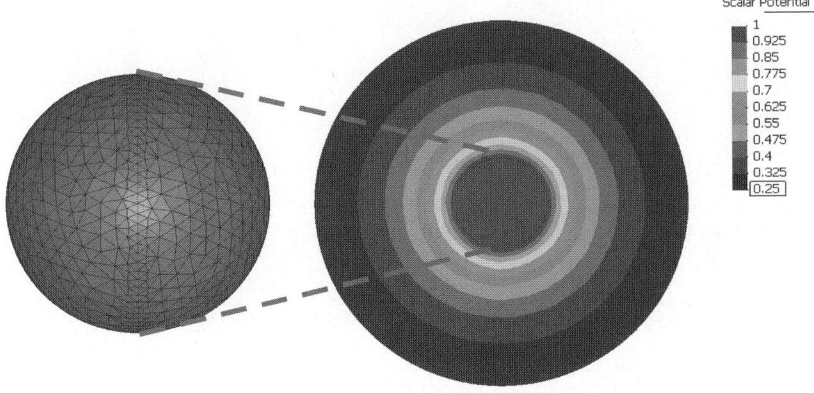

Fig. 2: Geometry (left) and scalar electric potential (right)

potential is constant, while outside it shows the expected $O\left(|\mathbf{r}|^{-1}\right)$ characteristic. In Fig. 3 the simulation result of this example is compared to its related analytical solution. The relative error of the simulation result is found to be of the order of 10^{-3} which could be diminished further on by an enhanced accuracy for the ACA matrix compression (the ACA accuracy used here is set to 10^{-3}) and a more refined discretization.

5.2 High-Voltage Surge Arrester

A 3D high-voltage surge arrester, a common technical device in the high-voltage technology for outdoor installations is a structure which mainly consists of metal, insulators and varistor material, is presented. In this EQS simulation, the electrical conductivity of the metal-oxide varistor material is taken into account by a nonlinear characteristic given in [12]. As excitation, the top of the device is set to a sinusoidal high-voltage of 471 kV with an angular frequency of 50 Hz while its stand is set to 0 V. The discrete model consists of 33.964 nodes in the domain Ω_{FEM} while the coupling boundary is discretized with 5.972 nodes and 2.988 triangles. The

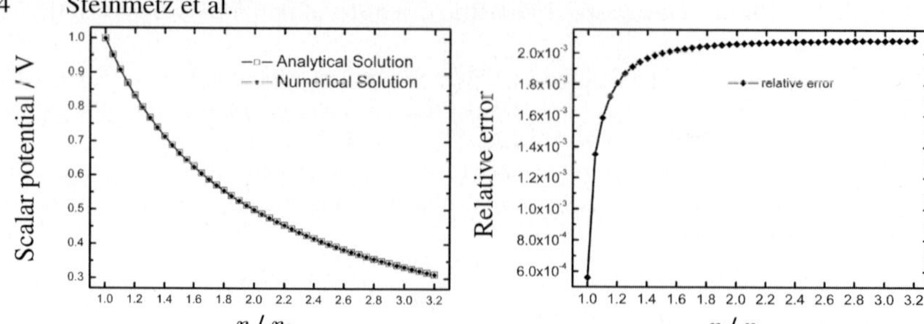

Fig. 3: Numerical and analytical computed solution (left), relative error of the numerical solution (right)

application of the ACA results in a reduction of the memory requirement to 10-25% of the uncompressed matrix blocks. Fig. 4 shows the geometry of this problem as well as the scalar

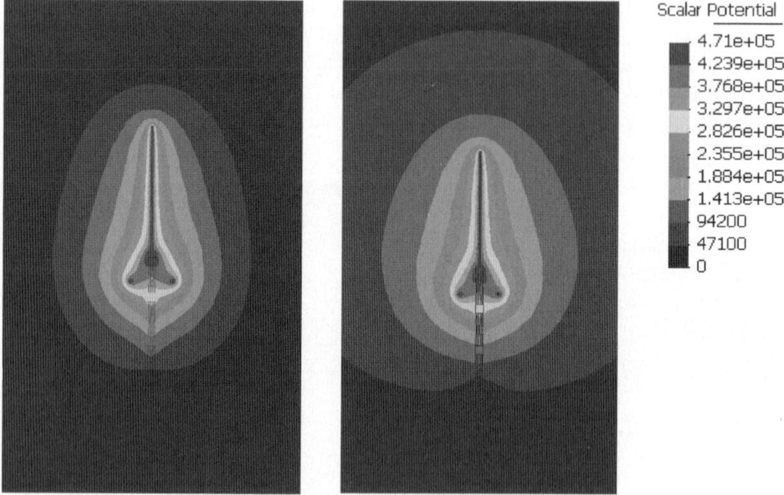

Fig. 4: Geometry, scalar electric potential computed by FEM and by FEM-BEM (from left). While the boundary is set to 0 V in the FEM simulation, the effect of the open boundary in the FEM-BEM simulation is obvious.

potential distribution at a time instant of maximum excitation voltage computed by a pure FEM simulation as well as by a simulation based on the proposed FEM-BEM coupling. For the FEM simulation, the boundary is provided with a Dirichlet boundary condition of 0 V [7].

6 Conclusion

A symmetric FEM-BEM coupled formulation for transient simulations of electro-quasistatic fields in the time domain was proposed. This formulation takes local nonlinear material behavior as well as the physical attenuation of the scalar potential condition for regular charge distributions into account. Discretization leads to symmetric, but in general indefinite linear systems of equations. Numerical results of a validation example and a realistic three dimensional technical application are presented.

References

1. Bebendorf, M., Rjasanow, S.: Adaptive low-rank approximation of collocation matrices. Computing **70** (2003) 1–24
2. Clemens, M., Wilke, M., Benderskaya, G., De Gersem, H., Koch, W., Weiland, T.: Transient Electro-Quasi-Static Adaptive Simulation Schemes. IEEE Trans. Mag. **240** (2004) 1294–1297
3. Costabel, M.: Symmetric Methods for the Coupling of Finite Elements and Boundary Elements. Boundary Elements IX Springer (1987) 411–420
4. Steinbach, O.: Numerische Nherungsverfahren fr elliptische Randwertprobleme. Teubner, Stuttgart (2003)
5. Hackbusch, W.: A sparse matrix arithmetic based on H-matrices. Part I: Introduction to H-matrices. Computing **62** (1999) 89–108
6. Hiptmair, R.: Symmetric coupling for eddy current problems. Report 148 (2000), SFB 382, University Tbingen
7. IEC 60099c-4-am2: Surge Arresters - Part 4: Metal-Oxide Surge Arresters Without Gaps for a.c. Systems
8. Kurz, S., Rain, O., Rjasanow, S.: The Adaptive Cross-Approximation Technique for the 3-D Boundary-Element Method. IEEE Trans. Mag. **38** (2002) 421–424
9. Preis, K.: Numerical Computation of Transient Quasistatic Electric Fields. Proc. CEM 2002, IEE professional network 2002, Bournemouth, UK
10. Ostrowski, J., Andjelic, Z., Bebendorf, M., Cranganu-Cretu, B., Smajic, J.: Fast BEM-Solution of Laplace Problems with H-Matrices and ACA. IEEE Trans. Mag. **42** (2006) 627–630
11. Kuhn, M., Steinbach, O.: Symmetric coupling of finite and boundary elements for exterior magnetic field problems. Math. Meth. Appl. Sci. **25** (2002) 357–371
12. Steinmetz, T., Helias, M., Wimmer, G. Fichte, L.O., Clemens, M.: Electro-Quasistatic Field Simulations Based on a Discrete Electromagnetism Formulation. IEEE Trans. Mag. **42** (2006) 755–758

Computational Errors in Hysteresis Preisach Modelling

Valentin Ionita[1] and Lucian Petrescu[1]

Univ. "Politehnica" of Bucharest, Electrical Eng. Dept.
Spl. Independentei-313, Bucharest, 060042, Romania **vali@mag.pub.ro**

Abstract - The paper analyzes the influence of the computational errors on the accuracy of magnetic material modelling with scalar Preisach model. The numerical tests on magnetic recording media allow the correct choosing of numerical algorithms for model parameter identification.
Keywords - Hysteresis modelling, Preisach model, computational errors, magnetic materials.

1 Introduction

The magnetic excitation systems are very useful in technical applications. Their design starts from the required distribution (in time and space) of the magnetic field, in order to produce the desired effects. But, this behaviour depends on the magnetic properties of the target object and requires a material model that includes the hysteresis phenomenon.
For technical applications, the Preisach hysteresis model [May91], [Tor99] offers a good rate between the computational efficiency and the result accuracy [LPA00]. The classical Preisach model considers that a ferromagnetic material is made up of dipoles (hysterons) having a magnetic behavior described by a rectangular hysteresis cycle. The distribution of these elementary operators with respect to their up- and down-switching values *(a,b)* is represented by Preisach function *P(a,b)* and it identifies the modeled material. The magnetization (model output) is computed by the superposition of the hysteron contributions. In this way, the material evolution can be followed in the Preisach triangle ($-H_s \leq b \leq a \leq +H_s$), H_s being the saturation magnetic field. This evolution corresponds to a moving staircase line between the areas corresponding to the positive and the negative saturated hysterons, in the Preisach triangle; the staircase line depends on all the previous values of the magnetic field *H* (model input).
The Preisach function identification may be done by analytical or numerical approximation. In the first case, one can determine the Preisach function by identifying the parameters of particular density functions (e.g. a factored -Lorentzian or a lognormal-Gauss distributions [Tor99]); it presents unpredictable modeling errors because there is not a real justification for assuming a particular distribution function. The numerical approximation involves a step-function defined on the meshed Preisach triangle and may use a limited set of experimental data [HR02].
The modelling errors may be: intrinsic model errors (according to the Preisach' hysteresis theory), experimental errors (e.g. measurement noise) or intrinsic computational errors. The computational errors of the model parameter identification influence the model accuracy in any electromagnetic field computation that uses it [DSA06]. One will assume that the Preisach function is identified in a numerical form, started from a set of experimental FORCs (first-

order reversal curves) which are obtained for bank cards, subway tickets, floppy disks and hard disks, with a vibrating sample magnetometer (VSM 7304 LakeShore$^{(R)}$). The FORCs number imposes the number of cells in the Preisach triangle mesh. The non-intrinsic experimental influences are controlled by choosing the maximum applied magnetic field and the field step, which determine the Preisach triangle boundary and meshing. Then, the numerical Preisach distribution is used in the computational model by superposing each cell contribution (identified from FORCs by solving the equation system using sequential substitution or direct methods) or by using the Everett integrals. Each procedure involves specific numerical errors. The impact of these experimental and numerical errors is presented for different histories of the applied magnetic field.

2 Computational Errors Generated by Experimental Data

The paper is focused on the computational errors of the classical Preisach identification procedure, in order to minimize them, but the origin of these errors may be in the used experimental data. The experimental FORCs are obtained by a vibrating sample magnetometer (VSM) for magnetic recording materials: bank cards, access card tapes, floppy disks. In these cases, the scalar Preisach model can be used, but the conclusions are also valid for any generalized model [IP06].
The experimental setup uses a thin sample, having (4x4) mm^2, which is vibrated into the airgap between the two poles of the VSM electromagnet. The applied magnetic field, which is parallel with the sample surface, is controlled by the current passing through the electromagnet coils and is measured with a Hall probe. For each field value, the corresponding magnetic moment of the sample is measured. The all measurement process is automated, the desired succession of the applied field values being built before the experiment, with a dedicated software on the computer that controls the VSM. The FORCs' measurement starts from the positive saturation. The applied field is decreased to the starting point of the first FORC; then, it is gradually increased to the saturation value, obtaining the first FORC. The process is repeated for the next FORC, which has the starting point for a smaller value of the magnetic field, and so on. The quasi-stationary variation of the magnetic field uses the desired field step. The measurement parameters influence the model identification. A correct value of the saturation magnetic field is usually high for magnetic recording materials and imposes a large number of FORCs. Our tests show that identification with 80 FORCs is more sensitive to the numerical errors, comparing to identification with 40 FORCs: for floppy disk sample, the conditioning number of the algebraic system matrix that must be solved is nc=1020 vs. nc=287 for 40 FORCs. At the same time, the equipment errors in fixing each applied magnetic field value have a greater influence if the field step is smaller.
A second problem is the presence of the reptation phenomenon and the saturation of the VSM yoke [Fio04], which affects the FORCs measurement. Indeed, the measurement of 5 consecutive cycles shows a relative mean increasing of the reversal field value, which depends of the magnetization level: for bank card sample, one obtains 0.4 % for saturation zone and 11.3 % for small negative magnetization. A solution could be the normalization of each FORC by its maximum value, before proceeding to the identification.
The equipment noise could also affect the experimental data accuracy, especially for small samples with weak magnetic moment. The FORCs smoothing can improve the numerical identification, but it changes the numerical hysteresis curves. For example, Figure 1 shows the effect of a low-pass filtering by calculating an average of adjacent points, for a subway magnetic ticket sample having (4x4) mm^2. The filtering reduces the local peaks of the Preisach function - see Figure 2 for unfiltered experimental data and Figure 3 for filtered data. The distribution shape shows that a numerical Preisach function is more accurate than an analytical distribution function [LPA00].

3 Numerical Errors

The identification assumes that the Preisach function is constant in each cell of the meshed triangle. These values are computed from a linear algebraic equation system, either by sequential substitution, or by direct methods (e.g. pivoting Gauss) because the FORCs number is small (less than 100) in order to have an efficient computational model [DSA06]. The obtained Preisach function presents small differences (10^{-12} % in only 6 cells) between the two methods of the system solving. The use of Everett function instead of Preisach function diminishes, by double integration, the errors. A good compromise can be obtained if the computed unknowns are the values of the Preisach function integral on each cell of the meshed Preisach triangle. The local peaks can produce local anomaly in numerical curves, like in Figure 4.

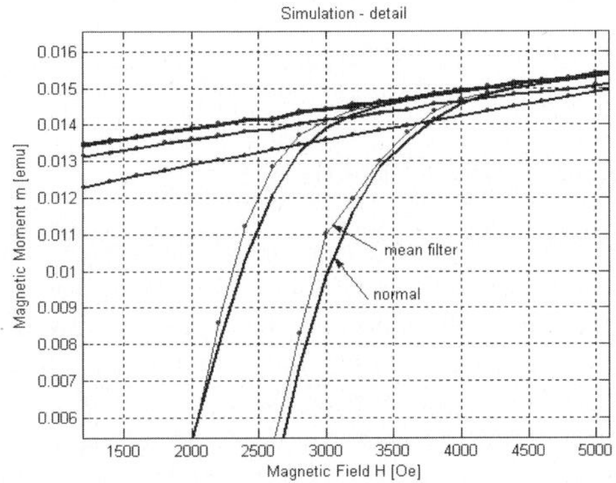

Fig. 1: Effect of the experimental data filtering on numerical simulation of subway magnetic ticket

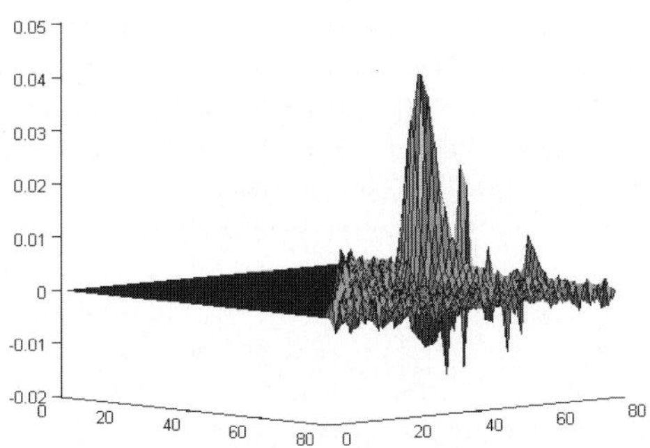

Fig. 2: Preisach function for identification with 80 FORCs

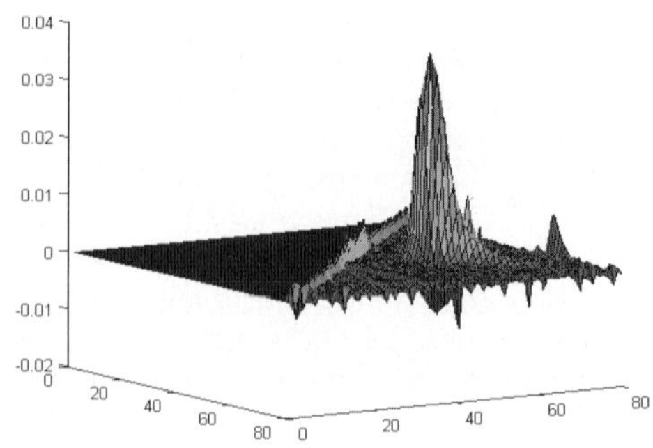

Fig. 3: Preisach function for identification with 80 filtered FORCs

4 Testing of Preisach Model Accuracy

The numerical simulation follows different histories of the applied magnetic field: evolution on first-order asymmetrical cycles (with constant or variable field step) and evolution on ascending and descending curves of different order. The experimental and numerical curves for a magnetic bank card sample are presented in Figure 4, 5 and 6. The model accuracy is sufficient for a technical purpose, like the electromagnetic field computation in devices with hysteretic magnetic materials. Significant errors could arise in the hysteresis cycle reversal points if the magnetic field step is variable and the magnetization (model output) must be interpolated on the Preisach triangle mesh (see Fig.5). A variable step meshing is better, but it could be prohibitive from the computation point of view.
Indeed, the errors are bigger for the hysteresis curves that start from the magnetic fields values which are close to the coercive one (see also [DSA06]), especially for hard magnetic materials. The explanation is that the slope of the major hysteresis cycle around the coercive field is big and two FORCs, starting from points having the magnetic field values close to the coercive field, are very different. At the same time, for the saturation zone, the use of the same field step generates FORCs that are very close to each other and are affected by the reptation phenomenon. The conclusion is that the FORCs' number is not so important as their distribution: the starting points of FORCs must have the magnetization values that are approximately equidistant. Unfortunately, this constraint leads to a more difficult experimental setup (the field step is variable) and to a non-uniform mesh of the Preisach triangle. Perhaps, these inconveniences could be overcome if a greater accuracy is imposed in a particular application.

5 Conclusions

The study outlines the sources of errors that could occur in hysteresis modelling. The experimental data used in model identification (FORCs in our case) must be carefully analyzed and filtered for generating a smooth numerical distribution function. The FORCs number must be correlated with the saturation magnetization, but it must not be very high, because the experimental noise and the reptation influence are amplified in the ill-conditioning identification procedure. It is more convenient, from the computation point of view, to use the Everett functions and a constant mesh step.

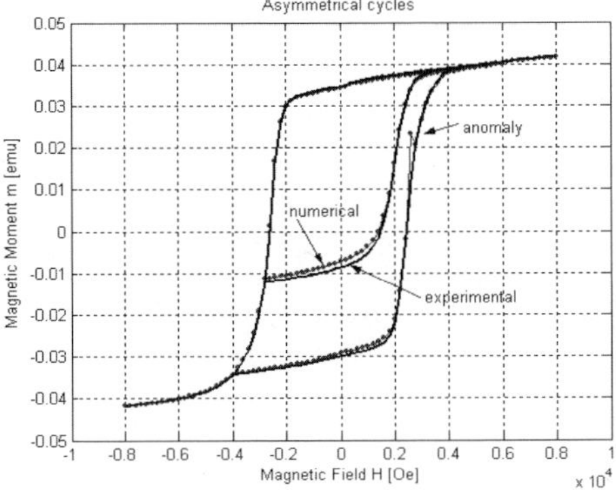

Fig. 4: Experimental and numerical asymmetrical hysteresis cycles for a bank card.

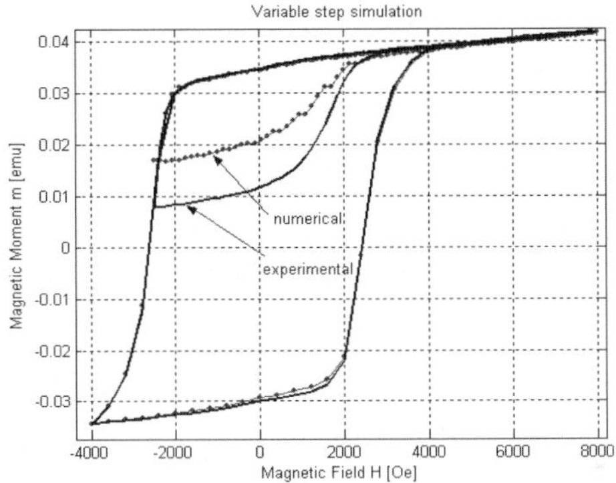

Fig. 5: Experimental and numerical hysteresis curves with variable step field for a magnetic bank card.

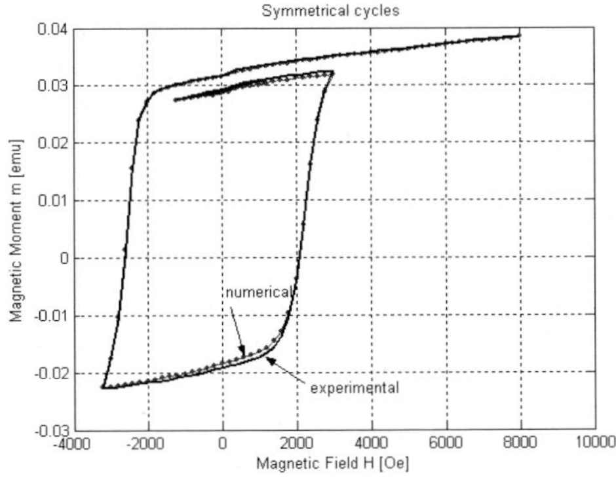

Fig. 6: Experimental and numerical symmetrical cycles for a magnetic bank card.

References

[DSA06] Dlala, E., Saitz, J., Arkkio, A.: Inverted and forward Preisach models for numerical analysis of electromagnetic field problems. IEEE Trans. on Mag., **42**, 1963-1973 (2006)

[Fio04] Fiorillo, F.: Measurement and Characterisation of Magnetic Materials. Elsevier, San Diego (2004)

[HR02] Henze, O., Rucker, W.: Identification procedures of Preisach model. IEEE Trans. on Mag., **38**, 833-836 (2002)

[IP06] Ionita, V., Petrescu, L.: Numerical advanced characterisation of magnetic recording media. J. Optoelectronics and Advanced Mat., **8**, 998-1000 (2006)

[LPA00] Liorzou, F., Phelps, B., Atherton, D.L. : Macroscopic models of magnetisation, IEEE Trans. on Mag., **36**, 418-428 (2000)

[May91] Mayergoyz, I.D.: Mathematical Models of Hysteresis. Springer Verlag, New York (1991)

[Tor99] Torre, E.D.: Magnetic Hysteresis. IEEE Press, Piscataway (1999)

Acknowledgements: This study was partially supported by A-362/2006 (MEC-CNCSIS) grant, CEEX MAGME/2006 (MEC-AMCSIT) contract and by CEEX MATHYS/2006 (MEC-MATNANTECH) contract.

Part IV

Mathematical and Computational Methods

Manifold Mapping for Multilevel Optimization*

Pieter W. Hemker and David Echeverría

Centrum voor Wiskunde en Informatica
Kruislaan 413, NL 1098 SJ Amsterdam, The Netherlands
{P.W.Hemker,D.Echeverria}@cwi.nl

Summary. We first show the idea behind a space-mapping iteration technique for the efficient solution of optimization problems. Then we show how space-mapping optimization can be understood in the framework of defect correction. We observe a difference between the solution of the optimization problem and the computed space-mapping solutions. We repair this discrepancy by exploiting the correspondence with defect correction iteration and we construct the manifold-mapping algorithm, which is as efficient as the space-mapping algorithm but converges to the accurate solution.

1 Introduction

Space mapping (Bandler et al. [1, 2]) is a technique to reduce the computing time in demanding optimization procedures by means of simple surrogate models. Space mapping makes use of both accurate (and time-consuming) models and less accurate (but cheaper) ones.
The original space-mapping procedure corresponds with right-preconditioning the coarse (inaccurate) model in order to accelerate the iterative procedure for the optimization of the fine (accurate) one. The iterative procedure used in space mapping for optimization can be understood as a defect correction iteration [3] and the convergence can be analyzed accordingly. We show that, right-preconditioning is generally insufficient and (also) left-preconditioning is needed. This leads to the improved space-mapping or 'manifold-mapping' procedure. This manifold mapping is shown in some detail in Section 5

2 Fine and coarse models in optimization

The optimization problem.

The specifications of an optimization problem are denoted by $(\mathbf{t}, \mathbf{y}) \equiv (t_i, y_i)_{i=1,\ldots,m}$. The independent variable is $\mathbf{t} \in \mathbb{R}^m$. The dependent variable $\mathbf{y} \in Y \subset \mathbb{R}^m$ represents the quantities that describe the behavior of the phenomena under study. The set Y is the *set of possible aims*.
The variable $\mathbf{y}$ does not only depend on $\mathbf{t}$ but also on control/design variables, $\mathbf{x}$. The difference between the measured data y_i and the values $y(t_i, \mathbf{x})$ may be the result of, e.g., measurement errors or the imperfection of the mathematical description.
Models that describe reality appear in several degrees of sophistication. Space mapping exploits the combination of the simplicity of the less sophisticated methods with the accuracy of the more complex ones. Therefore we distinguish the fine and the coarse model.

* Invited Paper at SCEE-2006

The fine model.

The *fine model* response is denoted by $\mathbf{f}(\mathbf{x}) \in \mathbb{R}^m$, with $\mathbf{x} \in X \subset \mathbb{R}^n$ the *fine model control variable*. The set $\mathbf{f}(X) \subset \mathbb{R}^m$ represents the fine model reachable aims. Notice that, with $n < m$, $\mathbf{f}(X)$ is an n-dimensional manifold in $Y \subset \mathbb{R}^m$. The fine model is assumed to be *accurate* but *expensive* to evaluate. For the optimization problem a *fine model cost function* $\| \mathbf{f}(\mathbf{x}) - \mathbf{y} \|$ should be minimized. So we look for

$$\mathbf{x}^* = \underset{\mathbf{x} \in X}{\operatorname{argmin}} \| \mathbf{f}(\mathbf{x}) - \mathbf{y} \| \,. \tag{1}$$

A design problem, characterized by the model $\mathbf{f}(\mathbf{x})$, the aim $\mathbf{y} \in Y$, and the space of possible controls $X \subset \mathbb{R}^n$, is a *reachable design* if the equality $\mathbf{f}(\mathbf{x}^*) = \mathbf{y}$ can be achieved for some $\mathbf{x}^* \in X$.

The coarse model.

The *coarse model* is denoted by $\mathbf{c}(\mathbf{z}) \in \mathbb{R}^m$, with $\mathbf{z} \in Z \subset \mathbb{R}^n$ the *coarse model control variable*. This model is assumed to be *cheap* to evaluate but *less accurate* than the fine model. The set $\mathbf{c}(Z) \subset \mathbb{R}^m$ is the set of *coarse model reachable aims reachable aims! coarse model*. For the coarse model we have the *coarse model cost function* $\| \mathbf{c}(\mathbf{z}) - \mathbf{y} \|$ and we denote its minimizer by $\mathbf{z}^*$,

$$\mathbf{z}^* = \underset{\mathbf{z} \in Z}{\operatorname{argmin}} \| \mathbf{c}(\mathbf{z}) - \mathbf{y} \| \,. \tag{2}$$

The space-mapping function.

The similarity or discrepancy between the responses of two models is expressed by the *misalignment function* $r(\mathbf{z}, \mathbf{x}) = \| \mathbf{c}(\mathbf{z}) - \mathbf{f}(\mathbf{x}) \|$. For a given $\mathbf{x} \in X$ it is useful to know which $\mathbf{z} \in Z$ yields the smallest discrepancy. Therefore, the *space-mapping function* $\mathbf{p} : X \subset \mathbb{R}^n \to Z \subset \mathbb{R}^n$ is introduced,

$$\mathbf{p}(\mathbf{x}) = \underset{\mathbf{z} \in Z}{\operatorname{argmin}}\, r(\mathbf{z}, \mathbf{x}) = \underset{\mathbf{z} \in Z}{\operatorname{argmin}} \| \mathbf{c}(\mathbf{z}) - \mathbf{f}(\mathbf{x}) \| \,. \tag{3}$$

Perfect mapping.

To identify the cases where the accurate solution $\mathbf{x}^*$ is related with the less accurate solution $\mathbf{z}^*$ by the space mapping function, a space-mapping function $\mathbf{p}$ is called a *perfect mapping* iff $\mathbf{z}^* = \mathbf{p}(\mathbf{x}^*)$.
We notice that *perfection* is not a property of the space-mapping function alone, but it also depends on the data $\mathbf{y}$ considered. A space-mapping function can be perfect for one data set but imperfect for a different data set, and if a design is reachable a space mapping is always perfect irrespective of the coarse model used.

3 Primal and dual space-mapping solutions

In literature many space mapping based algorithms can be found [1, 2], where two types can be distinguished: the primal and the dual.
The *primal* space-mapping approach seeks for a solution of the minimization problem

$$\mathbf{x}_p^* = \underset{\mathbf{x} \in X}{\operatorname{argmin}} \|\mathbf{p}(\mathbf{x}) - \mathbf{z}^*\| \,. \tag{4}$$

The dual determines

$$\mathbf{x}_d^* = \underset{\mathbf{x}\in X}{\operatorname{argmin}} \; |\!|\!| \, \mathbf{c}(\mathbf{p}(\mathbf{x})) - \mathbf{y} \, |\!|\!| \, , \tag{5}$$

where we can recognize $\mathbf{c}(\mathbf{p}(\mathbf{x}))$ as a *surrogate model.*
Both approaches coincide when $\mathbf{z}^* \in \mathbf{p}(X)$ and $\mathbf{p}$ is injective. If, in addition, the mapping is perfect both $\mathbf{x}_p^*$ and $\mathbf{x}_d^*$ are equal to $\mathbf{x}^*$. However, in general the space-mapping function $\mathbf{p}$ will not be perfect, and hence, a space-mapping based algorithm may *not* yield the solution of the fine model optimization. The principle of the approach is summarized in Figure 1.

The surrogate model: $\mathbf{c}(\mathbf{p}(\mathbf{x})) \approx \mathbf{f}(\mathbf{x})$.

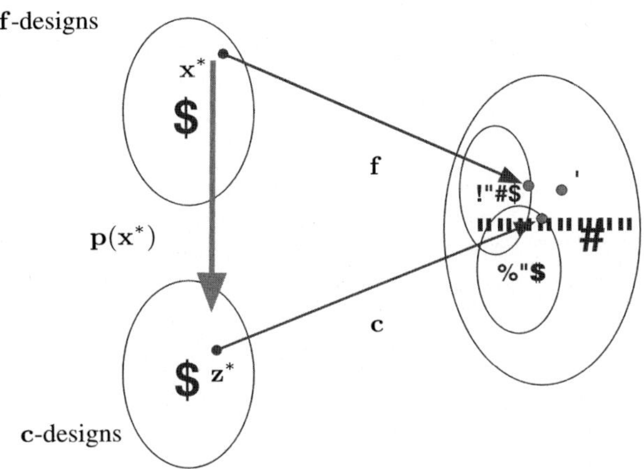

Fig. 1: The space mapping function $\mathbf{p}(\mathbf{x}) = \underset{\mathbf{z}\in Z}{\operatorname{argmin}} \; |\!|\!| \, \mathbf{c}(\mathbf{z}) - \mathbf{f}(\mathbf{x}) \, |\!|\!| \, .$

4 Defect correction iteration

The efficient solution of a complex problem by the iterative use of a simpler one, is known since long in computational mathematics as defect correction iteration [3].
To solve a nonlinear operator equation

$$\mathcal{F}\mathbf{x} = \mathbf{y}, \tag{6}$$

where $\mathcal{F} : D \subset E \to \widehat{D} \subset \widehat{E}$ is a continuous, generally nonlinear operator and E and $\widehat{E}$ are Banach spaces, defect correction iteration reads

$$\begin{cases} \mathbf{x}_0 &= \widetilde{\mathcal{G}}_0 \, \mathbf{y} \, , \\ \mathbf{x}_{k+1} &= \widetilde{\mathcal{G}}_{k+1} \left(\widetilde{\mathcal{F}}_k \, \mathbf{x}_k \; - \; \mathcal{F} \, \mathbf{x}_k + \mathbf{y} \right) \, , \end{cases} \tag{7}$$

where $\widetilde{\mathcal{F}}_k$ is a simpler version of $\mathcal{F}$ and $\widetilde{\mathcal{G}}_k$ is the (simple-to-evaluate) left-inverse of $\widetilde{\mathcal{F}}_k$.
For our optimization problems, where the design may be not reachable, $\mathbf{y} \in \widehat{D}$ but $\mathbf{y} \notin \mathcal{F}(D)$, so that no solution for (6) exists. We want to find the solution of (1), or

$$\mathbf{x}^* = \operatorname{argmin}_{\mathbf{x}\in D} \, \|\mathcal{F}\mathbf{x} - \mathbf{y}\|_{\widehat{E}} \, ,$$

which we associate with it a defect correction process for iterative optimization by taking $E = \mathbb{R}^n$, $\widehat{E} = \mathbb{R}^m$, $D = X$, $\widehat{D} = Y$ and by substitution of the operators:

$$
\begin{aligned}
\mathcal{F}\mathbf{x} = \mathbf{y} \quad &\Leftrightarrow \mathbf{f}(\mathbf{x}) = \mathbf{y}\,,\\
\mathbf{x} = \mathcal{G}\mathbf{y} \quad &\Leftrightarrow \mathbf{x} = \underset{\xi\in E}{\operatorname{argmin}}\ \|\mathbf{f}(\xi) - \mathbf{y}\|_{\widehat{E}}\,,\\
\widetilde{\mathcal{F}}_k\mathbf{x} = \mathbf{y} \quad &\Leftrightarrow \mathbf{c}(\overline{\mathbf{p}}_k(\mathbf{x})) = \mathbf{y}\,,\\
\mathbf{x} = \widetilde{\mathcal{G}}_k\mathbf{y} \quad &\Leftrightarrow \mathbf{x} = \underset{\xi\in E}{\operatorname{argmin}}\ \|\mathbf{c}(\overline{\mathbf{p}}_k(\xi)) - \mathbf{y}\|_{\widehat{E}}\,.
\end{aligned}
\tag{8}
$$

Here $\overline{\mathbf{p}}_k$ is *not* the space-mapping function but an arbitrary (easy to compute) bijection, e.g., the identity if $X = Z$. Notice that, in principle, also $\mathbf{c} = \mathbf{c}_k$ might be adapted during the iteration.
With (8) we derive from (7) the defect-correction iteration scheme for optimization:

$$\mathbf{x}_0 = \operatorname{argmin}_{\mathbf{x}\in X}\ \|\mathbf{c}(\overline{\mathbf{p}}_0(\mathbf{x})) - \mathbf{y}\|\,, \tag{9}$$

$$\mathbf{x}_{k+1} = \operatorname{argmin}_{\mathbf{x}\in X}\ \|\mathbf{c}(\overline{\mathbf{p}}_{k+1}(\mathbf{x})) - \mathbf{c}(\overline{\mathbf{p}}_k(\mathbf{x}_k)) + \mathbf{f}(\mathbf{x}_k) - \mathbf{y}\|\,. \tag{10}$$

In this iteration every minimization involves the surrogate model, $\mathbf{c}\circ\overline{\mathbf{p}}_k$.

Orthogonality and the need for left-preconditioning.

For the stationary points of the above process, $\lim_{k\to\infty}\mathbf{x}_k = \overline{\mathbf{x}}$, we can derive [5]:

$$\mathbf{f}(\overline{\mathbf{x}}) - \mathbf{y} \in \mathbf{c}(Z)^{\perp}(\mathbf{z}^*)\,. \tag{11}$$

Like the space-mapping methods, the above iteration has the disadvantage that, in general, the fixed point does not coincide with the solution of the fine model minimization problem. This is due to the fact that the approximate solution $\overline{\mathbf{x}}$ satisfies (11), whereas the (local) minimum $\mathbf{x}^*$ satisfies

$$\mathbf{f}(\mathbf{x}^*) - \mathbf{y} \in \mathbf{f}(X)^{\perp}(\mathbf{x}^*)\,.$$

Hence, differences between $\overline{\mathbf{x}}$ and $\mathbf{x}^*$ will be larger for larger distances between $\mathbf{y}$ and the sets $\mathbf{f}(X)$ and $\mathbf{c}(Z)$ and for larger angles between the linear manifolds tangential at $\mathbf{c}(Z)$ and $\mathbf{f}(X)$ near the optima.
By these orthogonality relations we see that it is advantageous, both for the conditioning of the problem and for the minimization of the residual, if the manifolds $\mathbf{f}(X)$ and $\mathbf{c}(Z)$ are found parallel in the neighborhood of the solution. However, by space mapping or by right-preconditioning, the relation between $\mathbf{f}(X)$ and $\mathbf{c}(Z)$ remains unchanged. This causes that the fixed point of traditional space mapping does, generally, not correspond with $\mathbf{x}^*$. This, however, can be improved by the introduction of an additional left-preconditioner. Therefore we consider such a preconditioner $\mathbf{S}$ so that near $\mathbf{c}(\mathbf{z}^*) \in Y$ the manifold $\mathbf{c}(Z) \subset Y$ is mapped onto $\mathbf{f}(X) \subset Y$:

$$\mathbf{f}(\mathbf{x}) \approx \mathbf{S}(\mathbf{c}(\overline{\mathbf{p}}(\mathbf{x})))\,.$$

In the next section we propose our *manifold-mapping* algorithm, where an affine operator maps $\mathbf{c}(Z)$ onto $\mathbf{f}(X)$ in the neighborhood of the solution. More precisely: it maps $\mathbf{c}(\overline{\mathbf{p}}(\mathbf{x}_k))$ to $\mathbf{f}(\mathbf{x}_k)$ and it approximately maps one tangential linear manifold onto the other. This restores the orthogonality relation $\mathbf{f}(\overline{\mathbf{x}}) - \mathbf{y} \in \mathbf{f}(X)^{\perp}(\overline{\mathbf{x}})$. Thus it improves significantly the traditional space-mapping approach and makes the solution $\mathbf{x}^*$ a stationary point of the iteration.

5 Manifold Mapping, the improved space mapping algorithm

We introduce the affine mapping $\mathbf{S}: Y \to Y$ such that $\mathbf{S}\,\mathbf{c}(\overline{\mathbf{z}}) = \mathbf{f}(\mathbf{x}^*)$ for a proper $\overline{\mathbf{z}} \in Z$, and the linear manifold tangential to $\mathbf{c}(Z)$ in $\mathbf{c}(\overline{\mathbf{z}})$ maps onto the one tangential to $\mathbf{f}(X)$ in $\mathbf{f}(\mathbf{x}^*)$. Because both $\mathbf{f}(X)$ and $\mathbf{c}(Z)$ are n-dimensional manifolds in $\mathbb{R}^m$, the mapping $\mathbf{S}$ can be described by

$$\mathbf{S}\,\mathbf{v} = \mathbf{f}(\mathbf{x}^*) + S\ (\mathbf{v} - \mathbf{c}(\overline{\mathbf{z}}))\,,$$

where S is an $m\times m$-matrix of rank n. A full rank $m\times m$-matrix S can be constructed, which has a well-determined part of rank n, while a remaining part of rank $m-n$ is free to choose. Because of the supposed similarity between the models $\mathbf{f}$ and $\mathbf{c}$ we keep the latter part close to the identity. The meaning of the mapping $\mathbf{S}$ is illustrated in Figure 2.
So we propose the following algorithm, where the optional right-preconditioner $\overline{\mathbf{p}}: X \to Z$ is still an arbitrary non-singular operator, which can be adapted to the problem. Often we will simply take $\overline{\mathbf{p}} = I$, the identity.

1. Set $k=0$, set $S_0 = I$ the $m\times m$ identity matrix, and compute
$$\mathbf{x}_0 = \text{argmin}_{\mathbf{x}\in X} \,\| \mathbf{c}(\overline{\mathbf{p}}(\mathbf{x})) - \mathbf{y}\| \,.$$
2. Compute $\mathbf{f}(\mathbf{x}_k)$ and $\mathbf{c}(\overline{\mathbf{p}}(\mathbf{x}_k))$.
3. If $k>0$, with $\Delta\mathbf{c}_i = \mathbf{c}(\overline{\mathbf{p}}(\mathbf{x}_{k-i})) - \mathbf{c}(\overline{\mathbf{p}}(\mathbf{x}_k))$ and $\Delta\mathbf{f}_i = \mathbf{f}(\mathbf{x}_{k-i}) - \mathbf{f}(\mathbf{x}_k)$, $i = 1,\cdots,\min(n,k)$, we define ΔC and ΔF to be the rectangular $m\times\min(n,k)$-matrices with respectively $\Delta\mathbf{c}_i$ and $\Delta\mathbf{f}_i$ as columns.
The generalized singular value decomposition of these (rectangular) matrices is $\Delta C = U_c\Sigma_c V^T$ and $\Delta F = U_f\Sigma_f V^T$, with U_c, U_f orthonormal, Σ_c and Σ_f diagonal and V non-singular.
4. The next iterant is computed as
$$\begin{aligned}\mathbf{x}_{k+1} = \text{argmin}_{\mathbf{x}\in X} \,\| &\mathbf{c}(\overline{\mathbf{p}}(\mathbf{x})) - \mathbf{c}(\overline{\mathbf{p}}(\mathbf{x}_k)) + \\ &\left[U_c\Sigma_c\Sigma_f^{\dagger}U_f^T + (I-U_cU_c^T)(I-U_fU_f^T)\right](\mathbf{f}(\mathbf{x}_k)-\mathbf{y})\| \,.\end{aligned} \tag{12}$$
5. Set $k := k+1$ and goto 2.

Here, $\Sigma_f^{\dagger}$ denotes the pseudo-inverse of Σ_f. It can be shown that (12) is asymtotically equivalent to
$$\mathbf{x}_{k+1} = \text{argmin}_{\mathbf{x}\in X} \,\| \mathbf{S}_k(\mathbf{c}(\overline{\mathbf{p}}(\mathbf{x}))) - \mathbf{y}\| \,, \tag{13}$$
where the approximate affine mapping is
$$\mathbf{S}_k\,\mathbf{v} = \mathbf{f}(\mathbf{x}_k) + S_k(\mathbf{v} - \mathbf{c}(\overline{\mathbf{p}}(\mathbf{x}_k))\,,$$
with $S_k = U_f\Sigma_f\Sigma_c^{\dagger}U_c^T + (I-U_f\,U_f^T)\,(I-U_c\,U_c^T)$.
If the above iteration converges with fixed point $\overline{\mathbf{x}}$ and mapping $\overline{\mathbf{S}}$, we have
$$\mathbf{f}(\overline{\mathbf{x}}) - \mathbf{y} \in \overline{\mathbf{S}}(\mathbf{c}(\overline{\mathbf{p}}(X)))^{\perp}(\overline{\mathbf{x}}) = \mathbf{f}(X)^{\perp}(\overline{\mathbf{x}})\,.$$

This, and the fact that $\mathbf{S}_k(\mathbf{c}(\overline{\mathbf{p}}(\mathbf{x}_k))) = \mathbf{f}(\mathbf{x}_k)$, makes that, under convergence to $\overline{\mathbf{x}}$, the fixed point is a (local) optimum of the fine model minimization.
The improved space-mapping scheme
$$\mathbf{x}_{k+1} = \underset{\mathbf{x}}{\text{argmin}} \,\| \mathbf{S}_k(\mathbf{c}(\overline{\mathbf{p}}_k(\mathbf{x})))) - \mathbf{y}\| \,,$$
can also be recognized as defect correction iteration with either $\widetilde{\mathcal{F}}_k = \mathbf{S}_k\circ\mathbf{c}\circ\overline{\mathbf{p}}_k$ and $\mathcal{F} = \mathbf{f}$ or with $\widetilde{\mathcal{F}}_k = \mathbf{S}_k\circ\mathbf{c}$ and $\mathcal{F} = \mathbf{f}\circ\overline{\mathbf{p}}_k^{-1}$.
An analysis and conditions for convergence of manifold mapping are found in [7]. To make the the algorithm more robust for ill-conditioned models, regularization can be used by means of the generalized singular value decomposition [8]. Notice that the singular value decomposition is applied to relatively small matrices so that the time for its computation is negligiable. Reports showing results of the manifold mapping technique for problems from practice can be found, e.g., in [4, 6]

The surrogate model: $\mathbf{S}_k \circ \mathbf{c} \circ \overline{\mathbf{p}} \approx \mathbf{f}$.

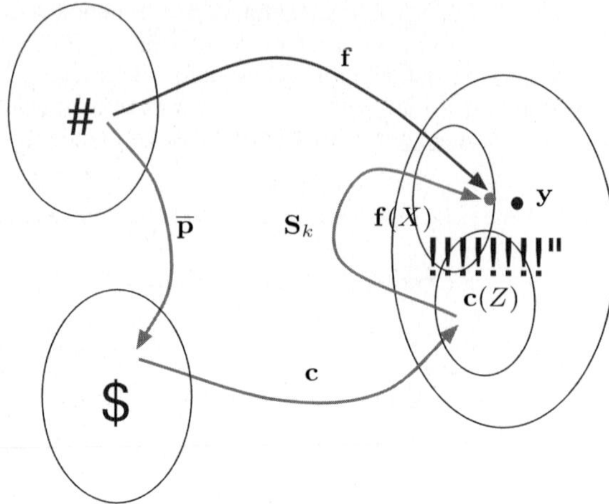

Fig. 2: Manifold Mapping.

6 Conclusion

By left preconditioning, manifold mapping improves traditional space mapping because it delivers the accurate optimum with the same computational efficiency as the space mapping algorithm.

References

1. M.H. Bakr, J.W. Bandler, K. Madsen, and J. Søndergaard. Review of the space mapping approach to engineering optimization and modeling. *Optimization and Engineering*, 1(3):241–276, 2000.
2. J.W. Bandler, Q.S. Cheng, A.S. Dakroury, A.S. Mohamed, M.H. Bakr, K. Madsen, and J. Søndergaard. Space mapping: The state of the art. *IEEE Transactions on Microwave Theory and Techniques*, 52:337–360, 2004.
3. K. Böhmer, P.W. Hemker, and H.J. Stetter. The defect correction approach. In K. Böhmer and H.J. Stetter, editors, *Defect Correction Methods: Theory and Applications*, volume 5 of *Computing Suppl.*, pages 1–32. Springer-Verlag, Berlin, Heidelberg, New York, Tokyo, 1984.
4. D. Echeverría. *Multi-level optimization: Space Mapping and Manifold Mapping*. PhD thesis, University of Amsterdam, 2007.
5. D. Echeverría and P.W. Hemker. Space mapping and defect correction. *Comp. Methods in Appl. Math.*, 5(2):107–136, 2005.
6. D. Echeverría, D. Lahaye, L. Encica, P.W. Hemker E.A. Lomonova, and A.J.A. Vandenput. Manifold-mapping optimization applied to linear actuator design. *IEEE Transactions on Magnetics*, 42(4):1183–1186, 2006.
7. P.W. Hemker and D. Echeverría. On the manifold mapping optimization technique. Technical Report MAS-R0612, CWI, Amsterdam, 2006.
8. P.W. Hemker and D. Echeverría. A trust-region strategy for manifold mapping optimization. Technical Report MAS-R0617, CWI, Amsterdam, 2006. Submitted to JCP.

Software Package for Multi-Objective Optimal Design of Electromagnetic Devices

Calin Munteanu, Gheorghe Mates, and Vasile Topa

Department of Electrotechnics, Technical University of Cluj-Napoca, C. Daicoviciu 15, 400020 Cluj-Napoca, Romania, Calin.Munteanu@et.utcluj.ro

1 Introduction

Resistive, capacitive and inductive parasitic coupling effects are very important in integrated circuit technology development and must be taken into account in the global circuit analysis and design. These values are seldom possible to be computed analytically because often we deal with complex multi-route layout geometries. Optimizing the placement of the routes inside the integrated design, in accordance with imposed constraints, one may lead to the decrease of the coupling effects.
Due to the high number of the design parameters involved, stochastic optimization methods based on Genetic Algorithms (GA) have been previously involved. The first solution proposed consisted in using a basic GA in which a single global fitness function is built up from the partial objectives. The main drawback of this method is that even if the global fitness function decreases during the optimal design process, there are partial objectives that get an important increase.
A first trial to eliminate this drawback has been done using the method of objective weighting [TMM03]. The drawback of this method is the fact that it requires some prior knowledge about the partial objectives behaviour in order to proper select the weighting values. This is quite difficult to be done in practice. Therefore the best solution would be to built up an algorithm that takes into account the information from partial objectives behaviour during the optimization process without any prior knowledge and also to take into account the global fitness function behaviour. The solution consists of setting-up a multi-objective optimal design algorithm based on Strength Pareto Evolutionary Algorithms (SPEA). The software package developed based on the above mentioned method, together with a practical application example will be presented in this paper.

2 Elitist Multi-Objective Optimal Design

Many industrial problems involve simultaneous optimization of several competing objectives. Usually, there is no single optimal solution, but rather a set of alternative solutions. These solutions are optimal in the wider sense that no other solutions in the search space are superior to them when all objectives are considered. They are known as Pareto optimal solutions. Mathematically, the concept of Pareto optimality can be defined as follows.
If one considers a multi objective minimisation problem with m parameters (decision variables) and n objectives:

$$Minimize: \quad \mathbf{y} = f(\mathbf{x}) = (f_1(x), f_1(x), ..., f_n(x)) \tag{1}$$

where $\mathbf{x} = (x_1, x_2, ..., x_n) \in X, \mathbf{y} = (y_1, y_2, ..., y_n) \in Y$.
A decision vector $\mathbf{a} \in X$ is said to dominate a decision vector $\mathbf{b} \in Y$ (also written as $a \succ b$) if:

$$\forall i \in \left\{\overline{1, n}\right\} : f_i(a) \leq f_i(b) \wedge \exists j \in \left\{\overline{1, n}\right\} : f_j(a) < f_j(b) \tag{2}$$

All decision vectors that are not dominated by any other decision vector are called non-dominated or Pareto - optimal. Often, there is a special interest in finding or approximating the Pareto optimal set, mainly to gain deeper insight into the problem and knowledge about alternate solutions respectively. Evolutionary Algorithms (EA) seem to be especially suited for this task, because they process a set of solutions in parallel, eventually exploiting similarities of solutions by crossover.
The algorithm implemented is based on the Strength Pareto Evolutionary Algorithms (SPEA). This algorithm introduces elitism by explicitly maintaining an external population $\overline{P}$. This population stores a fixed number of the non-dominated solutions that are found until the beginning of a simulation. At every generation, newly found non-dominated solutions are compared with the existing external population and the resulting non-dominated solutions are preserved. The SPEA does more than just preserving the elites; it also uses these elites to participate in the genetic operations along with the current population in the hope of influencing the population to steer towards good regions in the search space.
In this light, the general structure of the algorithm implemented is briefly described below [Deb01]:
Step 1. Find the best non-dominated set $A_1(P_t)$ of P_t. Copy these solutions to $\overline{P}_t$ or perform $\overline{P}_t = \overline{P}_t \cup A_1(P_t)$, where $\overline{P}_t$ is the external population at iteration *t*;
Step 2. Find the best non-dominated solutions $A_1\left(\overline{P}_t\right)$ of the modified population $\overline{P}_t$ and delete all dominated solutions or perform $\overline{P}_t = A_1(P_t)$;
Step 3. If the size of $\overline{P}_t > \overline{N}$, where $\overline{N}$ is the bounding limit of the external population size, use a clustering technique to reduce the size to $\overline{N}$. Otherwise, keep $\overline{P}_t$ unchanged. The resulting population is the external population $\overline{P}_{t+1}$ of the next generation;
Step 4. Assign fitness to each elite solution $i \in \overline{P}_{t+1}$ by using the strength $S_i = \frac{n_i}{N+1}$, where n_i is the number of the current population members that an external solution i dominates and N is the size of the population P. Then, assign fitness to each population member $j \in P_t$ by using $F_j = 1 + \sum_{i \in \overline{P}_t \wedge i \prec j} S_i$. The addition of 1 makes the fitness of any current population member P_t to be better than the fitness of any external population member $\overline{P}_t$. In this way, a solution with smaller fitness is the best;
Step 5. Apply a binary tournament selection with these fitness values (in a minimization sense), a crossover and a mutation operator to create the new population P_{t+1} of size N from the combined population $\left(\overline{P}_{t+1} \cup P_t\right)$.
There exist a number of real - parameter GA implementations, where crossover and mutation operators are applied directly to real parameter values. Since real parameters are used directly (without any string coding), solving real-parameter optimization problems is a step easier when compared to the binary-coded GA. For the optimization module has being used Simulated Binary Crossover (SBX) and Polynomial Mutation.

3 The MOOP Integrated Software Package

The integrated SPEA software package developed was called MOOP (Multi Objective Optimization Package) and it was written in C# language [Mar05].
In order to calculate the objective functions, the application uses a numerical analysis module built as an external component library. The package flowchart is presented in Figure 1.
For better usability, the input parameters can be kept into a file, MOOP providing new, open, and save functions. A screenshot of the MOOP main menu is shown in Figure 2.

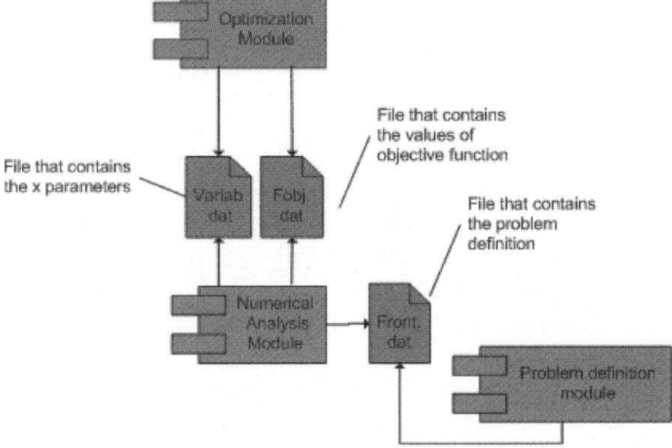

Fig. 1: Flowchart of the MOOP Software package

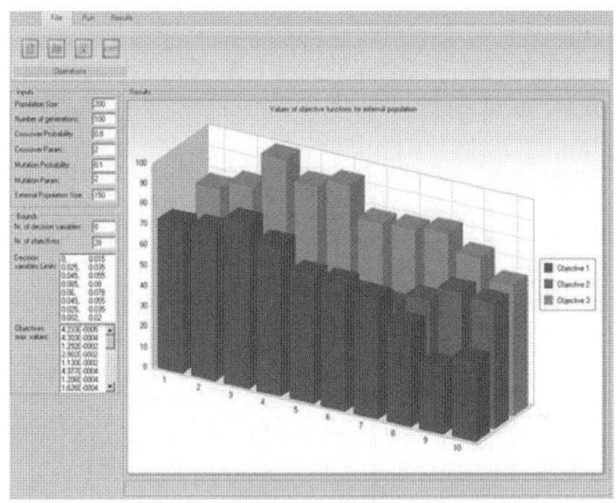

Fig. 2: Screenshot of the MOOP software package

After the optimization has completed, the user can view the charts from the last generation, all the values from the external population, the execution time, minimal and maximal values for all objectives and a proposed solution. The proposed solution is the solution with the smallest global fitness function. These values are saved in a Microsoft Excel type file.

4 Numerical Example

The SPEA algorithm developed was used for the shape optimization of the multi-terminal resistor presented in Figure 3 in order to decrease the partial resistances between its terminals [WY92]. The resistor has 8 terminals, represented by the thick segments numbered with the white bullets in Figure 3. The optimal design problem contains 28 partial objectives (the partial resistances between each pair of terminals) and uses 8 design parameters (the y coordinates of the black-marked nodes in Figure 3). One has to remark that by modifying the y coordinates of one of the 'movable' nodes during the optimization process, the y coordinates of the nearby segments' nodes are determined using parabolic interpolation.

	GA	SPEA
Chromosome length	32	–
Main population size	40	200
Ext. population size	–	150
Number of generations	1000	100
Crossover probability	0.8	0.8
Mutation probability	0.05	0.1
Crossover parameter	–	2
Mutation parameter	–	2
Total running time	7h 44 min	33h 43 min

Table 1: The optimal design algorithm settings

Using 204 boundary elements for the discretization of the whole geometry [Mun97], considering the resistor made by copper with 1 mm thickness, the partial resistance values for the initial geometry from in Figure 3 have been computed. In order to have a better relevance about the shape optimization process and results, the resistance values corresponding to the initial shape of the resistor are considered as reference values (unit values). Thus one can notice that for the starting process, the sum of all objectives is $F_{SUM} = 28$.

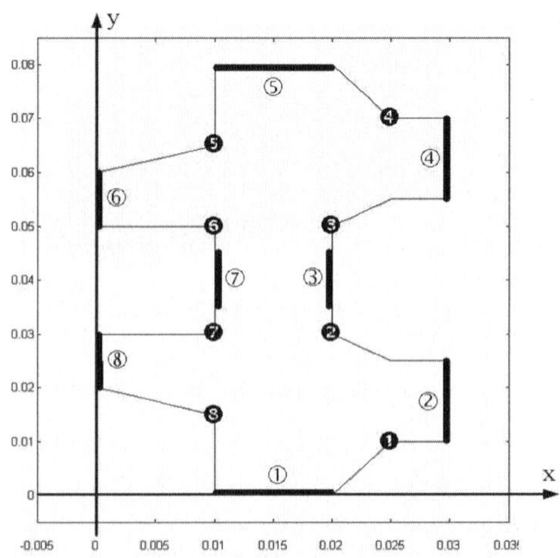

Fig. 3: The initial shape of the multi-terminal resistor

In order to emphasize the advantages of the optimization algorithm proposed, the numerical results are compared with those obtained by using an ordinary GA that looks for the minimisation of global objective function defined as suggested in [TMD00]. Table 1 presents the general settings for the optimal design algorithms.
The results using ordinary GA are presented in Table 2 while the corresponding optimal shape of the resistor is presented in Figure 4. One can notice that the overall sum decreases to $F_{SUM} = 24.98$, which means a global reduction of 11%. There are significant reduction of few partial resistances, for instance R_{58} and R_{68} with a factor of 0.6. In spite of these very good results, there are two resistances R_{12} and R_{45} with a high increasing factor of 1.48, while another 8 resistances have also increased values but with smaller factors.

		②	③	④	
		1.48	0.76	0.80	①
⑦	0.73		1.06	1.15	②
⑥	0.60	0.73		1.06	③
⑤	0.60	0.73	0.67		
④	0.84	1.13	1.17	1.48	
③	0.81	1.05	0.82	0.76	
②	1.12	1.10	0.83	0.81	
①	0.69	0.76	0.62	0.62	
	⑧	⑦	⑥	⑤	

Table 2: Partial resistance ratios, results using simple GA

Thus, as expected, the simple GA optimal design algorithm is suitable for a global optimization process but it does not take into account the variation of the partial objectives. For the actual type of applications this fact represents an important drawback.
Using the MOOP optimal design software, the interpretation of the results at the end of the optimization process is very much depending on the user interests. Of course, the overall decrease of the objectives sum value is a general indicator that the objectives are decreasing in ensemble but as it was proven in the case of the simple GA, this indicator is not relevant for partial objectives behaviour during optimization process.

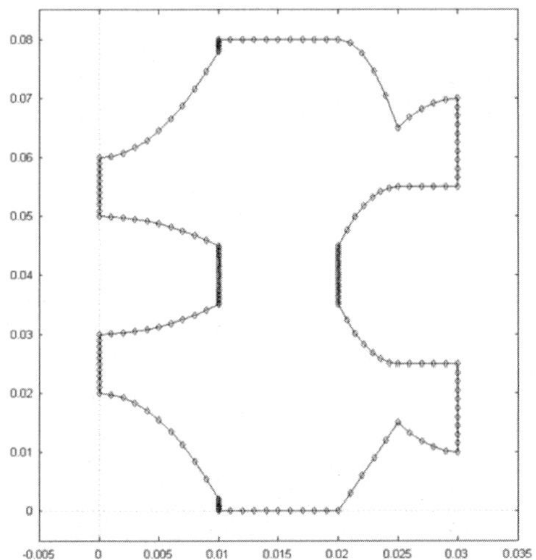

Fig. 4: Resistor optimal shape, results using ordinary GA

Of course the best result would be if one could get an as much as possible low overall sum, a high decrease of partial resistances and as much as possible low number and low values of resistances increase. In this light, the best result obtained is the one presented in Figure 5 and Table 3, where $F_{SUM} = 24.395$. As it can be noticed in this case, 27 of the partial resistances decreased with factors up to 0.63, while a single resistance has a very small increase, with factor 1.05.

		②	③	④	
		0.95	0.93	0.94	①
⑦	0.76		0.97	0.93	②
⑥	0.63	0.75		0.97	③
⑤	0.75	0.87	0.78		
④	0.76	0.98	0.95	0.8	
③	0.84	1.05	0.84	0.88	
②	0.96	0.98	0.76	0.83	
①	0.88	0.92	0.76	0.90	
	⑧	⑦	⑥	⑤	

Table 3: Partial resistance ratios, results using MOOP

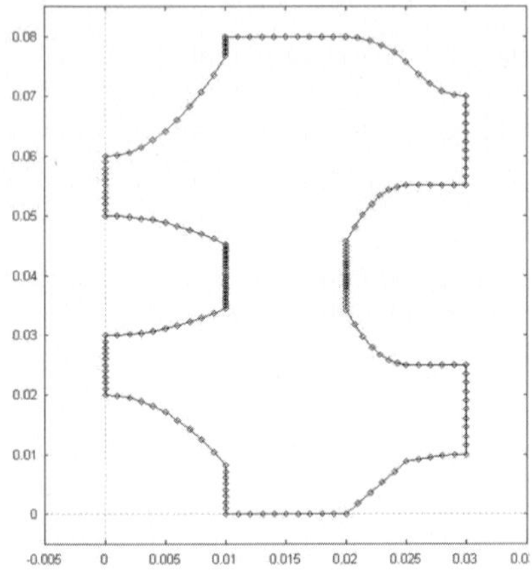

Fig. 5: Resistor optimal shape, results using MOOP

5 Conclusions

The paper emphasizes a multi-objective optimal design software package named MOOP that has been developed by the authors. The advantages of the new optimization technique with respect to the ordinary GA are outlined by a numerical example. The algorithm is suitable and effective for applications that require the control of the partial objectives evolution during the optimization process. The software package allows coupling with external solvers, being useful for a wide range of electromagnetic field optimal design applications.

Acknowledgement. The authors are grateful to the Ministry of Education and Research for the support within the frame of the Project 9 CEEX I 03 / 06.10.2005.

References

[Deb01] Deb, K.: Multi Objective Optimization using Evolutionary Algorithms, John Wiley & Sons (2001)

[Mar05] Marshall, D.: Programming Microsoft Visual C# 2005: The Language, Microsoft Press (2005)

[Mun97] Munteanu, C.: Metode numerice de analiză a câmpului electromagnetic. Metoda elementelor de frontieră, Ed. Casa Cărţii de ştiinţă, Cluj-Napoca, Romania (1997)

[WY92] Wang, Z., Yu, Q.: A Two-Dimensional Resistance Simulator Using the Boundary Element Method, IEEE Transactions on Computer-Aided Design, Vol. 11, No. 4, pp. 497 504 (1992)

[TMD00] Topa, V., Munteanu, C., De Mey, G., Deconinck, J., Simion, E.: Optimal Design of the Electromagnetic Devices using Numerical Methods, VUB University Press, Brussels, Belgium (2000)

[TMM03] Munteanu, C., Topa, V., Mates, Gh., Purcar, M., Grindei, L., Simion, E., De Mey, G.: Optimal Design of Electromagnetic Devices by Multi-Objective Optimization, Proceedings of the 6th International Workshop on Electric and Magnetic Fields, From Numerical Models to Industrial Applications, Aachen, Germany, pp. 133 137 (2003)

Optimal Design of Monolithic ESBT® Device carried out by Multiobjective Optimization.

Salvatore Spinella[1], Vincenzo Enea[2], Daniele Kroell[2], Michele Messina[2], and Cesare Ronsisvalle[2]

[1] Consorzio Catania Ricerche, Via A. Sangiuliano, 262, I95124 Catania, Italy
spins@unical.it
[2] STMicroelectronics, Stradale Primosole 50, I-95121 Catania, Italy
michele.messina@st.com

Summary. This work concerns the multiobjective optimization of an monolithic ESBT® device aimed to get a characterization of the best design. The optimization will select the epitaxial specifications (thickness, doping concentration) which minimize the energy dissipation, maximize the current flow and keep a breakdown voltage of 1000V. Since these goals are in conflict with each other the best solution must be characterized with respect to all trade-offs. The search was carried out with an extension of the DIRECT algorithm to the multiobjective case.

Keywords—Multiobjective optimization, ESBT®, process simulation, device simulation, Mixed-mode simulation.

1 Introduction

On-state voltage, breakdown voltage and switching losses represent the key points in the design of power devices devoted to high voltage and high frequency applications. In order to achieve significant efficiency improvements in DC-DC converter applications, which demand high currents and high switching frequencies, both conduction and switching energy losses need to be minimized.

ESBT® (Emitter Switching Bipolar Transistors) is an innovative power device particularly suitable for high voltage and high frequency applications [1]. The epitaxial structure of the collector region is a critical parameter of the ESBT® design:

- it characterizes the highest voltage sustainable during the off-state,
- it characterizes the current which flows into the device during the on-state,
- it characterizes the energy dissipation during a on-off cycle

The above specifications consist of the pair given by the collector region *thickness* and the doping *concentration* of the region (it must be noticed that the doping concentration is strongly related to the resistivity). A multiobjective problem formulation is necessary in order to achieve an optimal design with respect to the trade-offs of the operational performances.

2 The ESBT® Device

ESBT® consists of a high-voltage power BJT and low-voltage power MOSFET that are connected in cascode connection (see figure 1). It is a monolithic solution achieved through the integration of the MOSFET inside the emitter fingers of the BJT (see figure 2). It has been created a family of devices which can reach high breakdown voltage (up to 1.7 kV) with high switching frequency, while a low forward voltage drop is maintained. The driving of the bipolar transistor in a cascode connection is realized by the switching of a MOSFET connected in series with the emitter of the BJT. As a matter of fact by switching off the MOSFET, the emitter current of the BJT is immediately cut-off and then the whole collector current is diverted to the base terminal. By this way the bipolar transistor is turned-off very quickly because the charge stored in the base and collector is fast removed. In this way the BJT can operate at very high operating frequencies (up to 200 KHz). This device is useful in many applications as lighting and power supply.

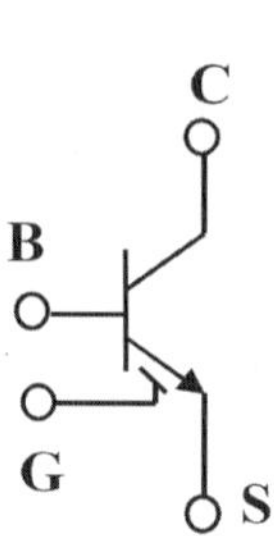

Fig. 1: The ESBT® symbol.

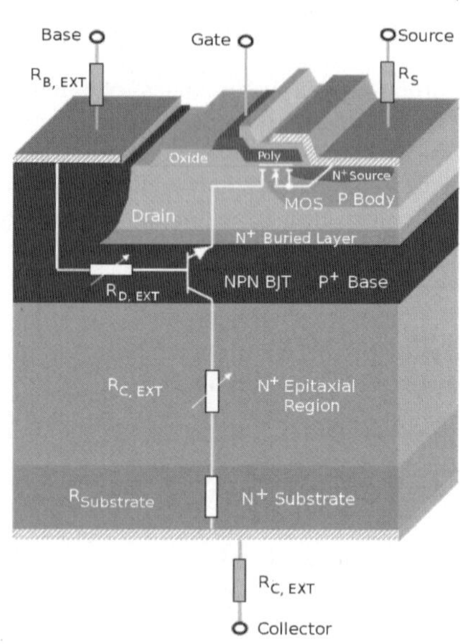

Fig. 2: Half elementary cell of the ESBT® device with superimposed the equivalent electrical circuit.

3 MultiDIRECT optimization Algorithm

A multiobjective problem is defined as

$$\min_{\mathbf{x}\in S}\{f_1(\mathbf{x}), f_2(\mathbf{x}), \ldots, f_k(\mathbf{x})\} \tag{1}$$

where we have $k \geq 2$ *objective functions* $f_i : \mathbb{R}^N \rightarrow \mathbb{R}$. S is called *decision space* and defines an *objective space* $Z \subseteq \mathbb{R}^k$ through the objective functions [2].

Minimization process follows the *Pareto optimality* criterion:

A decision vector $\mathbf{x}^*$ is Pareto optimal if there does not exist another decision vector $\mathbf{x} \in S$ such that $f_i(\mathbf{x}) \leq f_i(\mathbf{x}^*)$ for all $i = 1, \ldots, k$ and $f_j(\mathbf{x}) < f_j(\mathbf{x}^*)$ for at least one index j.

The MODirect method is an extension to the multiobjective case [3] of the DIRECT algorithm [4]. The method is based on three operations:

- Lipschitz constant estimation,
- choice for potential optimality of domain subregions,
- domain subdivision.

The choice for potential optimality is based on the estimation of Lipschitz constant for the objective function in a partition of the domain. This partition is built by hyperrectangles which are sampled in their centers in order to evaluate the value of the objective function. Therefore the estimation of Lipschitz constant leads to a possible choice of the hyperrectangles in the partition for a further sampling which exploits the estimation to balance global and local search and reaches a quasi-global solution in a large domain. In the main loop of the algorithm 1, hyperrectangles are selected for sampling if they have a large area, an high Lipschitz constant estimation, and a good value of the function in their center. Formally it is possible to give the following definition for the single objective problem in one variable:

Definition 1. ***[Potential optimality relative to the objective** i***]*** *Let S be the set of hyperrectangles generated by the algorithm after k iterations, and let f^{min} and f^{max} be respectively the ideal and nadir points of the cone centered in $f(c_{\widetilde{R}})$. An hyperrectangle $\widetilde{R} \in S$ with center $c_{\widetilde{R}}$ and measure $\alpha(\widetilde{R})$ is said potentially partial optimal relative to the i-th objective if there exists at least a Lipschitz constant $K_i^{lower} > 0$ such that*

$$f_i(\mathbf{c}_{\widetilde{R}}) - K_i^{lower}\alpha(\widetilde{R}) \leq f_i(\mathbf{c}_R) - K_i^{lower}\alpha(R) \tag{2}$$

$$f_i(\mathbf{c}_{\widetilde{R}}) - K_i^{lower}\alpha(\widetilde{R}) \leq f_i^{min} - \varepsilon|f_i^{min}|.\ \forall R \in S \tag{3}$$

or a constant $K_i^{upper} > 0$ such that

$$f_i(\mathbf{c}_{\widetilde{R}}) + K_i^{upper}\alpha(\widetilde{R}) \leq f_i(\mathbf{c}_R) + K_i^{upper}\alpha(R) \tag{4}$$

$$f_i(\mathbf{c}_{\widetilde{R}}) + K_i^{upper}\alpha(\widetilde{R}) \leq f_i^{max} - \varepsilon|f_i^{max}|.\ \forall R \in S \tag{5}$$

where $\varepsilon \sim 10^{-4}$ is a constant to control the clustering *during the search [4].*

This definition is easily extendible to the case of n variables.

Algorithm 1 DIRECT pseudocode

Require: Set of rectangles S
  $n \leftarrow 0$ {number of function calls}
  while $n < TotCalls$ **do**
    Choose $P \subseteq S$, set of potential optimal rectangles;
    Sample the rectangles in P updating the counter n;
    Subdivide the rectangles of P. Let subdivision be $D_P = \{R_1, R_2, \ldots, R_m\}$
    $S = S \setminus P \cup D_P$
  end while
  return the best minimum;

In order to obtain the heuristic which extends the above definition to the multiobjective case, let us redefine the Pareto optimality in general terms of efficiency [5].

Definition 2. ***[Efficiency criterion]*** *A decision vector $x^* \in X$ is* efficient *with respect to the convex cone D if there does not exist another decision vector $x \in X$ such that*

$$f(x^*) - f(x) \in D \tag{6}$$

The cone D is called *ordering cone* and if $D = R^n_+$ the efficiency criterion produces a partial ordering for the Pareto optimality criterion. This ordering is used by the algorithm as surrogate of linear ordering.

Remark 1. **[Multiple estimation of the Lipschitz constants]** Starting from the conditions 2 and 4 in Definition 1 it is possible to define the multiobjective optimality in terms of expected efficiency. For every objective i, from the above conditions we obtain estimates for K_i^{lower} in the form of an upper bound $\overline{K}_i^{lower} \geq 0$ and a lower bound $\underline{K}_i^{lower} \geq 0$ for K_i^{lower}. Analogously, for K_i^{upper} there will be an upper bound $\overline{K}_i^{upper} \geq 0$ and a lower bound $\underline{K}_i^{upper} \geq 0$.

The heuristic criterion leading to the choice of the optimal hyperrectangles in the multiobjective case is motivated by the potential increase of the expected efficiency.

Definition 3. ***[Multiobjective potential optimality]*** *Given the estimations of the upper bounds and the lower bounds for the Lipschitz constant of every objective i in the cone centered in $f(c_{\widetilde{R}})$, the hyperrectangle $\widetilde{R}$ is said potentially optimal if*

$$\sqrt{\sum_{i=1}^{k} [\underline{K}_i^{lower}]^2} \leq \sqrt{\sum_{i=1}^{k} [\overline{K}_i^{lower}]^2} \tag{7}$$

or

$$\sqrt{\sum_{i=1}^{k} [\underline{K}_i^{upper}]^2} \leq \sqrt{\sum_{i=1}^{k} [\overline{K}_i^{upper}]^2} \tag{8}$$

Moreover, let f^{min} and f^{max} be respectively the ideal and nadir points of the cone centered in $f(c_{\widetilde{R}})$. The choice of hyperrectangle $\widetilde{R}$ leads to a non trivial improvement of objective functions

$$\sum_{i=1}^{k} [f_i(\mathbf{c}_{\widetilde{R}}) - K_i^{lower} \alpha(\widetilde{R})]^2 \leq \sum_{i=1}^{k} [f_i^{min} - \varepsilon |f_i^{min}|]^2 \tag{9}$$

or

$$\sum_{i=1}^{k} [f_i(\mathbf{c}_{\widetilde{R}}) + K_i^{upper} \alpha(\widetilde{R})]^2 \leq \sum_{i=1}^{k} [f_i^{max} - \varepsilon |f_i^{max}|]^2 \tag{10}$$

The above definition gives a heuristic rule to choose hyperrectangles which are potentially optimal in the sense of either increasing the efficiency of the objective vector or taking into account possible trade-off (the latter arises from considering both lower and upper bounds for the Lipschitz constant). Equations 9 and 10 can be interpreted as controlling the clustering nearby the optimal points. If an hyperrectangle is potential optimal then it will be sampled in the points $\mathbf{c} \pm \delta \mathbf{e}_i$, $i = 1 \ldots N$, where $\mathbf{c}$ is the center point of the hyperrectangle, δ is one-third the side length of the hyperrectangle, and $\mathbf{e}_i$ is the ith unit vector.

Afterwards the hyperrectangle will be subdivided in thirds along its widest sides based on a dominance sorting of function values $f(\mathbf{c} \pm \delta \mathbf{e}_i)$ with respect to their efficiency. This strategy increases the attractiveness of searching near points with good function values in the large hyperrectangles.

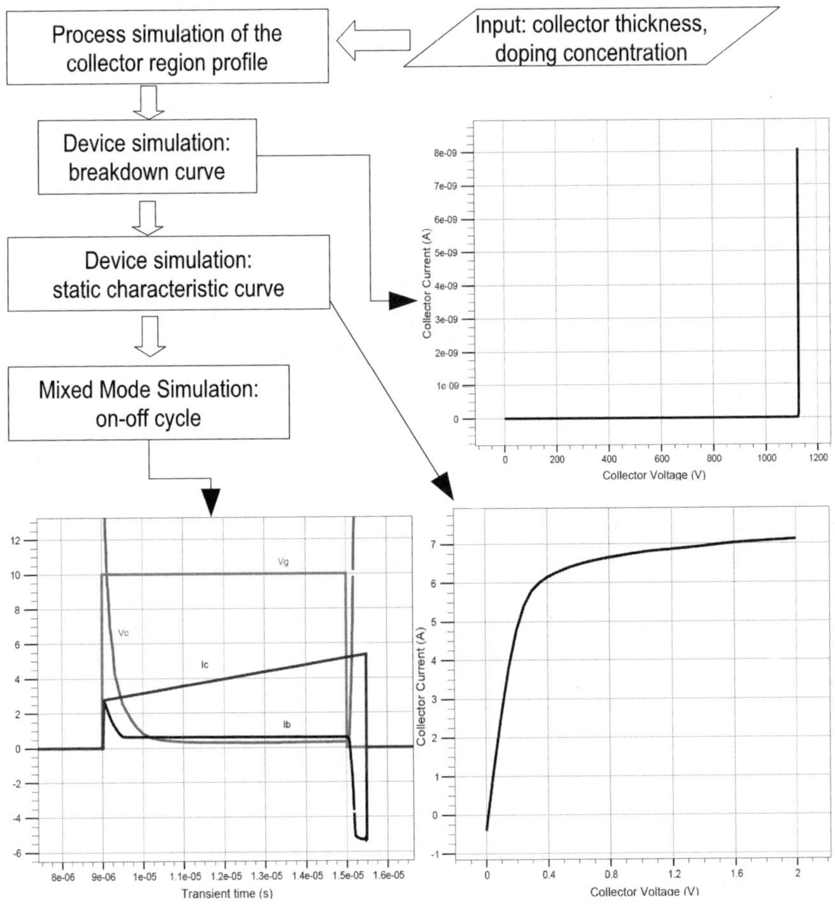

Fig. 3: The simulation flow.

4 Simulation Flow and Results

A simulation flow performed by ©SILVACO tools has been planned and it was used to evaluate the above target (see figure 3). The flow accepts as input collector thickness and doping concentration of the collector region, then a process simulation simulates the device structures. Then three device simulations extract the values of energy dissipation, current capability and breakdown voltage. Notice that the device simulations are independent and therefore can be performed in parallel.

The optimization has been carried out with respect to 3 targets:
- energy dissipation of a on-off cycle (minimizing),
- current capability (maximizing),
- breakdown voltage (to fix at 1130 Volt).

The last target constrains the optimization to functional solution which assure good process tolerances. A budget of 350 simulations has been established to perform the whole optimization.

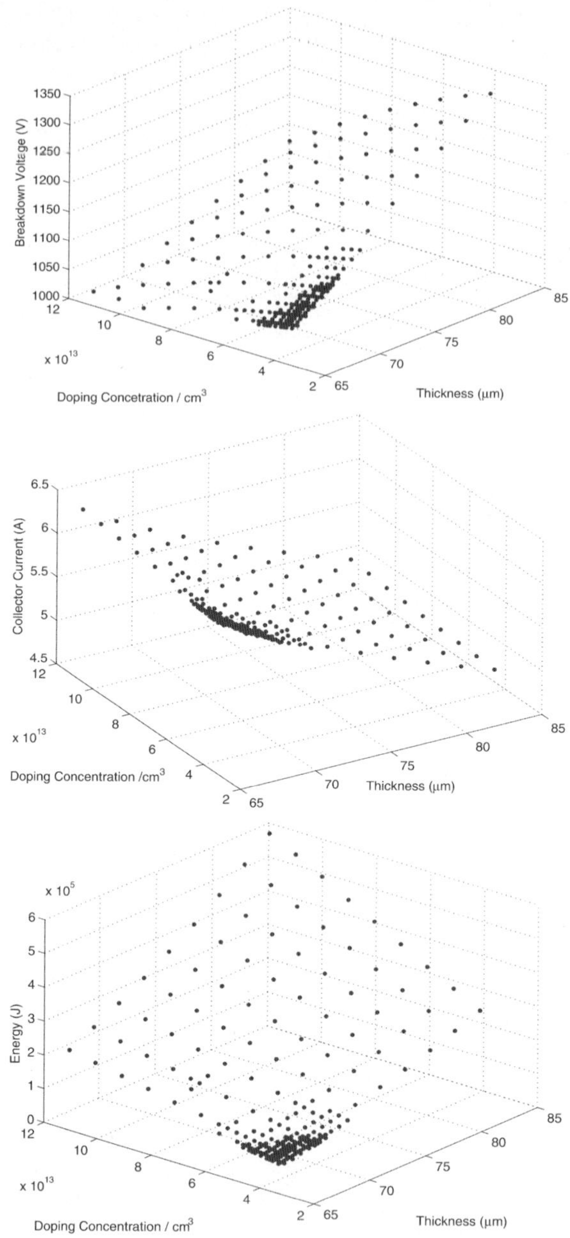

Fig. 4: The design variable space against performances.

The figure 4 shows the sampling in the design space of collector thickness and doping concentration of the collector region against each performance. The figure 5 shows the Pareto front which follows the optimization sampling. The sampling allows to characterize the optimal pair collector thickness-concentration and several alternative designs were found.

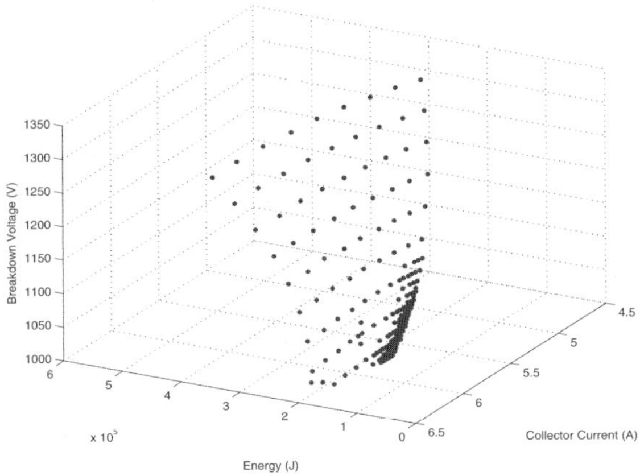

Fig. 5: Sampling in the objective space.

These results are also useful to evaluate the behaviour of the overall performances, their trade-offs, and the correlations with respect to the two design variables. For instance the following linear relations were discovered with respect to the design variable for energy (E), current flow (CF) and breakdown voltage (BV)

$$E(t,d) = 1.2972t + 2.2429 \cdot 10^{-19}d - 8.5939 \cdot 10^{-5}$$
$$CF(t,d) = -76477t + 1.7183 \cdot 10^{-15}d + 11.097$$
$$BV(t,d) = 12429000t - 1.4664 \cdot 10^{-12}d + 311.59$$

where t is the collector thickness (in μm) and d is the doping concentration (cm^{-3}) of the collector region. Also the correlations were computed and the results are shown in table 1

	Energy	Current Flow	Breakdown voltage
Energy	1	-0.83036	0.47727
Current Flow	-0.83036	1	-0.87562
Breakdown voltage	0.47727	-0.87562	1

Table 1: Table of correlation among objectives

5 Conclusion

A successful optimization test on a power device has been done. The multiobjective methodology was proved useful to guide device design. Furthermore the sampling could become a knowledge base for the future scaling of the power device towards higher breakdown voltages.

Acknowledgment

The authors want to thank Prof. A. M. Anile for the useful suggestions and discussions about this work.

References

1. S. Musumeci, R. Pagano, A. Raciti, S. Buonomo, C. Ronsisvalle, and R. Scollo, "A new monolithic emitter-switching bipolar trasistor (esbt) in high voltage converter applications," in *IAS '03*, 2003.
2. Kaisa M. Miettinen, *Nonlinear Multiobjective Optimization*, Kluwer Academic Publisher, 1998.
3. Spinella S. and A. M. Anile, "A posteriori multiobjective optimization," in *Applied and Industrial Mathematics in Italy*, 2005, pp. 520 – 529.
4. D. R. Jones, C.D. Perttunen, and B. E. Stuckman, "Lipschitzian optimization without the lipschitz constant," *J. Optim. Theory Appl.*, vol. 79, no. 1, pp. 157 – 181, 1993.
5. Yoshikazu Sawaragi, Hirotaka Nakayama, and Tetsuzo Tanino, *Theory of Multiobjective Optimization*, Academic Press Inc., 1985.

On Fast Optimal Control for Energy-Transport-based Semiconductor Design

C. R. Drago

Dipartimento di Matematica e Informatica, Università di Catania,
Viale A. Doria 6, I-95125 - Catania, drago@dmi.unict.it

Abstract. This papers deals with the optimal design of semiconductor devices, based on the adjoint method and the Energy Transport model. A partially decoupled adjoint system is obtained by considering the electrostatic potential as the new design variable and by interpreting the Poisson equation as a state equation for the doping profile. This leads to an efficient iterative optimization algorithm based on a variant of the Gummel iteration.

1 Introduction

The interest in optimal control for semiconductor design has attracted considerable recent attention in both the engineering and applied mathematics community. A major objective in the optimal design is to improve the current flow over some contacts, for fixed applied voltages, by a slight modification of the device doping profile. Besides standard black box optimization methods [9], [12], which, in general, requiring many solvers of the forward model, have an high computational cost, the adjoint method, recently proposed in the field of optimal semiconductor design [5], [6], provides good results by drastically reducing the amount of computational costs. Using the adjoint calculus, the evaluation of the complete gradient vector of the objective functional requires a single run of the (non-linear) forward model and one solve of the (linear) adjoint system, independently of the dimension of the parameters space.
The simulations of semiconductor devices, on the other hand, have been addressed, during the last years, by using different types of models, which range from microscopic models, like the Boltzmann-Poisson model to the macroscopic ones, like the hydrodynamic, the standard Drift-Diffusion and the Energy Transport model.
The adjoint method based on Drift Diffusion model has been proposed and analyzed in [5],[6]. Herein, the doping profile, which enters as a source term into the state equations, was considered as a control variable. Meanwhile, the same approach was extended to the Energy Transport model in [2].
The Energy Transport model, unlike the drift diffusion one, that is based on the assumption of isothermal motion, takes into account also the thermal effects related to the electron flow through the semiconductor crystal, due to the ongoing miniaturization of semiconductor devices, those effects cannot be neglected any longer, if one wants to improve the physical description of the device.
The dimensionless stationary energy-transport (ET) model for charge carriers in a semiconductor enclosed in a bounded domain $\Omega \subset \mathbb{R}^d$, $d = 1, 2, 3$, in the unipolar case and by using the dual entropy variables $w = (w_1 = (\mu - V)/T, w_2 = -1/T)$, is given by the following

balance equations for the electron density and the energy, coupled to the Poisson equation for the electrostatic potential V (see [8] for a complete overview):

$$\mathrm{div} I_1 = 0, \quad (1)$$

$$\mathrm{div} I_2 = Q(w,V), \quad (2)$$

$$\lambda^2 \Delta V = N(w,V) - C(x), \quad (3)$$

where

$$I_1 = -\sum_{k=1}^{2} D_{1k}(w,V)\nabla w_k, \quad I_2 = -\sum_{k=1}^{2} D_{2k}(w,V)\nabla w_k. \quad (4)$$

I_1 is the carrier flux density, μ is the chemical potential, T the temperature, $C(x)$ the doping concentration, $\lambda = \sqrt{\frac{\varepsilon_s U_T}{q C_m L^2}}$ the Debye length.
Assuming the parabolic band approximation one has for the electron density $N(w,V)$ $= (-1/w_2)^{3/2} \exp(w_1 - w_2 V)$. Moreover the energy relaxation term is given by: $Q(w,V) = -\frac{3}{2} N(w,V)(-\frac{1}{w_2} - 1)/\tau_w$, where $\tau_w = \tau_0 \mu_0 U_t / L^2$ is the scaled energy relaxation time. For the physical parameters see Table 1. The diffusion matrix (Dij) is symmetric and positive definite [8].
To get a well posed problem, system (1)-(4) has to be supplemented with appropriate boundary conditions. We assume that the boundary $\partial\Omega$ of the domain Ω splits into two disjoint parts Γ_D and Γ_N, where Γ_D models the Ohmic contacts of the device and Γ_N represents the insulating parts of the boundary. Let ν denote the unit outward normal vector along the boundary, we consider the following mixed boundary conditions

$$w_1 = w_{1D}, \quad w_2 = w_{2D}, \quad V = V_D \qquad \text{on } \Gamma_D, \quad (5)$$

$$I_i \cdot \nu = \nabla V \cdot \nu = 0 \qquad i = 1,2 \qquad \text{on } \Gamma_N, \quad (6)$$

where w_{1D}, w_{2D} and V_D are the $H^1(\Omega)$–extensions of fixed functions defined on Γ_D.
In [5], [6], [2] an optimal control approach have been presented, where the natural design variable was the doping profile and a penalty term related to C was introduced in the cost functional to stabilize the system.
A different approach, which leads to fast optimization algorithm, was investigated in [1] for the drift diffusion model. The main idea was to consider the doping profile C as a state variable and the electrostatic potential V as a control variable. The Poisson equation was reinterpreted as a state equation for the state variable C and a penalty dependence of the functional on

$$W = \triangle(V - V^*), \quad (7)$$

rather than on $C - C^*$ was introduced.
In the present communication we are addressing some analytical and numerical results concerning the ET model, using and extending previous results from [1].

2 The optimal design problem and the analytical setting

In order to introduce a functional analytic framework, we consider the following minimization problem:

$$\min_{\mathcal{D}} F_\gamma(w,V,W) \quad (8)$$

with the admissible domain

$$\mathcal{D} = \left\{(w,V,W) \in H^1(\Omega)^2 \times \left(H^1(\Omega) \cap L^\infty(\Omega)\right) \times L^2(\Omega) \text{ satisfying(1)} - (2), (7)\right\}$$

and functionals of the type

$$F_\gamma(w,V,W) = \frac{1}{2}\left[\int_\Gamma I_1 d\nu - \bar{I}\right]^2 + \frac{\gamma}{2}\int_\Omega |W(x)|^2 dx, \tag{9}$$

where $\gamma > 0$ balances the effective cost.
As initial guess V^*, we assume the one obtained as solution of the ET model, with C^* as reference doping profile. Moreover we assume $\bar{I} = 1.5 \cdot I^*$, i.e we try to gain an amplification by 50% of the reference current I^*, corresponding to the reference doping profile C^*.

Remark 1. The above objective functional F_γ is weak lower semicontinuous in $H^1(\Omega)^3 \times L^2(\Omega)$ and the admissible domain $\mathcal{D}$ weakly closed, if $\nabla(V - V*)$ remains in $L^2(\Omega)$.

Theorem 1. *Let* $\gamma > 0$ *then the constrained minimization problem* (8) *admits a solution*

$$(\bar{w}, \bar{V}, \bar{W}) \in H^1(\Omega)^2 \times \left(H^1(\Omega) \cap L^\infty(\Omega)\right) \times L^2(\Omega)$$

Proof. Let $\{w_n, V_n, W_n\}$ a minimizing sequence, then $\{W_n\}$ is bounded in $L^2(\Omega)$, and by standard elliptic regularity, $V_n - V^*$ is uniformly bounded in $H^2(\Omega) \hookrightarrow C(\bar{\Omega})$. since $V^* \in L^\infty(\Omega)$, we may conclude that V_n is uniformly bounded in $L^\infty(\Omega)$. On the other hand, in [8], Jüngel proves that $\|w_n\|_{H^1(\Omega)} \leq c_1$, where $c_1 > 0$ depends on the $L^\infty(\Omega)$-norm of V_n and the $H^1(\Omega)$-norm of w_D. Hence, from the previous estimates, there exists a subsequence, again denoted by $\{w_n, V_n\}$, such that

$$(w_n, V_n) \rightarrow \left(\bar{w}, \bar{V}\right) \text{ weakly in } H^1(\Omega)^3$$

which, by Rellich theorem, implies strong convergence in $L^2(\Omega)^3$. For the uniform L^∞-bound, we also have $V_n \rightarrow \bar{V}$ weakly-* in $L^\infty(\Omega)$. Finally $\{W_n\}$ converges weakly in $L^2(\Omega)$. For the weak closedness of the admissible domain and the weak lower semicontinuity of the objective function one can conclude that the weak limit of the subsequence is a minimizer of problem (8).

3 The Karush Kuhn Tucker conditions

Since we want to tackle a constrained optimization problem [7], we write the Karush Kuhn Tucker system by using the Lagrangian associated to the minimization problem

$$\mathcal{L}(w_1,w_2,V,W;\mu_1,\mu_2,\mu_3) \stackrel{\text{def}}{=} F_\gamma(w_1,w_2,V,W) + \int_\Omega \sum_{i,j=1}^{2} D_{ij}(w,V)\nabla w_j \cdot \nabla \mu_i dx -$$

$$- \int_\Omega Q\mu_2 dx + \int_\Omega W\mu_3 dx + \int_\Omega \nabla(V - V^*)\cdot\nabla\mu_3 dx$$

assuming $(\mu_1, \mu_2, \mu_3) \in H^1_{0,D}(\Omega)^3$, where $H^1_{0,D}(\Omega) = \left\{\varphi \in H^1(\Omega) |\ \varphi|_{\partial\Omega_D} = 0\right\}$ [13]. The first-order optimality conditions are given by imposing the variations of the Lagrangian with respect to all the state variables and the dual variables (μ_1, μ_2, μ_3) equal to zero. It is a simple matter to show that the variation with respect to the dual variables gives the equality constraint system itself (1)-(2),(7). Moreover the variation with respect to the state variable (w_1, w_2, V, W) yields the following conditions

$$\frac{\partial F_\gamma}{\partial w_1}\widehat{w}_1 + \int_\Omega \sum_{i,j=1}^{2} \frac{\partial D_{ij}(w,V)}{\partial w_1}\widehat{w}_1 \nabla w_j \cdot \nabla\mu_i dx + \tag{10}$$

$$+ \int_\Omega \sum_{k=1}^{2} D_{k1}(w,V)\nabla\widehat{w}_1 \cdot \nabla\mu_k dx - \int_\Omega \frac{\partial Q(w,V)}{\partial w_1}\widehat{w}_1\mu_2 dx = 0,$$

$$\frac{\partial F_\gamma}{\partial w_2}\widehat{w}_2 + \int_\Omega \sum_{i,j=1}^{2} \frac{\partial D_{ij}(w,V)}{\partial w_2}\widehat{w}_2 \nabla w_j \cdot \nabla \mu_i dx + \tag{11}$$

$$+\int_\Omega \sum_{k=1}^{2} D_{k2}(w,V)\nabla \widehat{w}_2 \cdot \nabla \mu_k dx - \int_\Omega \frac{\partial Q(w,V)}{\partial w_2}\widehat{w}_2 \mu_2 dx = 0,$$

$$\frac{\partial F_\gamma}{\partial V}\widehat{V} + \int_\Omega \sum_{i,j=1}^{2} \frac{\partial D_{ij}(w,V)}{\partial V}\widehat{V}\nabla w_j \cdot \nabla \mu_i dx - \int_\Omega \frac{\partial Q(w,V)}{\partial V}\widehat{V}\mu_2 dx +$$

$$+\int_\Omega \nabla \widehat{V} \cdot \nabla \mu_3 dx = 0 \tag{12}$$

$$\int_\Omega \widehat{W}(\gamma W + \mu_3)dx = 0, \tag{13}$$

for all the variations of $(\widehat{w}_1, \widehat{w}_2, \widehat{V}, \widehat{W}) \in H^1(\Omega)^3 \times L^2(\Omega)$. From (13) it follows $\mu_3 = -\gamma W$ and eliminating the μ_3 from (12) one gets, finally

$$\frac{\partial F_\gamma}{\partial V}\widehat{V} + \int_\Omega \sum_{i,j=1}^{2} \frac{\partial D_{ij}(w,V)}{\partial V}\widehat{V}\nabla w_j \cdot \nabla \mu_i dx - \int_\Omega \frac{\partial Q(w,V)}{\partial V}\widehat{V}\mu_2 dx -$$

$$-\int_\Omega \gamma \nabla \widehat{V} \cdot \nabla W dx = 0 \tag{14}$$

which will be taken as the optimality condition for the design variable W, corresponding to the minimization problem. From now on, we will proceed, assuming the particular functionals F_γ in (9), which is of particular interest in the following numerical results.
For every $(\widehat{w}_1, \widehat{w}_2, \widehat{V}) \in H^1_{0,D}(\Omega)$, it is a simple matter to show that

$$F'_\gamma(w_1, w_2, V)(\widehat{w}_1, \widehat{w}_2, \widehat{V}) = \left(\int_\Gamma I_1 d\nu - I^*\right)\int_\Gamma \left(-\sum_{k=1}^{2} D_{1k}(w,V)\nabla \widehat{w}_k \cdot \nu\right) d\sigma$$

If we choose the lagrangian μ_1 such that $\mu_1 = 0$ only on $\partial\Omega_D \setminus \Gamma$ and $\mu_1 = \eta$ on Γ the optimality system will assume a simpler form.
First of all the Lagrangian becomes

$$\mathcal{L}(w_1, w_2, V, W; \mu_1, \mu_2, \mu_3) \stackrel{\text{def}}{=} F_\gamma(w_1, w_2, V, W) + \int_\Omega \sum_{i,j=1}^{2} D_{ij}(w,V)\nabla w_j \cdot \nabla \mu_i dx$$

$$-\int_\Omega Q\mu_2 dx + \int_\Omega W\mu_3 dx + \int_\Omega \nabla(V - V^*) \cdot \nabla \mu_3 dx + \eta \int_\Gamma I_1 d\sigma$$

and the optimality with respect to w_1 yields

$$\left(\int_\Gamma I_1 \cdot d\nu - I^* + \eta\right)\int_\Gamma -D_{11}(w,V)\nabla \widehat{w}_1 \cdot \nu d\sigma + \int_\Omega \sum_{i,j=1}^{2} \frac{\partial D_{ij}(w,V)}{\partial w_1}\widehat{w}_1 \nabla w_j \cdot \nabla \mu_i dx$$

$$+\int_\Omega \sum_{k=1}^{2} D_{k1}(w,V)\nabla \widehat{w}_1 \cdot \nabla \mu_k dx - \int_\Omega \frac{\partial Q(w,V)}{\partial w_1}\widehat{w}_1 \mu_2 dx = 0.$$

Then assuming finally $\eta = -\int_\Gamma I_1 \cdot d\nu + I^*$, this reduces to the weak form corresponding to the following elliptic partial differential equations

$$-\sum_{k=1}^{2} \text{div}\ (D_{k1}(w,V)\nabla\mu_k) + \sum_{ij=1}^{2}\left(\frac{\partial D_{ij}(w,V)}{\partial w_1}\nabla w_j\right)\cdot\nabla\mu_i - \\ -\frac{\partial Q(w,V)}{\partial w_1}\mu_2 = 0, \tag{15}$$

subject to the following boundary conditions

$$\begin{aligned} \mu_1 - \int_\Gamma J\cdot\nu + I^* &= 0 \quad \text{on } \Gamma \\ \mu_1 &= 0 \quad \text{on } \partial\Omega_D \setminus \Gamma \\ \frac{\partial\mu_1}{\partial\nu} &= 0 \quad \text{on } \partial\Omega_N \end{aligned}$$

Analogously one gets for the optimality with respect to μ_2

$$-\sum_{k=1}^{2} \text{div}\ (D_{k2}(w,V)\nabla\mu_k) + \sum_{i,j=1}^{2}\left(\frac{\partial D_{ij}(w,V)}{\partial w_2}\nabla w_j\right)\cdot\nabla\mu_i - \\ -\frac{\partial Q(w,V)}{\partial w_2}\mu_2 = 0, \tag{16}$$

subject to the following boundary conditions

$$\begin{aligned} \mu_2 &= 0 \quad \text{on } \partial\Omega_D \\ \frac{\partial\mu_2}{\partial\nu} &= 0 \quad \text{on } \partial\Omega_N \end{aligned}$$

Finally, the optimality condition with respect to W in strong form reads as

$$\gamma \triangle W + \sum_{i,j=1}^{2}\left(\frac{\partial D_{ij}(w,V)}{\partial V}\nabla w_j\right)\cdot\nabla\mu_i - \frac{\partial Q(w,V)}{\partial V}\mu_2 = 0 \tag{17}$$

subject to homogeneous Dirichlet condition on $\partial\Omega_D$ and homogeneous Neumann boundary condition on $\partial\Omega_N$.
The equations (15)-(16) can be written in the simplified form

$$\text{div}\left(-\sum_{k=1}^{2} D_{ki}(w,V)\nabla\mu_k\right) + \sum_{k=1}^{2}\mathbf{b}_{ki}\cdot\nabla\mu_k - \mathbf{c}_i\cdot\mu = 0, \tag{18}$$

where $i = 1,2$ and

$$\mathbf{b}_{ki} = \sum_{j=1}^{2}\frac{\partial D_{kj}}{\partial w_i}\nabla w_j, \quad \mathbf{c}_i = \left(0, \frac{\partial Q}{\partial w_i}\right), \quad \mu = (\mu_1, \mu_2).$$

Remark 2. The matrix (D_{ki}) is symmetric positive definite and there exists a $\delta = \delta(V) > 0$ such that

$$\sum_{i,k=1}^{2} D_{ki}\xi_k\xi_i \geq \delta(V)|\xi|^2 \quad \text{for all } \xi \in \mathbb{R}^2.$$

Moreover, taking into account the $L^\infty(\Omega)$–bound on V, there exists some $\delta_0 > 0$ such that $\delta(V) \geq \delta_0$ (see [8]).
If we define

$$\mathbf{h} = \left(\sum_{k=1}^{2}\left(\frac{\partial D_{1k}}{\partial V}\nabla w_k\right), \sum_{k=1}^{2}\left(\frac{\partial D_{2k}}{\partial V}\nabla w_k\right)\right) \quad \text{and} \quad \mathbf{g} = (0, \frac{\partial Q}{\partial V}),$$

equation (17) can be written as

$$-\gamma \triangle W = \mathbf{h} \cdot \nabla \mu - \mathbf{g} \cdot \mu \tag{19}$$

where $\nabla \mu = (\nabla \mu_1, \nabla \mu_2)$.

Remark 3. With respect to the direct optimal control approach (cf. [2]), where analyzing the adjoint system was a difficult task, here the adjoint system (10)-(12) has a partially decoupled structure with respect to the Lagrangian variables, then proving existence and uniqueness of Lagrangian variables $(\mu_1, \mu_2, \mu_3) \in H^1_{0,D}(\Omega)^3$, for given primal variables, consists of the analysis of two subsequently variational problems, which in turn are coercive in $H^1_{0,D}(\Omega)$.

Theorem 2. *Assuming all the coefficients* $D_{ij}, b_{ij}, c_i, h_i, g_i \in L^\infty(\Omega)$. *There exists a constant* $l = l(\Omega, C, \|\mathbf{b}_{ik}\|_{L^\infty(\Omega)}, \delta_0) > 0$ *such that for each* $(w_1, w_2, V) \in \mathcal{D}$ *with*

$$\sum_{i=1}^{2} \|\mathbf{c}_i\|_{L^\infty(\Omega)} \leq l \quad \text{and} \quad \sum_{i,k=1}^{2} \|\mathbf{b}_{ik}\|_{L^\infty(\Omega)} < \frac{\delta_0}{C}$$

(where $C = C(\Omega) > 0$ *is the Poincaré constant), system* (18)-(19) *admits a unique solution* $(\mu_1, \mu_2, W) \in H^1_{0,D}(\Omega)^3$.

Proof. It is a straight-forward application of the Lax-Milgram Theorem.

4 A Fast Optimization Algorithm and Numerical Optimal Designs

The partially decoupling in the adjoint systems suggests a fast iterative optimization algorithm, based on a variant of the Gummel iteration [1], [4], [11]. This avoids solving the fully coupled ET system, but only needs solving the continuity equations and their adjoints.
Namely, one first solves (7) with given W, for the potential V, then the coupled continuity equations (1)-(2) with given potential V, for w_1 and w_2. Furthermore for given potential V and given w_1 and w_2, one solves the coupled adjoint equations (15)-(16) to obtain the lagrangian variables μ_1 μ_2. Finally, a gradient step will be performed with respect to the design variable W by using the optimality equation (17). In fact, by introducing an additional damping parameter τ one can interpret this iteration as a descent algorithm

$$-\gamma \triangle W + \tau W = \tau W^* + \sum_{i,j=1}^{2} \left(\frac{\partial D_{ij}(w, V)}{\partial V} \nabla w_j \right) \cdot \nabla \mu_i - \frac{\partial Q(w, V)}{\partial V} \mu_2,$$

here, W^* is the old value of W.
Altogether we will perform the following algorithm.

ALGORITHM 1

1. *Let* $W^0 = 0$.
2. *For* $n = 1, 2, \ldots$ *solve* $\Delta V^n = \Delta V^* + W^{n-1}$.
3. *Solve the continuity equation* (1)- (2), *with given* V^n, *for* w_1^n *and* w_2^n.
4. *Solve the adjoints equations* (15)-(16), *with given* V^n, w_1^n *and* w_2^n.
5. *Perform a gradient step by updating* W:

$$-\gamma \triangle W^n + \tau W^n = \tau W^{n-1} + \sum_{i,j=1}^{2} \left(\frac{\partial D_{ij}(w^n, V^n)}{\partial V} \nabla w_j^n \right) \cdot \nabla \mu_i^n - \frac{\partial Q(w^n, V^n)}{\partial V} \mu_2^n,$$

6. *Independent update of C by:* $C^n - C^* = -\lambda^2 W^n + N^n - N^*$.

Now we are showing some numerical results of a one-dimensional $n^+ - n - n^+$ ballistic diode, which is a simple model for the channel of a MOS transistor. The semiconductor domain is given by the interval $\Omega = (0, L)$, with $L > 0$. In the n^+-regions a maximal doping concentration of $C_m = 5 \cdot 10^{17}$ cm^{-3} is prescribed. In the n–channel the minimal doping density is $2 \cdot 10^{15}$ cm^{-3}. The length of the n^+–regions is 0.1μm, whereas the length of the channel region equals 0.4μm. The numerical values of the physical parameters are given in Table 1. The scaled parameters were set to $\lambda^2 = 9.0278 \cdot 10^{-5}$ and $\tau_w = 4 \cdot 10^{-3}$. We

Parameter	Physical meaning	Numerical value
q	elementary charge	$1.60219 \cdot 10^{-19}$As
ε_s	permittivity constant	$1.04479442 \cdot 10^{-12}$AsV^{-1}cm^{-1}
μ_0	(low field) mobility constant	$1.5 \cdot 10^3$cm^{2}V^{-1}s^{-1}
U_T	thermal voltage at $T_0 = 300$K	0.025852V
τ_0	energy relaxation time	$0.4 \cdot 10^{-12}$s
L	length of the device	0.6μm

Table 1: Physical Parameters

solve the constrained optimization problem (8) by using Algorithm 1. For the parameter γ we have chosen 10^{-8} and $\tau = 10^{-1}$ for the constant damping parameter. The state system was discretized by a variant of the well–known exponentially fitted Scharfetter–Gummel scheme [11, 4]. The computations were performed on a uniform grid of 301 points to ensure to have no grid effects, but the same results can be already obtained for 101 points. For the biasing voltage at the working point we chose $\bar{U} = 1V$ and tried to gain an amplification of the current I^* by 50%, i.e. we set $\bar{I} = 1.5 \cdot I^*$. In Figure 1 we present the optimal doping profile as well as the reference doping C^*. Further, we depict the densities, velocities and temperatures before and after the optimization. Note, that the change in the potential, as well as in the velocities and temperatures before and after the optimization is quite small while a more significant change happens just in the electron density. Nevertheless, we reach our objective as can be also seen from the given current–voltage characteristics (IVC). The overall performance of the algorithm is promising, since already 10 gradient steps are sufficient to reach the optimum.

5 Conclusion and future works

In this work a fast optimization method was performed and validate in the framework of energy transport-based optimal semiconductor design. An inspection of the evolution of the objective demonstrates the efficiency of the approach, since the minimum is obtained with only few iterations. Since in each iteration, we only have to solve two systems of elliptic partial differential equations, the numerical effort per iteration is similar to two Gummel-type iterations per step. The overall performance of the algorithm is already very promising for a one-dimensional numerical test. Applications of the algorithm for two-dimensional cases will be the aim of future works.

Acknowledgments. The author would like to thank prof. R. Pinnau and A. M. Anile for helpful discussions.

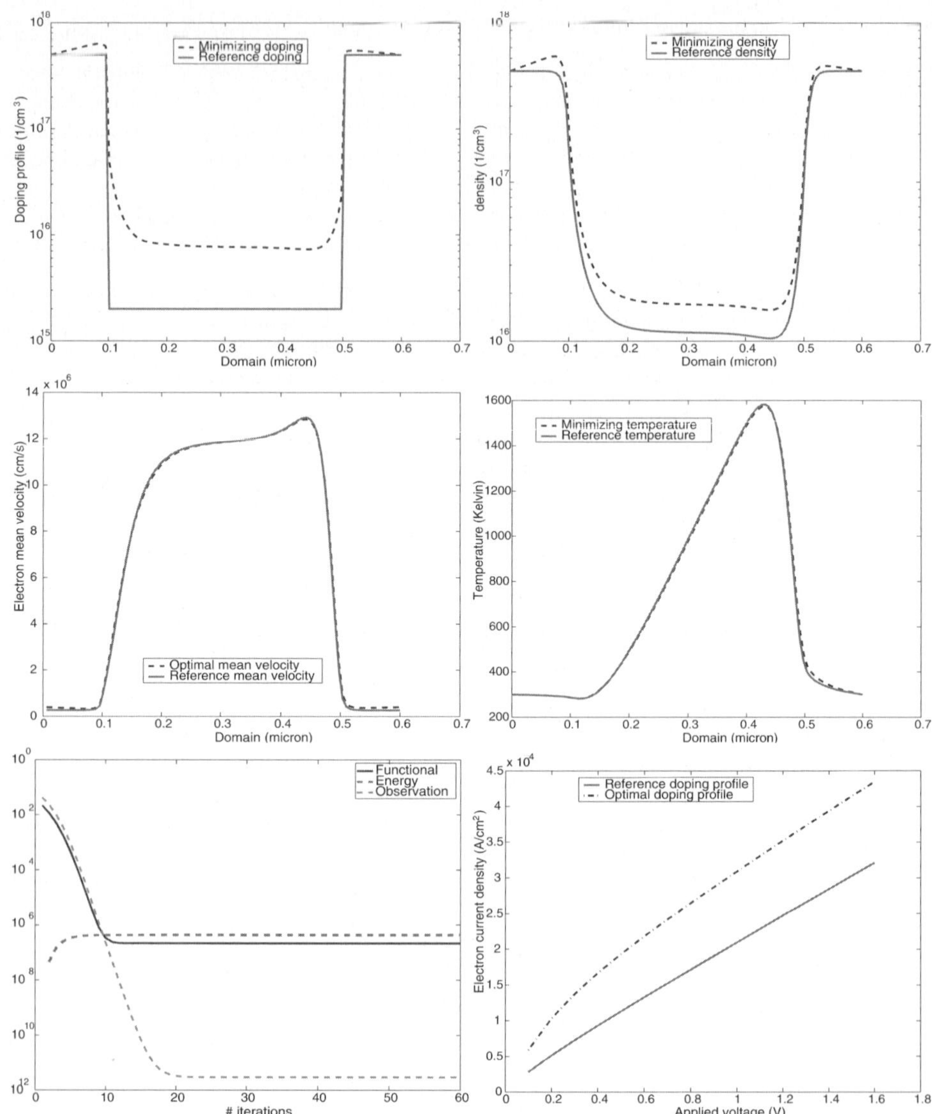

Fig. 1: Optimized doping profile, electron density, electron mean velocity, temperature, evolution of the cost functional for a biasing voltage of 1 V, and the corresponding IVCs

References

[1] M. Burger, R. Pinnau, Fast optimal design for semiconductor devices. SIAM J. Appl. Math., 64 (1) pp. 108-126, 2003

[2] C. R. Drago, A. M. Anile, An optimal Control approach for an Energy Transport Model in Semiconductor Design. *Scientific Computing in Electrical Engineering, SCEE'04, Mathematics in Industry*, Vol. 9:323–331, Springer Verlag, 2006.

[3] C. R. Drago, R. Pinnau, The Semicondutor Model Hierarchy in Optimal Dopant Profiling, submitted.

[4] H.K. Gummel. A self-consistent iterative scheme for one-dimensional steady state transistor calculations. IEEE Trans. Elec. Dev., ED-11:455-465, 1964.

[5] M. Hinze, R. Pinnau, An optimal control approach to semiconductor design. Math. Mod. Meth. Appl. Sc., 12 (1) pp.. 89-107, 2002

[6] M. Hinze, R. Pinnau, Mathematical Tools in Optimal Semiconductor Design, (to appear on TTSP 2006)

[7] K. Ito and K. Kunish. Augmented Lagrangian-SQP-methods for nonlinear optimal control problems of tracking type SIAM J Control and Optimization, 34:874-891,1996

[8] A. Jüngel, Quasi-hydrodynamic Semiconductor Equations, Progress in Nonlinear Differential Equation and Their Application (2001) Birkhäuser

[9] W. R. Lee, S. Wang, K. L. Teo, An optimization approach to a finite dimensional parameter estimation problem in semiconductor device design. Journal of Computational Physics, 156 pp. 241-256, 1999

[10] J.L. Lions Optimal Control of Systems Governed by Partial Differential Equations. Springer-Verlag, New York, 1971. Translated by S.K. Mitter

[11] D.L. Scharfetter and H.K. Gummel. Large signal analysis of a silicon read diode oscillator. IEEE Trans. Electr. Dev., 15:6477, 1969.

[12] M. Stockinger, R. Strasser, R. Plasun, A. Wild, S. Selberherr. A qualitative study on optimized MOSFET doping profiles. In Proocedings SISPAD 98 Conf., pp 77-80, 1998

[13] G. M. Troianello, Elliptic Differential Equations and Obstacle Problems, Plenum Press, New York, first edition, 1987

Extended Hydrodynamical Models for Charge Transport in Si

Roberto Beneduci[1], Giovanni Mascali[1], and Vittorio Romano[2]

[1] University of Calabria and INFN-Gruppo c.Cosenza, rbeneduci@unical.it, g.mascali@unical.it
[2] University of Catania, romano@dmi.unict.it

Summary. In this paper we present a hierarchy of extended hydrodynamical models for electron transport in Silicon, which differ from each other for the number of scalar and vector moments of the electron distribution function used as state variables. The closure of the moment equations is achieved by means of the Maximum Entropy Principle. The main scattering mechanisms between electrons and phonons are taken into account. An application to the case of bulk Silicon is presented.

1 Introduction

The description of the functioning of modern electron devices requires an increasingly accurate physical modeling of carrier transport in semiconductors [AMR03], since the presence of very high and rapidly varying electric fields produces phenomena which cannot be described by means of the standard drift-diffusion or energy transport models. This has led to the construction of new models that are, loosely speaking, called hydrodynamical models. These models are usually derived from the infinite hierarchy of the moment equations of the Boltzmann transport equation by suitable truncation procedures. However most of them suffered from serious theoretical drawbacks due to the ad hoc treatment of the closure problem. Recently [Rom00, MR02, AMR03], these drawbacks have been overcome by means of an appropriate method based on the Extended Thermodynamics of moments [MuR98, Lev96]. This method allows one to obtain approximate distribution functions, the so called Maximum entropy distribution functions, by means of which it is possible to solve the closure problem in a physically grounded way.
In many hydrodynamical models [AMR03], only two scalar moments and two vector moments of the electron distribution function are used: the number density, the average energy, the velocity and the energy flux. These moments are the only scalar and vector moments which have an immediate physical meaning. However in situations where very high or very rapidly varying fields are present, those models may fail and a higher number of moments may be required to have a satisfactory description of the physical situation [MuR98, Str00].
In this paper we construct models where the numbers of scalar and vector moments can be arbitrarily great and different from each other. The main difficulty which occurs in the construction of these models is the inversion of the constraint relations which, however, can be tackled by means of suitable approximations and numerical procedures. We consider the problem of electron transport in Si, but the models can be easily extended to other semiconductors. A non parabolic approximation, the Kane dispersion relation, is used for the electron energy in the conduction band and all the main scattering mechanisms of electrons with phonons are

treated. We also try to investigate how the number of moments affects the results at least in the case of bulk Silicon.

2 The electron transport in Si, the semiclassical Boltzmann equation

We consider the case of Silicon unipolar devices for which the charge transport is due to the electrons in the six equivalent valleys around the six minima of the conduction band [JR83]. We assume that, for those electrons, the relation between the energy $\mathcal{E}$ and the quasi-wave vector $\mathbf{k}$, both measured from the bottom of the valley, is given by the Kane dispersion relation

$$\mathcal{E}(k)\left[1+\alpha\,\mathcal{E}(k)\right]=\frac{\hbar^2k^2}{2m^*},\qquad \mathbf{k}\in\mathbb{R}^3, \tag{1}$$

which involves a parameter α, called the non-parabolicity factor, while m^* is the electron effective mass. In the semiclassical kinetic approach, the charge transport is described by the *Boltzmann equation* [Tom93], which reads [3]

$$\frac{\partial f}{\partial t}+v^i(\mathbf{k})\frac{\partial f}{\partial x^i}-\frac{qE^i}{\hbar}\frac{\partial f}{\partial k^i}=\mathcal{C}[f], \tag{2}$$

where $f(\mathbf{x},\mathbf{k},t)$ is the electron distribution function, $\mathbf{v}$ is the electron group velocity related to the energy $\mathcal{E}$ by $\mathbf{v}=\frac{1}{\hbar}\nabla_{\mathbf{k}}\mathcal{E}$, $\hbar$ is the Planck reduced constant, q is the absolute value of the electron charge and $\mathcal{C}[f]$, the collision term, represents the effects due to scatterings with phonons and impurities. The electric field $\mathbf{E}$ is calculated by solving the Poisson equation for the electric potential ϕ

$$\mathbf{E}=-\nabla_{\mathbf{x}}\phi,\qquad \nabla_{\mathbf{x}}(\epsilon\nabla_{\mathbf{x}}\phi)=-q(N_+-N_--n), \tag{3}$$

N_+ and N_- respectively being the donor and acceptor densities (which depend only on the position), ϵ the dielectric constant and n the electron number density

$$n=\int_{\mathbb{R}^3} f d\mathbf{k}.$$

The equations (2)-(3) constitute the Boltzmann-Poisson system that is the basic semiclassical model of electron transport in semiconductors.
As said, $\mathcal{C}[f]$ reflects the various scattering mechanisms the electrons undergo in a semiconductor [Tom93, AMR03]. Some of them leave the electrons in the same valley as they are before the collision (intravalley transitions), while other scatterings can drive the electrons into a different valley (intervalley transitions) according to suitable selection rules. In the non-degenerate case the form of $\mathcal{C}[f]$ is

$$\mathcal{C}[f]=\int_{\mathbb{R}^3}\left[w(\mathbf{k}',\mathbf{k})\,f(\mathbf{x},\mathbf{k}',t)-w(\mathbf{k},\mathbf{k}')f(\mathbf{x},\mathbf{k},t)\right]\,d\mathbf{k}',$$

where $w(\mathbf{k},\mathbf{k}')$ represents the sum of the various electron scattering rates from a state with wave vector $\mathbf{k}$ to one with wave vector $\mathbf{k}'$. We take into account the following scattering mechanisms for Silicon

[3] summation over repeated indices is understood.

- electron - acoustical phonon intravalley scattering for which the transition rate, in its elastic approximation (valid when the thermal energy is much greater than that of the phonon involved in the scattering), reads

$$w_{ac}(\mathbf{k},\mathbf{k}') = \mathcal{K}_{ac}\,\delta(\mathcal{E}' - \mathcal{E})\,,$$

with $\mathcal{K}_{ac}$ acoustical intravalley scattering kernel coefficient and δ Dirac function,
- electron - phonon intervalley scattering, for which there are six contributions

$$w_{\alpha}(\mathbf{k},\mathbf{k}') = \mathcal{K}_{\alpha}\left[n_{\alpha}\,\delta(\mathcal{E}' - \mathcal{E} - \hbar\omega_{\alpha}) + (n_{\alpha}+1)\delta(\mathcal{E}' - \mathcal{E} + \hbar\omega_{\alpha})\right]\,,$$

where α runs over the three g_1, g_2, g_3 and the three f_1, f_2, f_3 intervalley scatterings [JR83], $\mathcal{K}_{\alpha}$ are the correspondent optical or acoustical intervalley scattering kernel coefficients and

$$n_{\alpha} = \frac{1}{exp\left(\frac{\hbar\omega_{\alpha}}{K_B T_L}\right) - 1}$$

is the occupation number of phonons with frequency ω_{α}, K_B and T_L respectively being the Boltzmann constant and the lattice temperature.

3 Moment equations and maximum entropy principle

Starting from the transport equation (2), it is possible to get balance equations for macroscopic quantities associated with the electron flow. In fact, multiplying eq. (2) by a sufficiently regular function $\psi(\mathbf{k})$ and integrating over $\mathcal{B} = \mathbb{R}^3$, the first Brillouin zone in the Kane approximation, the generic *moment equation* [AMR03] is obtained

$$\frac{\partial M_{\psi}}{\partial t} + \int_{\mathbb{R}^3} \psi(\mathbf{k}) v^i(\mathbf{k}) \frac{\partial f}{\partial x^i} d\mathbf{k} - \frac{q}{\hbar} E^j \int_{\mathbb{R}^3} \psi(\mathbf{k}) \frac{\partial f}{\partial k^j} d\mathbf{k} = \int_{\mathbb{R}^3} \psi(\mathbf{k}) \mathcal{C}[f] d\mathbf{k}, \tag{4}$$

with

$$M_{\psi} = \int_{\mathbb{R}^3} \psi(\mathbf{k}) f d\mathbf{k},$$

moment relative to the weight function ψ.
In this paper, at difference with [Rom00, MR02, AMR03], we use as ψ the functions $1, \mathcal{E}, \mathcal{E}^2, \ldots, \mathcal{E}^N, \mathbf{v}, \mathcal{E}\mathbf{v}, \ldots, \mathcal{E}^M\mathbf{v}$, with N and M to be suitably chosen according to the physical situation under study. The resulting moment equations are

$$\frac{\partial (n\,F_A)}{\partial t} + \frac{\partial (n\,F_A^i)}{\partial x^i} + qE_i\,A\,n\,F_{A-1}^i = n\,C_{F_A}, \quad A = 0, \ldots, N, \tag{5}$$

$$\frac{\partial (nF_B^i)}{\partial t} + \frac{\partial (nF_B^{ij})}{\partial x^j} + \frac{q}{\hbar} E_j n\left(G_B^{ij} + B\,F_{B-1}^{ij}\right) = n\,C_{F_B^i}, \; B = 0, \ldots, M, \tag{6}$$

where

$$n\,F_A = \int_{\mathbb{R}^3} \mathcal{E}^A f\,d^3k, \quad n\,F_B^i = \int_{\mathbb{R}^3} \mathcal{E}^B v^i f\,d^3k,$$

$$n\,F_B^{ij} = \int_{\mathbb{R}^3} \mathcal{E}^B v^i v^j f\,d^3k, \quad n\,G_B^{ij} = \frac{1}{\hbar}\int_{\mathbb{R}^3} \mathcal{E}^B \frac{\partial v^i}{\partial k^j} f\,d^3k, \tag{7}$$

$$n\,C_{F_A} = \int_{\mathbb{R}^3} \mathcal{E}^A C[f]\,d^3k, \quad n\,C_{F_B^i} = \int_{\mathbb{R}^3} \mathcal{E}^B v^i C[f]\,d^3k.$$

n, F_A ($A \geq 1$) and F_B^i are the variables used to describe the electrons, while F_B^{ij}, G_B^{ij}, C_{F_A} and $C_{F_B^i}$ are extra-variables which respectively are fluxes, the former two sets, and production terms the latter ones.
For $A = 1$ and $B = 0, 1$, F_A and F_B^i are respectively equal to W, $\mathbf{V}$, and $\mathbf{S}$ which, together with the number density n, represent the usual fundamental variables used in the models described in [Rom00, MR02, AMR03]: the average energy, the velocity and the energy flux. Therefore, if one takes $N = M = 1$, those models are recovered. The higher moments do not have an immediate physical meaning. The moment equations (5), (6) do not form a closed system, since the number of the unknowns is greater than that of the equations. It is necessary to find closure relations for the fluxes $(7)_2$ and the production terms $(7)_3$, that is to express them as functions of the fundamental variables $(7)_1$.
A way to get constitutive relations, founded on sound physical bases, consists of using the maximum entropy principle [AMR03, MuR98, Lev96]. This principle furnishes the form of the distribution function which makes the best use of the knowledge of a finite number of moments. In particular, in the present case the maximum entropy distribution is the one which makes the electron entropy extremal under the constraints of fixed values of the fundamental variables. The electron entropy can be written as

$$s[f] = -k_B \int_{\mathbb{R}^3} (f \log f - f)\, d^3 k\,,$$

therefore the maximum entropy distribution function is

$$f_{ME} = \exp\left[-\frac{1}{k_B}(\Lambda_A \mathcal{E}^A + \Lambda_B^i \mathcal{E}^B v_i)\right],$$

where the Λ's are the Lagrange multipliers which take care of the constraints $(7)_1$.
In order to determine the Lagrange multipliers in terms of n, F_A, F_B^i, $A = 1, \ldots, N$, $B = 0, \ldots, M$, one has to insert the expression of the maximum entropy distribution function into $(7)_1$ and solve the resulting system. After that the closure relations can be obtained by evaluating the appropriate moments of f, and $\mathcal{C}[f]$, with f replaced by the corresponding maximum entropy function. However, on account of the algebraic difficulties, we can get only approximate expressions for the Lagrange multipliers under reasonable physical assumptions on the distribution function.
At equilibrium the distribution function is isotropic

$$f^{(eq)} = \exp\left[-\left(\frac{1}{k_B}\Lambda_0^{(eq)} + \frac{\mathcal{E}}{k_B\, T_L}\right)\right],$$

that is at equilibrium

$$\Lambda_1^{(eq)} = \frac{1}{T_L}, \qquad \Lambda_A^{(eq)} = 0, \quad A = 2, \ldots, N\,,$$
$$\Lambda_B^{i\,(eq)} = 0, \qquad B = 0, \ldots, M\,.$$

On the basis of Monte Carlo results, we assume that the anisotropy of f^{ME}, remains small even out of equilibrium. We formally introduce a *small* anisotropy parameter δ, assume that the Lagrange multipliers are analytic in δ and expand them around $\delta = 0$ up to the first order. By taking into account the representation theorems for isotropic functions, consistently with the small anisotropy assumption, one has that the Λ_A's are of order zero in δ, while the Λ_B^i's are of the first order in δ. Therefore the maximum entropy distribution function can be approximated as

$$f_{ME} \approx \exp\left(-\frac{1}{k_B}\Lambda_A \mathcal{E}^A\right)\left[1 - \frac{\delta}{k_B}\Lambda_B^i\, v_i \mathcal{E}^B\right]. \tag{8}$$

4 Inversion of the constraint relations

In order to express the Lagrange multipliers in terms of the fundamental moments, we have to invert the following two systems of equations

$$n\,F_A = \int_{\mathbb{R}^3} \exp(-\frac{1}{k_B}\Lambda_C \mathcal{E}^C)\mathcal{E}^A\, d^3k, \quad A = 0,\ldots,N\,, \tag{9}$$

$$n\,F_B^i = C_{BD}\Lambda_D^i\,, \qquad B, D = 0,\ldots,M\,, \tag{10}$$

where

$$C_{BD} = -\frac{8\pi}{3}\frac{\sqrt{2m^*}}{\hbar^3 k_B}\int_0^\infty \mathcal{E}^{B+D}\frac{\left[\mathcal{E}(1+\alpha\mathcal{E})\right]^{\frac{3}{2}}}{1+2\alpha\mathcal{E}}\exp(-\frac{1}{k_B}\Lambda_C\mathcal{E}^C)\,d\mathcal{E},$$

The first equation in the first system, corresponding to $A = 0$, immediately gives Λ_0 in terms of n and Λ_A, $A = 1,\ldots,N$, while the inversion of the remaining equations in the first system has to be performed numerically. Once this is done, the inversion of the second system of equations only requires the computation of $C^{-1}_{A\,B}$, whose elements are functions of n and F_A, $A = 1,\ldots,N$

$$\Lambda_A^i = C^{-1}_{AB} n F_B^i\,.$$

5 Fluxes

Once the Lagrangian multipliers are expressed as functions of the fundamental variables, the constitutive equations for the fluxes can be obtained. Up to the first order terms in δ, one has

$$F_B^{ij} = \frac{8\pi\sqrt{2m^*}}{3\hbar^3}\delta^{ij}\int_0^\infty \mathcal{E}^B\frac{\left[\mathcal{E}(1+\alpha\mathcal{E})\right]^{\frac{3}{2}}}{(1+2\alpha\mathcal{E})}\exp(-\frac{1}{k_B}\Lambda_C\mathcal{E}^C)\,d\mathcal{E}\,,$$

$$G_B^{ij} = \frac{4\pi\sqrt{2m^*}}{\hbar^3}\delta^{ij}\int_0^\infty \left[1 - \frac{4}{3}\frac{\alpha\mathcal{E}(1+\alpha\mathcal{E})}{(1+2\alpha\mathcal{E})^2}\right]\mathcal{E}^B\sqrt{\mathcal{E}(1+\alpha\mathcal{E})}$$

$$\times\exp(-\frac{1}{k_B}\Lambda_C\mathcal{E}^C)\,d\mathcal{E}\,, \quad B = 0,\ldots,M\,,$$

δ^{ij} being the Kronecker delta.

6 Production terms

By using the same procedure as before, also the expressions of the moments of the collision term can be found.

6.1 Acoustic phonon intravalley scattering

Since we are considering this scattering in its elastic approximation, we get

$$C_{F_A} = 0\,, \quad A = 0,\ldots,N \tag{11}$$

$$C^{(ac)}_{F_B^i} = \frac{64}{3}\pi^2\frac{{m^*}^2}{\hbar^6}\mathcal{K}_{(ac)}\,C^{-1}_{ED}F_D^i\int_0^\infty \mathcal{E}^{(B+E+2)}\,(1+\alpha\,\mathcal{E})^2$$

$$\times\exp\left(-\frac{1}{k_B}\Lambda_C\mathcal{E}^C\right)\,d\mathcal{E}\,, \quad B, D, E = 0,\ldots,M. \tag{12}$$

6.2 Intervalley scatterings

These scattering mechanisms are inelastic, therefore we obtain

$$n\,C^{(\alpha)}_{F_A} = 32\frac{m^{*3}}{\hbar^6}\pi^2\mathcal{K}_\alpha\left[n_\alpha\psi^{(A)}_{1,\alpha}(\Lambda_C) - (n_\alpha+1)\psi^{(A)}_{2,\alpha}(\Lambda_C)\right], \tag{13}$$

$$C^{(\alpha)}_{F_B{}^i} = \frac{64}{3}\frac{\pi^2 m^{*2}}{\hbar^6}\mathcal{K}_\alpha C^{-1}_{ED}F_D{}^i\left[(n_\alpha+1)\psi^{(B,E)}_{3,\alpha} + n_\alpha\psi^{(B,E)}_{4,\alpha}\right], \tag{14}$$

where, defining $\mathcal{E}^\alpha_+ = \mathcal{E} + \hbar\omega_\alpha$,

$$\psi^{(A)}_{1,\alpha}(\Lambda_C) = \int_0^\infty \exp\left(-\frac{1}{k_B}\Lambda_C\mathcal{E}^C\right)\sqrt{\mathcal{E}\mathcal{E}^\alpha_+(1+\alpha\mathcal{E}^\alpha_+)(1+\alpha\mathcal{E})}$$
$$\times\left[(\mathcal{E}^\alpha_+)^A - \mathcal{E}^A\right](1+2\alpha\mathcal{E})(1+2\alpha\mathcal{E}^\alpha_+)\,d\mathcal{E},$$

$$\psi^{(A)}_{2,\alpha}(\Lambda_C) = \int_0^\infty \exp\left(-\frac{1}{k_B}\Lambda_C(\mathcal{E}^\alpha_+)^C\right)\sqrt{\mathcal{E}\mathcal{E}^\alpha_+(1+\alpha\mathcal{E}^\alpha_+)(1+\alpha\mathcal{E})}$$
$$\times\left[(\mathcal{E}^\alpha_+)^A - \mathcal{E}^A\right](1+2\alpha\mathcal{E})(1+2\alpha\mathcal{E}^\alpha_+)\,d\mathcal{E},$$

$$\psi^{(B,E)}_{3,\alpha}(\Lambda_C) = \int_0^\infty \frac{\left[\mathcal{E}^\alpha_+(1+\alpha\mathcal{E}^\alpha_+)\right]^{\frac{3}{2}}}{(1+2\alpha\mathcal{E}^\alpha_+)}\sqrt{\mathcal{E}(1+\alpha\mathcal{E})}\exp\left(-\frac{1}{k_B}\Lambda_C(\mathcal{E}^\alpha_+)^C\right)$$
$$\times(\mathcal{E}^\alpha_+)^{B+E}(1+2\alpha\mathcal{E})\,d\mathcal{E},$$

$$\psi^{(B,E)}_{4,\alpha}(\Lambda_C) = \int_0^\infty \mathcal{E}^{B+E}\frac{\left[\mathcal{E}(1+\alpha\mathcal{E})\right]^{\frac{3}{2}}}{(1+2\alpha\mathcal{E})}\exp\left(-\frac{1}{k_B}\Lambda_C\mathcal{E}^C\right)\sqrt{\mathcal{E}^\alpha_+(1+\alpha\mathcal{E}^\alpha_+)}$$
$$\times(1+2\alpha\mathcal{E}^\alpha_+)\,d\mathcal{E}.$$

7 Application to bulk Silicon

We test the hierarchy of models in a one-dimensional homogeneous case in a semiconductor with constant doping and applied electric field, using various values of N and M. In figure 1, we show the drift velocity versus the electric field. Fixing N=2 and changing M, one can see that the results at high electric fields vary up to M=10, for greater values of M practically no more changes are present. Taking N=3 and changing M, the variations are very small. Comparing the results with the Monte Carlo ones shown in [Tom93] together with the experimental data, we can conclude that the best results are obtained by using a higher number of scalar moments. We believe that this is due to the fact that no expansion is done with respect to the scalar Lagrange multipliers at variance with the vector ones. The drawback is that when a higher number of scalar moments is used, the Maxwellian is at the boundary of the realizability region in the Lagrange multipliers space so that instabilities arise in the numerical resolution at low fields.
In order to get an idea about the computational costs, in table 1 we report the run time with a pentium 3 pc, for a single value of the electric field, at the varying of the number of moments. The model with $N = M = 3$ requires a CPU time about a factor 2.4 greater than that of the model with $N = M = 2$. However the computing effort, even in the most accurate case, is still adequate for CAD purposes considering the very good agreement with the much more CPU time consuming Monte Carlo simulations.

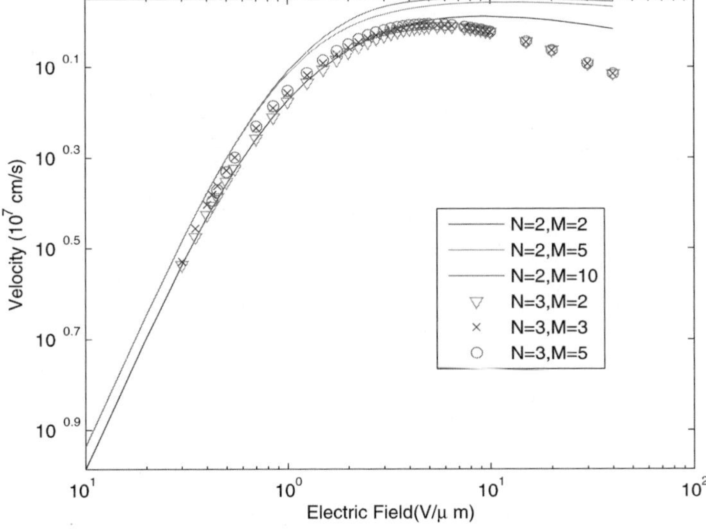

Fig. 1: Drift velocity vs the electric field

N	M	time(s)	N	M	time(s)
2	2	17	3	2	29
2	5	57	3	3	40
2	10	200	3	5	72

Table 1: Run times

References

[AMR03] Anile, A. M., Mascali, G., Romano, V.: Recent developments in hydrodynamical modeling of semiconductors. In: Lecture Notes in Mathematics **1823**. Springer, Berlin Heidelberg New York (2003)

[JR83] Jacoboni, C., Reggiani, L.: The Monte Carlo method for the solution of charge transport in semiconductors with application to covalent materials. Rev. Mod. Phys., **55**, 645–705 (1983)

[Lev96] Levermore, C. D.: Moment Closure Hierarchies for Kinetic Theories. J. Stat. Phys, **83**, 331–407 (1996)

[MR02] Mascali, G., Romano, V.: Hydrodynamical model of charge transport in GaAs based on the maximum entropy principle. Cont. Mech. Thermodyn., **14**, 405–423 (2002)

[MuR98] Müller, I., Ruggeri, T.: Rational Extended Thermodynamics, Springer-Verlag, Berlin (1998)

[Rom00] Romano, V.: Non parabolic band transport in semiconductors: closure of the production terms in the moment equations. Cont. Mech. Thermodyn., **12**, 31–51 (2000)

[Str00] Struchtrup, H.: Extended Moment Method for Electrons in Semiconductors. Phys. A, **275**, 229–255 (2000)

[Tom93] Tomizawa, K.: Numerical simulation of sub micron semiconductor devices. Artech House, Boston (1993)

On the Implementation of a Delaunay-based 3-dimensional Mesh Generator

K.J. van der Kolk and N.P. van der Meijs

Delft University of Technology, EEMCS, Circuits and Systems Group,
Mekelweg 4, NL-2628 CD Delft
{keesjan,nick}@cas.et.tudelft.nl

1 Introduction

A typical problem in engineering is to find a numerical solution to a partial differential equation (or a coupled set thereof), given a number of boundary conditions, and the usual approach of solving the problem starts by discretizing the domain into elementary volumes. In this paper, we focus on mesh generation suitable for the solution of field problems in arbitrary VLSI structures. We assume that the problem cannot be easily reduced to a lower dimensionality by exploiting symmetry or regularity, so that the problem-domain is intrinsically three-dimensional. Also, we assume that the selected numerical technique (e.g., the finite element method) requires a three-dimensional discretization (as opposed to a surface-discretization).
Surveys on mesh generation are given in [2] and [5]. The mesh generator described in this paper is based on techniques from the Delaunay-based mesh generation literature. The main benefit of these techniques is that the quality of the resulting meshes can be *guaranteed*, and, equally important, that the meshes are still small enough to be *practically useful*. An additional advantage is that computation of the mesh is efficient in practice. In general, mesh computation is much faster than solving the subsequent numerical problems. An example mesh generated by our implementation is shown in Figure 1.
Although the principles of Delaunay-based mesh generation are well-known, implementing a mesh generator of this kind is a real challenge. Besides the fact that coding the topological manipulation of the three-dimensional basic elements (tetrahedra in our case) is tedious, one must make sure that the algorithm is robust against floating point errors. Furthermore, the Delaunay-based mesh generation theory allows fast construction of the mesh using only local operations, and it is not trivial to exploit this fact.

2 Delaunay Refinement

Our mesh-generator follows the traditional approach of Delaunay-based mesh refinement. Limitation of space prohibits us to give a full account of the method and its theoretical underpinnings, hence we are obliged to refer the reader to the literature. See, e.g., [5], or [10] for especially good expositions.
For later reference, the global Delaunay refinement algorithm is depicted in Figure 2. The terms SPLIT_1, SPLIT_2, and SPLIT_3 denote the operations of subsegment, subfacet and tetrahedron splitting, respectively. Given a tetrahedron t, $\gamma(t)$ denotes its circumradius-to-shortest-edge ratio. Further, the constant B reflects the "minimum tetrahedron quality," and should be selected greater than 2.

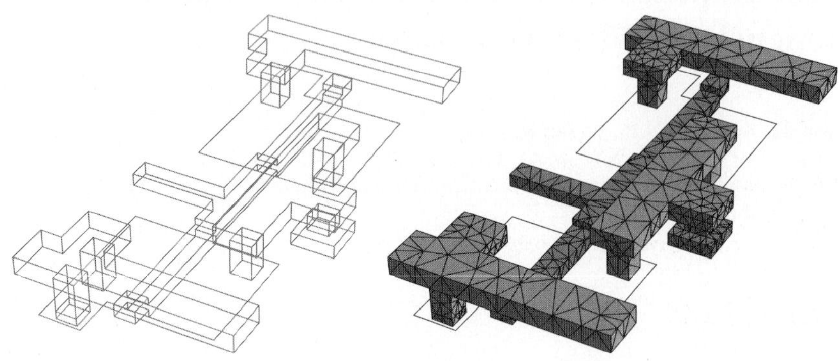

Fig. 1: Example PLC and corresponding mesh. The structure is contained in a bounding box (not shown) and the exterior of the structure is meshed as well in this case.

The input of the algorithm is restricted to a piecewise linear complex (PLC) with angles greater than or equal to $\pi/2$ (this holds both for inter-edge angles and dihedral angles). In our case of VLSI structures, this forms no real limitation. In cases where smaller input-angles are unavoidable, a remedy is to clip-off the offending angles, thereby introducing a slight modification of the original geometry, but one could even mesh the clipped domains separately, in some cases.

3 Delaunay Tetrahedrization

An essential data-structure maintained by the mesh-generator is the Delaunay tetrahedrization (abbreviated DT; we use the same abbreviation for the Delaunay *triangulation*, but add the prefix '2d' or '3d' when confusion may occur) [4]. The DT is constructed incrementally: we start with a basic DT (in fact a single tetrahedron), and as new points are added to the mesh, the representation of the DT is updated. Mathematical tools described in Section 4 guarantee that the DT is uniquely defined.
For incremental point-insertion, we use the Bowyer-Watson method [3, 11]. In essence, given a point p to be inserted into the DT, the method computes the so called Bowyer-Watson polyhedron, which is the union of tetrahedra which have p in their circumsphere (it can be shown that this polyhedron can be computed in $\mathcal{O}(t)$, where t is the number of tetrahedra in the polyhedron; a simple breadth-first search can be used). Then, the polyhedron is emptied (its constituent tetrahedra are removed from the mesh), and new tetrahedra are formed between p and the triangular faces of the polyhedron. It is theoretically guaranteed that the resulting complex is the DT.
Note that the initial DT, consisting exclusively of the points in the input PLC, can be constructed more efficiently by using a sweepline algorithm, as mentioned in [10].

```
while True:
    (STEP 1) if some subsegment s is encroached:
        SPLIT1 s

    else (STEP 2) if some subfacet u is encroached:
        let c be the circumcenter of u

        if c encroaches upon a subsegment s:
            SPLIT1 s
        else:
            SPLIT2 u

    else (STEP 3) if there exists a tetrahedron t with γ(t) > B:
        let c be the circumcenter of t

        if c encroaches upon some subsegment s:
            SPLIT1 s
        else if c encroaches upon some subfacet u:
            SPLIT2 u
        else:
            SPLIT3 t
    else:
        break
```

Fig. 2: Pseudo-code for the three-dimensional mesh-refinement algorithm.

4 Geometric Predicates

The mesh generator depends on geometric predicates for making elementary decisions. Essentially, geometric predicates form the connection between topological information (how elements are connected together) and geometric information (where elements are physically located). Reducing the number of geometrical predicates to a minimum creates the best opportunities for making our algorithm robust w.r.t. degeneracies and round-off error.

We require only two different predicates. One predicate, ORIENT3D, determines the relative orientation of four points in 3d space. Another predicate, INSPHERE, determines, given four points in 3d space, whether a fifth points lies inside or outside the circumscribing sphere of those four points. Both predicates can be computed by evaluating the sign of a determinant. See [10] for precise definitions of these predicates. Note that the algorithm does not utilize two-dimensional equivalents of the predicates (see also Section 8).

In order to uniquely define the Delaunay tetrahedrization, it is necessary and sufficient that any degeneracies are expelled from the geometric predicates. This means that the corresponding determinants must be either positive or negative, but never exactly zero.

We implement non-degeneracy by using the Simulation of Simplicity (SoS) method devised by Edelsbrunner and Mücke [6]. However, it would be highly inefficient to use the SoS method directly for all geometric predicates that need to be evaluated, since the method relies on exact arithmetic. Therefore, we compute first, for every predicate we encounter, the sign of the determinant without the symbolic perturbation exerted by SoS. Only if a degeneracy is found, we resort to SoS.

For the computation of regular signs of determinants, we rely on Shewchuk's adaptive floating point predicate library [9]. For SoS, we have implemented a module capable of performing the necessary symbolic manipulations, where the handling of exact floating-point arithmetic is delegated to the GNU multiprecision library (`libgmp` [1]). In order to reduce the probability of degeneracies, we (physically) perturb the inserted points slightly. The result is that in practice, the SoS-module is used only in a small fraction of the cases, and thus its efficiency is marginal. Most of the optimizations mentioned in [6] can be ignored (at least in our case) since they do not result in any appreciable advantage.

5 Elementary Data-structures

A fair amount of research was needed to come up with the actual data-structures to be used in the mesh-generator. Of course, having proper data-structures is of paramount importance to the efficiency of the algorithm. Here we give a overview of the types of object that the mesh generator handles. This overview is a necessary aid in understanding the algorithm and in seeing why it is efficient.

For every point inserted in the mesh, a node-object is allocated, which contains the physical x, y, z-coordinates of the point. The address of the node is used as the perturbation-index for the SoS method (see Section 4). All other objects (subsegments, subfacets, tetrahedra, etc.) refer to a point by a pointer to the corresponding node-object. Every subsegment-object records an (arbitrarily large) set of subfacet-objects to which it is attached (its 'wings'). Every subfacet records a set of (at most two) abutting tetrahedra. Each subfacet contains three pointers to neighboring subfacets (a pointer can be null in case a neighbor does not exist). Similarly, every tetrahedron contains four pointers to its neighbors.

As a local optimization, a subfacet stores the orientation of each neighboring subfacet, by storing the neighbor's edge-number. Similarly, a tetrahedron stores the orientation of each neighboring tetrahedron, by storing its face-number and by storing the orientation of that face. Note however, that this information can be easily obtained by a local topological investigation based on node-comparisons.

6 Detached Elements

In the final mesh, every subfacet has two abutting tetrahedra associated with it (assuming the subfacet is not at the boundary of the mesh). However, during mesh refinement, a subfacet may of course be detached from its two potentially abutting tetrahedra.

We keep subfacets attached to tetrahedra as much as possible. Thus, when inserting a point into the mesh (modifying the three-dimensional DT, and/or two-dimensional DTs), we examine every face of every *modified* tetrahedron, and see if it matches some subfacet which is in a detached state. If we find such a subfacet, we simply attach it to the tetrahedron. The table used for attaching subfacets and tetrahedra is implemented as a hash-table. Indexing is done using the set of the three corresponding node pointers.

A similar mechanism is used to attach subsegments to subfacets. In our implementation, we also explicitly attach subsegments to tetrahedra (thus requiring an extra table).

7 Encroached Elements

Throughout the refinement process, certain subsegments and subfacets may become encroached [8, 10] and they remain encroached until the corresponding subsegment or subfacet is split (note that a subfacet may also disappear from the mesh due to some other splitting).

Subfacets which are (possibly) encroached are kept in a list. Any time a subfacet may have become encroached (this happens during point-insertion), we add it to the list. It turns out that the Bowyer-Watson insertion-polyhedron (see Section 3) used during point-insertion contains exactly those subfacets which may have become encroached by the point to be inserted, thus the amount of elements to be inserted into the encroachment-list is limited to the neighborhood of the given point.

The actual encroachment-test is delayed until the elements are fetched from the encroachment-list: whenever we need to select an arbitrary encroached subfacet, we pick one from the encroachment-list, and we check whether the subfacet still exists, and whether it is indeed encroached; encroachment is easily detected by examining the apices of the two tetrahedra abutting the subfacet. It remains to be said that for subsegments, the approach is similar.

8 Representation of Facets

For every facet of the PLC, we need to keep a separate (2d) Delaunay triangulation (all nodes contained in the facet are inserted into both the corresponding 2d DT and the global 3d DT, but not vice versa). Each 2d DT is actually implemented using three-dimensional predicates (we define an auxiliary node at some distance from the plane of the facet, and use this node to augment the set of parameters of our three-dimensional predicates).
Most operations on 2d DT's are straighforward modifications of the corresponding operations on 3d DT's. In 2d DT's, however, we remove subfacets at the exterior of the facet, for efficiency. The boundaries of a facet thus form the constraining edges in a constrained Delaunay triangulation (CDT), see, e.g., [5]. Using the fact that these edges form a set of contiguous boundaries, one can show that (in 2d) the Bowyer-Watson insertion scheme is still applicable and retains its efficiency (with only some trivial modifications).

9 Zone Bookkeeping

For each tetrahedron, we record whether it is inside, or outside the part of the domain to be meshed, and this information is updated dynamically. Whenever it is unknown in what part of the domain a tetrahedron resides, its zone is marked as 'unknown'.
If, when querying the zone of a tetrahedron τ, its zone is not 'unknown', we can be sure that we have the correct zone. Otherwise, we perform a straight walk through the mesh towards some point at infinity, and we count the number of walls crossed. When we run into the boundary of the mesh, or if we encounter some tetrahedron with a known zone, we can reconstruct the zone information of τ. After a query, in order to optimize future queries, we record the zone information with τ, and also recursively with its neightbors.
Note from Figure 2 that the concept of the zone in which a tetrahedron lies is properly defined only at STEP 3, since then all subfacets are unencroached and guaranteed to be part of the mesh (in this case no tetrahedron can pierce through a wall of the domain). Fortunately, we need the zone-information exclusively in this step: when splitting skinny tetrahedra, we need to consider just the tetrahedra which are on the 'inside' of the domain, as refinement of the exterior would be wasteful.
Note that in the context of physical VLSI layout, the domain is often bounded by a large cuboid; in that case, determining the zone of a tetrahedron is quite simple. However, when analyzing a design in parts, one can easily end up with different types of geometry.

10 Point-location

We frequently need to determine the tetrahedron in which a given point lies. For example, when splitting a tetrahedron τ_1, we need to insert a node into the tetrahedron τ_2 containing the circumcenter of τ_1. Finding a tetrahedron containing a given point, or 'point location', is implemented using a linear walk (see, e.g., [7]).
A linear walk takes time proportional to the number of tetrahedra visited, and thus the main trick is to always make sure that the initial tetrahedron is topologically close to the final tetrahedron. In the case of splitting a skinny tetrahedron, an obvious candidate for the initial tetrahedron is of course the skinny one, and fortunately, in practice, with this choice one observes that the amount of intermediate tetrahedra visited is generally below a small constant.

There are other situations in which point-location is required. For example, when splitting a subfacet f, we need to find the tetrahedron containing the circumcenter of f (because we need to insert a node into it). Now since the subfacet is not guaranteed to be attached to a tetrahedron (in fact, it is encroached so it is likely not attached), we have no simple initial tetrahedron. To remedy this, we record with each detached subfacet a tetrahedron which is in its proximity (basically, the tetrahedron to which it was attached most recently) so that we can use this tetrahedron as our starting point. Of course, once a subfacet is detached, it can 'drift' away from this tetrahedron (due to insertions elsewhere), but practice shows that point-location still approximately takes constant time. Intuitively, the effect of drifting is only small, since all operations on the mesh are designed to be as localized as possible for additional efficiency reasons.
We also need point-location in two dimensions, i.e., we need the ability to find the subfacet in which a given point lies. The operation is similar to that described above, except that in 2d, the elements at the exterior of the domain are not present. Thus, a linear walk may prematurely halt at the border of the domain. In such cases, we perform an additional breadth-first search from the subfacet at which the linear walk ended. This search is expected to be relatively inexpensive (practice confirms this) since we already assume that the starting subfacet is close to the target subfacet.

11 Conclusion

Although limitation of space precluded an in-depth treatment, we have described the most important ingredients of an efficient three-dimensional mesh generator. Our implementation works and produces around 15.000 tetrahedra per second on a 2.4GHz Intel-based workstation. The reader may obtain a copy of the implementation by contacting the authors.

References

1. The GNU multi-precision arithmetic library, `http://www.swox.com/gmp`.
2. Bern and Eppstein. Mesh generation and optimal triangulation. In *Computing in Euclidean Geometry, Edited by Ding-Zhu Du and Frank Hwang, World Scientific, Lecture Notes Series on Computing – Vol. 1*. 1992.
3. A. Bowyer. Computing Dirichlet tessellations. *Computer J.*, 24:162–166, 1981.
4. B. Delaunay. Sur la sphère vide. *Bull. Acad. Sci. USSR(VII)*, pages 793–800, 1934. Classe Sci. Mat. Nat.
5. H. Edelsbrunner. *Geometry and Topology for Mesh Generation.* Cambridge University Press, 2001.
6. H. Edelsbrunner and E. P. Mücke. Simulation of simplicity, a technique to cope with degenerate cases in geometric computations. *ACM Trans. Graphics*, 9:66–104, 1990.
7. E. P. Mücke, I. Saias, and B. Zhu. Fast randomized point location without preprocessing in two- and three-dimensional delaunay triangulations. In *Proceedings of the 11th Annual Symposium on Computational Geometry*, pages 274–283, 1996.
8. J. Ruppert. A new and simple algorithm for quality 2-dimensional mesh generation. In *Proc. 4th ACM-SIAM Symp. on Disc. Algorithms*, pages 83–92, 1993.
9. Jonathan Richard Shewchuk. Adaptive Precision Floating-Point Arithmetic and Fast Robust Geometric Predicates. *Discrete & Computational Geometry*, 18(3):305–363, October 1997.
10. Jonathan Richard Shewchuk. *Delaunay refinement mesh generation*. PhD thesis, School of Computer Science, Carnegie Mellon University, 1997.
11. D. F. Watson. Computing the n-dimensional Delaunay tessellation with application to Voronoi polytopes. *Computer J.*, 24:167–171, 1981.

Coupled FETI/BETI Solvers for Nonlinear Potential Problems in (Un)Bounded Domains

Ulrich Langer[1, 2] and Clemens Pechstein[2]

[1] Institute of Computational Mathematics, Johannes Kepler University, Altenberger Str. 69, 4040 Linz, Austria ulanger@numa.uni-linz.ac.at
[2] Special Research Program SFB F013, Johannes Kepler University, Altenberger Str. 69, 4040 Linz, Austria clemens.pechstein@numa.uni-linz.ac.at

Summary. In nonlinear electromagnetic field computations, one is not only faced with large jumps of material coefficients across material interfaces but also with high variation in these coefficients even inside homogeneous materials due to the nonlinearity. The radiation condition can conveniently be taken into account by a coupled boundary integral and domain integral variational formulation. The coupled finite and boundary element discretization leads to large-scale nonlinear algebraic systems. In this paper we propose special inexact Newton methods where the Jacobi systems arising in every step of the Newton method are solved by a special preconditioned finite and boundary element tearing and interconnecting solver. The numerical experiments show that the preconditioner proposed in the paper can handle large jumps in the coefficients across the material interfaces as well as high variation in these coefficients on the subdomains. Furthermore, the convergence does not deteriorate if many inner subdomains touch the unbounded exterior subdomain.

Keywords—Nonlinear problems, electromagnetic field computations, domain decomposition, FEM, BEM, Newton's methods, unbounded computational domains

1 Introduction

Domain decomposition (DD) methods of iterative substructuring type, such as the widely used classical finite element tearing and interconnecting (FETI) methods [3], dual-primal FETI (FETI-DP) methods [4] and balanced domain decomposition by constraints (BDDC) [5], are powerful methods for large-scale field computations on parallel computers. In the classical FETI method, the finite element subspaces are treated separately on each subdomain including its boundary. The global continuity across the interfaces is enforced by Lagrange multipliers, which leads to a saddle point problem that can be solved iteratively via its dual problem. From the Lagrange multipliers, the solution can easily be computed. The iteration process is nothing but a preconditioned conjugate gradient (PCG) subspace iteration.
To obtain a fast method, a careful choice of the preconditioner is essential. For the case that the coefficients of the underlying elliptic partial differential equation (PDE) are constant in each subdomain, quasi-optimal preconditioners are available. It is proved that the condition number grows proportionally to $(1+\log(H/h))^2$, where h is the mesh size and H the average diameter of the subdomains. Moreover, the preconditioners are robust with respect to jumps in the coefficients across subdomain interfaces [7, 13]. The PCG subspace iteration involves the use of standard Dirichlet and Neumann solvers.

To summarize, the main success of FETI, FETI-DP and BDDC methods is certainly due to their rather general structure, wide applicability, moderate complexity, robustness, and finally their scalability with respect to parallel computing. For a comprehensive introduction to domain decomposition methods, FETI and FETI-DP methods, we refer to the monograph by Toselli and Widlund [18].

Recently, Langer and Steinbach have introduced the boundary element tearing and interconnecting (BETI) methods [11] as a boundary element counterpart of the FETI methods, as well as the coupled FETI/BETI methods [12]. The BETI method uses boundary-element-based analogs of the FETI operators. Due to spectral arguments, all the properties of FETI methods mentioned above remain valid for BETI methods. Furthermore, inexact and data-sparse techniques are available, cf. [9].
Coupling boundary element and finite element discretizations, one can benefit from the advantages of both discretization techniques. For instance, in electromagnetics, source terms and nonlinearities can be treated more efficiently by the finite element method (FEM) than by the boundary element method (BEM), whereas unbounded domains, moving parts and air regions can efficiently be handled by BEM. We refer to [2] for the symmetric coupling of finite and boundary elements, and to [6, 8] for using this coupling technique to construct domain decomposition solvers.

In this contribution, we use coupled FETI/BETI methods to solve nonlinear potential problems as they appear in nonlinear magnetostatics in two dimensions,

$$\operatorname{div}(\nu(x, |\nabla u(x)|)\nabla u(x)) = f(x) \qquad \text{for } x \in \Omega\,. \tag{1}$$

Additionally we may have suitable transmission conditions, Dirichlet boundary conditions and/or–in case Ω is unbounded–a radiation condition.
In Section 2 we give a review of the coupled FETI/BETI methods for the case that the coefficient ν is constant in the subdomains Ω_i, considering a bounded domain with Dirichlet boundary conditions.
Solving the nonlinear problem (1) in a bounded domain Ω with FETI/BETI methods has been investigated by the present authors in [10]. We give an outline of the main issues in Section 3. Applying Newton's method to the global formulation, the spectrum of the Jacobi matrices in the nonlinear subdomains may show high variation, especially if there are singularities in the solution. We propose a special preconditioner to overcome these difficulties and show its good numerical behavior in a typical magnetostatic model problem.
Section 4 is devoted to unbounded domains. We show how the BETI operators change and give some analytic results. Up to now, we can only prove a suboptimal condition number estimate whereas the performance of the numerical experiments is far more promising.

2 Coupled FETI/BETI Methods

Let $\Omega \subset \mathbb{R}^d$ (where $d = 2, 3$) be a bounded, connected domain with a Lipschitz boundary Γ and the outward unit normal vector n. We assume that Ω is decomposed into p non-overlapping simply-connected Lipschitz domains Ω_i, i. e. $\overline{\Omega} = \bigcup_{i=1}^{p} \overline{\Omega}_i$. It is assumed that the diameters $H_i = \operatorname{diam} \Omega_i$ are all of comparable size and bounded by the maximal diameter H. We define the local boundaries $\Gamma_i = \partial\Omega_i$ and the interfaces $\Gamma_{ij} = \overline{\Omega}_i \cap \overline{\Omega}_j$, and denote the outward unit normal vector on Γ_i by n_i. In the following, we consider the Poisson problem with homogeneous Dirichlet boundary conditions and piecewise constant coefficients, to find u satisfying

$$\begin{aligned} -\operatorname{div}(\alpha_i \nabla u) &= f \quad \text{in } \Omega_i\,, \\ u &= 0 \quad \text{on } \Gamma\,, \qquad\qquad \alpha_i \frac{\partial u}{\partial n_i} + \alpha_j \frac{\partial u}{\partial n_j} = 0 \quad \text{on } \Gamma_{ij}\,, \end{aligned} \tag{2}$$

with α_i = const. The generalization to inhomogeneous and mixed boundary conditions is straightforward.

2.1 Dirichlet-to-Neumann maps

The solution u_i of the local subproblem

$$-\text{div}(\alpha_i \nabla u_i) = 0 \qquad \text{in } \Omega_i\,, \qquad\qquad u_i = g_i \qquad \text{on } \Gamma_i\,, \tag{3}$$

defines the *Steklov-Poincaré operator* $S_i\, g_i := \alpha_i \frac{\partial u_i}{\partial n_i}$, mapping the Dirichlet trace g_i to the corresponding Neumann trace. The contribution of a right hand side f_i to the Neumann trace of the solution is described by the *Newton potential* $N_i\, f_i = -\alpha_i \frac{\partial v_i}{\partial n_i}$, where v_i solves

$$-\text{div}(\alpha_i \nabla v_i) = f_i \qquad \text{in } \Omega_i\,, \qquad\qquad v_i = 0 \qquad \text{on } \Gamma_i\,. \tag{4}$$

These operators can be approximated by the FEM. Fixing a triangulation $\mathcal{T}_{i,\,h}$ of Ω_i with a mesh size h, we denote by K_i the FEM stiffness matrix and partition it according to boundary unknowns (subscript Γ) and inner unknowns (subscript I). The Schur complement

$$S_{i,\,h}^{\mathit{FEM}} := K_{II,\,i} - K_{\Gamma I,\,i}\, K_{II,\,i}^{-1}\, K_{I\Gamma,\,i} \tag{5}$$

is a symmetric and stable approximation of S_i, and

$$N_{i,\,h}^{\mathit{FEM}} \underline{f}_i := \underline{f}_{\Gamma,\,i} - K_{\Gamma I,\,i}\, K_{II,\,i}^{-1}\, \underline{f}_{I\,i} \tag{6}$$

is a stable approximation of the Newton potential N_i, cf. [17].
On the other hand, one can approximate the Steklov-Poincaré operator by means of the BEM, cf. [6, 17]. The solution of (3) satisfies the Caldéron equation

$$\begin{pmatrix} g_i \\ t_i \end{pmatrix} = \begin{pmatrix} \frac{\alpha_i}{2} I - \mathcal{K}_i & \mathcal{V}_i \\ \mathcal{D}_i & \frac{\alpha_i}{2} I + \mathcal{K}_i' \end{pmatrix} \begin{pmatrix} g_i \\ t_i \end{pmatrix} \tag{7}$$

where t_i is the co-normal derivative and $\mathcal{V}_i$, $\mathcal{K}_i$, $\mathcal{K}_i'$, $\mathcal{D}_i$ are the usual boundary integral operators, the single layer potential operator, the double layer potential operator, its adjoint, and the hypersingular integral operator, respectively. In three dimensions the single layer potential operator $\mathcal{V}_i$ is always elliptic, whereas in two dimensions, due to the logarithm in the fundamental solution, it is only elliptic if $\operatorname{diam} \Omega_i < 1$. This property can always be achieved by a suitable coordinate scaling.
After discretizing and eliminating t_i, one obtains the symmetric and stable approximation

$$S_{i,\,h}^{\mathit{BEM}} := D_{i,\,h} + \left(\tfrac{\alpha_i}{2} M_{i,\,h}^{\top} + K_{i,\,h}^{\top}\right) V_{i,\,h}^{-1} \left(\tfrac{\alpha_i}{2} M_{i,\,h} + K_{i,\,h}\right), \tag{8}$$

where $V_{i,\,h}$, $K_{i,\,h}$, $D_{i,\,h}$ are the boundary element matrices corresponding to $\mathcal{V}_i$, $\mathcal{K}_i$, $\mathcal{D}_i$, respectively, and $M_{i,\,h}$ is a mass matrix.
Note that the two approximations $S_{i,\,h}^{\mathit{FEM}}$ and $S_{i,\,h}^{\mathit{BEM}}$ are compatible and both spectrally equivalent to the Galerkin matrices of the exact Steklov-Poincaré operators S_i. The application of $S_{i,\,h}^{\mathit{FEM}}$ or $S_{i,\,h}^{\mathit{BEM}}$ simply corresponds to the solution of local Dirichlet problems. For details we refer to [11, 12].

2.2 Tearing and Interconnecting

Introducing separate variables u_i on the local subdomains, one can re-enforce the continuity of the solution u across interfaces Γ_{ij} by the constraints

$$\sum\nolimits_{i=1}^{p} B_i\, u_i = 0\,, \tag{9}$$

where the B_i are incidence matrices.
Problem (2) can be written as a constraint minimization problem, as well as a saddle point problem involving Lagrange multipliers. Using the notion of the pseudo-inverse (†) and a special projection P addressing the kernels of the subproblems, it is possible to eliminate the primal unknowns u_i. Finally, one obtains the discrete dual FETI/BETI formulation, to find the Lagrange multiplier λ such that

$$P^T F \lambda = d\,, \tag{10}$$

where the FETI/BETI operator F is defined by

$$F = B\left[S_h^{\mathit{FEM/BEM}}\right]^\dagger B^\top = \sum\nolimits_{i=1}^{p} B_i \left[S_{i,h}^{\mathit{FEM/BEM}}\right]^\dagger B_i^\top\,, \tag{11}$$

where $B = \left[B_i\right]_{i=1}^{p}$, $S_h^{\mathit{FEM/BEM}} := \left[S_{i,h}^{\mathit{FEM/BEM}}\right]_{i=1}^{p}$. The application of the pseudo-inverses $[S_{i,h}^{\mathit{FEM/BEM}}]^\dagger$ can be realized by the simple solution of regularized local Neumann problems. Since F is symmetric positive definite on range(P), one can solve the dual problem (10) by a preconditioned conjugate gradient subspace iteration. The preconditioner

$$M_{S,\alpha}^{-1} = (BD_\alpha^{-1}B^\top)^{-1} BD_\alpha\, S_h^{\mathit{FEM/BEM}}\, D_\alpha B^\top (BD_\alpha^{-1}B^\top)^{-1}\,, \tag{12}$$

first introduced and fully analyzed by Klawonn and Widlund [7], satisfies the quasi-optimal condition number estimate

$$\kappa(PM_{S,\alpha}^{-1}P^T P^T F P) \le C(1 + \log(H/h))^2\,, \tag{13}$$

independent of the values–and therefore possible jumps–of the coefficients α_i. It is well known that an appropriate norm of the iteration error of the conjugate gradient method will decrease at least by a factor $2\left(\frac{\sqrt{\kappa}-1}{\sqrt{\kappa}+1}\right)^n$ in n steps. The robustness with respect to the jumps in the coefficients α_i is due to the special diagonal scaling matrix D_α, involving weighted mean values of α_i on cross points, interface lines and interfaces between the subdomains. We further point out that one step of the PCG subspace iteration is performed by the solution of one local Dirichlet and one local Neumann problem on each of the subdomains, and the application of the projection and some scaling matrices, which are both global operations but of a rather small dimension. This is why these tearing and interconnecting methods are most suitable for parallelization. For a more detailed description see e. g. [7, 10, 12]

2.3 Varying coefficients

In this subsection, we consider a varying matrix coefficient $A_i(x)$ instead of a constant coefficient α_i on some of the subdomains discretized by the FEM. We assume $A_i(x)$ to be constant on the finite elements $T \in \mathcal{T}_{i,h}$. In order to determine the amount of variance, we introduce the *spectral variance measure*

$$m_{SV}(A_i) := \frac{\sup_{x\in\Omega_i} \overline{\alpha}_i(x)}{\inf_{x\in\Omega_i} \underline{\alpha}_i(x)}\,, \tag{14}$$

where $\overline{\alpha}_i(x)$ and $\underline{\alpha}_i(x)$ denote the maximal and minimal local eigenvalues of $A_i(x)$, respectively. The application of a preconditioner with Steklov-Poincaré operators corresponding to constant coefficients leads in the worst case to a condition number proportional to $\max_i m_{SV}(A_i)$, which is not at all acceptable in magnetostatic applications. In [10] the present authors have proposed a new preconditioner $\widehat{M}_{S,A}^{-1}$ based on a varying scalar coefficient $\widehat{\alpha}_i(x)$ which can be easily computed from $A_i(x)$, together with a suitable diagonal scaling matrix $D_{\widehat{\alpha}}$. If the local *anisotropy measure*

$$m_{\mathit{anis}}(A_i) := \sup\nolimits_{x\in\Omega_i} \frac{\overline{\alpha}_i(x)}{\underline{\alpha}_i(x)} \tag{15}$$

is moderate, our new preconditioner works fine, cf. Table 1 and [10].

d.o.f.	Lagr.	H/h	$m_{anis}(\zeta)$	$m_{SV}(\zeta)$	Newt.	PCG-steps	ref
806	408	6.3	12.0	187.4	6	14.0	12
3539	777	12.6	12.0	469.9	4	17.8	14
14357	1515	25.3	12.0	852.4	4	21.3	17
57833	2991	20.6	12.0	1496.4	3	25.3	19
232145	5943	101.2	12.4	2670.9	4	27.8	21

Table 1: Average number of FETI PCG iterations to get a reduction of 10^{-8} in the residual of linear subproblems during Newton's iteration, with 70 subdomains, compared to a linear reference problem (ref.) of the same size.

3 Nonlinear Problems

We now consider the following nonlinear magnetostatic model problem:

$$\begin{aligned} -\text{div}[\nu_i(|\nabla u|)\nabla u] &= f \quad \text{in } \Omega_i\,, \\ u &= 0 \quad \text{on } \Gamma\,, \qquad \nu_i(|\nabla u|)\tfrac{\partial u}{\partial n_i} + \nu_j(|\nabla u|)\tfrac{\partial u}{\partial n_j} = 0 \quad \text{on } \Gamma_{ij}\,. \end{aligned} \tag{16}$$

Assuming that the functions $t \mapsto \nu_i(t)\,t\,:\,[0,\infty) \to [0,\infty)$ are strongly monotonically increasing and piecewise C^2, (16) is uniquely solvable in the weak sense and the corresponding Newton iteration converges locally at a quadratic rate. For our computations, we generated these material curves from noisy measurements using the robust interproximation technique introduced by Pechstein and Jüttler in [15]. In the linearized problems a varying matrix coefficient $\zeta_i(\nabla u_h^{(k)}(x))$ appears. In our numerical experiments, it turns out that typically the anisotropy measure $m_{anis}(\zeta_i(\nabla u_h^{(k)}))$ is small, whereas the spectral variance measure $m_{SV}(\zeta_i(\nabla u_h^{(k)}))$ becomes rather large. As one can observe in Table 1, with our new preconditioner $\widehat{M}_{S,A}^{-1}$ such linearized Newton-problems can be solved satisfactorily. In order to get a good initial guess for Newton's iteration, it is convenient to set up a hierarchy of nested grids and use coarse grid solutions as initial guesses on finer levels. For details, see [10].

4 Unbounded domains

In this section we allow that one of the subdomains, namely Ω_0, is the unbounded exterior $\text{ext}(\Gamma_0)$ of its boundary Γ_0, where the interior $\text{int}(\Gamma_0)$ is bounded. Usually, for a typical domain decomposition, $H_0 := \text{diam}\,\text{int}(\Gamma_0) \text{g} H_i$ for $i \neq 0$. In Ω_0, we assume that the homogeneous Poisson equation is satisfied together with a suitable radiation condition, e. g. for $d = 3$,

$$-\alpha_0 \Delta u = 0 \quad \text{in } \Omega_0\,, \qquad |u(x)| = \mathcal{O}(|x|^{-1}) \quad \text{for } |x| \to \infty\,, \tag{17}$$

cf. [14, 16]. In magnetostatic field computations, α_0 could equal $1/\mu_0$, where μ_0 is the permeability of vacuum. The Dirichlet-to-Neumann map on Ω_0 reads

$$S_0^{\text{ext}} = \mathcal{D}_0 + \left(\tfrac{\alpha_0}{2} I - \mathcal{K}_0'\right) \mathcal{V}_0^{-1} \left(\tfrac{\alpha_0}{2} I - \mathcal{K}_0\right). \tag{18}$$

This means the difference between the interior and exterior Steklov-Poincaré operator is just the sign in front of the double layer potential operators $\mathcal{K}$ and $\mathcal{K}'$. As a consequence, S_0^{ext} is always a one-to-one mapping, and Ω_0 must be treated as a non-floating subdomain, cf. [7]. It can be shown that

$$\langle S_0^{\text{ext}} v,\, v\rangle \simeq |v|^2_{H^{1/2}(\Gamma_0)} + \tfrac{1}{H_0}\|v\|^2_{L_2(\Gamma_0)}\,, \tag{19}$$

where in two dimensions the coordinates have to be scaled such that $H_0 \simeq 1$. The BEM-approximation of S_0^{ext} works analogously to (8).

p	H_0/H_F	(H_F/h_F)	4	8	16	32
10	3	(PCG)	8	9	11	13
37	6		9	11	13	16
145	12		9	13	15	17
577	24		11	14	16	–
2305	48		12	–	–	–

Table 2: Iteration numbers of the PCG subspace iteration of the FETI/BETI method in presence of a twodimensional exterior domain. p: Number of subdomains.

By means of Sobolev interpolation techniques and the auxiliary results stated in [7, 18], one can show that the FETI/BETI preconditioner defined in (12) yields the following condition number estimate in the presence of one exterior domain Ω_0,

$$\kappa(P\, M_{S,\alpha}^{-1}\, P^\top\, P^\top\, F\, P) \leq C\Big(1 + \log\Big(\max_{F\subset\Gamma_S} \tfrac{H_F}{h_F}\Big)\Big)^2 \max_{F\subset\Gamma_0} \tfrac{H_0}{H_F}\,, \tag{20}$$

where $\Gamma_S = \bigcup_{i=1}^p \Gamma_i$ is the skeleton of the domain decomposition, and F runs over interfaces of subdomains, i. e. $\overline{F} = \Gamma_{ij}$ for some $i \neq j$, and $H_F := \operatorname{diam} F$. By h_F we denote the minimal mesh size on the interface F. As in the standard case, C is independent of the diameters H_i, H_F, the mesh size h_F and possible jumps in the coefficients α_i.
Note that $(\frac{H_0}{H_F})^{d-1}$ is proportional to the number of subdomains touching Γ_0. Our first numerical experiments show even better performance than expected from the estimate (20). In Table 2 one can observe a logarithmic instead of a linear growth in the condition numbers, when increasing $\frac{H_0}{H_F}$.
Whenever computing on parallel machines with such exterior domains, one has to keep the load balancing of the processors in mind. Our suggestion is either to use a special decomposition with different mesh sizes to obtain balance, or to treat the exterior subdomain on a group of processors and apply an inner parallelization of the corresponding local problem, e. g. using the techniques described in [1] and the references therein. In experiments, we have observed that for nonlinear problems in unbounded domains, a coarse solution of a homogeneous Dirichlet problem (without considering the exterior domain) may serve as a good initial guess for the Newton iteration on the fine level including the exterior domain.

5 Conclusion

In this work we gave an overview on the standard FETI and BETI methods for linear large-scale potential problems on bounded domains, and an outline on how to extend and use these techniques in the presence of nonlinearities and unbounded domains. Our numerical results are rather promising in both cases.
For a proof of the effectiveness of our new preconditioner that can be applied to nonlinear problems, an analysis of FETI/BETI methods for varying coefficients is needed, which is certainly a challenging subject of future research. For the unbounded case, we could give gave some analysis leading to a suboptimal condition number estimate, while our numerical experiments show an even better, quasioptimal behavior.

Acknowledgments

The authors would like to thank Olaf Steinbach and Günther Of (Graz University of Technology, Austria) for fruitful discussions, and gratefully acknowledge the financial support of the Austrian Grid project and the FWF (Austrian Science Funds) Special Research Program SFB F013. We remark that in our numerical experiments we have used PARDISO (`www.computational.unibas.ch/cs/scicomp`) as a sparse direct linear solver and Steinbach's BEM-package OSTBEM.

References

1. Ainsworth, M. and Guo, B.: Analysis of iterative sub-structuring techniques for boundary element approximation of the hypersingular operator in three dimensions, Appl. Anal. **81**(2), 241–280 (2002)
2. Costabel, M.: Symmetric coupling of finite elements and boundary elements. In: Brebbia, C.A., Kuhn, G., Wendland, W.L. (ed.) Boundary elements IX, 411–420, Springer, Berlin (1987)
3. Farhat, C. and Roux, F.-X.: A method of finite element tearing and interconnecting and its parallel solution algorithm, Int. J. Numer. Meth. Engrg., **32**, 1205–1227 (1991)
4. Farhat, C., Lesoinne, M., and Pierson, K.: A scalable dual-primal domain decomposition method, Numer. Linear Algebra Appl., **7**, 687–714 (2000)
5. Dohrmann, C.R.: A preconditioner for substructuring based on constrained energy minimization, SIAM J. Sci. Comput., **25**(1), 246–258 (2003)
6. Hsiao, G.C., Steinbach, O., Wendland, W.L.: Domain decomposition methods via boundary integral equations, J. Comput. Appl. Math., **125**, 521–537 (2000)
7. Klawonn, A., Widlund, O.B.: FETI and Neumann-Neumann iterative substructuring methods: Connections and new results, Comm. Pure Appl. Math., **54**, 57–90 (2001)
8. Langer, U.: Parallel iterative solution of symmetric coupled FE/BE-equations via domain decomposition, Contemp. Math., **157**, 335–344 (1994)
9. Langer, U., Of, G., Steinbach, O., Zulehner, W.: Inexact data-sparse boundary element tearing and interconnecting methods, SIAM J. Sci. Comp., **29**(1), 290–314 (2007)
10. Langer, U., Pechstein, C.: Coupled finite and boundary element tearing and interconnecting solvers for nonlinear potential problems, ZAMM - J. Appl. Math. Mech., **86**(12), 915–931 (2006)
11. Langer, U., Steinbach, O.: Boundary element tearing and interconnecting methods, Computing, **71**(3), 205–228 (2003)
12. Langer, U., Steinbach, O.: Coupled boundary and finite element tearing and interconnecting methods. In: Kornhuber, R., Hoppe, R., Periaux, J., Pironneau, O., Widlund, O., Xu, J. (ed.) Proceedings of the 15th int. conference on domain decomposition. LNCSE, **40**, 83–97, Springer, Heidelberg (2004)
13. Mandel, J., Dohrmann, C.R., Tezaur, R.: An algebraic theory for primal and dual substructuring methods by constraints, App. Num. Math., **54**, 167–193 (2005)
14. McLean, W.: Strongly elliptic systems and boundary integral equations, Cambridge University Press (2000)
15. Pechstein, C., Jüttler, B.: Monotonicity-preserving interproximation of *B*-*H*-curves, J. Comp. Appl. Math., **196**, 45–57 (2006)
16. Steinbach, O.: Numerische Näherungsverfahren für elliptische Randwertprobleme. Finite Elemente und Randelemente. B.G. Teubner, Stuttgart, Leipzig, Wiesbaden (2003)
17. Steinbach, O.: Stability estimates for hybrid coupled domain decomposition methods, Lecture Notes in Mathematics, **1809**, Springer, Berlin Heidelberg (2003)
18. Toselli, A., Widlund, O.: Domain decomposition methods – Algorithms and theory, Springer Series in Computational Mathematics, **34**, Springer, New York (2005)

A Hierarchical Preconditioner within Edge Based BE-FE Coupling in Electromagnetism

K. Straube[1], I. Ibragimov[2], V. Rischmüller[1], and S. Rjasanow[2]

[1] Robert Bosch GmbH, PF 10 60 50, 70049 Stuttgart, Germany
katharina.straube@de.bosch.com,
volker.rischmueller@de.bosch.com
[2] University of Saarland, PF 15 11 50, 66041 Saarbrücken, Germany
ilgis@num.uni-sb.de, rjasanow@num.uni-sb.de

In this paper, a numerical algorithm solving large sparse linear systems that arise in electromagnetic field computation will be presented. It is based on hierarchical partitioning of the matrix and uses block-wise low-rank approximation in combination with element dropping in order to construct a preconditioner for iterative solution. Within the BE-FE coupling, this approximate factorisation is applied as preconditioner for the FE system. The treatment of multiply connected domains will also be described. The efficiency of the presented solver will be shown by means of an electromagnetic valve.

1 Introduction

In the design of electromagnetic components, numerical field computation of three-dimensional problems plays an important role. Efficient solver concepts are necessary to retrieve information about the components behaviour already at an early stage of development. The spatial discretisation is done by coupling of the boundary element method (BEM) and the finite element method (FEM) both based on edge elements. Fine discretisation of complex problems leads to large systems of equations. The BEM part is solved with asymptotically optimal complexity by using block-wise adaptive cross approximation (ACA) [2]. In larger problems, the main cost is then caused by the FEM part. In this paper, an efficient preconditioner for the large sparse FE matrix will be investigated.
The use of BE-FE coupling for complicated geometry can lead to multiply connected subdomains. Therefore, the discrete space approximating the boundary data needs to be extended in order to consider those degrees of freedom corresponding to the holes. This was already described in [10] for the Galerkin BEM. In this work, these degrees of freedom will be considered for the edge collocation method as described in [9, 11] (cf. Section 2). The discretisation then yields a regular non-symmetric system of equations consisting of sparse FE matrices and dense BE matrices, which is solved iteratively.
Due to the ill conditioning of the FE matrix, a preconditioner needs to be constructed. In [6], different strategies of hierarchical concepts solving the sparse FE system were presented. Especially, a non-recursive algorithm computing a preconditioner was developed that combines a block Cholesky decomposition with low-rank approximation and element dropping (Section 3). In this work, the efficiency of this preconditioner will be shown within the solution of the coupled BE-FE problem.
For this, a component of the fuel injection system is simulated using BE-FE coupling and considering multiply connected domains (Section 4). The FE stiffness matrix will be precon-

ditioned by the method described in Section 3 and two other preconditioning concepts. With this, an evaluation of the preconditioners will be carried out.

2 Discretisation and solver

Magnetostatic field problems in $\mathbb{R}^3$ can be described in terms of the magnetic vector potential $\mathbf{A}$ by

$$\operatorname{curl} \frac{1}{\mu} \operatorname{curl} \mathbf{A} = \jmath, \tag{1}$$

where $\jmath$ is the electric current density. The material parameter μ describes the magnetic permeability and may depend on the magnetic field. It is assumed that $\operatorname{div} \jmath = 0$. In order to use BE-FE coupling, the domain is decomposed into an inner domain Ω^- containing all conducting and magnetic materials of the component and an exterior infinite domain $\Omega^+ = \mathbb{R}^3 \setminus \Omega^-$ (cf. Figure 1). The FEM will be applied in Ω^- while the BEM is used in Ω^+. At the coupling

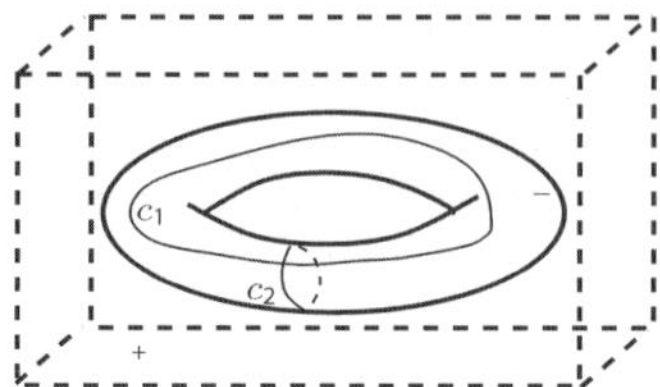

Fig. 1: Decomposition into Ω^+ and Ω^- and two homology cycles c_1 and c_2 of the boundary

boundary Γ of these two domains, the boundary data $\mathbf{A} \times \mathbf{n}$ and $\operatorname{curl} \mathbf{A} \times \mathbf{n}$ needs to be continuous. Here, $\mathbf{n}$ is the outer normal field on Γ.
The variational formulation of (1) is constructed in the Hilbert space $\mathbb{H}(\operatorname{curl}, \Omega^-)$ containing all square integrable functions with an existing curl in the weak sense. With $\mathbf{w} \in \mathbb{H}(\operatorname{curl}, \Omega^-)$ it reads

$$\int_{\Omega^-} \frac{1}{\mu} \operatorname{curl} \mathbf{A} \cdot \operatorname{curl} \mathbf{w} dx - \int_{\Gamma} \gamma_N \mathbf{A} \cdot \gamma_D \mathbf{w} dS_x = \int_{\Omega^-} \jmath \cdot \mathbf{w} dx, \tag{2}$$

with the Dirichlet and Neumann trace operators γ_D and γ_N. The respective trace spaces are $\mathbb{H}_{\perp}^{-1/2}(\operatorname{curl}_{\Gamma}, \Gamma)$ and $\mathbb{H}_{||}^{-1/2}(\operatorname{div}_{\Gamma} 0, \Gamma)$. The trace operators as well as the surface curl and the surface divergence operators are defined in [4]. The boundary integral in (2) prepares the coupling to Ω^+. With the help of the fundamental solution of the Laplace operator $\mathbf{A}^*(\mathbf{x}, \mathbf{y}) = \mathbf{I}/4\pi|\mathbf{x} - \mathbf{y}|$, one can derive a representation formula [9]. In case of $\jmath = 0$ in Ω^+, it only contains boundary integrals:

$$\begin{aligned} \mathbf{A}(\mathbf{y}) = & \int_{\Gamma} (\gamma_N \mathbf{A}^*(\mathbf{x}, \mathbf{y}))^T \gamma_D \mathbf{A}(\mathbf{x}) dS_x - \int_{\Gamma} \gamma_D \mathbf{A}^*(\mathbf{x}, \mathbf{y}) \gamma_N \mathbf{A}(\mathbf{x}) dS_x \\ & + \int_{\Gamma} \gamma (\operatorname{div} \mathbf{A}^*(\mathbf{x}, \mathbf{y}))^T \gamma_n \mathbf{A}(\mathbf{x}) dS_x . \end{aligned} \tag{3}$$

The last integral, where γ is a standard and γ_n a normal trace operator, corresponds to a gauge potential which will be eliminated by the discretisation with collocation over cycles [9].

Let Ω_h^- be a discretisation of Ω^- having k_E edges and k_N nodes. The respective boundary mesh Γ_h has k_E^Γ edges and k_N^Γ nodes. Equation (2) and (3) will be discretised by the use of Whitney p-forms $\mathcal{W}^p$, $p = 0, 1$. For $p = 0$ they are formed by the continuous Lagrangian nodal elements, and for $p = 1$ they are tangentially continuous vector fields defined along edges [3]. With the Whitney 1-forms $\boldsymbol{\omega}_i \in \mathcal{W}^1(\Omega_h^-)$, the approximation of $\mathbf{A}$ reads $\mathbf{A}_h = \sum_{i=1}^{k_E} \beta_i \boldsymbol{\omega}_i$. The approximation of the Dirichlet data contains the degrees of freedom β_i^Γ corresponding to the boundary, so that $(\gamma_D \mathbf{A})_h = \sum_{i=1}^{k_E^\Gamma} \beta_i^\Gamma \boldsymbol{\omega}_i^\Gamma$ with $\boldsymbol{\omega}_i^\Gamma \in \mathcal{W}^1(\Gamma_h)$. As explained in [9], the Neumann data has a zero surface divergence $\mathrm{div}_\Gamma \gamma_N \mathbf{A} = 0$. In order to exploit this, the discrete kernel space

$$\ker(\mathrm{div}_\Gamma) = \mathbf{curl}_\Gamma \mathcal{W}^0(\Gamma_h) \cup \mathcal{K}^1(\Gamma_h) \subset \mathcal{W}^1(\Gamma_h)$$

is used to discretise $\gamma_N \mathbf{A}$. The operator $\mathbf{curl}_\Gamma$ acts on scalar functions and is the adjoint operator of curl_Γ [4]. Here, $\mathcal{K}^1(\Gamma_h)$ is a finite-dimensional space discretising cohomology fields due to the holes in the domain. The Betti number $b = \dim(\mathcal{K}^1(\Gamma_h))$ denotes its dimension, which is given by twice the number of holes of Ω_h^-. The discrete representation of the cohomology group was described in [10]. Therefore, b representative cycles on Γ_h surrounding the holes need to be constructed (cf. c_1 and c_2 in Figure 1). The discrete space $\mathcal{K}^1(\Gamma_h)$ is spanned with the help of scalar functions ψ_k being piecewise linear and continuous on Γ_h except for a jump $[\psi_k]_{c_k} = 1$ across the corresponding homology cycle c_k. With $\widetilde{\mathbf{curl}}_\Gamma$ being the surface curl on $\Gamma \backslash c_k$, $\eta_k = \widetilde{\mathbf{curl}}_\Gamma \psi_k$, $k = 1, \ldots, b$ yields a basis of $\mathcal{K}^1(\Gamma_h)$. The discrete Neumann data reads

$$(\gamma_N \mathbf{A})_h = \sum_{i=1}^{k_N^\Gamma} \varphi_i \mathbf{curl}_\Gamma \lambda_i + \sum_{k=1}^{b} \varphi_{c_k} \sum_{(j,m)\in S_k} \mathbf{curl}_\Gamma \lambda_{j,m},$$

with $\lambda_i \in \mathcal{W}^0(\Gamma_h)$ and $\lambda_{j,m}$ being a restriction of λ_j on the m-th element. The pair (j, m) belongs to the index set S_k if the m-th element lies on one side of the oriented cycle c_k and the j-th node is contained in c_k.

The DeRham collocation method for boundary integral equations was considered in [9] for trivial domains. The evaluation of the discretised representation formula along k_N^Γ closed cycles yields a fully populated unsymmetric system of equations. For multiply connected domains, b additional cycles given by the homology paths of Γ_h are used for the collocation. With this, $k_N^\Gamma + b$ linearly independent equations form the BE system. This discretisation of (2) and (3) yields the system

$$\begin{pmatrix} Q_{\Omega\Omega} & Q_{\Omega\Gamma} & 0 \\ Q_{\Gamma\Omega} & Q_{\Gamma\Gamma} & T \\ 0 & H & G \end{pmatrix} \begin{pmatrix} \beta^\Omega \\ \beta^\Gamma \\ \varphi \end{pmatrix} = \begin{pmatrix} g_\Omega \\ g_\Gamma \\ 0 \end{pmatrix} \tag{4}$$

with $k_E + k_N^\Gamma + b$ unknowns and equations.

The vector potential ansatz is unique up to gradient and cohomology fields. Because of this ambiguity, Q is singular having a large kernel. The regularisation of the system by constructing a matrix of representative vectors spanning the kernel due to gradient fields is applied, so that $K = Q + UU^T$, $U \in \mathbb{R}^{k_E \times k_N}$. The kernel matrix contains discrete gradient fields given by the incidence matrix between edges and nodes of the mesh [3]. The kernel properties of the block matrices stated in [11] allow the regularisation of the system (4). The kernel due to the cohomology fields is eliminated by consideration of the cohomology in the BEM and therefore, the BE-FE system matrix is regular.

This regularised unsymmetric system is solved iteratively by the GMRES method. The BE matrices H and G are well conditioned, so that no preconditioning is required. In Section 3, the construction of a preconditioner for the FE stiffness matrix $K \in \mathbb{R}^{k_E \times k_E}$ will be described.

3 The hierarchical preconditioner

A recent development of numerical linear algebra is the application of hierarchical matrices ($\mathcal{H}$-matrices) to dense matrices arising from integral equations. $\mathcal{H}$-matrices are based on a geometrical clustering of the degrees of freedom so that the matrix can be partitioned into smaller blocks $A \in \mathbb{R}^{n \times m}$ where low-rank approximation $A = UV^T$ with $U \in \mathbb{R}^{n \times r}$ and $V \in \mathbb{R}^{m \times r}$, $r \ll n$ can be applied. In the context of sparse matrices, the idea of hierarchical approximation can be reused in order to approximate the much more populated matrix of the Cholesky decomposition. This was already done for general elliptic differential equations [1]. In [6], we investigated the $\mathcal{H}$-matrix based Cholesky decomposition under consideration of memory reducing clustering. The more promising method was given by a non-recursive way of an approximate decomposition, called HSILLT. It is based on low-rank approximation in combination with element dropping . For this algorithm, we will show the performance in Section 4 so that it is briefly explained.
By a hierarchical interface clustering of the degrees of freedom, a permuation is computed which reorders the system matrix $K \in \mathbb{R}^{k_E \times k_E}$ so that a block structure as shown in Figure 2 arises. The idea stems from a reordering strategy reducing the memory requirement of the Cholesky decomposition called nested-dissection [5]. This permutation is computed with the help of the geometry information corresponding to the degrees of freedom. The clustering algorithm consists of recursive repeats of the two steps: 1. Geometrical bisection, 2. Construction of the interface cluster.

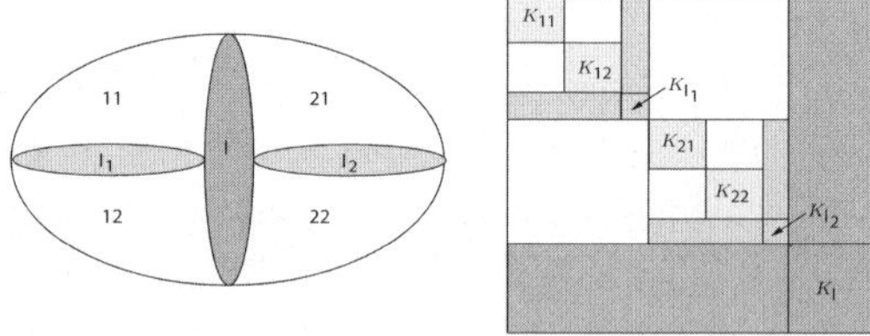

Fig. 2: Hierarchical interface clustering, geometry (left), matrix partitioning (right)

With the application of the clustering, the matrix can be organised in block rows as follows:

$$\left(\begin{array}{c|c|c|c|c}
K_{11} & \multicolumn{4}{c}{K_{*1}^T} \\
\hline
 & \ddots & \multicolumn{3}{c}{\cdots} \\
\cline{3-5}
K_{*1} & \vdots & K_{(i-1)(i-1)} & \multicolumn{2}{c}{K_{*(i-1)}^T} \\
\cline{4-5}
 & & K_{*(i-1)} & K_{ii} & K_{*i}^T \\
\cline{4-5}
 & & & K_{*i} & \ddots
\end{array}\right).$$

The Cholesky decomposition works block column wise with two accuracies $\varepsilon_{\text{drop}}$ and $\varepsilon_{\text{appr}}$. The first one controls the zero bound of small sub-diagonal block rows and the second one is the bound for low-rank approximation of the Schur complement.
The exact decomposition of the first block column would yield

$$\begin{pmatrix} K_{11} & K_{*1}^T \\ K_{*1} & K \end{pmatrix} = \begin{pmatrix} L_{11} & 0 \\ L_{*1} & I \end{pmatrix} \begin{pmatrix} I & 0 \\ 0 & K - L_{*1}L_{*1}^T \end{pmatrix} \begin{pmatrix} L_{11}^T & L_{*1}^T \\ 0 & I \end{pmatrix},$$

where $K_{11} \in \mathbb{R}^{k_1 \times k_1}$ and $K_{*1} \in \mathbb{R}^{(n-k_1) \times k_1}$. Here, $K_{11} = L_{11}L_{11}^T$ and $L_{*1}^T = L_{11}^{-1}K_{*1}^T$. The low-rank approximation of the matrix L_{*1}^T is given by a reduced QR-decomposition with accuracy $\varepsilon_{\text{appr}}$ so that $L_{*1}^T \approx U_1 V_1$. Here, $U_1 \in \mathbb{R}^{k_1 \times r}$ contains r orthonormal columns, i.e. $U_1^T U_1 = I \in \mathbb{R}^{r \times r}$ and $V_1 \in \mathbb{R}^{r \times (n-k_1)}$. This yields

$$\begin{pmatrix} K_{11} & K_{*1}^T \\ K_{*1} & K \end{pmatrix} \approx \begin{pmatrix} L_{11} & 0 \\ V_1^T U_1^T & I \end{pmatrix} \begin{pmatrix} I & 0 \\ 0 & K - V_1^T V_1 \end{pmatrix} \begin{pmatrix} L_{11}^T & U_1 V_1 \\ 0 & I \end{pmatrix}.$$

Since the block K_{*1}^T is column sparse with only k_2 non-zero columns, a column sparse structure of V_1 results respectively. Thus, only the non-zero columns of this matrix will be stored as a fully populated matrix together with an additional information about the indices of the non-zero columns.
The computation of the Schur complement $K - V_1^T V_1$ will be postponed until the decomposition of the ith column is done. There the additional fill-in appears, and it is important to explain how we deal with it.
In the ith elimination step, there are $(i - 1)$ matrices $V_1, \ldots, V_{i-1}$ which must be used to compute the Schur complement updates of the matrices K_{ii} and K_{*i}^T. Thus, some additional non-zero columns will arise in K_{*i}^T. Here, the second accuracy $\varepsilon_{\text{drop}}$ is used. It allows this fill-in only if the norm of the additional column is larger than the norm of the diagonal block times $\varepsilon_{\text{drop}}$.

Algorithm 1 HSILLT

1: **for all** block columns i **do**
2: Compute the updates for the matrices K_{ii} and K_{*i} arising from the previous Schur complements,
3: Compute the Cholesky decomposition of the diagonal block $K_{ii} = L_{ii} L_{ii}^T$,
4: Compute the sub-diagonal block
$k_2 < k_1 : L_{*i} = K_{*i} L_{ii}^{-T}$,
$k_2 \geq k_1 : \textit{with}\ K_{*i} \approx \widetilde{V}_i \widetilde{U}_i,\ \widetilde{V}_i^T \widetilde{V}_i = I,\ \textit{compute}\ L_{*i} = \widetilde{V}_i (\widetilde{U}_i L_{ii}^{-T})$,
5: $k_2 < k_1$: Compute a low-rank approximation of $L_{*i} \approx V_i U_i,\ U_i^T U_i = I$,
$k_2 \geq k_1$: Do postcompression for $\widetilde{V}_i (\widetilde{U}_i L_{ii}^{-T}) \approx V_i U_i,\ U_i^T U_i = I$.
6: **end for**

This algorithm has a memory complexity of $\mathcal{O}(rn \log_2 n)$ and the number of operations is $\mathcal{O}(r^3 n \log_2^2 n)$ [6].

4 Numerical example

In order to evaluate the presented method by an industrial application, a magnetic valve as an essential component of a fuel injection system is simulated by BE-FE coupling. It consists of a ringshaped coil to carry the exciting current, a core and a yoke as well as a moving armature. The discretisation of the three-dimensional domain is performed with the help of tetrahedral and prismal edge elements. Due to the toroidal geometry, its discretisation has one hole along the z-axis of the mesh (cf. Figure 3). Therefore, two additional degrees of freedom are added in order to construct the correct discretisation space for the Neumann data as explained in section 2. The domain has 4000 boundary elements and 31000 finite elements. The excitation is given by a current of $10A$ and the coil has 100 windings. We assume a non-linear material with an approximated magnetisation curve. Because of this, the Newton-Raphson method is applied. In every Newton step, the resulting linear system of equations is solved by the GMRES method.
For increasing problem size, the FE matrix K is getting more ill-conditioned. Therefore, a preconditioner is constructed by HSILLT (cf. Section 3). For the comparison, two other solver

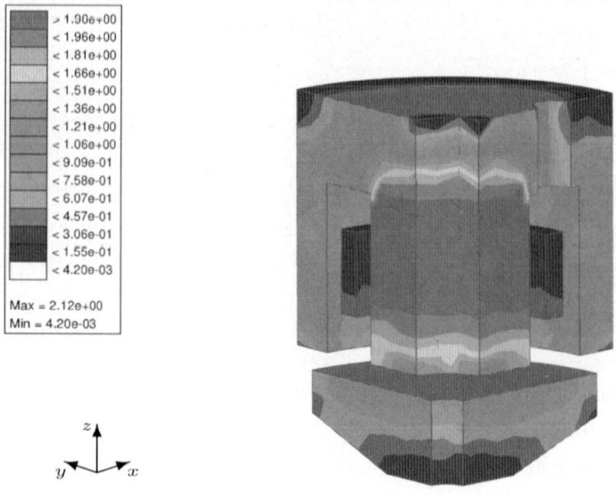

Fig. 3: A quarter of the valve geometry, where only material components and the coil are shown. The colour scale indicates the magnitude of the magnetic induction.

	GMRES-Iter	Mem_L [MB]	t_{LL^T} [min]	t_{tot} [min]
HSILLT	1255	232	9	34
Kaporin	1305	227	130	171
Taucs	1287	258	36	64

Table 1: Comparison of different preconditioner concepts for the magnetic valve.

concepts are also applied: an incomplete factorisation method by Kaporin [7] and one given by the Taucs library [12]. The preconditioner HSILLT uses the interface clustering for reordering the matrix. In the other cases, we choose a reordering also based on a nested-dissection strategy given by the Metis library [8]. The accuracies of the preconditioners are chosen in order to get similar convergence of the iterative method.

With all three preconditioners, the solution as shown in Figure 3 was computed. The magnitude of the magnetic induction in the material domain is shown in a quarter of the valve. The maximum field can be found in the centre of the core and its value is 2.1 Tesla.

In Table 1, the performance of the preconditioners is compared. The number of Newton iteration steps is 10 for all three preconditioners in order to reach a Newton residual of 10^{-9}. In every Newton step the GMRES method terminates with a residual of 10^{-12}. The complete number of required GMRES steps is stated in the first column. The memory required for storing the incomplete Cholesky factor L is given by Mem_L, and the required time for its computation is t_{LL^T}. Moreover, the total computation time t_{tot} can be seen.

The factorisation time of HSILLT is four times faster than that of Taucs. Taucs computes a decomposition column-wise by dropping elements via an accuracy criterion, whereas HSILLT uses a low-rank approximation of sub-diagonal blocks. The block structure of those sub-diagonal entries allows us to use fast level-3-blas matrix operations (cf. Algorithm 1). The Kaporin method is said to yield high quality preconditioners. However, it doesn't use a block structure and requires a lot of index searches, so that it operates very slowly. A comparison for more complicated materials should be done in the future.

5 Conclusions

In this paper, we have presented a method to compute magnetic field problems on multiply connected domains with edge based BE-FE coupling and the time efficient preconditioner HSILLT. The efficiency of HSILLT was succesfully demonstrated by means of the numerical example of the fuel injection system.

References

1. M. Bebendorf. Why approximate LU decomposition of finite element discretisations of elliptic operators can be computed with almost linear complexity. *preprint 8/2005, Max-Planck-Insitut MiS, Leipzig*, 2005.
2. M. Bebendorf and S. Rjasanow. Adaptive low-rank approximation of collocation matrices. *Computing*, 70(1):1–24, 2003.
3. Bossavit, A. *Computational Electromagnetism.* Academic Press series in Electromagnetism. Academic Press, 1997.
4. A. Buffa and P. Ciarlet. On traces for functional spaces related to Maxwell's equations. Part I: An integration by parts formula in Lipschitz polyhedra. *Meth. Appl. Sci.*, 24:9–30, 2001.
5. A. George. Nested dissection of a regular finite element mesh. *SIAM J. Numer. Anal.*, 10:345–363, 1973.
6. I. Ibragimov, S. Rjasanow, and K. Straube. Hierarchical Cholesky decomposition of sparse matrices arising from curl-curl-equation. *to appear in J. Numer. Math.*, 2006.
7. I. E. Kaporin. High quality preconditioning of a general symmetric positive definite matrix based on its $U^TU + U^TR + R^TU$-decomposition. *Numer. Linear Algebra Appl.*, 5:483–509, 1998.
8. G. Karypis and V. Kumar. *MeTis: Unstructured graph partitioning and sparse matrix ordering system, version 2.0*, 1995.
9. S. Kurz, O. Rain, V. Rischmüller, and S. Rjasanow. Discretization of boundary integral equations by differential forms on dual grids. *IEEE Trans. Mag.*, 40(2):826–829, 2004.
10. J. Ostrowski. *Boundary element methods for inductive hardening*. PhD thesis, Universität Tübingen, 2003.
11. O. Rain, B. Auchmann, S. Kurz, V. Rischmüller, and S. Rjasanow. Edge-based BE-FE coupling for electromagnetics. *IEEE Trans. Mag.*, 42(4):679–682, 2006.
12. S. Toledo. Taucs: A library of sparse linear solvers, http://www.tau.ac.il/∼stoledo/taucs. 2003.

Solution of Band Linear Systems in Model Reduction for VSLI Circuits*

Alfredo Remón[1], Enrique S. Quintana-Ortí[1], and Gregorio Quintana-Ortí[1]

Depto. de Ingeniería y Ciencia de Computadores
Universidad Jaume I
12.071–Castellón, Spain
{remon,quintana,gquintan}@icc.uji.es

Summary. We investigate the solution of linear systems that appear when model reduction via system balancing is applied in circuit simulation and design. We show how the properties and structure of the coefficient matrices allow the development and use of efficient parallel algorithms for the solution of the corresponding linear systems. Experimental results are reported on an Intel SMP multiprocessor.

Key words: Large linear systems, model order reduction, system balancing, VLSI circuits, Lyapunov equations, LR-ADI iteration, multithreaded BLAS, multicore and SMP architectures.

1 Introduction

We consider dynamical linear systems, given in generalized state-space form by

$$\begin{aligned} E\dot{x}(t) &= Ax(t) + Bu(t), \quad & t > 0, \\ y(t) &= Cx(t) + Du(t), \quad & t \geq 0, \end{aligned} \tag{1}$$

where $A, E \in \mathbb{R}^{n\times n}$, $B \in \mathbb{R}^{n\times m}$, $C \in \mathbb{R}^{p\times n}$, $D \in \mathbb{R}^{p\times m}$, and $x(0) = x^0 \in \mathbb{R}^n$ is the initial state. Here, n is the order of the system and the associated transfer function matrix (TFM) is $G(s) = C(sE - A)^{-1}B + D$. In model order reduction (MOR) we are interested in finding

$$\begin{aligned} \widehat{E}\dot{\widehat{x}}(t) &= \widehat{A}\widehat{x}(t) + \widehat{B}u(t), \quad & t > 0 \\ \widehat{y}(t) &= \widehat{C}\widehat{x}(t) + \widehat{D}u(t), \quad & t \geq 0, \end{aligned} \tag{2}$$

of order r, with $r \ll n$, and TFM $\widehat{G}(s) = \widehat{C}(s\widehat{E} - \widehat{A})^{-1}\widehat{B} + \widehat{D}$ which "approximates" $G(s)$. Systems of the form (1) arise in circuit simulation and design (CSD) [4]. When modeling the interconnect or the pin package of VLSI circuits, the order is often too large to allow simulation in an adequate time or to even tackle the model using differential equation solvers. Therefore, MOR is frequently used to replace the circuit model by one of much smaller order. Here we will only consider methods based on system balancing, which are specially appealing in that they provide global computable error bounds and can preserve the system properties. For a survey of different MOR techniques, see [1].

* This research was supported by the DAAD programme Acciones Integradas HA2005-0081, the CICYT project TIN2005-09037-C02-02 and FEDER, and project No. P1B-2004-6 of the *Fundación Caixa-Castelló/Bancaixa and UJI.*

There exist various methods for MOR which aim at balancing the system [1]. The core computation in many of these is the solution of the Lyapunov equations

$$\begin{array}{r} AW_oE^T + EW_oA^T + BB^T = 0, \\ A^TW_cE + E^TW_cA + C^TC = 0, \end{array} \tag{3}$$

for Cholesky or full rank factors of $W_o, W_c \in \mathbb{R}^{n \times n}$. Lyapunov solvers based on the LR-ADI iteration [7] are specially efficient when both A and E are sparse, and the constant terms in (3) are of low numerical rank (a usual case in CSD). The LR-ADI method requires, at a given iteration j, the solution of a linear system of the form

$$U_{j+1} := \widehat{A}_j^{-1} U_j = (A + \gamma_j E)^{-1} U_j, \tag{4}$$

where $\{\gamma_j\}_{j=0}^{\infty}$ are scalars with periodicity t_s, and $\{U_j\}_{j=0}^{\infty}$ have all m or p columns, depending respectively on whether the first or the second equation in (3) is being solved for. Therefore, the linear systems in iterations j and $j + t_s$ share the same coefficient matrix and the use of direct solvers is highly recommendable. For further details on the LR-ADI iteration and its parallelization on distributed-memory parallel architectures see, respectively, [7, 6] and [2, 3].
The coefficient matrices of the linear systems (4) associated with CSD models often present a band structure (or can be transformed to that form), allowing the use of band solvers in LAPACK. In this paper we describe how specialized band solvers can be designed to efficiently exploit the properties and structure of these matrices. In particular, we describe new algorithms for the factorization and backward/forward substitution stages with band symmetric positive definite matrices. Combined with a *multithreaded* implementation of BLAS, the codes allow the parallel solution of large-scale linear systems on current multicore and SMP architectures in reasonable time. Experimental results on a 14-way Intel Itanium2 multiprocessor provide evidence in support.

2 Efficient Solvers for Band Linear Systems

In this section we first review the codes in LAPACK for the factorization of band matrices. We then propose two new variants with the potential to attain higher efficiency. To illustrate this, we employ the routine in LAPACK for the Cholesky factorization of a band matrix, xPBTRF [5]. The same modifications carry over to the LU factorization of band matrices [8]. We conclude the section by describing how to improve the performance of the routine for the solution of triangular band systems in current implementations of BLAS, xTBSV.

2.1 LAPACK routine xPBTRF

Given a symmetric positive definite (s.p.d.) matrix $\widehat{A} \in \mathbb{R}^{n \times n}$ with bandwidth k_d, routine xPBTRF delivers a lower (or upper) triangular factor $L \in \mathbb{R}^{n \times n}$, of bandwidth k_d, such that $\widehat{A} = LL^T$. Upon completion, L overwrites the lower triangular part of $\widehat{A}$. Now, let $\widehat{A}$ be partitioned as

$$\left(\begin{array}{c|c} A_{TL} & A_{TR} \\ \hline A_{BL} & A_{BR} \end{array}\right) = \left(\begin{array}{c|cccc} A_{00} & A_{01} & A_{02} & & \\ \hline A_{10} & A_{11} & A_{12} & A_{13} & \\ A_{20} & A_{21} & A_{22} & A_{23} & A_{24} \\ & A_{31} & A_{32} & A_{33} & A_{34} \\ & & A_{42} & A_{43} & A_{44} \end{array}\right), \tag{5}$$

where $A_{TL}, A_{00} \in \mathbb{R}^{k \times k}$, $A_{11}, A_{33} \in \mathbb{R}^{n_b \times n_b}$, and $A_{22} \in \mathbb{R}^{k_d - n_b \times k_d - n_b}$. Here, the block size n_b is chosen to tune the performance of the routine and is related with the size of the

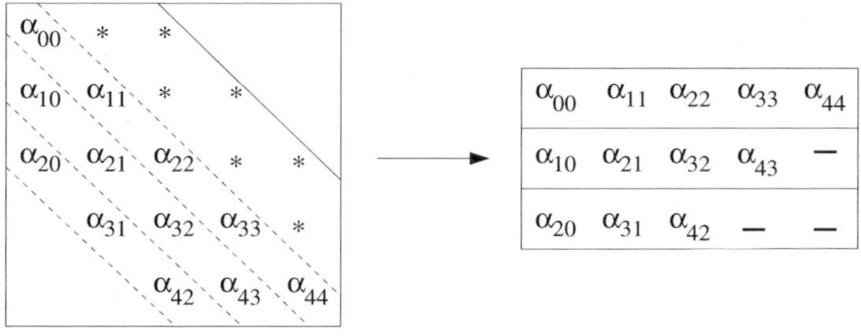

Fig. 1: Compact storage of a 5×5 symmetric band matrix.

cache of the architecture. At a given iteration of the right-looking algorithm in xPBTRF, A_{BL} and the lower triangular part of A_{TL} have been overwritten with the corresponding blocks of L, and A_{BR} has been updated conformally. The following operations are then performed on different parts of A_{BR} during the current iteration:

$$\begin{array}{rrcll}
1.1) & \text{Factorize } A_{11} & = & L_{11}L_{11}^T, & (\text{xPOTF2}) \\
2.1) & A_{21}\ (= L_{21}) & := & A_{21}L_{11}^{-T}, & (\text{xTRSM}) \\
2.2) & \text{TRIL}(A_{22}) & := & \text{TRIL}(A_{22}) - L_{21}L_{21}^T, & (\text{xSYRK}) \\
 & W & := & \text{TRIU}(A_{31}), & \\
3.1) & W & := & WL_{11}^{-T}, & (\text{xTRSM}) \\
3.2) & A_{32} & := & A_{32} - WL_{21}^T, & (\text{xGEMM}) \\
3.3) & \text{TRIL}(A_{33}) & := & \text{TRIL}(A_{33}) - WW^T, & (\text{xSYRK}) \\
 & \text{TRIU}(A_{31})\ (= L_{31}) & := & \text{TRIU}(W). &
\end{array} \quad (6)$$

Here $\text{TRIU}(A_{ij})$ and $\text{TRIL}(A_{ij})$ denote, respectively, the upper and lower triangular parts of A_{ij}, and the expressions are annotated with the LAPACK/BLAS routines that are employed for their computation. Provided $n_b \ll k_d$, a major part of the floating-point arithmetic operations (flops) are performed in terms of the BLAS-3 computation in 2.2), and high performance is to be expected if a tuned implementation of xSYRK is utilized. On the other hand, no attempt is made to exploit the upper triangular structure of A_{31} in 3.1)–3.3) as there is no appropriate kernel in BLAS. The packed storage scheme utilized for symmetric band matrices (illustrated in Fig. 1), and the use of BLAS kernels in 3.1)–3.3), results in the copies to/from the work space W.

Operations involving small blocks during the factorization stage in general do not attain high performance. While theoretically the influence (cost) of the computations involving A_{11}, A_{31}, and A_{33} on the overall process should be small, practice has shown us otherwise. In some experiments with s.p.d. matrices, the optimal block size n_b determines that as much as 20–30% of the time is spent in these small operations; the ratio is even higher when multiple processors and a multithreaded BLAS are employed.

In order to overcome this problem, we propose two different algorithms with a common goal: to *integrate the updates on the small blocks with those of larger operations*. The first algorithm, xPBTRF+A, requires padding the data structure containing $\widehat{A}$ with n_b rows of zeros in the bottom. In this way, the update of A_{31} can be combined with that of A_{21} in a single call to xTRSM, and a single call to xSYRK suffices to update A_{22}, A_{32}, and A_{33}. Access to the strictly lower triangular part of A_{31} in routines xTRSM and xSYRK only touches the zero entries in the padded zone (see Fig. 2 (left)) and therefore does not affect the result. This strategy requires space for an extra $n_b \times n$ block which, provided $n_b \ll k_d$, is small.

In the second algorithm, xPBTRF+B, no padding is required. As access to the strictly lower triangular of A_{31} in routines xTRSM and xSYRK would actually involve the elements in A_{11} in the compact storage (see Fig. 2 (right)), we first copy the latter block to an auxiliary $n_b \times n_b$

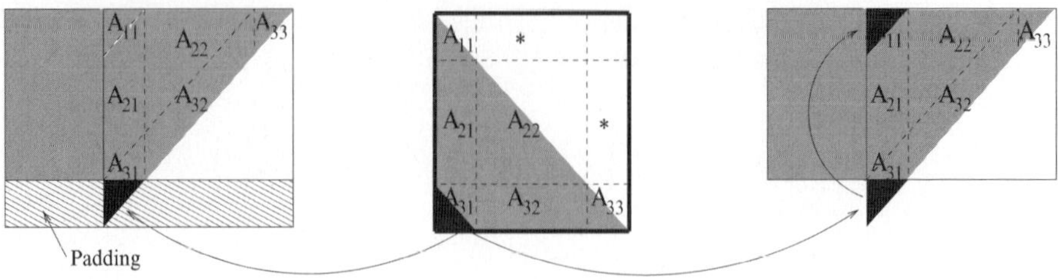

Fig. 2: Access to elements in the strictly lower triangle of A_{31} in the compact storage for the symmetric band matrix in xPBTRF+A (left) and xPBTRF+B (right).

workspace, W, while simultaneously setting its entries to zero. Now, the updates of A_{21} and A_{31} can be combined in a single call to xTRSM, and a single call to routine xSYRK suffices to update all three A_{22}, A_{32}, and A_{33}. When these operations have been completed, the elements of A_{11} are restored from those in W.

2.2 BLAS implementations of xTBSV

We have experimentally evaluated several implementations of the routine for the solution of linear systems with triangular band coefficient matrix, xTBSV. Our tests included some of the most efficient implementations of BLAS for the Intel family of processors, as MKL (`http://www.intel.com`), Goto BLAS (`http://www.tacc.utexas.edu/resources/software`), and ATLAS (`http://www.netlib.org/atlas`). Invariably, these implementations delivered low performance, probably due to the little interest of tuning such a routine to vendors.
As the performance of the xTBSV routine is crucial in our CSD problems, we have developed our own implementation. For that purpose, we started from the legacy code for xTBSV at `http://www.netlib.org/blas`. This implementation is a "direct" code, with no references to BLAS. We then substituted as many lines of this code by calls to BLAS. We expect that, depending on the actual bandwidth of the triangular band matrix, the use of BLAS outpeforms a direct code.
On the other hand, there is no specification in BLAS for a triangular band solver with multiple right-hand sides, although such routine is necessary, e.g., in the solution of linear systems arising in CSD. Besides, the presence of multiple right-hand side vectors enables the use of a BLAS-3 version instead of having to perform multiple calls to a BLAS-1/2 routine xTBSV, one per right-hand side. Certainly, such routine could benefit much our band linear system solvers and we therefore developed it. Following the LAPACK convention the new routine was named as xTBSM.

3 Experimental Results

The following experiments were performed using IEEE double-precision (real) arithmetic on a parallel SMP platform consisting of 14 Intel Itanium2 processors@1.5 GHz, with 6 MB of L3 cache per processor, and 30 GB of shared RAM. The multithreaded BLAS implementation in MKL 8.1 was employed. We consider three different examples from the Oberwolfach MOR benchmark collection (`http://www.imtek.uni-freiburg.de/simulation/benchmark`). Matrices $-A$ and E in the systems are s.p.d. allowing the use of the Cholesky factorization during the factorization of the corresponding linear systems. MATLAB routine `symrcm` was used to reduce the bandwidth in the second and third examples below. A brief description of the examples follows:

Example	Time (in secs.) $p\times$ DTBSV	Time (in secs.) DTBSM	Speed-up DTBSM
t2dah	2.09e−1	2.32e−2	9.37
chip_v0	4.13e−1	8.26e−2	5.00
gas	5.79e+0	1.29e+0	4.46

Table 1: Performance of the triangular band solvers.

Example `t2dah`**.** This is a model of a μ thruster array with n=11,445 states, $m = 1$ input, $p = 7$ outputs, and bandwidth k_d=231.
Example `chip_v0`**.** This model is used for 3D simulation of convective thermal flow in a chip with n=20,082, $m = 1, p = 5$, and k_d=1,226.
Example `gas`**.** This is a μ machined metal oxide gas sensor array with n=66,917, $m = 1$, $p = 28$, and k_d=1,957.
Although sparse (parallel) linear system solvers as SuperLU or MUMPS could be used, in some of these examples this leads to explosive fill-in so that memory is rapidly exhausted and the factorization is not possible and/or keeping the factors during the LR-ADI iteration becomes infeasible.
Our first experiment evaluates the performance of the (double-precision) factorization routine DPBTRF, and the two variants we described in the previous section, DPBTRF+A and DPBTRF+B, when linked with a multithreaded implementation of BLAS (MKL). The plots in the left-hand side of Fig. 3 report the execution times of the three routines for the narrow- (`t2dah`), medium- (`chip_v0`), and wide-band (`gas`) problems as the number of threads (and therefore processors) is increased. The bandwidth of the different problems determines the maximum number of threads that should be used in order to reduce the solution time of the linear systems. Thus, 2, 4, and 8 threads produce the lowest execution times for examples `t2dah`, `chip_v0`, and `gas`, respectively. Using more threads does produce an increase of the execution time due to the overhead of communications among processors.
The plots in the right-hand side of Fig. 3 show the speed-ups attained by modified variants, computed as the ratio between the execution time of DPBTRF and those of routines DPBTRF+A and DPBTRF+B. For the narrow-band case, the speed-up of the new variants using 2 threads is between 25% and 37%. However, we recognize that this is due to the poor performance of the LAPACK routine DPBTRF, which should have been executed using a single thread. For the other two examples, the speed-ups range from 5% and 10% except in very few cases (the use of three threads seems to be problematic here).
Our next experiment investigates the performance of triangular band solvers when multiple linear systems are to be solved. This is the case, e.g., for the Lyapunov equations associated with the three CSD examples considered in this section. Linear systems as in (4) with (p=)7, 5, and 28 right-hand sides need to be solved in the iteration for examples `t2dah`, `chip_v0`, and `gas`, respectively. Table 1 reports the execution routines obtained by solving the triangular band systems associated with the three examples using a single processor. We note that p calls to routine DTBSV are necessary while a single call to DTBSM suffices. The third column in the table reports remarkable speed-ups when solving the systems via routine DTBSM over DTBSV.

4 Conclusions

We have addressed the parallel solution of band linear systems as those arising in model order reduction of dynamical linear systems for circuit simulation and design. Current codes for band linear systems in LAPACK, and implementations in BLAS are not optimal when executed on SMP parallel architectures. We have presented two new routines for the factorization of s.p.d. band matrices that outperform the routine in LAPACK by 5–10%, depending

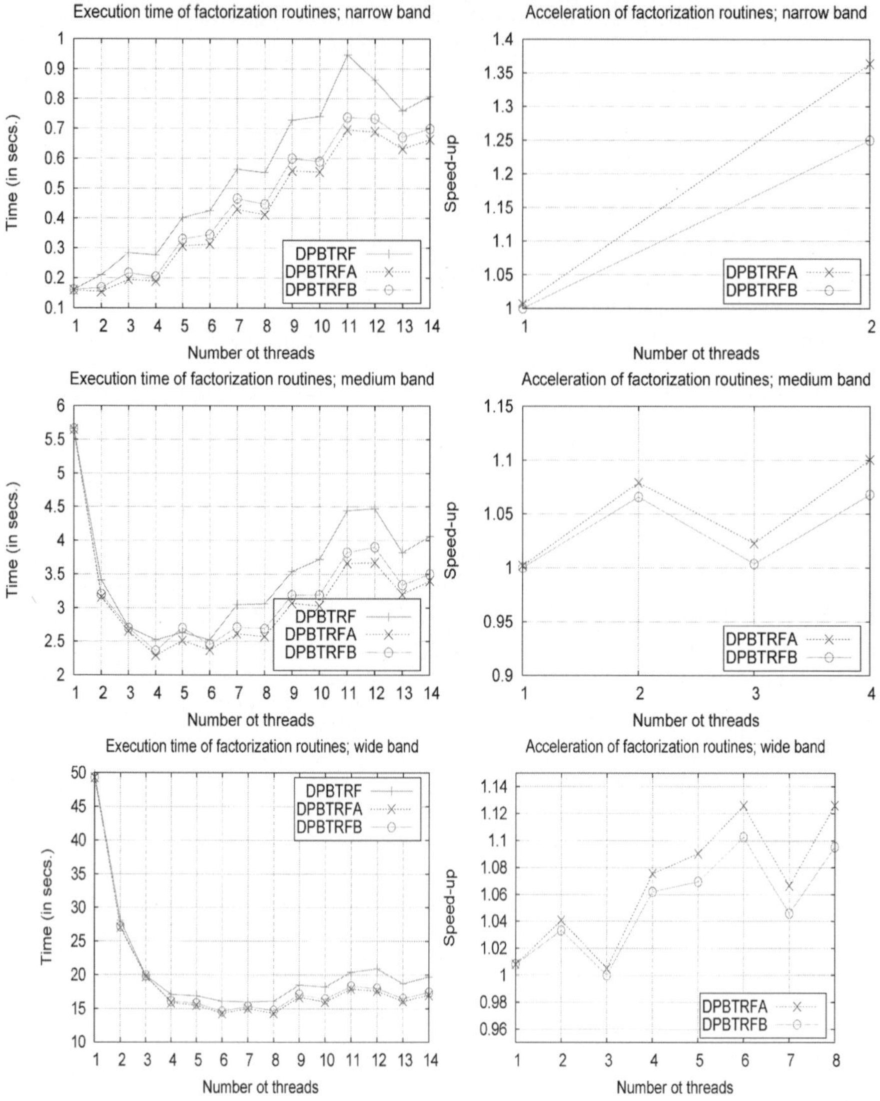

Fig. 3: Performance of the parallel factorization routines.

on the bandwidth of the problem and the number of threads/processors that are employed. We have also coded a BLAS-like routine for the solution of triangular band systems with multiple right-hand sides. By enabling the introduction of BLAS-3 kernels, the new routine clearly outperforms implementations of routine DTBSV in all BLAS that were tested. The codes allow the solution of large-scale linear systems on current parallel multicore and SMP architectures in a reduced time while avoiding explosive fill-in of sparse linear system solvers in certain examples.

References

1. A.C. Antoulas. *Approximation of Large-Scale Dynamical Systems*. SIAM Publications, Philadelphia, PA, 2005.
2. J.M. Badía, P. Benner, R. Mayo, and E.S. Quintana-Ortí. Solving large sparse Lyapunov equations on parallel computers. In B. Monien and R. Feldmann, editors, *Euro-Par 2002 – Parallel Processing*, number 2400 in Lecture Notes in Computer Science, pages 687–690. Springer-Verlag, Berlin, Heidelberg, New York, 2002.
3. J.M. Badía, P. Benner, R. Mayo, and E.S. Quintana-Ortí. Parallel algorithms for balanced truncation model reduction of sparse systems. In J. Dongarra, K. Madsen, and J. Wasniesky, editors, *PARA'04 – Workshop on state-of-the-art in Scientific Computing*, number 3732 in Lecture Notes in Computer Science, pages 267–275. Springer-Verlag, Berlin, Heidelberg, New York, 2004.
4. C.-K. Cheng, J. Lillis, S. Lin, and N.H. Chang. *Interconnect Analysis and Synthesis*. John Wiley & Sons, Inc., New York, NY, 2000.
5. J. Du Croz, P. Mayes, and G. Radicati. Factorizations of band matrices using level 3 BLAS. In *CONPAR 90/VAPP IV: Proc. of the Joint Int. Conf. on Vector and Parallel Processing*, pages 222–231, London, UK, 1990. Springer-Verlag.
6. J.-R. Li and J. White. Low rank solution of Lyapunov equations. *SIAM J. Matrix Anal. Appl.*, 24(1):260–280, 2002.
7. T. Penzl. A cyclic low rank Smith method for large sparse Lyapunov equations. *SIAM J. Sci. Comput.*, 21(4):1401–1418, 2000.
8. A. Remón, E.S. Quintana-Ortí, and G. Quintana-Ortí. Parallel LU factorization of band matrices on SMP systems. In M. Gerndt and D. Kranzmüller, editors, *HPCC'06*, Lecture Notes in Computer Science. Springer-Verlag, Berlin, Heidelberg, New York, 2006. To appear.

MOESP Algorithm for Converting One-dimensional Maxwell Equation into a Linear System

E. F. Yetkin[1], H. Dağ[2], and W. H. A. Schilders[3]

[1] Istanbul Technical University, Informatics Institute, Istanbul, Turkey
fatih@be.itu.edu.tr
[2] Isk University, Information Technologies Department, Istanbul, Turkey
dag@isikun.edu.tr
[3] Technical University of Eindhoven, CASA, Eindhoven, The Netherlands
w.h.a.schilders@tue.nl

Summary. We present a method for converting 1-D Maxwell equation into a linear system using the Multivariable Output Error State Space (MOESP) method, a subspace system identification method. To show the efficiency of the method, we first apply it to a set of ordinary differantial equations. Input and output from the equation set are computed by numerical methods and the obtained data is used for building the required matrices. An appropriate Single Input Single Output (SISO) linear system is estimated by MOESP algorithm for the equation at hand. The goal of the research is to build a low order linear state space system model for the Maxwell equation. On the other hand the order estimation for the system can be used in other way. For example, with this estimation one can determine an appropriate order for the physical system, for which one of the well-known model order reduction techniques can be used to obtain a reduced order model.

1 Introduction

In general, system identification methods are mainly developed in the area of automatic control to determine the best model (in the sense of input-output relationship) from a given observed input-output data set. In this study, a 1-D Maxwell equation is converted into a set of state-space equations using MOESP algorithm, which is a member of subspace system identifacation family of algorithms. The idea can be useful when simulation of the VLSI interconnections are considered. The computation of the effects of VLSI interconnections is mainly based on the solution of the Maxwell equations on chip geometries. The RLC parasitic circuits are realized with the solution of Maxwell equations. Finally, the model order reduction algorithms are employed to reduce the dimension of the linear subsystem of these RLC circuits [ANT05]. In this study, 1-D Maxwell equation is directly converted into a small order SISO system without using any model reduction algorithm. Therefore, it can be also useful for finding an appropriate reduction order of the model order reduction process. Before dealing with the Maxwell equations however, let us use an ordinary differential equation set to show the usage and the details of the method.
The remaining of the paper is organized as follows. In section 2, the methodology and the MOESP algorithm are briefly explained, whereas section 3 contains some numerical results and discussions. We present, in section 4 some concluding remarks and the future work.

2 Definition of the Problem

2.1 Introduction

To explain the basics of and the implementations details of the MOESP algorithm, a general n^{th} order ordinary differential equation (ODE) is considered. We also present numerical results for this case in the paper. Then the method is applied to the partial differential equations (PDE), more specifically to the Maxwell equations.

2.2 n^{th} Ordinary Differential Equation as a Discrete Linear System

A general linear differential equation of order n with zero inital values on an interval I is defined as,

$$y^{(n)}(t) + a_{n-1}y^{(n-1)}(t) + \cdots + a_0 y(t) = u(t), \quad t \in I \tag{1}$$

This system can be reduced to an associated first order ordinary differential equation system.

$$\frac{d}{dt}X = AX + Bu \tag{2}$$

where;

$$A = \begin{bmatrix} 0 & 1 & 0 & \dots & 0 \\ 0 & 0 & 1 & \dots & 0 \\ \vdots & \vdots & \vdots & \dots & \vdots \\ 0 & 0 & 0 & \dots & 1 \\ -a_0 & -a_1 & -a_2 & \dots & -a_{n-1} \end{bmatrix}, \quad X = \begin{bmatrix} x_1 \\ x_2 \\ \vdots \\ x_{n-1} \\ x_n \end{bmatrix}, \quad B = \begin{bmatrix} 0 \\ 0 \\ \vdots \\ 0 \\ 1 \end{bmatrix} \tag{3}$$

Matrix **A** is called *companion matrix*. Components of the **X** vector are called state variables. Furthermore, it is always possible to define the desired output as a linear combination of these state variables [CARL97].

$$Y = CX + Du \tag{4}$$

It is also possible to convert this continuos system into a discrete system with the help of any numerical integration algorithm. For example, if we choose Euler method for integration we obtain below difference equations for x_n at time t_{k+1},

$$\frac{x_n^{k+1} - x_n^k}{h} = a_{n1}x_1^k + a_{n2}x_2^k + \cdots + a_{nn}x_n^k + b_n u^k \tag{5}$$

or in matrix form,

$$\begin{aligned} X_{k+1} &= (I + hA)X_k + Bu_k \\ Y_{k+1} &= CX_{k+1} + Du_{k+1} \end{aligned} \tag{6}$$

Using (6), one can write the input-output formulas for each data point;

$$Y_{k+j} = C(I + hA)^j X_k + \sum_{i=1}^{j} C(I + hA)^{i-1} Bu_{k+j-i} + Du_{k+j} \tag{7}$$

Using (7) we can derive matrix input-output equations which play a fundamental role in subspace identification,

$$\begin{bmatrix} y_k \\ y_{k+1} \\ \vdots \\ y_{k+j} \end{bmatrix} = \begin{bmatrix} C \\ C\widehat{A} \\ \vdots \\ C\widehat{A}^j \end{bmatrix} X_k + \begin{bmatrix} D & & & \\ CB & D & & \\ \vdots & \ddots & \ddots & \\ C\widehat{A}^{j-1}B & \dots & CB & D \end{bmatrix} \begin{bmatrix} u_k \\ u_{k+1} \\ \vdots \\ u_{k+j} \end{bmatrix} \tag{8}$$

where $\widehat{A} = (I + hA)$. One can define two Hankel matrices in terms of u_k and y_k to generalize the structure. These Hankel matrices are called as $U_{0|k-1}$ and $Y_{0|k-1}$ respectively.

$$\begin{bmatrix} u(0) & u(1) \dots & u(N-1) \\ u(1) & u(2) \dots & u(N) \\ \vdots & \vdots & \vdots \\ u(k-1) & u(k) \dots & u(k+N-2) \end{bmatrix} \in \mathcal{R}^{kmxN} \begin{bmatrix} y(0) & y(1) \dots & y(N-1) \\ y(1) & y(2) \dots & y(N) \\ \vdots & \vdots & \vdots \\ y(k-1) & y(k) \dots & y(k+N-2) \end{bmatrix} \in \mathcal{R}^{kpxN} \tag{9}$$

where k is strictly greater than the order of the system n, p is the number of the outputs of the system, m is the number of the inputs and finally N is a sufficiently large number for fixing the Hankel matrix. 0 and $k-1$ values in the definitions of Hankel matrices are used for determination of the upper-left and lower-left elements respectively. Using this Hankel matrix definitions one can write below equations for the nth order ordinary differential system.

$$\begin{aligned} Y_{0|k-1} &= \mathcal{O}_k X_0 + \Phi_k U_{0|k-1} \\ Y_{k|2k-1} &= \mathcal{O}_k X_k + \Phi_k U_{k|2k-1} \end{aligned} \tag{10}$$

where

$$\mathcal{O}_k = \begin{bmatrix} C \\ C\widehat{A} \\ \vdots \\ C\widehat{A}^j \end{bmatrix}, \quad \Phi_k = \begin{bmatrix} D & & & \\ CB & D & & \\ \vdots & \ddots & \ddots & \\ C\widehat{A}^{j-1}B & \dots & CB & D \end{bmatrix} \tag{11}$$

Here, X_0 and X_k are the initial states respectively. $U_{0|k-1}$ and $Y_{0|k-1}$ are called past inputs and outputs and $U_{k|2k-1}$ and $Y_{k|2k-1}$ are called future inputs and outputs [KAT05].
Data matrices $U_{0|k-1}$, $Y_{0|k-1}$ can be written in more compact form as:

$$\begin{bmatrix} U_{0|k-1} \\ Y_{0|k-1} \end{bmatrix} = \begin{bmatrix} I_{km} & 0_{kmxn} \\ \Phi_k & \mathcal{O}_k \end{bmatrix} \begin{bmatrix} U_{0|k-1} \\ X_0 \end{bmatrix} \tag{12}$$

Finally, it can be said that it is always possible to rewrite (1) as a matrix equation as given in (12).

2.3 MOESP Algorithm

LQ decomposition, which is the dual of the QR decomposition, is used to make the upper-right block of the data matrix zero. LQ decomposition of a matrix can be given as,

$$\begin{bmatrix} U_{0|k-1} \\ Y_{0|k-1} \end{bmatrix} = \begin{bmatrix} L_{11} & 0 \\ L_{21} & L_{22} \end{bmatrix} \begin{bmatrix} Q_1^T \\ Q_2^T \end{bmatrix} \tag{13}$$

where $L_{11} \in \mathcal{R}^{kmxkm}$, $L_{22} \in \mathcal{R}^{kpxkp}$, $Q_1 \in \mathcal{R}^{Nxkm}$ $Q_2 \in \mathcal{R}^{Nxkp}$.
The actual computation of LQ decomposition is performed by taking transpose of the QR decomposition of the matrix.
Using orthogonality conditions on the input output spaces, below equation can be obtained for L_{22},

$$\mathcal{O}_k X_0 Q_2 = L_{22} \tag{14}$$

where $\mathcal{O}_k$ is extended observability matrix, X_0 is the initial states. If we take the SVD of the L_{22} matrix we get,

$$L_{22} = [U_1 U_2] \begin{bmatrix} \Sigma_1 & 0 \\ 0 & 0 \end{bmatrix} \begin{bmatrix} V_1^T \\ V_2^T \end{bmatrix} = U_1 \Sigma_1 V_1^T \tag{15}$$

In MOESP algorithm, the system dimension is determined by the singular values of the L_{22} matrix and with this decomposition we have,

$$\mathcal{O}_k X_0 Q_2 = U_1 \Sigma_1 V_1^T. \tag{16}$$

From last identity, we can define the extended observability matrix as

$$\mathcal{O}_k = U_1 \Sigma_1^{1/2}. \tag{17}$$

With (17) we have the C matrix of the estimated system as $C = \mathcal{O}_k(1:p, 1:n)$ and the A matrix as a a solution of below least square equation $\mathcal{O}_k(1:p(k-1), 1:n)A = \mathcal{O}_k(p+1:kp, 1:n)$.
Computation of the B and the D matrices are more complex. We refer the reader to work in [CIG98], [VD92-1], and [VD92-2] for further information. Algorithm of the method is given in Fig. (1).

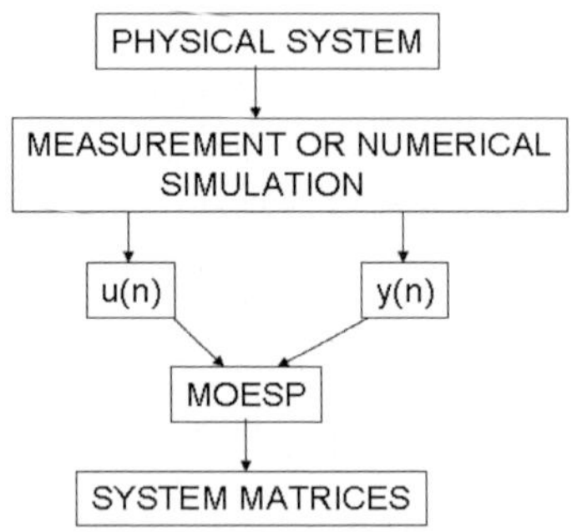

Fig. 1: Main MOESP Algorithm

3 Numerical Examples

3.1 Example ODE System

A second order ODE equation is selected for estimation. The equation is

$$\frac{d^2\varphi(t)}{dt^2} + \frac{d\varphi(t)}{dt} - 10 = 0 \tag{18}$$

Here, the input u is constant and equals to 10 and initial values of equation taken as zero. The output y is computed by a Runge-Kutta algorithm. Data matrices are created after the input and output data collected. Then the SMI Toolbox employed to produce the estimation [SMI]. The singular value distribution of L_{22} matrix is shown in Fig. (2).

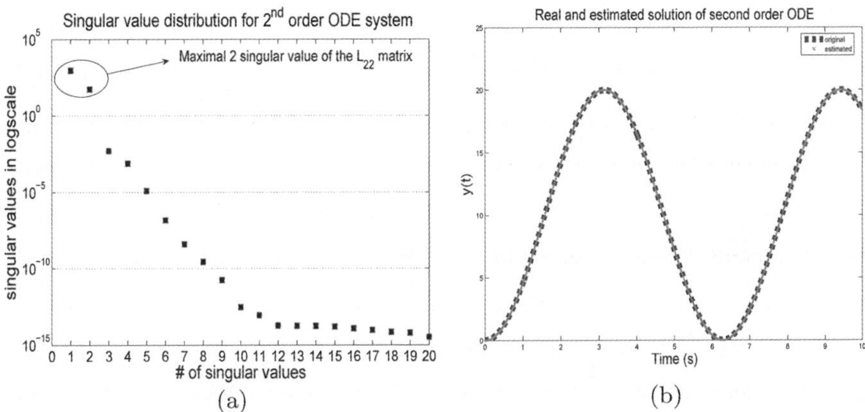

Fig. 2: (a) Singular value distribution of data matrix for u(t)=10 and estimated system order $n = 2$, (b) Original and estimated outputs for a estimated system order $n = 2$

3.2 Maxwell Equation

Consider a one-dimensional space where there are only variations in the x dimension. Assume that the electric field has only a z component. With Faraday and Ampere's laws we can write 1-D Maxwell equations as

$$\begin{aligned} \mu\frac{\partial H_y}{\partial t} &= \frac{\partial E_z}{\partial x}, \\ \epsilon\frac{\partial E_z}{\partial t} &= \frac{\partial H_y}{\partial x}. \end{aligned} \tag{19}$$

The source function is applied to the 0th node of the computational domain and data is collected as the electrical field of 50th node. After discretization, FDTD (Finite Difference Time Domain) algorithm is employed to obtain the input data u_k and output data y_k. The singular value distribution of the L_{22} matrix and the original and estimated outputs are given in Figs. 3 and 4. Here, two source functions are considered. First one is a sinusoidal and second one is an exponential function . For exponential source function MOESP algorithm works more accurately. Estimated order n, is selected as 2 in both cases.

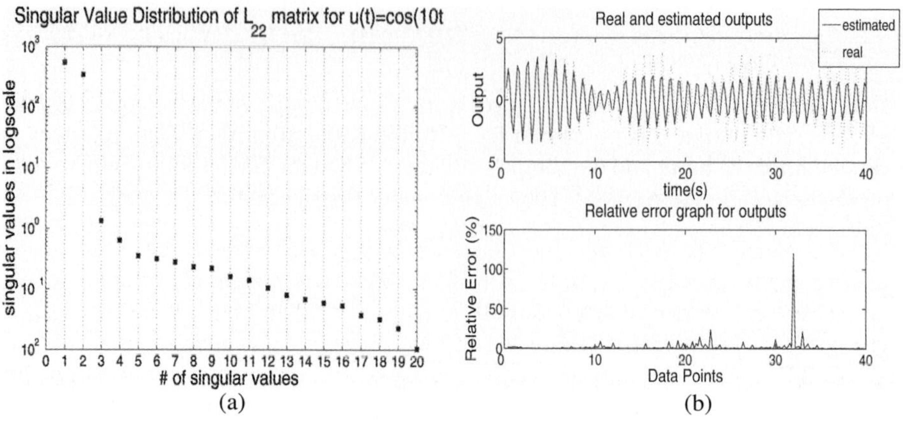

Fig. 3: (a) Singular value distribution of L_{22} matrix for $u(t) = cos(10t)$ (b) Original and estimated outputs and relative error for $u(t) = cos(10t)$ where the estimated system order $n = 2$

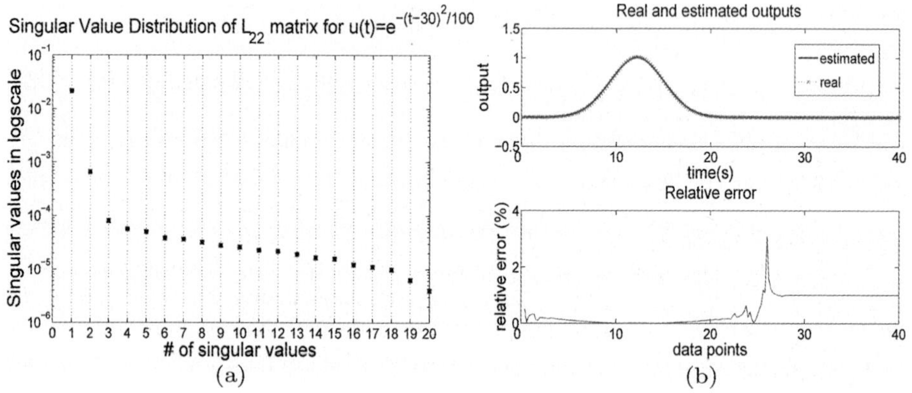

Fig. 4: (a) Singular value distribution of L_{22} for $u(t) = exp^{-(t-30)^2/100}$, (b) Original and estimated outputs and relative error for estimated system order $n = 2$

3.3 Comparison of Estimations

For ODE systems, singular value distributions of the L_{22} matrix is reasonable if one considers that a second order differential equation is estimated. Ratios of its maximum two singular values to the other singular values are sufficiently small and the singular values except first two largest ones can be neglected. This situation can be seen from the Fig. 2, the outputs are exactly matched.
This can be verified with one of the possible measures of accuracy which named as VAF (Variance According For) [SMI]. It is defined as,

$$VAF = 1 - \frac{variance(y - y_{est})}{variance(y)} * 100\% \tag{20}$$

where y is the original output and y_{est} is the estimated output. The VAF of two signals that are the same is 100%. If they differ, the VAF will be lower and if the two signals are completly different then VAF gets value of -1000.

For ODE example, VAF is equal to 100%. It means that the estimation works very successfully for this set of equation.
In the Figs. 3 and 4, singular values of the L_{22} matrix are relatively close to each other and we cannot select an exact estimation order like for the ODE system in Fig. 2. Therefore, it can be said that, for Maxwell equations the accuracy of the estimations is more sensitive to the selection of the estimation order. VAF values for these equations are 86% and 96.5% respectively. It can be also said that for exponential source functions, MOESP algorithm produces more accurate result. This fact can be observed from Figs 3 and 4. Its possible reason is the periodicity of the input and output vectors. Linear dependency of the columns of data matrices are determined by the input output vectors. Here we can say that, for non-periodic input sources Maxwell equations also can be modeled as a linear system with high accuracy. But in the case of the periodical input sources some other methodologies have to be used to improve the accuracy of the method.

4 Conclusion and Future Work

We examined the algorithm MOESP to convert a 1-D Maxwell equation into a SISO linear discrete state-space system.
Method is applied to an ordinary differential equation first and it is observed that the method produces a linear system quadruple (A,B,C,D) with high accuracy. However, when applying the same method to 1D-Maxwell equation accuracy of the method varies depending on the input source. With non-periodical input signals results are more accurate than those of the periodical input signal case. There can be a relationship between the periodicity of the input-output data and the behaviour of the algorithm. The future work will be focused on finding this relationship, i.e., the relationship between the order of the estimated system and the properties of data matrices.
We so far have studied the SISO modeling of the equations. In SISO models one has to define only one output point. On the other hand, in realistic systems more than one output point are required for modelling. Therefore, the method has to be extended to MIMO (Multiple Input Multiple Output) cases for extending the implementation area.

References

[ANT05] Antoluas, A. C.: Approximation of Large-Scale Dynamical Systems, SIAM (2005).
[KAT05] Katayama, T.: Subspace Methods for System Identification, Springer-Verlag (2005).
[VD92-1] Verhaegen, M., Dewilde P.: Subspace Model Identification Part 1. The Output-Error State-Space Model Identification Class of Algorithms, Int. J. Control, vol:56, No:5, pp:1187-1210, Taylor&Francis (1992).
[VD92-2] Verhaegen, M., Dewilde P.: Subspace Model Identification Part 2. Analysis of the Elementary Output-Error State-Space Model Identification Algortihm, Int. J. Control, vol:56, No:5, pp:1211-1241, Taylor&Francis (1992).
[CIG98] Ciggaar, E.: The MOESP Algorithm for Time Domain Modelling, Nat. Lab. Tchnical Note, TN 123/98 (1998).

[CARL97] Coddington E. A., Carlson R., Linear Ordinary Differential Equations, SIAM, Philedelphia, 1997.

[SMI] Haverkamp B., Verhaegen M.: State Space Model Identification Software for Multivariable Dynamical Systems, Delft University of Technology, Technical Notes, TUD/ET/SCE96.015 (1996).

Adaptive Methods for Transient Noise Analysis

Thorsten Sickenberger and Renate Winkler

Humboldt-Universität zu Berlin, Institut für Mathematik, 10099 Berlin
sickenberger/winkler@math.hu-berlin.de

Stochastic differential algebraic equations (SDAEs) arise as a mathematical model for electrical network equations that are influenced by additional sources of Gaussian white noise. In this paper we discuss adaptive linear multi-step methods for the numerical integration of SDAEs, in particular stochastic analogues of the trapezoidal rule and the two-step backward differentiation formula, together with a new step-size control strategy. Test results illustrate the performance of the presented methods.

1 Transient noise analysis in circuit simulation

Transient analysis is often performed without taking noise effects into account. But due to the parasitic effects, this is no longer possible. The increasing scale of integration, high clock frequencies and low supply voltages cause smaller signal-to-noise ratios. In several applications the noise influences the system behaviour in an essentially nonlinear way such that linear noise analysis is no longer satisfactory and transient noise analysis, i.e., the simulation of noisy systems in the time domain, becomes necessary (see [DeWi03, Wi04]). Here we deal with the thermal noise of resistors as well as the shot noise of semiconductors that are modelled by additional sources of additive or multiplicative Gaussian white noise currents that are shunt in parallel to the noise-free elements [DS98].
Thermal noise of resistors having a resistance R is caused by the thermal motion of electrons and is described by Nyquist's theorem. Hence, the associated current is modelled by additive noise,

$$i_{th} = \sqrt{\frac{2kT}{R}}\xi(t), \qquad k = 1.3806 \times 10^{-23},$$

where T is the temperature, k is Boltzmann's constant and $\xi(t)$ is a standard Gaussian white noise process. Shot noise of pn-junctions, caused by the discrete nature of currents due to the elementary charge, is modelled by multiplicative noise. If the noise-free current through the pn-junction is described by a characteristic $i = g(u)$ in dependence on a voltage u, the associated Gaussian white noise current is modelled by

$$i_{shot} = \sqrt{q_e |g(u)|}\xi(t), \qquad q_e = 1.602 \times 10^{-19},$$

where $\xi(t)$ again is a standard Gaussian white noise process and q_e is the elementary charge. Combining Kirchhoff's Current law with the element characteristics and using the charge-oriented formulation yields a stochastic differential algebraic equation (SDAE) of the form (see [GF99] for the deterministic case)

$$A\frac{d}{dt}q(x(t)) + f(x(t),t) + \sum_{r=1}^{m} g_r(x(t),t)\xi_r(t) = 0\,, \tag{1}$$

where A is a constant singular incidence matrix determined by the topology of the dynamic circuit parts, the vector $q(x)$ consists of the charges and the fluxes, and x is the vector of unknowns consisting of the nodal potentials and the branch currents through voltage-defining elements. The term $f(x,t)$ describes the impact of the static elements, $g_r(x,t)$ denotes the vector of noise intensities for the r-th noise source, and ξ is an m-dimensional vector of independent Gaussian white noise sources (see e.g. [DeWi03, Wi04]). Hence, one has to deal with a large number of equations as well as of noise sources. Compared to the other quantities the noise intensities $g_r(x,t)$ are small.
We understand (1) as a stochastic integral equation

$$Aq(X(s))\Big|_{t_0}^{t} + \int_{t_0}^{t} f(X(s),s)ds + \sum_{r=1}^{m}\int_{t_0}^{t} g_r(X(s),s)dW_r(s) = 0,\;\; t\in[t_0,T]\,, \tag{2}$$

where the second integral is an Itô-integral, and W denotes an m-dimensional Wiener process (or Brownian motion) given on the probability space $(\Omega,\mathcal{F},P)$ with a filtration $(\mathcal{F}_t)_{t\geq t_0}$. The solution is a stochastic process depending on the time t and on the random sample ω. For a fixed sample ω representing a fixed realization of the driving Wiener noise, the function $X(\cdot,\omega)$ is called a realization or a path of the solution. Due to the influence of the Gaussian white noise, typical paths are nowhere differentiable.
Especially for oscillating solutions in circuit simulation one is interested in the phase noise. We aim at the simulation of solution paths that reveal the phase noise. From the solution paths statistical data of the phase as well as of moments of the solution can be computed in a post-processing step. We therefore use the concept of strong solutions and strong (mean-square) convergence of approximations.
Using techniques from the theory of DAEs as well as of the theory of stochastic differential equations (SDEs) one derives existence and uniqueness for the solutions as well as convergence results for certain drift-implicit methods for systems with index 1 DAE [Wi03].

2 Adaptive numerical methods

An efficient integrator must be able to change the step-size. We present adaptations of known schemes for SDEs that are implicit in the deterministic part (drift) and explicit in the stochastic part (diffusion) of the SDAE. Designing the methods such that the iterates have to fulfill the constraints of the SDAE at the current time-point is the key idea to adapt known methods for the SDEs to (2).
We consider stochastic analogues of the two-step backward differentiation formula (BDF_2) and the trapezoidal rule, where only the increments of the driving Wiener

process are used to discretize the diffusion part. Analogously to the Euler-Maruyama scheme we call such methods multi-step Maruyama methods. The variable step-size BDF_2 Maruyama method for the SDAE (2) has the form (see [Si05] and e.g. [BuWi05] in the case of constant step-sizes)

$$A\frac{q(X_\ell) - \frac{(\kappa_\ell+1)^2}{2\kappa_\ell+1}q(X_{\ell-1}) + \frac{\kappa_\ell^2}{2\kappa_\ell+1}q(X_{\ell-2})}{h_\ell} + \frac{\kappa_\ell+1}{2\kappa_\ell+1}f(X_\ell,t_\ell)$$
$$+\sum_{r=1}^{m} g_r(X_{\ell-1},t_{\ell-1})\frac{\Delta W_r^\ell}{h_\ell} - \frac{\kappa_\ell^2}{2\kappa_\ell+1}\sum_{r=1}^{m} g_r(X_{\ell-2},t_{\ell-2})\frac{\Delta W_r^{\ell-1}}{h_\ell} = 0, \quad (3)$$

$\ell = 2,\ldots,N$. Here, X_ℓ denotes the approximation to $X(t_\ell)$, $h_\ell = t_\ell - t_{\ell-1}$, and $\Delta W_r^\ell = W_r(t_\ell) - W_r(t_{\ell-1}) \sim N(0,h_\ell)$ on the grid $0 = t_0 < t_1 < \cdots < t_N = T$. The coefficients of the two-step scheme (3) depend on the step-size ratio $\kappa_\ell = h_\ell/h_{\ell-1}$ and satisfy the conditions for consistency of order one and two in the deterministic case and of order $1/2$ in the stochastic case (see [Si05]).
A correct formulation of the stochastic trapezoidal rule for SDAEs requires more structural information (see [SiWi06]). It should implicitly realize the stochastic trapezoidal rule for the so called inherent regular SDE of (2) that governs the dynamical components. One possibility is to discretize the constraints differently, which requires the explicit knowledge of the constraints or, equivalently, a projector R along im$A$. The discrete equations

$$A\frac{q(X_\ell) - q(X_{\ell-1})}{h_\ell} + \frac{1}{2}(I-R)\big(f(X_\ell,t_\ell) + f(X_{\ell-1},t_{\ell-1})\big)$$
$$+Rf(X_\ell,t_\ell) + \sum_{r=1}^{m} g_r(X_{\ell-1},t_{\ell-1})\frac{\Delta W_r^\ell}{h_\ell} = 0, \quad (4)$$

$\ell = 1,\ldots,N$, imply the correct constraints and realize the trapezoidal rule for the inherent regular SDE.
Both the BDF_2 (3) and the trapezoidal rule (4) have only an asymptotic order of strong convergence of $1/2$, i.e.,

$$\|X(t_\ell) - X_\ell\|_{L_2(\Omega)} := \max_{\ell=1,\ldots,N}(E|X(t_\ell) - X_\ell|^2)^{1/2} \leq c \cdot h^{1/2}, \quad (5)$$

where $h := \max_{\ell=1,\ldots,N} h_\ell$ is the maximal step-size of the grid. (For additive noise the order may be 1.) This holds true for all numerical schemes that include only information on the increments of the Wiener process.
However, the noise densities given in Sec. 1 contain small parameters and the error behaviour is much better. In fact, the errors are dominated by the deterministic terms as long as the step-size is large enough [BuWi05]. In more detail, the error of the given methods behaves like $\mathcal{O}(h^2 + \varepsilon h + \varepsilon^2 h^{1/2})$, when ε is used to measure the smallness of the noise $(g_r(x,t) = \varepsilon\widehat{g}_r(x,t),\ r = 1,\ldots,m,\ \varepsilon \ll 1)$. Thus we can expect order 2 behaviour if $h \gg \varepsilon$.
The smallness of the noise also allows special estimates of the local error terms, which can be used to control the step-size. In [RoWi05] the authors presented a step-size control for the drift-implicit Euler scheme in the case of small noise that leads to adaptive step-size sequences that are uniform for all paths, see also [DeWi03, Wi04]. The estimates of the dominating local error term are based on values of the drift

term and do not cost additional evaluations of the coefficients of the SDE or their derivatives. In [SWW06, SWW07] we extend this strategy to stochastic linear multi-step methods with deterministic order 2 and present an estimate of the mean-square local errors. Again it is based on divided differences of values of the drift term and leads to step-size sequences that are identical for all computed paths.

3 Numerical results

Here, we illustrate the potential of the step-size control strategy by simulation results for the stochastic BDF_2 applied to three test problems. For the first and the second example we use an implementation of the adaptive methods discussed in the previous section in fortran code. To be able to handle real-life problems, a slightly modified version of the schemes for MNA together with the new step-size control has been implemented in Qimonda's in-house simulator TITAN. The third example shows the performance of this industrial implementation.

A nonlinear test-SDE

First, we consider a nonlinear scalar SDE with known explicit solution. The drift and diffusion coefficients are tunable by real parameters α and β, which we have chosen as $\alpha = -10$ and $\beta = 0.01$:

$$X(t) = \int_0^t -(\alpha+\beta^2 X(s))(1-X(s)^2)ds + \int_0^t \beta(1-X(s)^2)dW(s), \quad t \in [0,T]\,. \tag{6}$$

The solution is given by

$$X(t) = \frac{\exp(-2\alpha t + 2\beta W(t)) - 1}{\exp(-2\alpha t + 2\beta W(t)) + 1}\,. \tag{7}$$

In Figure 1 we present a work-precision diagram. We plotted the tolerance ($\triangle$) and the mean-square norm of the errors for adaptively chosen (+) and constant ($\times$) step-sizes for 100 computed paths vs. the number of steps in logarithmic scale. Lines with slopes -2 and -0.5 are provided to enable comparisons with convergence of order 2 or $1/2$. We observe order 2 behaviour up to accuracies of 10^{-4}. The results show that the proposed step-size control works very well for step-sizes above this threshold and provides considerably more accurate results than the computation with the same number of constant steps.

A MOSFET inverter circuit

Secondly, we consider a model of an inverter circuit with a MOSFET-transistor under the influence of thermal noise. The equivalent circuit diagram is given in Figure 2. The MOSFET is modelled as a current source from source to drain that is controlled by the nodal potentials at gate, source and drain.
The thermal noise of the resistor and of the MOSFET is modelled by additional white noise current sources that are shunt in parallel to the original, noise-free elements. To make the effect of the noise more visible we scaled the noise intensities by a factor of 1000. For the simulation we used the BDF_2 with adaptively chosen step-sizes.

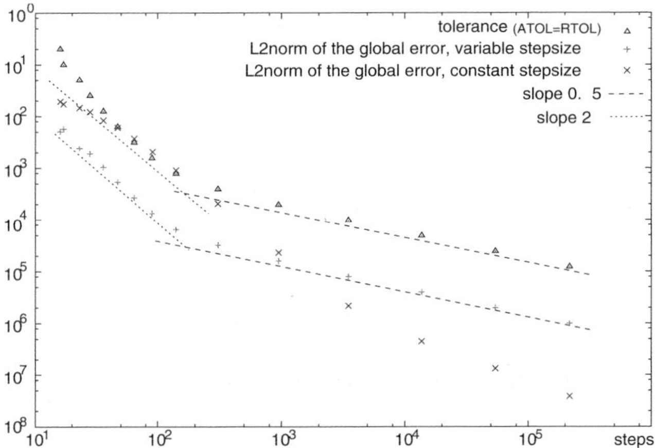

Fig. 1: Tolerance and accuracy versus steps for a test-SDE.

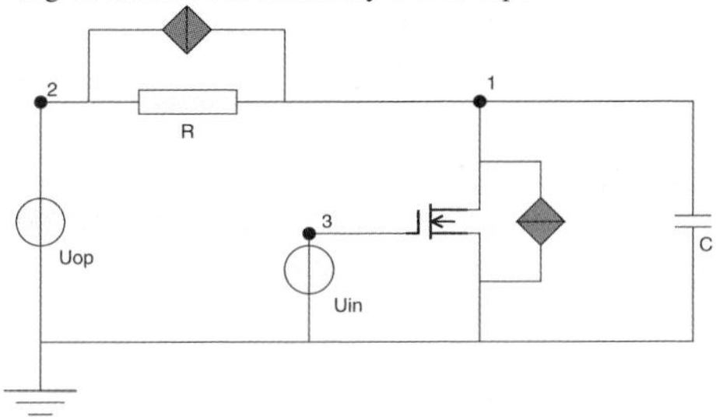

Fig. 2: Thermal noise sources in a MOSFET inverter circuit

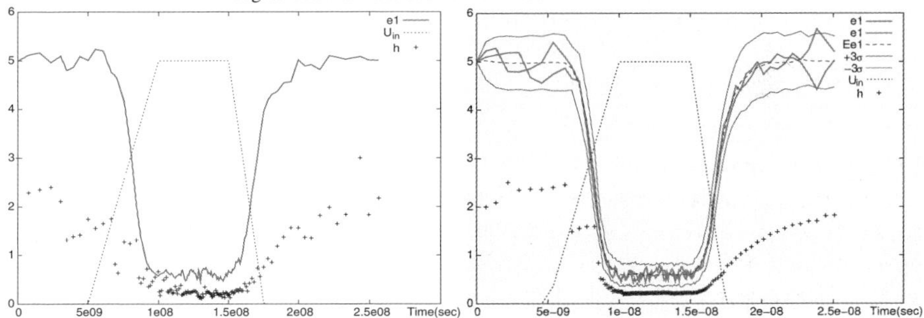

Fig. 3: Simulation results for the noisy inverter circuit:
1 path 127(+29 rejected) steps; 100 paths 134(+11 rejected) steps

In Figure 3 we present simulation results, where we plotted the input voltage U_{in} and values of the output voltage e_1 versus time. We compare the results for the computation of a single path (left picture) with those for the computation of 100 simultaneously computed solution paths (right picture), where the dark lines additionally show the values of two different solution paths, the dotted line gives the mean of 100 paths and the gray lines the 3σ-confidence interval for the output voltage e_1. Moreover, the applied step-sizes, suitably scaled, are shown by means of single crosses. Using the information of an ensemble of simultaneously computed solution paths smoothes the step-size sequence and reduces the number of rejected steps considerably, compared to the simulation of a single path. Also the computational cost mainly determined by the number of integration steps is reduced.

A voltage controlled oscillator

Finally, we present simulation results for a voltage controlled oscillator that has been used as a test application. It is a simplified version of a fully integrated 1.3 GHz VCO for GSM in 0.25 μm standard CMOS (see [Ti00]). For simulation, the oscillator is embedded in a test environment, using a virtual output buffer load and tuning voltage as well as core current modelled as independent DC sources. The VCO is tunable from about 1.2 GHz up to 1.4 GHz. The unknowns of the VCO in the MNA system are the charges of the six capacities, the fluxes of the four inductors, the 15 nodal potentials and the currents through the voltage sources. This circuit contains 5 resistors and 6 MOSFETs, which induce 53 sources of thermal or shot noise. To make the differences between the solutions of the noisy and the noise-free model more visible, the noise intensities had been scaled by a factor of 500.
Numerical results obtained with a combination of the BDF$_2$ and the trapezoidal rule are shown in Fig. 4, where we plotted the difference of the nodal potential $V(7) - V(8)$ of node 7 and 8 versus time. The solution of the noise-free system is given by a dashed line. Four sample paths (dark solid lines) are shown. They cannot be considered as small perturbations of the deterministic solution, phase noise is highly visible. To analyze the phase noise we repeated the simulation ten times with

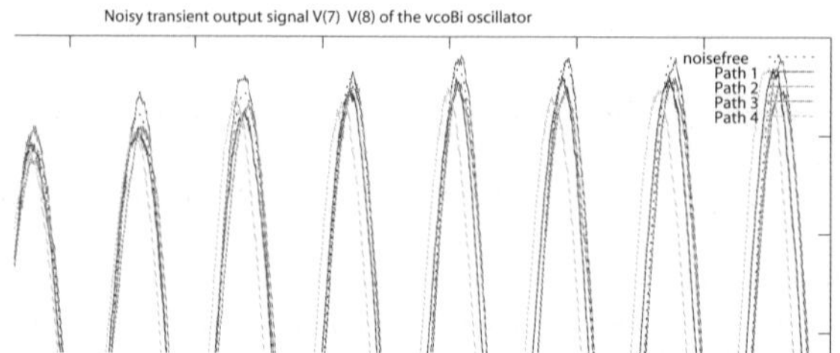

Fig. 4: Noisy transient output signal of a VCO.

different initialization of the pseudo-random numbers. Then we computed the length of the first 50 periods for each solution path. On Fig. 5 the mean μ of the frequencies (horizontal lines), the smallest and the largest frequencies (boundaries of the vertical thin lines) and the boundaries of the confidence interval $\mu \pm \sigma$ (the plump lines) are presented, where σ was computed as the empirical estimate of the standard deviation. The mean appears increased and differs by about +0.25% from the noiseless, deterministic solution. Further on, the frequencies vacillate from 1.18 GHz (-0.95%) up to 1.21 GHz (+1.55%). So the transient noise analysis shows that the voltage con-

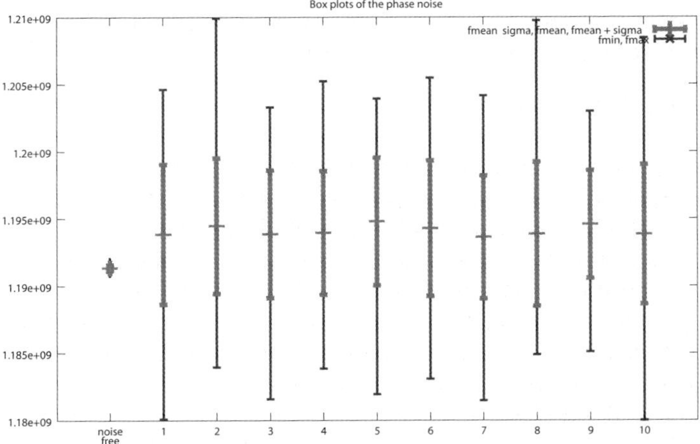

Fig. 5: Boxplots of the phase noise, scaled by a factor of 500

trolled oscillator runs in a noisy environment with increased frequencies and smaller phases, respectively.

References

[BuWi05] Buckwar, E., Winkler, R.: Multi-step methods for SDEs and their application to problems with small noise. SIAM J. Num. Anal., **44**(2), 779–803 (2006)

[DS98] Demir, A., Sangiovanni-Vincentelli, A.: Analysis and simulation of noise in nonlinear electronic circuits and systems. Kluwer Academic Publishers (1998)

[DeWi03] Denk, G., Winkler, R.: Modeling and simulation of transient noise in circuit simulation. Mathematical and Computer Modelling of Dynamical Systems, to appear

[GF99] Günther, F., Feldmann, U.: CAD-based electric-circuit modeling in industry I. mathematical structure and index of network equations. Surv. Math. Ind., **8**, 97–129 (1999)

[Hi01] Higham, D.J.: An algorithmic introduction to numerical simulation of stochastic differential equations. SIAM Review, **43**, 525–546 (2001)

[Ti00] Tiebout, M.: A fully integrated 1.3 GHz VCO for GSM in 0.25 μm standard CMOS with a phasenoise of -142 dBc/Hz at 3MHz offset. In: Proceedings 30th European Microwave Conference, Paris (2000)

[RoWi05] Römisch, W., Winkler, R.: Stepsize control for mean-square numerical methods for stochastic differential equations with small noise. SIAM J. Sci. Comp., **28**(2), 604–625 (2006)

[Si05] Sickenberger, T.: Mean-square convergence of stochastic multi-step methods with variable step-size. J. Comput. Appl. Math., to appear

[SiWi06] Sickenberger, T., Winkler, R.: Efficient transient noise analysis in circuit simulation. In: Proceedings of the GAMM Annual Meeting 2006, Berlin, Proc. Appl. Math. Mech. **6**(1), 55–58 (2006)

[SWW06] Sickenberger, T., Weinmüller, E., Winkler, R.: Local error estimates for moderately smooth problems: Part I - ODEs and DAEs. BIT Numerical Mathematics, to appear

[SWW07] Sickenberger, T., Weinmüller, E., Winkler, R.: Local error estimates for moderately smooth problems: Part II - SDEs and SDAEs. In preparation

[Wi03] Winkler, R.: Stochastic differential algebraic equations of index 1 and applications in circuit simulation. J. Comput. Appl. Math., **157**(2), 477–505 (2003)

[Wi04] Winkler, R.: Stochastic differential algebraic equations in transient noise analysis. In: Proceedings of Scientific Computing in Electrical Engineering, September, 5th - 9th, 2004, Capo D'Orlando , Springer Series Mathematics in Industry, 151–158 (2006)

Efficient Execution of Loosely Coupled Tasks in Grid Platforms

Felicia Ionescu[1], Stefan Diaconescu, Alexandru Gherega, and Gabriel Dimitriu

University Politehnica Bucharest, Splaiul Independentei Nr. 313, Postal Code: 060042, Romania fionescu@tech.pub.ro

Abstract
Grid technologies offer powerful computing resources for all domains, but the highly heterogeneous and dynamic nature of the Grids needs adaptable, scalable and extensible scheduling systems. In this paper we describe a dynamic, centralized scheduling mechanism based on Master-Worker paradigm for efficient execution of a set of loosely coupled tasks in a Grid environment. This mechanism offers high programmability features, adaptability and reliability towards processor failure. Experiments are presented that demonstrate the effectiveness of our approach.
Key-words: parallel and distributed systems, grid computing, service-oriented architecture, dynamic scheduling, master-worker model

1 Introduction

The grid computing paradigm aggregates the view on existing hardware and software resources, coordinating resource sharing and problem solving in dynamic, multi-institutional virtual organizations [FKT01]. Grid platforms are now developed in a service-oriented architecture, defined as Open Grid Service Architecture (OGSA) [FKN02], which standardizes all the services one finds in a grid application. There are two implementation specifications for OGSA: Open Grid Services Infrastructure (OGSI), released in 2003 and Web Services Resource Framework (WSRF) [OASIS] introduced in 2004 by a team from IBM and the Globus Alliance [GLOBS]. WSRF is an attempt to re-factor many of the concepts in OGSI to be more consistent with today's Web Services, allowing the manipulation of state, with no modifications to Web Services tooling.
Currently the Globus project is the highly favourite grid toolkit, having been adopted by IBM, HP, etc.
Another interesting grid toolset is WSRF.NET, which is an open-source implementation of the WSRF suite of specifications developed at Virginia University [WSRFN] for Windows systems under .NET platform. For this toolset, a remote job execution service, which allows remote execution and data movement between Windows machines across the grid platform, and a Scheduler Service, based on WS-Notification services were developed [WHE05]. In this work we developed a dynamic scheduling mechanism based on Master-Worker paradigm [MWWIS] in a WSRF.NET grid, in

order to study the efficiency of the execution of a set of loosely coupled tasks in a grid-aware approach.
The outline of this paper is as follows. Section 2 describes a Master-Worker model for execution of distributed applications composed of a set of loosely coupled tasks in a grid platform. Section 3 presents a discussion of the implementation of the Master-Worker model in a WSRF.NET grid. Section 4 contains experimental results and conclusions.

2 Master-Worker execution model for a set of loosely coupled tasks

The Master-Worker (MW) model (also known as Master-Slave model) has been widely used for developing parallel applications. In the MW model there are two distinct types of processes: master and workers. The master process assigns the tasks to the workers taking into account the dependencies between them. The workers typically perform most of computational work by just executing those tasks. The MW model has proved to be efficient in developing applications with different degrees of granularity of parallelism (grain size) and is particularly useful when the dependencies between tasks are low.
The aim of our work is to demonstrate the viability of the MW paradigm for efficient execution of a set of loosely coupled tasks in a Grid environment. To this end, we have written a Grid implementation prototype of the model using the WSRF.NET toolset and have deployed it on the local network platform consisting of 18 PC nodes (Pentium IV 2GHz, 80GB HDD, 512 MB RAM) connected with one Gigabit Ethernet switch. Each node runs .NET framework under Windows XP operation system, offering the grid fabric needed for experiments.
For this experiment, we used a generic parallel application in the form of an acyclic task dependence graph (TDG), represented by the couple G = (V,E), where V is the set of vertices in the graph, corresponding to the tasks, and E is the set of directed edges, indicating the precedence relations between tasks. There is no need for communication between tasks and the synchronizations took place only at the start and the end of tasks. The communications on the network involve only short messages exchanged between master and workers, containing parameters and returned values of the methods invoked by the master process on the workers, so that loosely coupled tasks and bandwidth unlimited communications can be assumed.
The tasks in the graph are simple loop (vector) operations, and the grain size can be estimated from the vector length (iteration count) and can be stored as a computation cost (execution time) of the graph node. The estimated grain size of tasks (computation cost) is not needed for Master-Worker algorithm and is used only for analysis of the algorithm behavior.
The graphs used for experiments were generated off-line and stored in files available on the disk, each task containing its execution parameters (vector length and operation), a list of all precedent tasks (parent tasks) and a list of all dependent tasks (child tasks).
The master and worker components are developed as WSRF Grid Services (Master Service and Worker Service), deployed in different nodes of the grid. The Master Service receives an execution command (containing a TDG) from a client application

and distributes the tasks to all available Worker Services, hosted in Worker Nodes of the grid.
The worker Grid service publishes an interface with at least one operation (doTask()) that the Master Service calls in order to dispatch a task to a worker. The master Grid Service publishes an interface with different operations, such as doGraph() function that the client calls in order to start the distributed execution of the program represented as a TDG. The client is just a simple C# application with a graphical interface, which collects different parameters of the tasks (vector length and operations) and passes these parameters to the Master Service operation (doGraph()).

3 Implementation of the Master-Worker model

The master maintains a list of all available workers (AW), a list of all running workers (RW) and processes the received TDG, creating a list of ready tasks (RT) and a list of waiting tasks (WT). A task is ready to be executed if it has no parents or all its parents were already executed. The distributed execution of a TDG is accomplished as a loop in the main thread of the master service, which runs until all tasks are processed, as is presented in Fig. 1.
In this loop, the available workers list is checked and, if the list is not empty, an available worker is selected; else, the main thread is waiting until a worker becomes available. When an available worker is selected, the main thread checks the list of ready tasks and, if this list is not empty, a ready task is selected.
At this point of the execution, the main thread creates an auxiliary thread and loops back to available workers list check. The auxiliary thread calls the doTask() function of the selected Worker Service, with the parameters of the selected task, runs until the function returns, updates the list of available workers, the list of ready tasks and the list of waiting tasks and exits. Every worker task is invocated in a new thread, so that they can execute concurrently, while the main thread returns and tests for waiting tasks to be distributed.
When the main thread checks the list of ready tasks, if this list is empty and the waiting tasks list is not empty, the main thread is waiting until a task becomes ready; if both waiting tasks list (WT) and running workers list (RW) are empty, the execution of all tasks of the TDG is accomplished, the final execution time is measured and total execution time is returned to the client.
All lists described before are shared variables, concurrently accessed by the main thread and auxiliary threads: the main thread extracts workers from AW list and tasks from RT list; the auxiliary threads insert available workers in the AW list and move tasks from the WT list into the RT list. All these operations are implemented as synchronized thread-safe operations available in C#.

4 Experimental results and conclusions

This section contains several experiments, showing the dependence of the efficiency of parallel applications on the characteristics of the task dependence graph. The experiments are not meant to be comprehensive, but only to give the reader an image of aspects that application programmers must consider when building master-worker

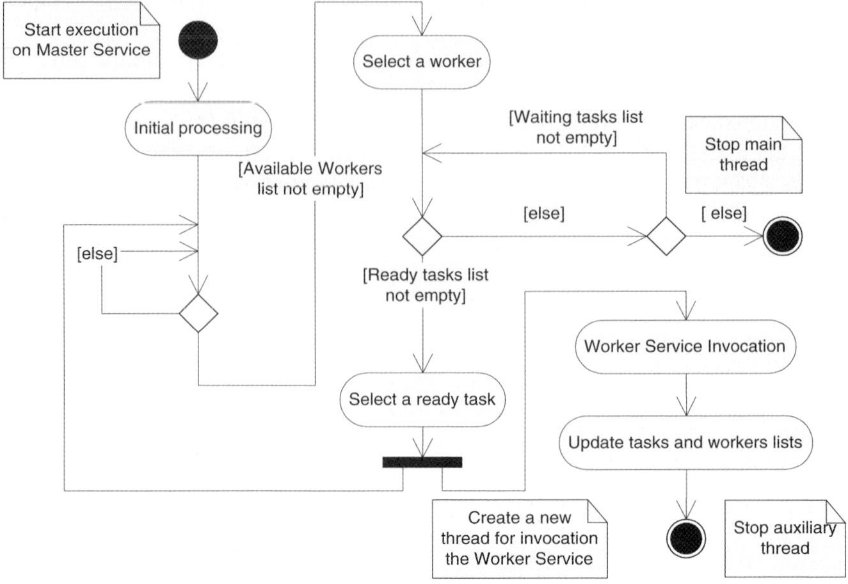

Fig. 1: Activity diagram of the Master Service

applications for grid environments. The computing environment (a WSRF.NET grid prototype) was used in dedicated mode for these experiments.
For each TDG, several executions, with different task size and different number of workers (grid nodes) were accomplished. The sequential execution time (T_S) of a TDG with a given size of the tasks, was measured executing only in the Master Service, by replacing the call of doTask() function on a worker service, with a local call. The parallel execution time (T_P) of a TDG with a given size of the tasks, on a given number of workers (p) was measured in the Master Service in a normal operation of master-worker mechanism, described in the previous section. The efficiency E is computed with the well-known expression: $E = T_S/(p * T_P)$.
The results obtained for a TDG with 64 tasks on 1, 2, 4, 8, 16 workers (grid nodes), for different task medium grain size (given in seconds) are presented in Fig. 2.
These results show that the efficiency decreases when the number of workers (processors) increases and when the task size decreases. This behavior is mainly caused by the centralized control of scheduling operation (executed in master), which represents a fraction α of unparallelizable computation. As Amdahls law establishes, a fraction α of inherently sequential or unparallelizable computation limits the efficiency that can be achieved with p processors to a value given by the expression $1/(\alpha * (p - 1) + 1)$.
Indeed, when the medium task grain size decreases, the total execution time decreases, the unparalellizable computations (consumed in Master Service for tasks distribution) represent a greater fraction α from this execution, and the efficiency decreases; the same effect is determined by the increasing of the number of workers (p).

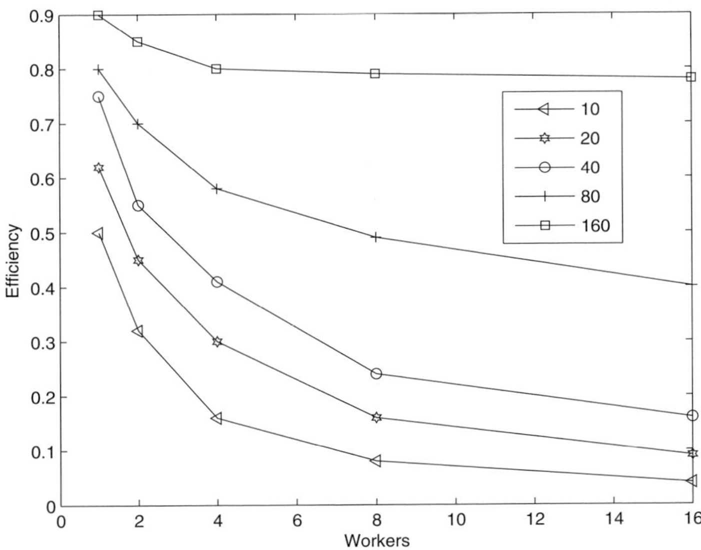

Fig. 2: Efficiency vs number of workers for different task grain size

Many other parameters of the system, which are not comprised here, can also influence the efficiency of the distributed execution: the number of tasks that can be concurrently executed (that depends on the total number of tasks and the degree of task dependences of the TDG), the unbalance of tasks size, the heterogeneous computing power of workers etc.

In our approach, master-worker paradigm offers a good scalability as long as task size is high enough, but the efficiency decreases with increassing number of workers for small task size. However, this paradigm offers a lot of interesting features in a grid environment, which can be easily obtained with some extensions of the base implementation presented in this paper: programmability (users should easily be able to take an existing application code and integrate it with the system), adaptability (the system should transparently adapt to the dynamic and heterogeneous execution environment) and reliability (the system should perform the correct computations in the presence of worker processors failure).

Acknowlegement: This work was supported by CNCSIS Project 36-GR/ 2.06.2006 Code A44 (UPB Part 14) started in 2006.

References

[FKT01] I. Foster, C. Kesselman, S. Tuecke. The Anatomy of the Grid: Enabling Scalable Virtual Organizations, International Journal of Supercomputer Applications and High Performance Computing, 15(3), 2001, pp. 200-222.

[FKN02] I. Foster, C. Kesselman, J.M. Nick, S. Tuecke. The Physiology of the Grid: An Open Grid ServiceArchitecture for Distributed System Integration, Proc. of Open Grid Service Infrastructure WG, Global Grid Forum, June 2002

[OASIS] OASIS Web Services Resource Framework (WSRF), http://www.oasis-open/org

[GLOBS] Globus Alliance, www.globus.org

[WSRFN] WSRF.NET: http://www.ws-rf.net, http://www.cs.virginia.edu/gsw2c/wsrf.net.html

[WHE05] G. Wasson and M. Humphrey. Exploiting WSRF and WSRF.NET for Remote Job Execution in Grid Environments. International Parallel and Distributed Processing Symposium (IPDPS 2005), Denver CO, April 4-8, 2005.

[MWWIS] Master-Worker Home Page: http://www.cs.wisc.edu/condor/

Colour Figures

Comparison of model reduction methods with applications to circuit simulation

Roxana Ionutiu, Sanda Lefteriu, Athanasios C. Antoulas

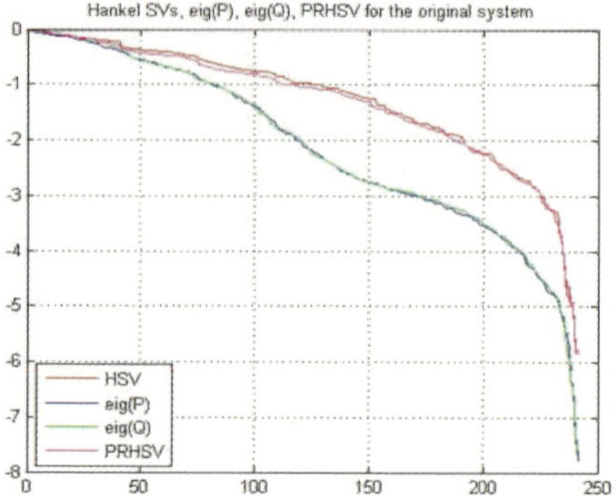

Fig. 2 (p. 14) Hankel singular values, positive real Hankel singular values, eigenvalues of $\mathcal{P}$, eigenvalues of $\mathcal{Q}$.

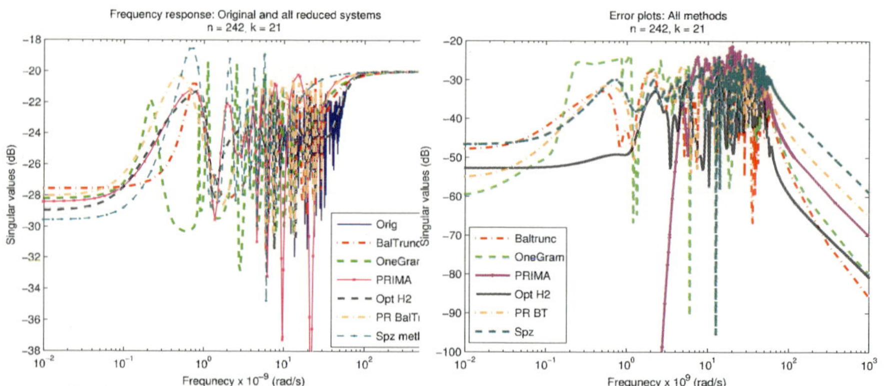

Fig. 3 (p. 18) Frequency response of original and all reduced systems

Fig. 4 (p. 18) Error for all reduced systems

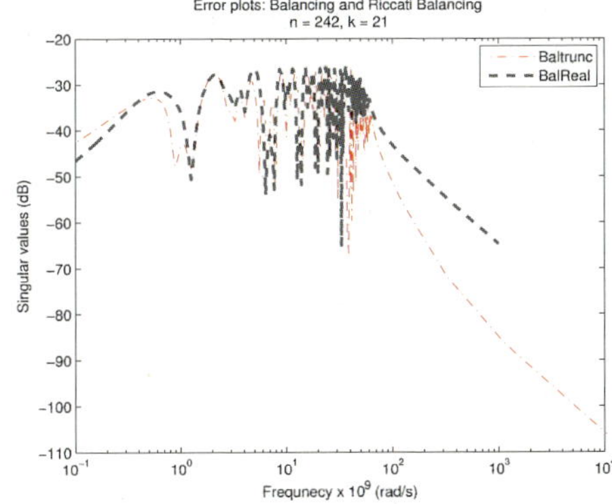

Fig. 5 (**p. 18**) Error systems: Balanced truncation and Positive Real Balanced Truncation

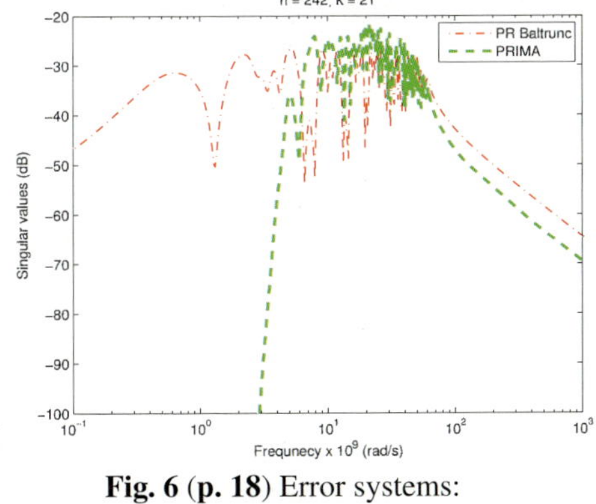

Fig. 6 (**p. 18**) Error systems: Positive Real Balanced Truncation and PRIMA

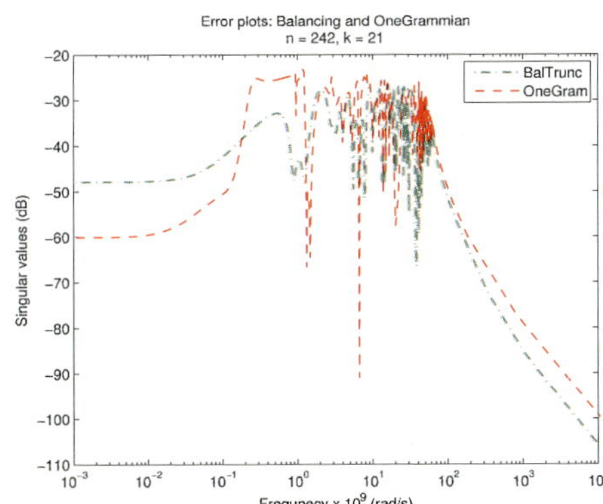

Fig. 7 (**p. 18**) Error systems: Balanced truncation and One Gramian

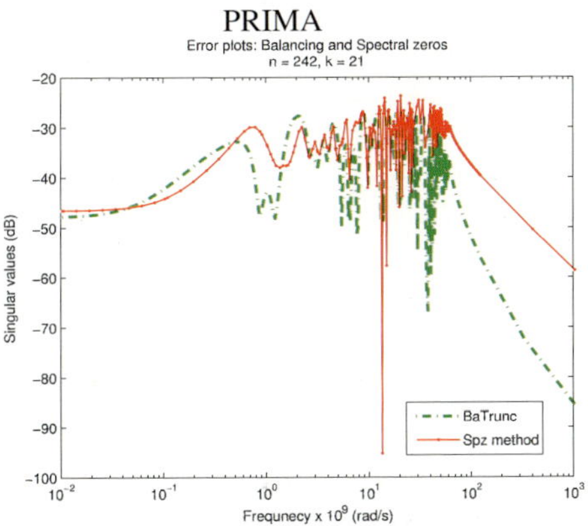

Fig. 8 (**p. 18**) Error systems: Balanced truncation and Projection using spectral zeros

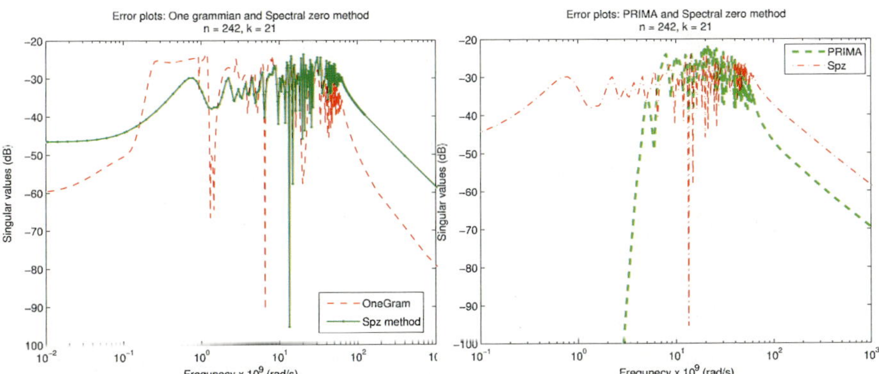

Fig. 9 (**p. 19**) Error systems: One Gramian and Projection using spectral zeros

Fig. 10 (**p. 19**) Error systems: PRIMA and Projection using spectral zeros

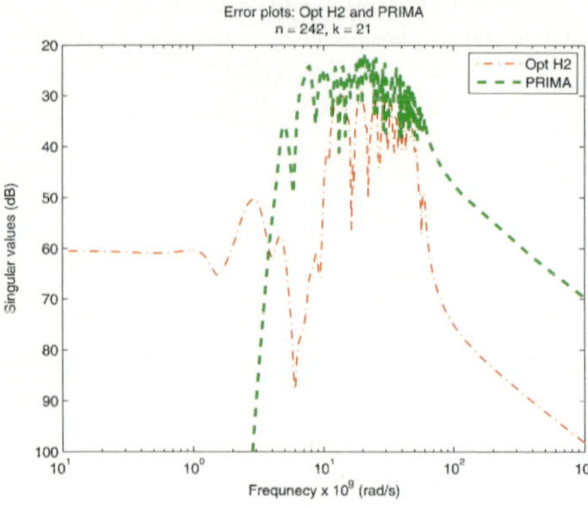

Fig. 11 (**p. 19**) Error systems: Optimal $\mathcal{H}_2$ and PRIMA

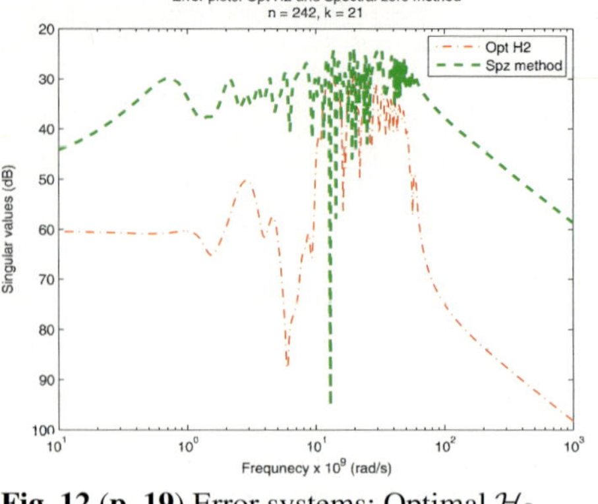

Fig. 12 (**p. 19**) Error systems: Optimal $\mathcal{H}_2$ and Spz method

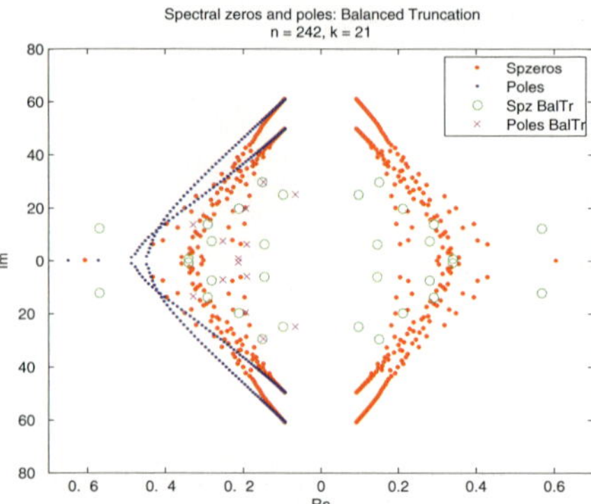

Fig. 13 (**p. 20**) Spectral zeros and poles for original system and reduced with Balanced truncation

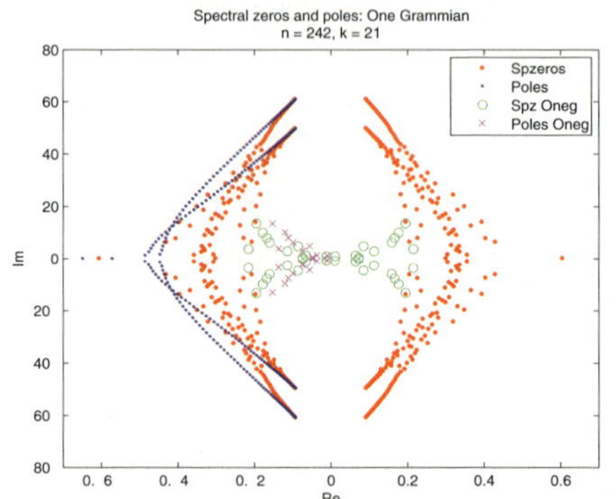

Fig. 14 (**p. 20**) Spectral zeros and poles for original system and reduced with One Gramian

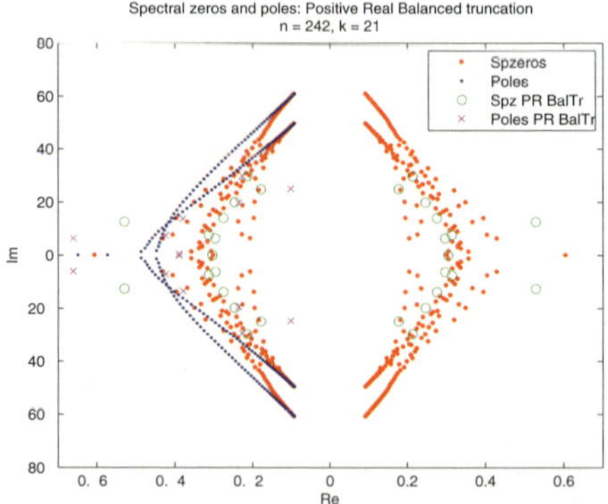

Fig. 15 (**p. 20**) Spectral zeros and poles for original system and reduced with Riccati Balanced truncation

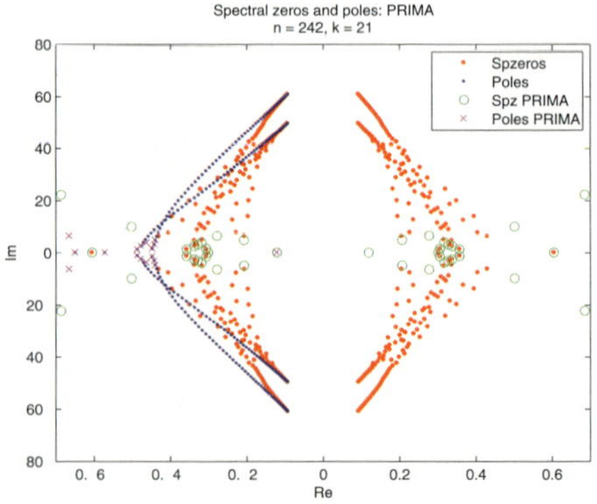

Fig. 16 (**p. 20**) Spectral zeros and poles for original system and reduced with PRIMA

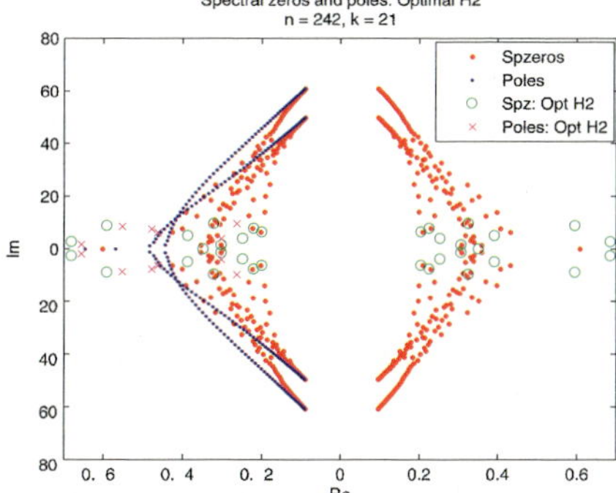

Fig.17 (p. 21) Random initial shifts: Poles and Spectral zeros of original system and reduced with Optimal $\mathcal{H}_2$

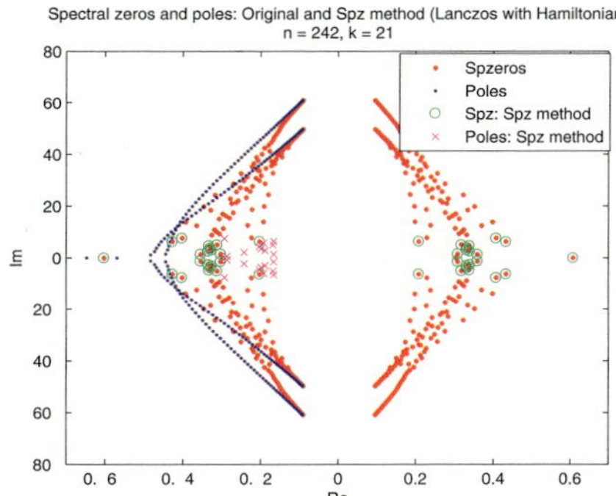

Fig. 18 (p. 21) Spectral zeros and poles of original and reduced with projection using spectral zero selection

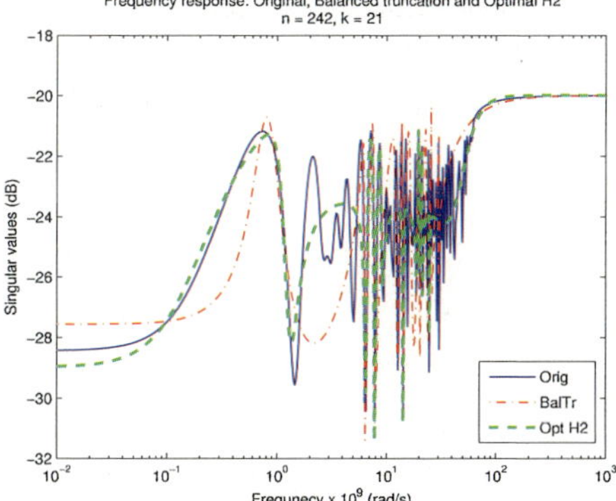

Fig. 19 (p. 22) BT poles as initial shifts: Frequency response of original and reduced systems

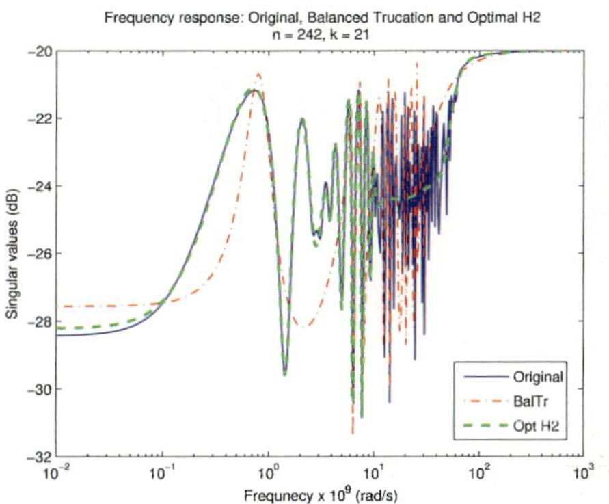

Fig. 20 (p. 22) Random intial shifts: Frequency response of original and reduced systems

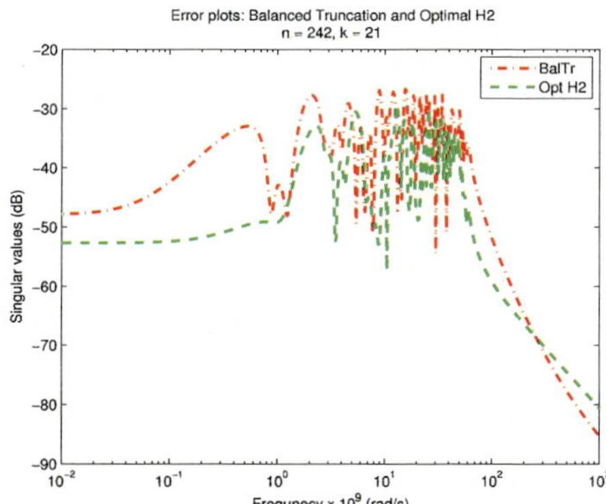

Fig. 21 (p. 22) BT poles as initial shifts: Error systems

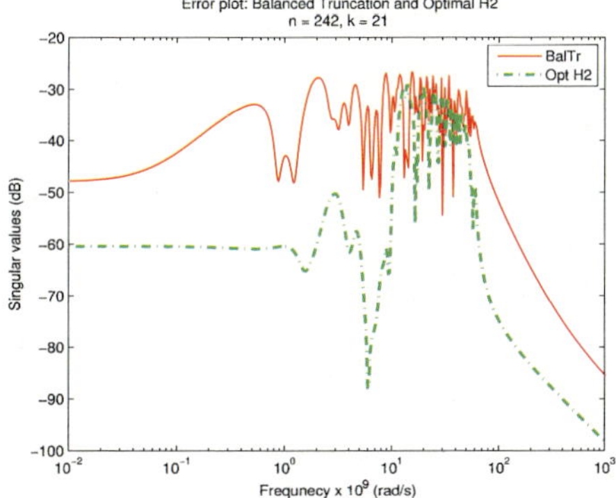

Fig. 22 (p. 22) Random initial shifts: Error systems

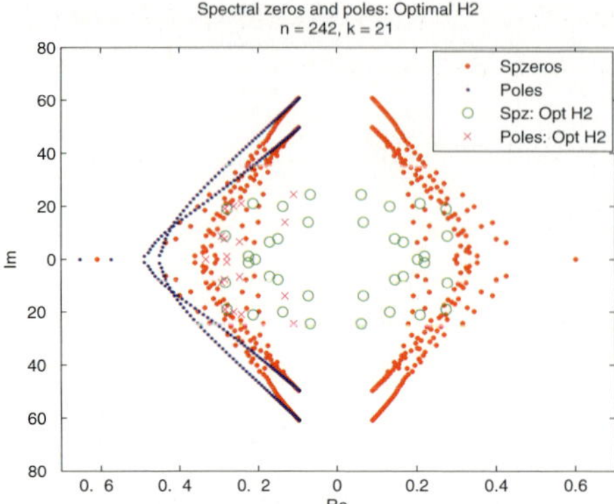

Fig. 23 (**p. 22**) BT poles as initial shifts: Poles and Spectral zeros of original system and reduced with Optimal $\mathcal{H}_2$

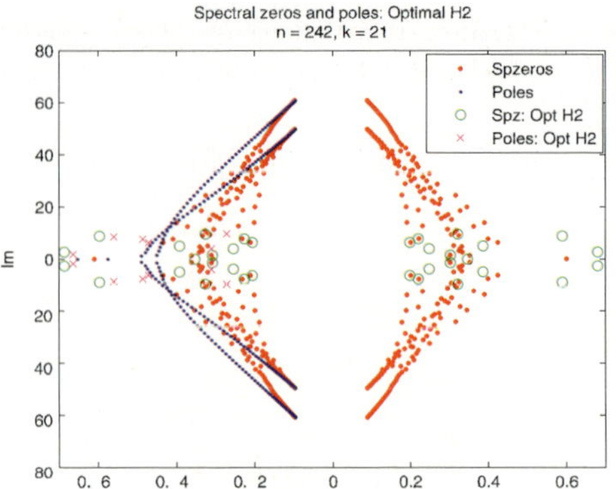

Fig. 24 (**p. 22**) Random initial shifts: Poles and Spectral zeros of original system and reduced with Optimal $\mathcal{H}_2$

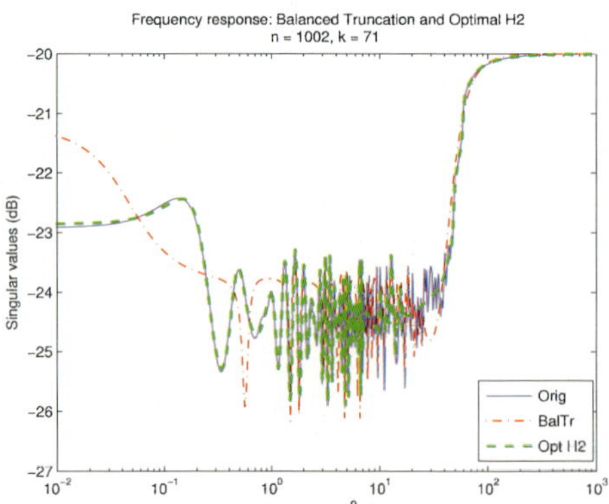

Fig. 25 (**p. 23**) Random initial shifts: Frequency response: original, reduced with balanced truncation and Optimal $\mathcal{H}_2$

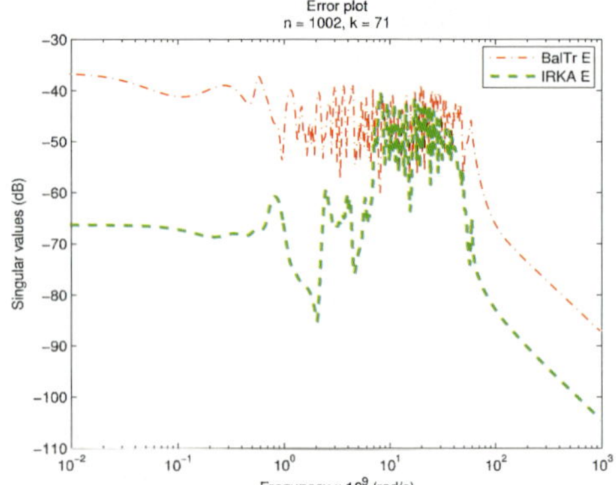

Fig. 26 (**p. 23**) Random initial shifts: Error systems for balanced truncation and Optimal $\mathcal{H}_2$

Transient Field-Circuit Coupled Models with Switching Elements for the Simulation of Electric Energy Transducers

Herbert De Gersem, Galina Benderskaya, Thomas Weiland

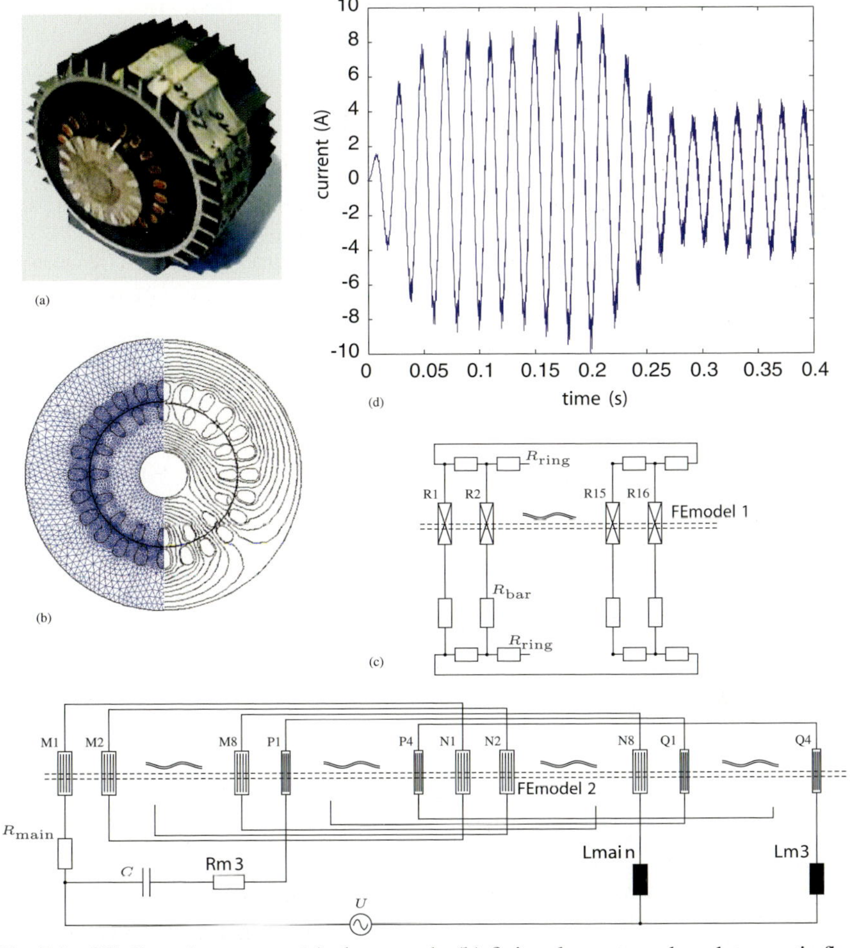

Fig. 5 (p. 36) Capacitor motor: (a) photograph; (b) finite-element mesh and magnetic flux lines at no-load operation; (c) external circuit with the applied sinusoidal voltage U, the capacitance C, the resistances R_{main} and R_{aux} and inductances L_{main} and L_{aux} modelling the end winding parts and the resistances R_{bar} and R_{ring} modelling the rotor ring and rotor-bar parts outside the finite-element model; (d) current through the main stator winding during start-up.

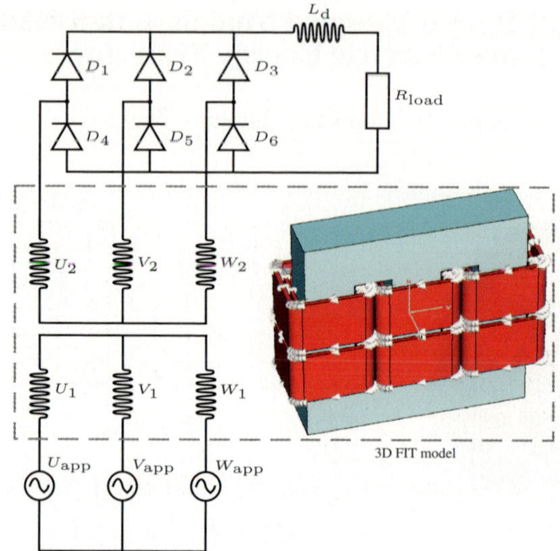

Fig. 6 (p. 37) 3D finite-integration model of a three-phase transformer connected to an external electric circuit for the power grid, diode rectifier and inductive load.

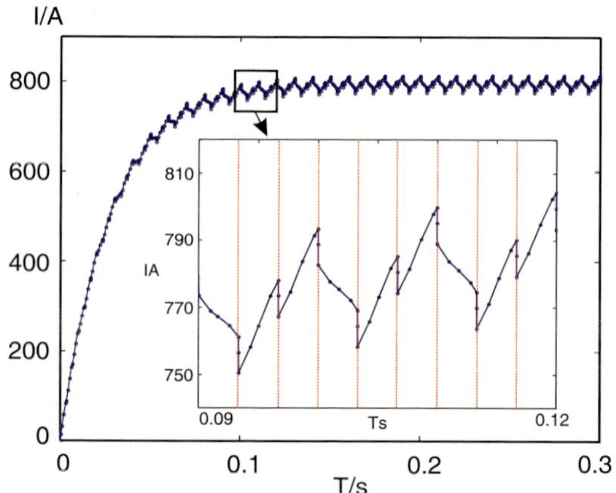

Fig. 7 (p. 37) Current through the first coil at the high-voltage side of the three-phase transformer.

Technology and Device Modeling in Micro and Nano-electronics: Current and Future Challenges

Andrea Marmiroli, Gianpietro Carnevale, Andrea Ghetti

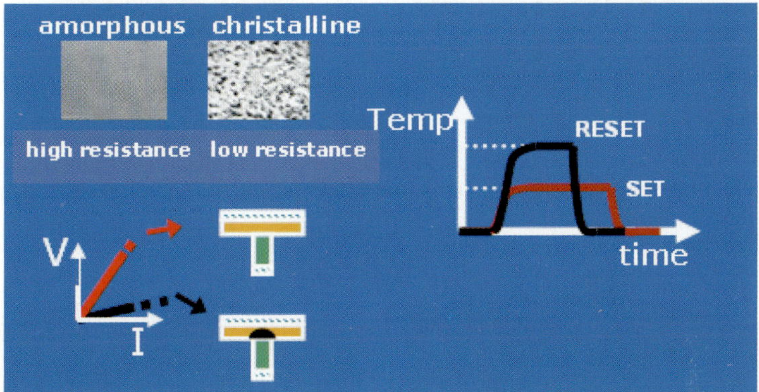

Fig. 1 (p. 42) Phase Change Memory (PCM) description: low and high resistivity of PCM material is associated to the crystalline and amorphous state, respectively. At the left side, bottom part of the figure the schematic pictures of a cross section of the bit architecture are sketched: the "T" shape at the lower side correspond to the low resistance state, with the material partially modified in the crystalline configuration. The upper "T" shape corresponds to the high resistance state, with the material in the amorphous phase.

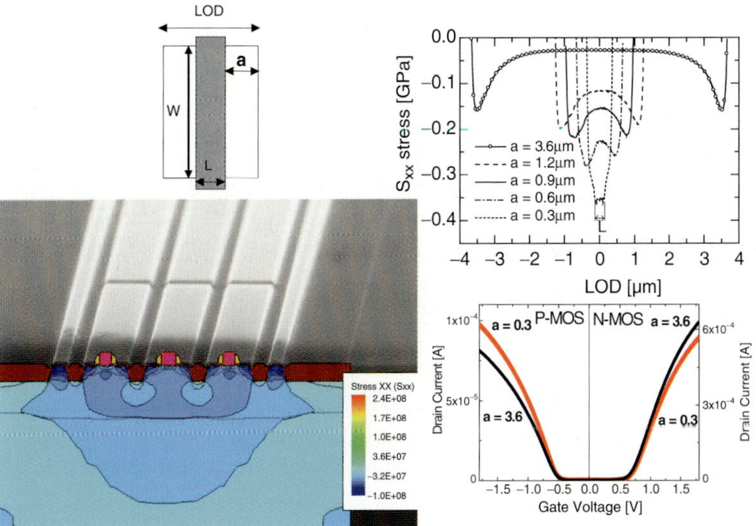

Fig. 2 (p. 43) Strain effect in mosfet transistors

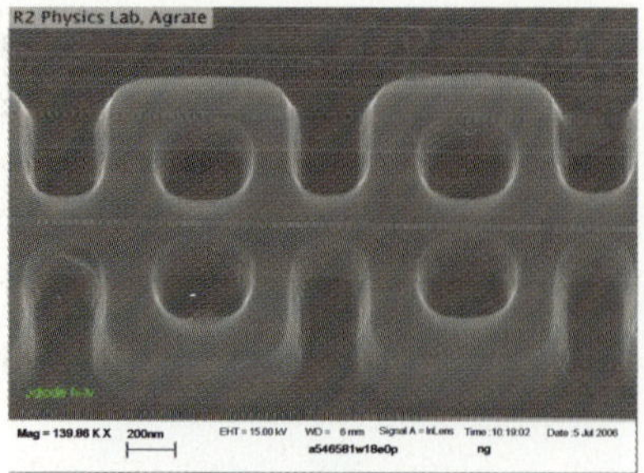

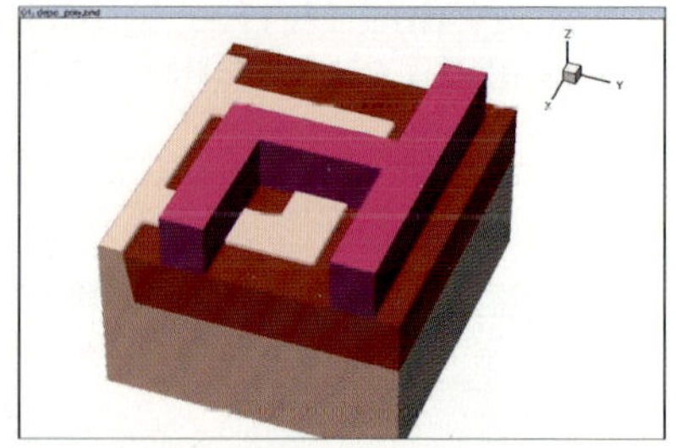

Fig. 4 (**p. 45**) A SEM picture of a silicon active area (oxide has been removed during stripping operation) and the corresponding 3D structure (pink region are silicon, brown is oxide and violet is poly-silicon material).

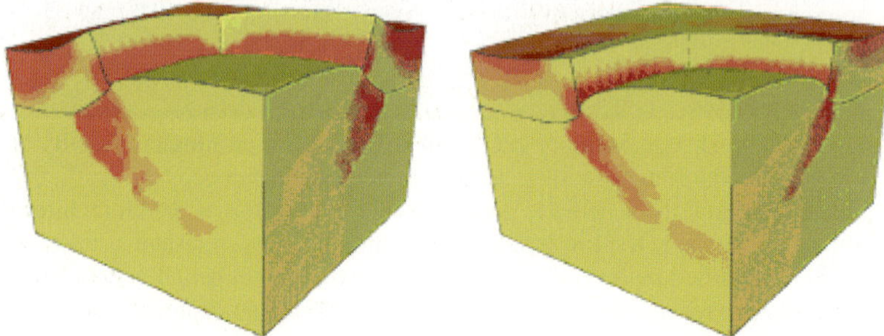

Fig. 5 (**p. 46**) At the left, the pressure distribution calculated with FEDOS without stress dependent oxidation. At the right stress dependent oxidation is included [Hol05].

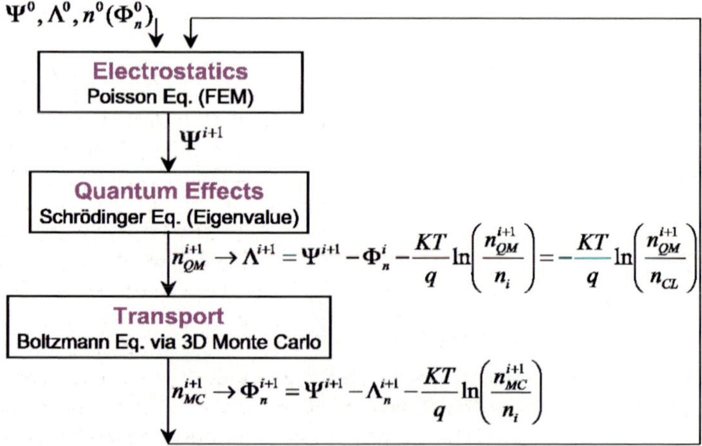

Fig. 6 (**p. 47**) Main blocks of the simulation program and their interactions. Simulation starts by reading an initial guess computed with conventional programs.

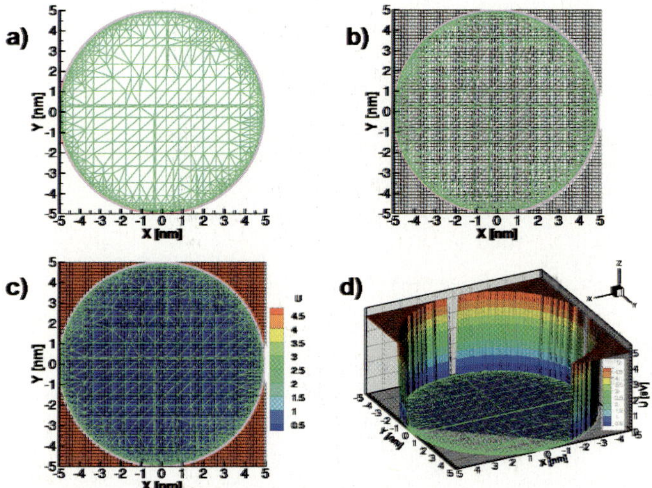

Fig. 7 (**p. 48**) Numerical solution of the 2D SE in the case of a circular well with R=5nm. a) initial finite element mesh; b) domain map to a uniform tensor product grid; c) contour plot of the energy profile; d) partial 3D view of the energy profile.

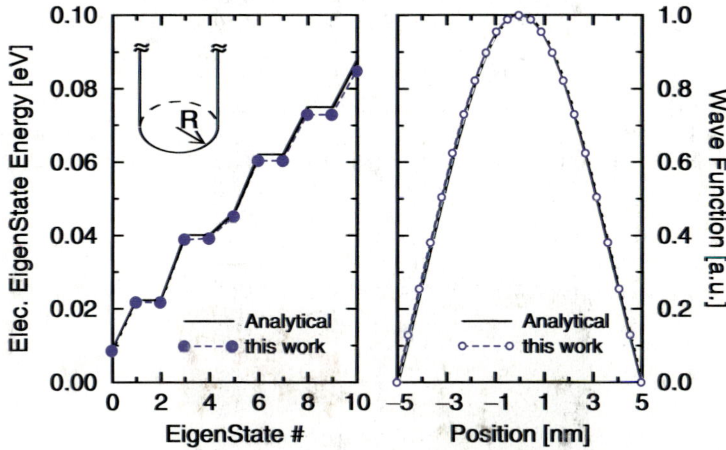

Fig. 8 (**p. 49**) Validation of the numerical solution of the 2D SE in the case of a circular well with R=5nm. Solid line: analytical solution; symbols: simulation. Left: eigenstate energy; right: first eigenstate wave function.

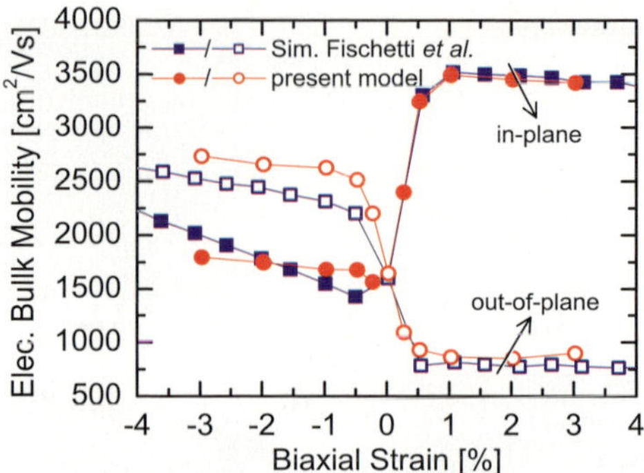

Fig. 9 (p. 49) Simulated electron bulk mobility (□/■) in comparison with calculation of [Fis96] (○/●) for un-doped silicon under biaxial strain. Closed/open symbols refer to in-plane/out-of-plane mobility.

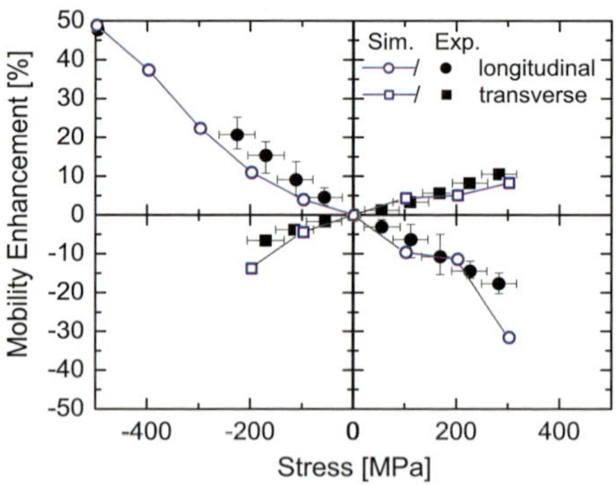

Fig. 10 (p. 50) Monte Carlo simulation of hole mobility enhancement in comparison with experimental data from wafer bending experiments of [Tho04] under uniaxial stress.

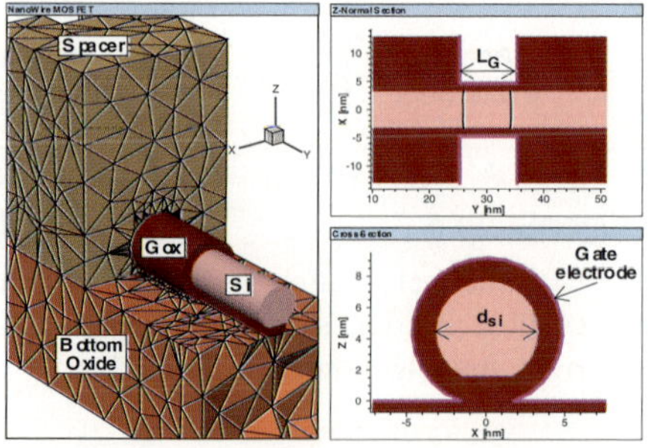

Fig. 11 (p. 50) Silicon NanoWire n-MOSFET reported in [Yan04] and simulated in this work. a) partial 3D view; b) horizontal (z-normal) section; c) channel cross-section. $L_G = 10nm$, $t_{ox} - 1.5nm$, $d_{si} = 6.5nm$. Not all dielectrics are shown.

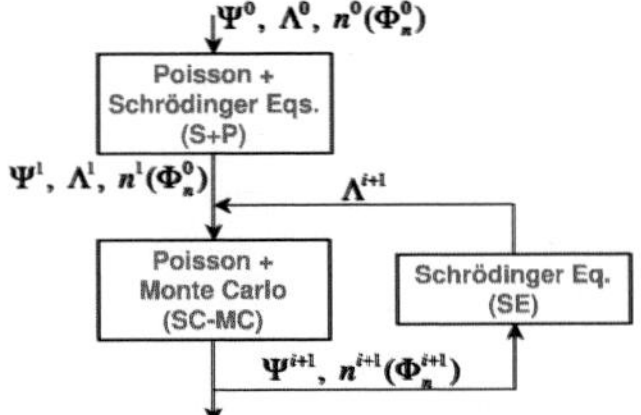

Fig. 12 (p. 51) Schematic representation of the iteration scheme. Convergence is reached after a few iterations. (Notice that $\Phi^1 = \Phi^0$)

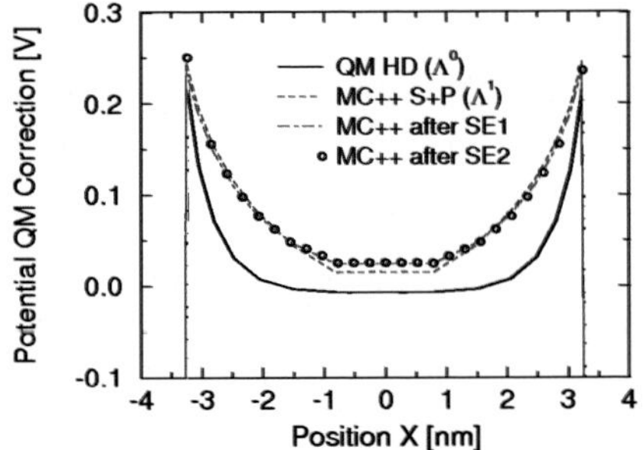

Fig. 13 (p. 51) Evolution of the potential QM correction during iterations. Λ^0 is the initial profile computed with conventional density-gradient hydrodynamic simulation (QM-HD). Λ^1 is the first guess provided by the self-consistent solution of the Schrödinger-Poisson Eq. (S+P).

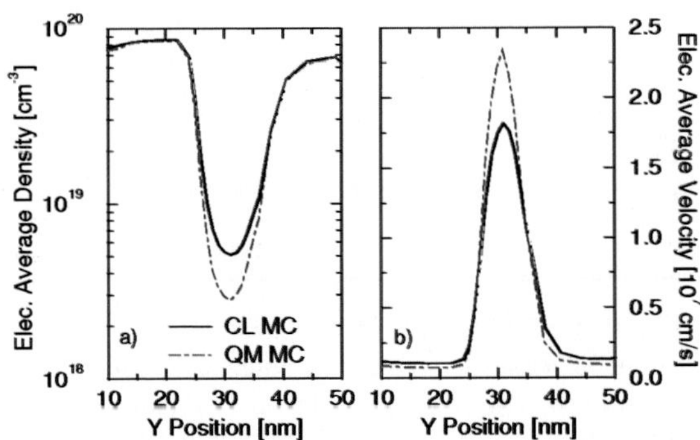

Fig. 15 (p. 52) Simulated electron density (a) and velocity (b) averaged on a channel cross-section as a function of the position for $V_G = 0.5V$, $V_{DS} = 1V$ with (QM MC, dot-dashed line) and without (CL MC, solid line) QM correction.

A Demonstrator Platform for Coupled Multiscale Simulation

Carlo de Falco, Georg Denk

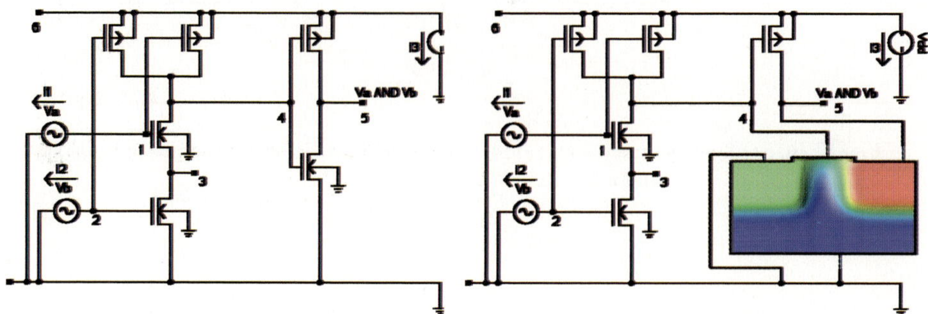

Fig. 2(a) (p. 70) The CMOS AND gate simulated

Fig. 2(b) (p. 70) The set-up for coupled simulation of the AND gate

Accurate Modeling of Complete Functional RF Blocks: CHAMELEON RF

H.H.J.M. Janssen, J. Niehof and W.H.A. Schilders

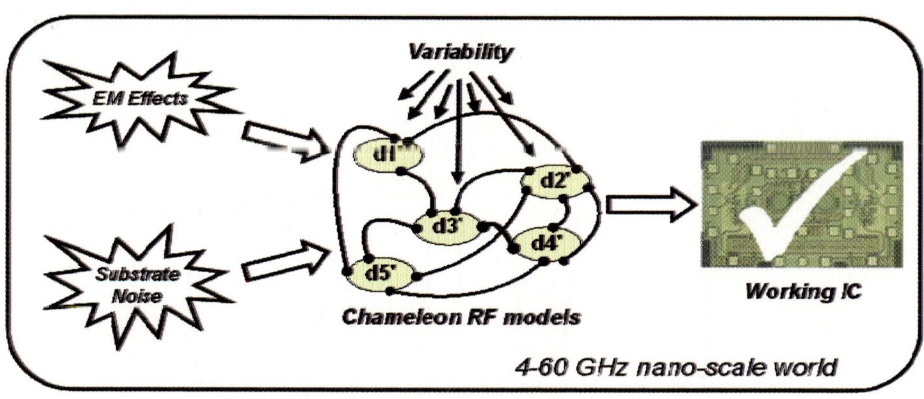

Fig. 1 (p. 83) Overview of the Chameleon RF system.
For 4-60 GHz frequencies made possible by nano-scale integration technologies, electromagnetic and substrate noise effects require hierarchical connector-equipped models of full RF functional blocks in order to enable creation of working chips. The models will be variability-aware to account for relatively increasing effects of manufacturing tolerances.

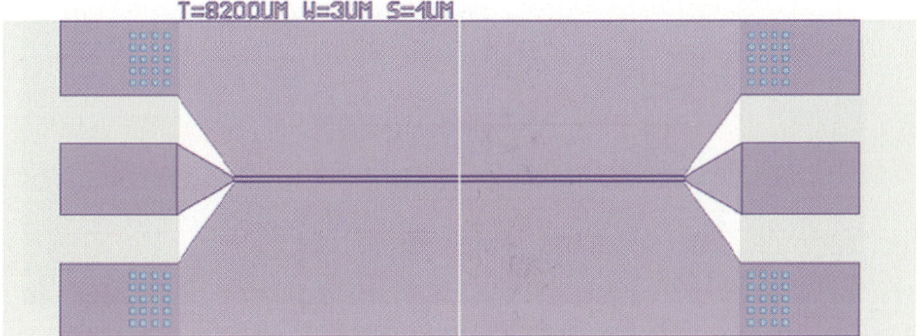

Fig. 2 (p. 84) Layout of coplanar line, 3 μm wide and 8.2 mm long, with 1 μm spacing between the line and ground.

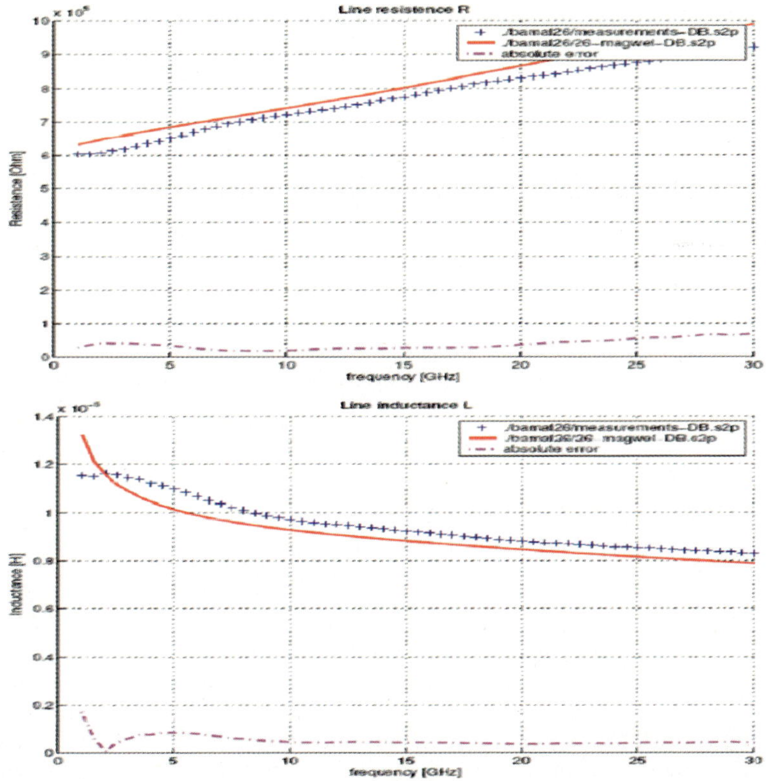

Fig. 3 (p. 85) Comparison of simulations with measurementsof the line parameters R and L (calculated from the impedance Z=R=jωL) for the coplanar line.

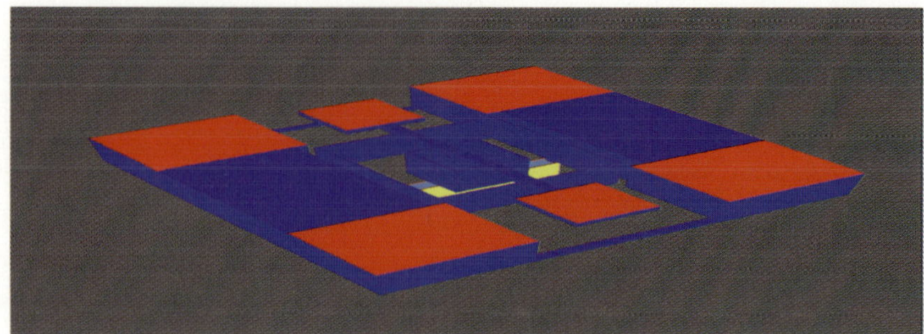

Fig. 4 (p. 86) Substrate isolation structure layout.

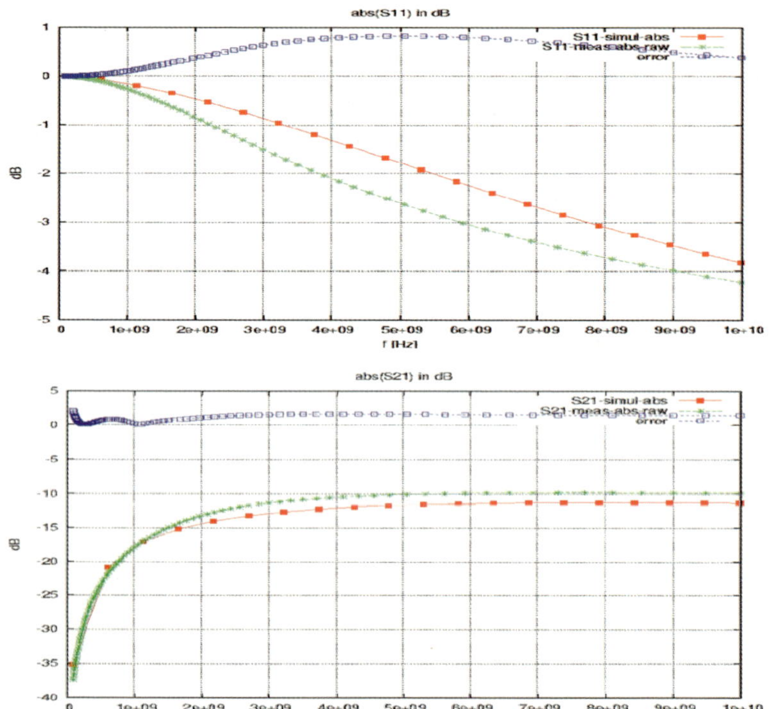

Fig. 5 (p. 87) S11 and S12 curves: the comparison between measurement and simulation shows good agreement (¡1 dB error) for frequencies up to 10 GHz.

Finite Volume Method Applied To Symmetrical Structures in Coupled Problems

Ioana - Gabriela Sîrbu

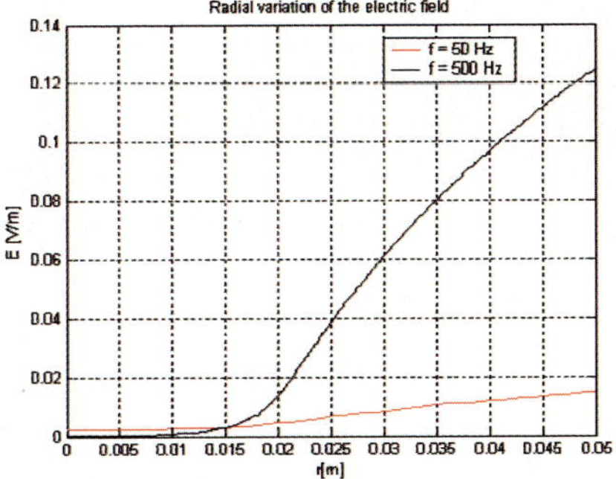

Fig. 3 (**p. 111**) The radial variation of the electric field

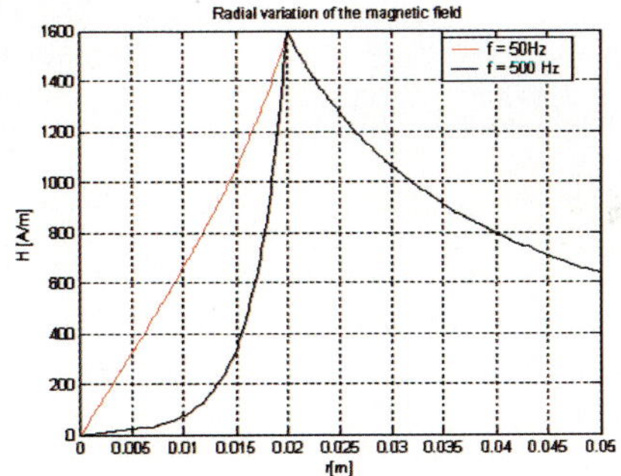

Fig. 4 (**p. 112**) The radial variation of the magnetic field

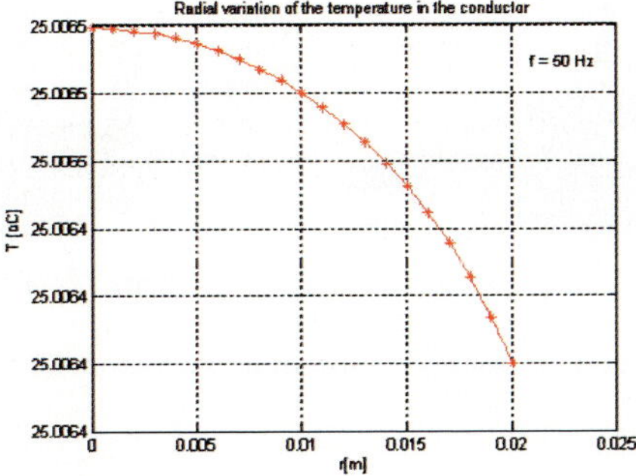

Fig. 7 (**p. 113**) The radial variation of the temperature for f = 50 Hz

Scattering Matrix Analysis of Cascaded Periodic Surfaces

Adriana Savin, Raimond Grimberg, Rozina Steigmann

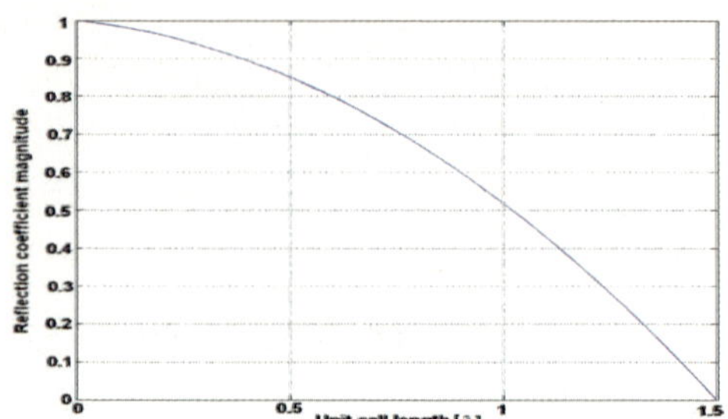

Fig. 2 (**p. 118**) Reflection coefficient magnitude vs. unit cell size

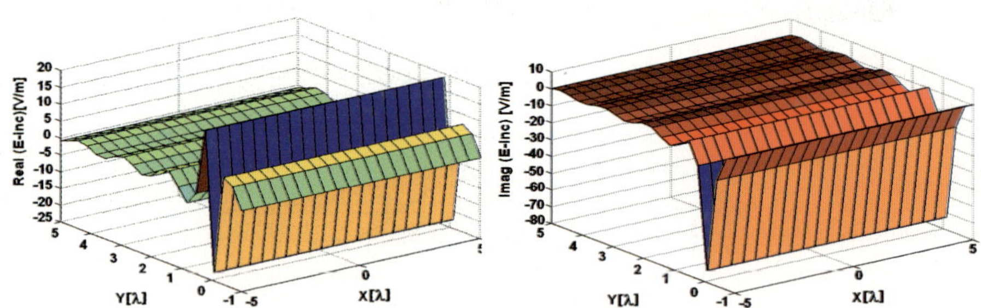

Fig. 3 (**p. 120**) The electromagnetic incident field a) Real component of TM electric field; b) Imaginary component of TM electric field

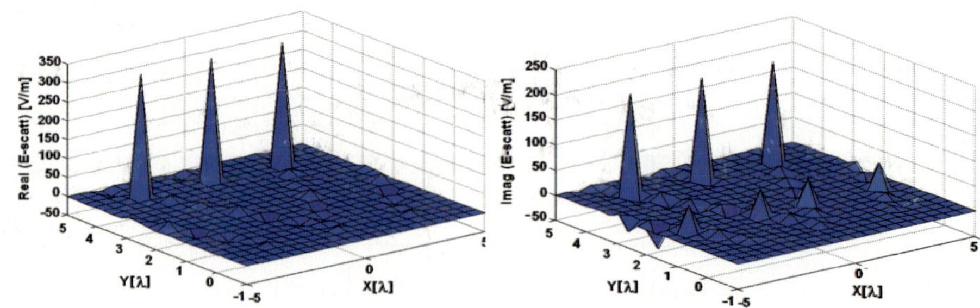

Fig. 5 (**p. 120**) The scattered field for $a = 1.5\lambda$. a) real component; b) imaginary component

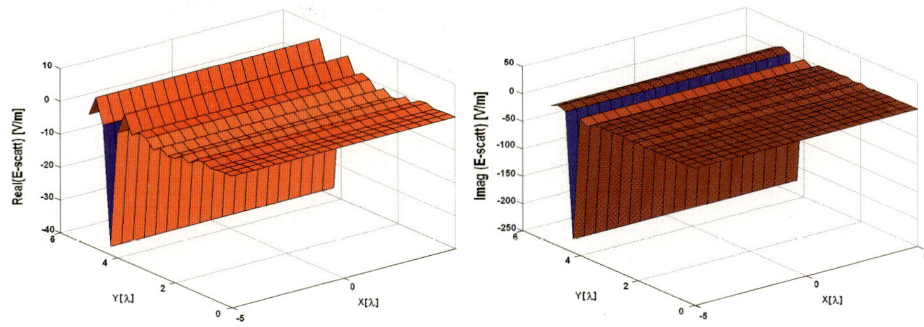

Fig. 6 (p. 121) The scattered field for $a = \lambda/2$. a) real component; b) imaginary component

Outstanding Issues in Model Order Reduction

João M. S. Silva, Jorge Fernández Villena, Paulo Flores, L. Miguel Silveira

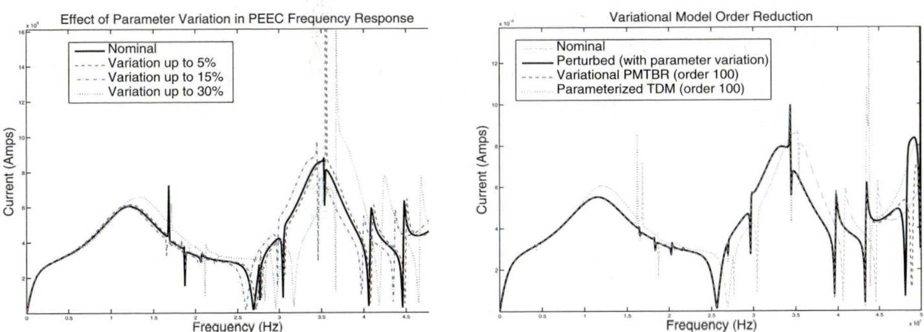

Fig. 3 (p. 150) Variational PEEC: effects on the frequency response (left) and performance of parametric MOR methods (right).

Adjoint Transient Sensitivity Analysis in Circuit Simulation

Z. Ilievski, H. Xu, A. Verhoeven, E.J.W.ter Maten, W.H.A. Schilders, R.M.M. Mattheij

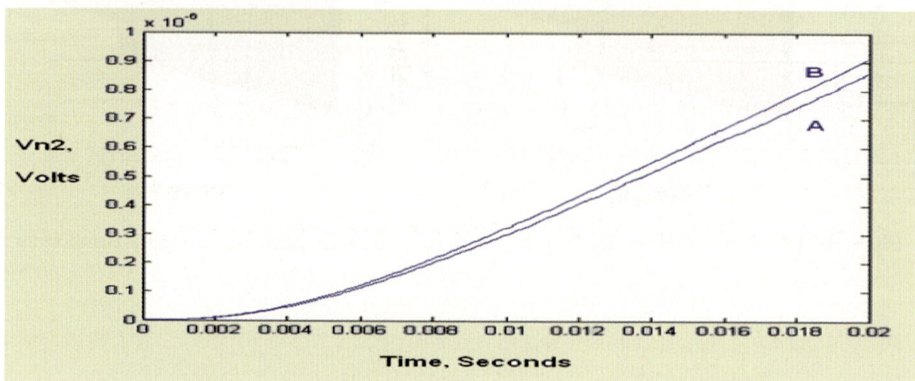

Fig. 1 (p. 188) Voltage differences for successive sensitivity values, node 2

Index Reduction by Element-Replacement for Electrical Circuits

Simone Bächle and Falk Ebert

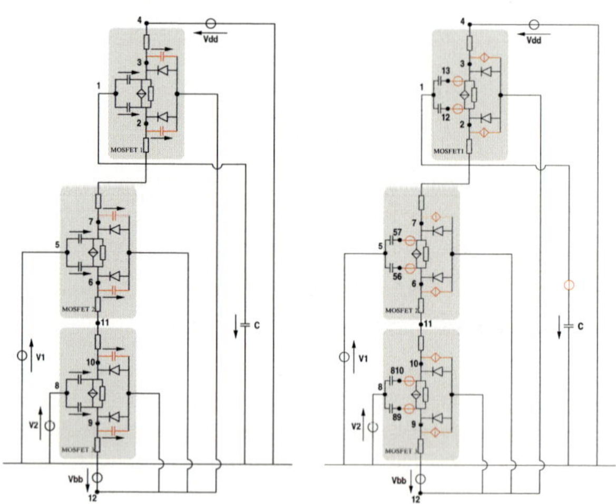

Fig. 1 (p. 195) NAND gate replacement circuit, index 2 circuit (left), index 1 circuit with additional voltage sources (right)

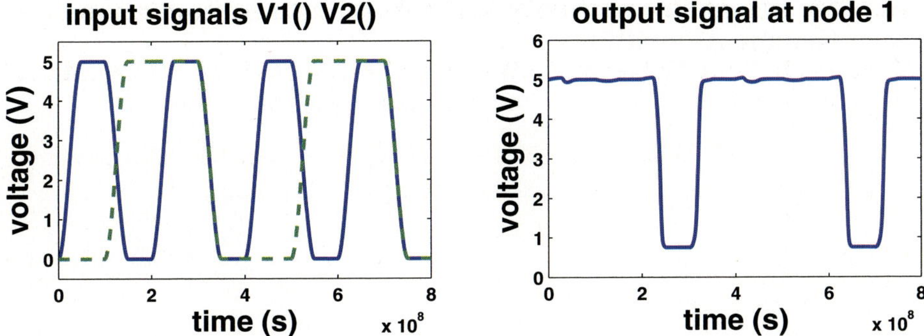

Fig. 2 (p. 195) Input signals (left) and reference solution for the output signal (right)

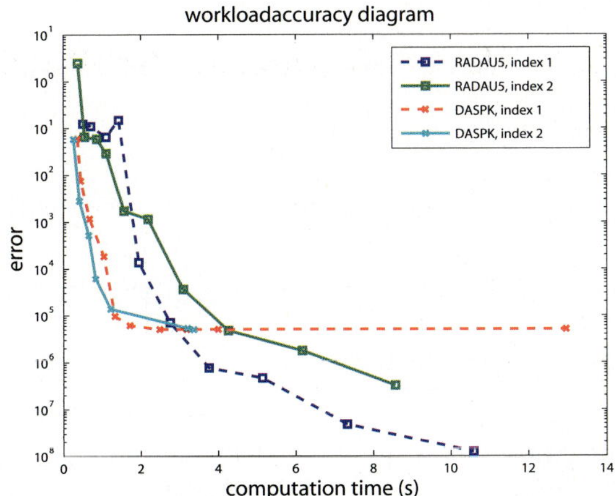

Fig. 3 (p. 196) workload vs. error for the NAND gate example

A Filter Design Framework with Multicriteria Optimization Based on a Genetic Algorithm

Neag Marius, Marina Topa, Liviu Nedelea, Lelia Festila, Vasile Topa

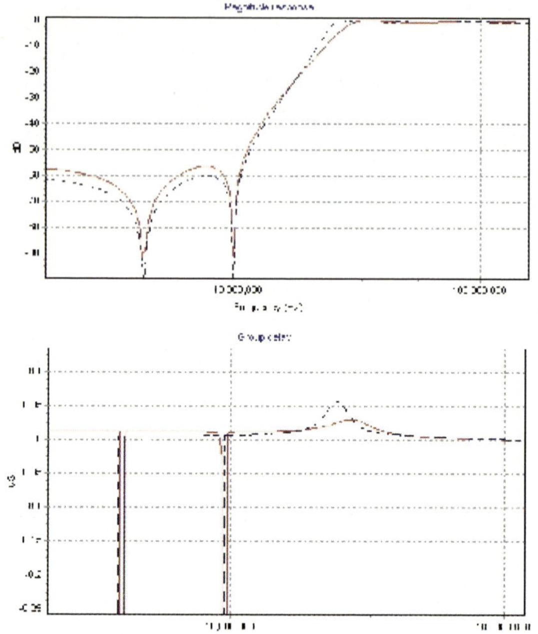

Fig. 2 (p. 211) Magnitude response and group delay of the multicriteria optimized filter

Thermal Network Method in the Design of Power Equipment

C. Gramsch, A. Blaszczyk, H. Lbl, S. Grossmann

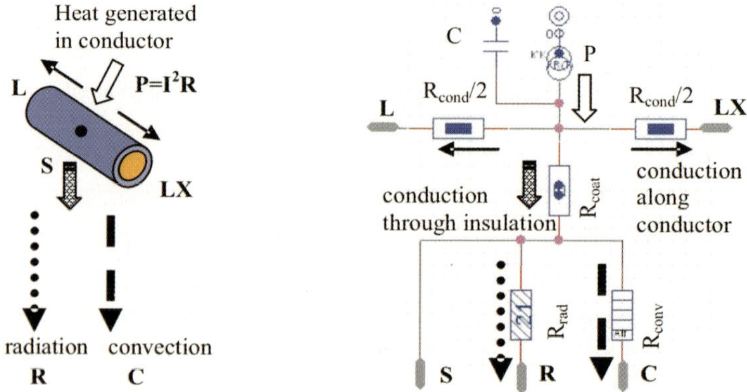

Fig. 1 (p. 214) Thermal network model of coated conductor

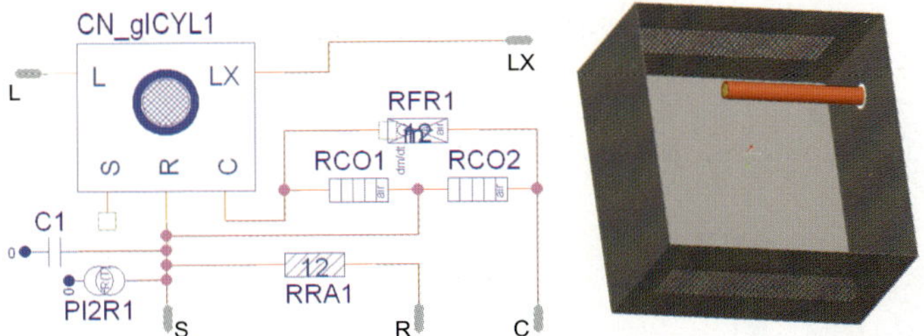

Fig. 2 (p. 215) Example of a hierarchical thermal network representing coated conductor in a ventilated enclosure

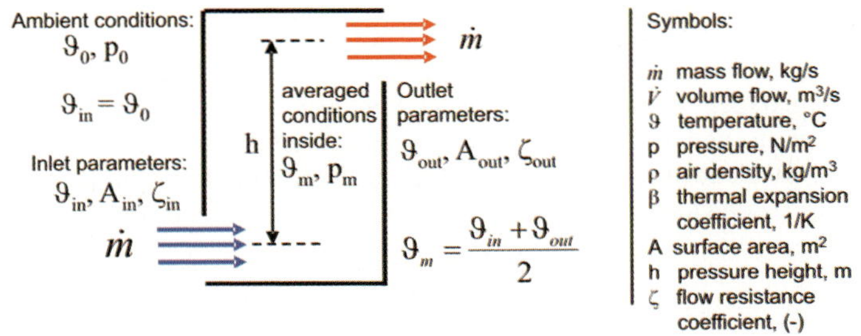

Fig. 3 (p. 216) Basic relationships for the natural ventilation of a compartment

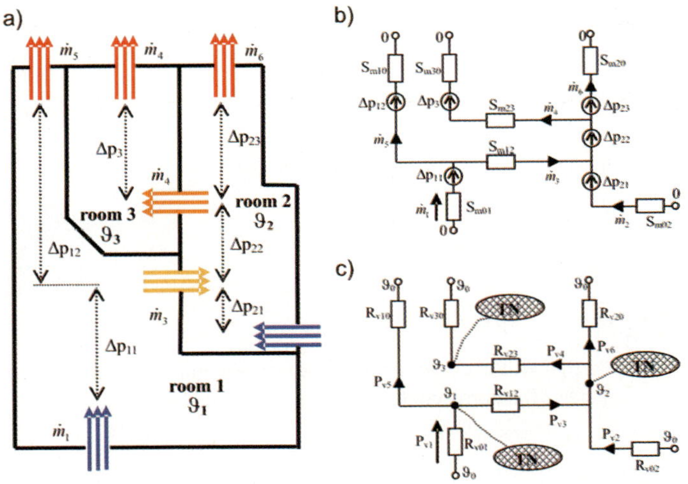

Fig. 4 (p. 218) Example of an arrangement with air flow between compartments (a) and the corresponding models of (b) ventilation network and (c) thermal network.
(The configuration of compartments and the ventilation openings is the same as in the arrangement shown in Fig. 5.)

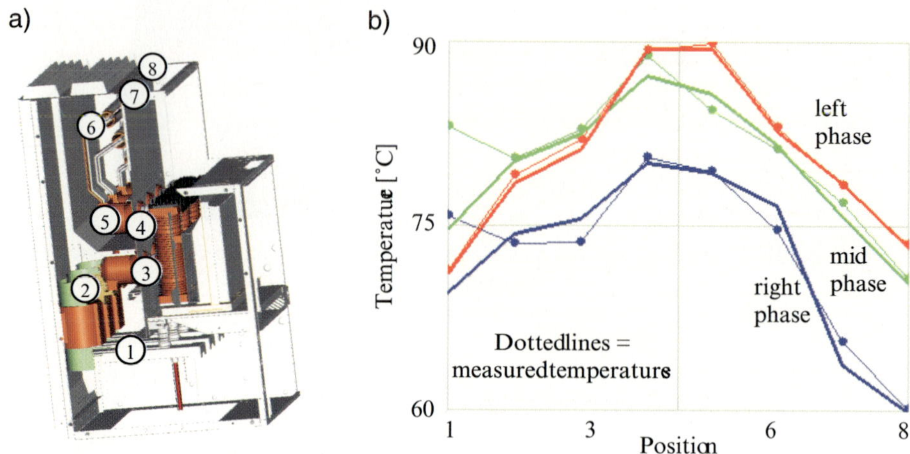

Fig. 5 (p. 218) Air insulated medium voltage switchgear arrangement
Comment: the numbers shown in circles on geometry view (a) denote positions of measurement/calculation points at the metal surface of the current carrying conductors. These positions correspond to the values shown on the X-axis of the temperature distribution (b)

Automatic partitioning for multirate methods

A. Verhoeven, B. Tasić, T.G.J. Beelen, E.J.W. ter Maten, R.M.M. Mattheij

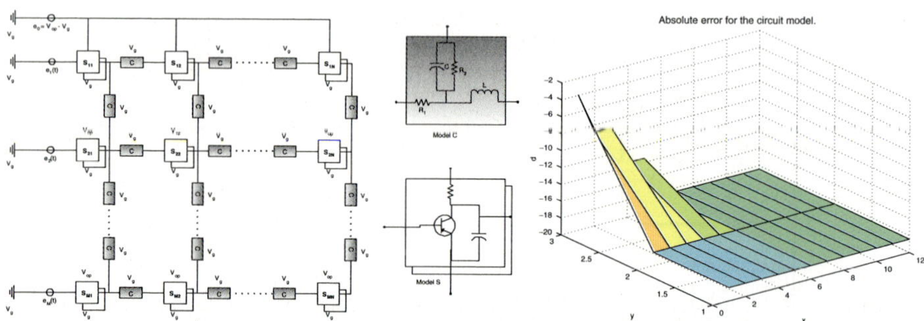

Fig. 1 (p. 229) At the left the circuit diagram of test example and at the right the typical shape of thc corresponding error vector for $M = 3, N = 6$.

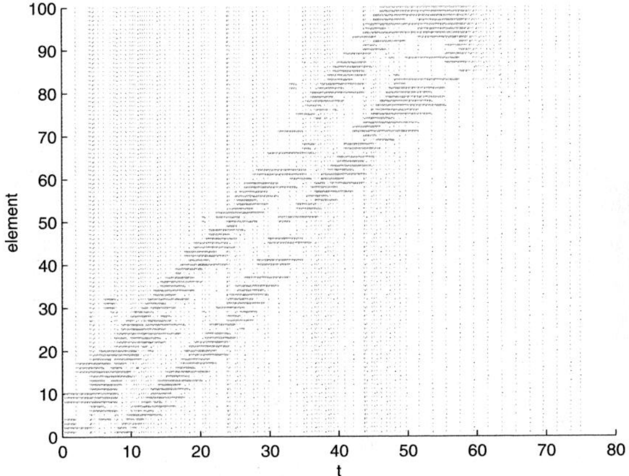

Fig. 3 (p. 235) Timepoints per element for the inverter chain (case 1 with $\alpha = \frac{3}{2}$).

Newton and Approximate Newton Methods in Combination with the Orthogonal Finite Integration Technique

H. De Gersem, I. Munteanu, T. Weiland

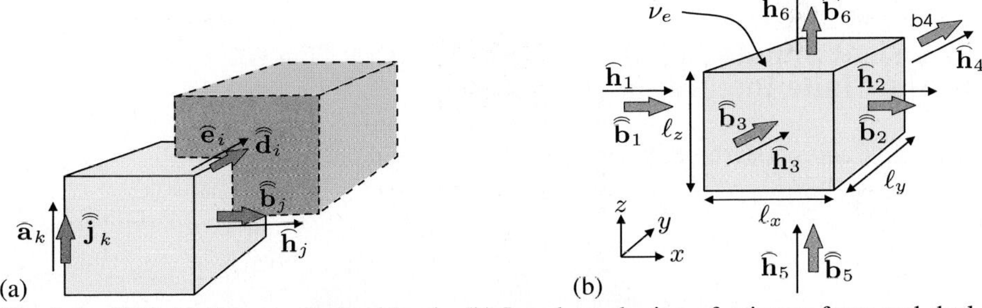

Fig. 1 (p. 276) (a) Primary-dual grid pair. (b) Local numbering of primary faces and dual edges associated at a primary grid cell

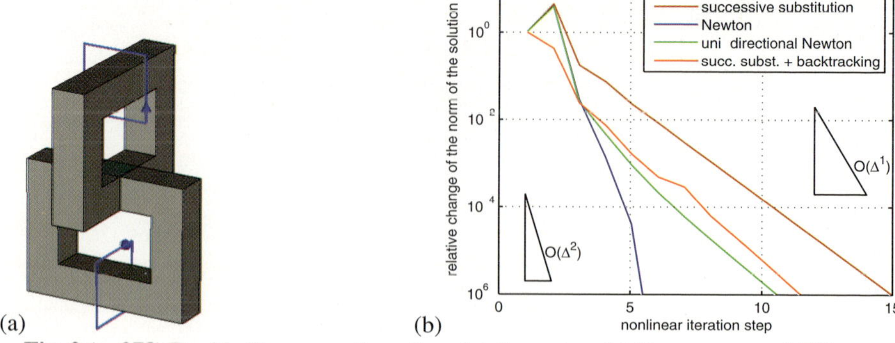

Fig. 2 (p. 278) Double C-core configuration: (a) Geometry, (b) Convergence of different linearization techniques

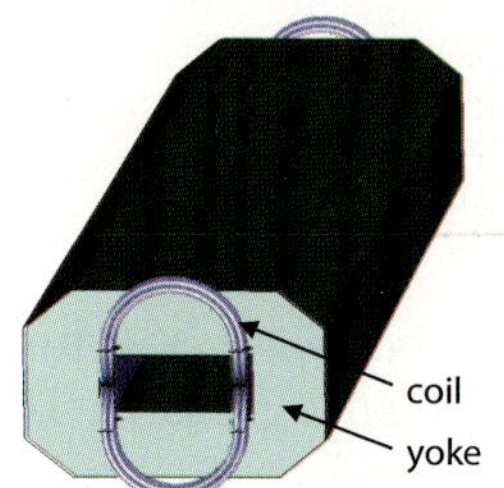

Fig. 3 (p. 279) Geometry of the Nuclotron magnet device.

Transient Simulation of a Linear Actuator Discretized by the Finite Integration Technique

Mariana Funieru, Herbert De Gersem, Thomas Weiland

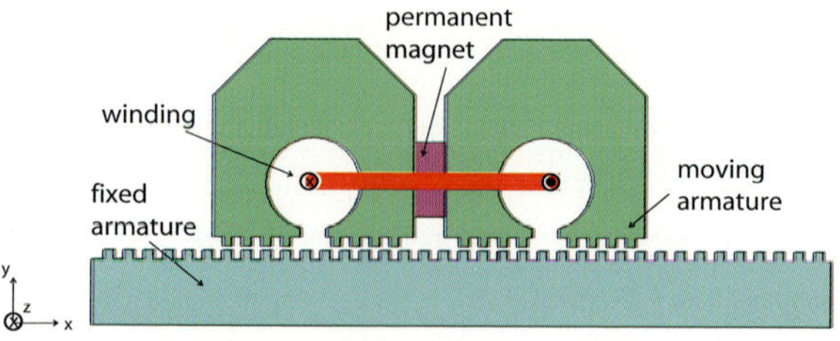

Fig. 1 (p. 282) Lincar actuator

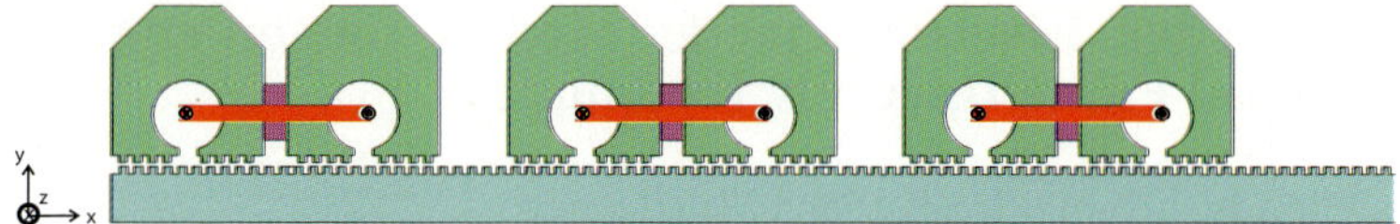

Fig. 2 (**p. 284**) A 3-phased linear actuator.

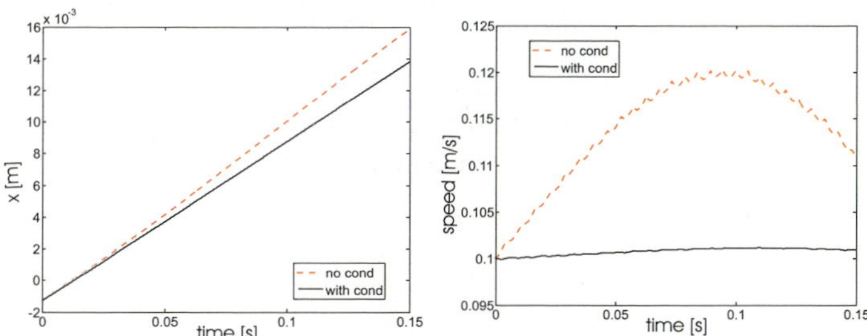

Fig. 3 (**p. 284**) Position and speed for a block current.

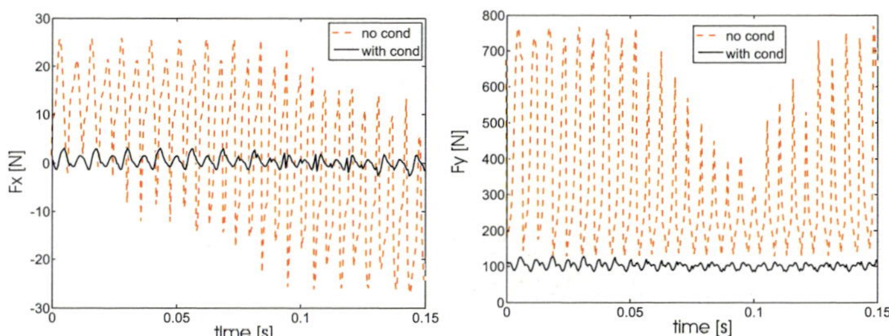

Fig. 4 (**p. 284**) Displacement force F_x and vertical attraction force F_y for a block current.

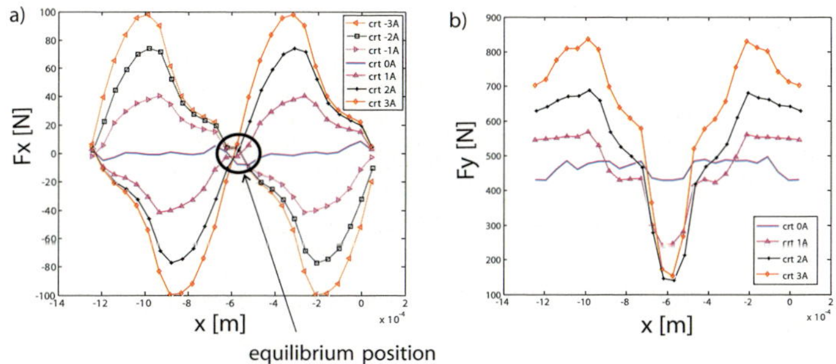

Fig. 5 (**p. 285**) Forces versus position: F_x a) and F_y b).

Reduced Order Electromagnetic Models for on-chip passives based on dual Finite Integrals Technique

Gabriela Ciuprina, Daniel Ioan, Diana Mihalache

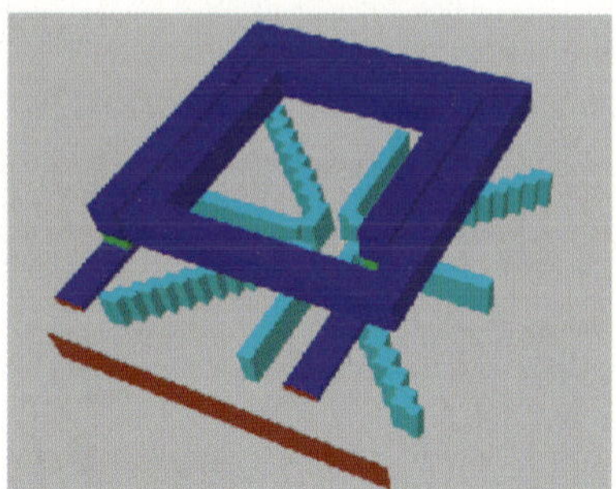

Fig. 1 (p. 288) A typical on-chip component (www.imec.be/codestar)

Techniques to Reduce the Equivalent Parallel Capacitance for EMI Filters Integration

Adina Racasan, Calin Munteanu, Vasile Topa, Claudia Racasan

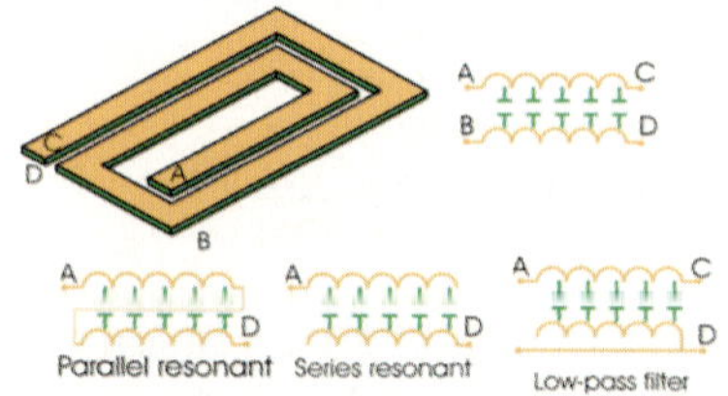

Fig. 1 (p. 295) The integrated LC structure

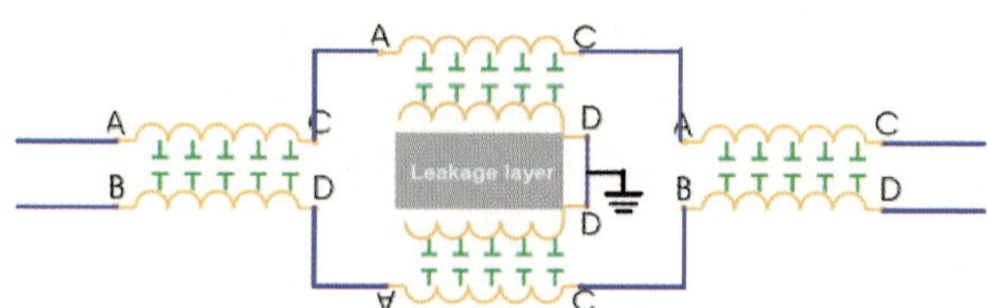

Fig. 2 (p. 296) Integrated EMI filter composition

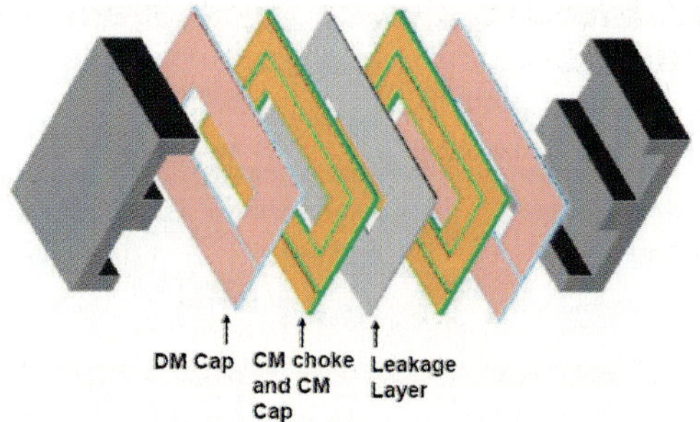

Fig. 3 (p. 296) Physical structure of integrated EMI filter

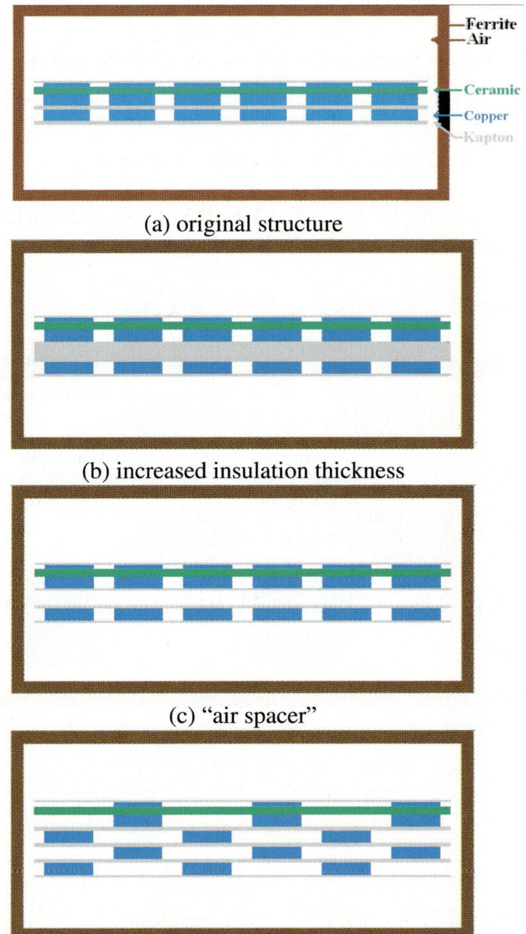

Fig. 4 (p. 298) FEA Simulation models of different winding structures

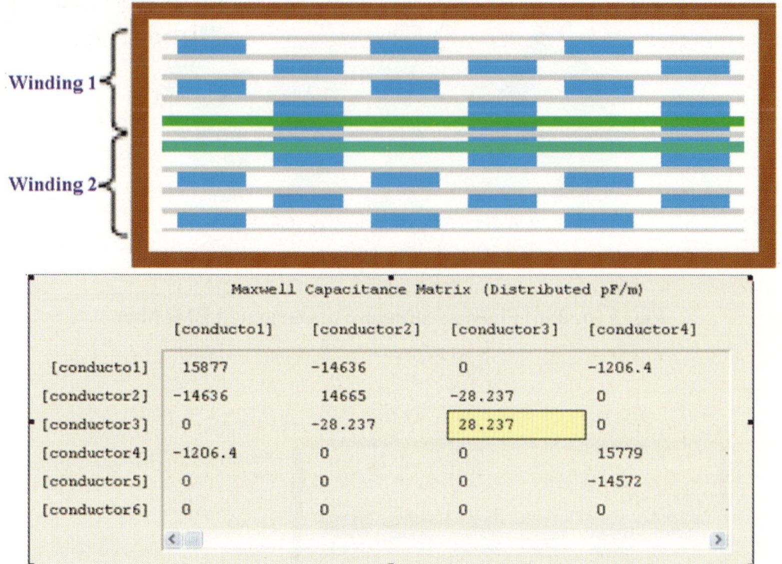

Maxwell Capacitance Matrix (Distributed pF/m)

	[conducto1]	[conductor2]	[conductor3]	[conductor4]
[conducto1]	15877	-14636	0	-1206.4
[conductor2]	-14636	14665	-28.237	0
[conductor3]	0	-28.237	28.237	0
[conductor4]	-1206.4	0	0	15779
[conductor5]	0	0	0	-14572
[conductor6]	0	0	0	0

Fig. 7 (**p. 299**) Two staggered windings not interleaved

Fig. 8 (**p. 299**) Staggered and interleaved windings

Symmetric Coupling of the Finite-Element and the Boundary-Element Method for Electro-Quasistatic Field Simulations

T. Steinmetz, N. Gödel, G. Wimmer, M. Clemens, S. Kurz, M. Bebendorf, S. Rjasanow

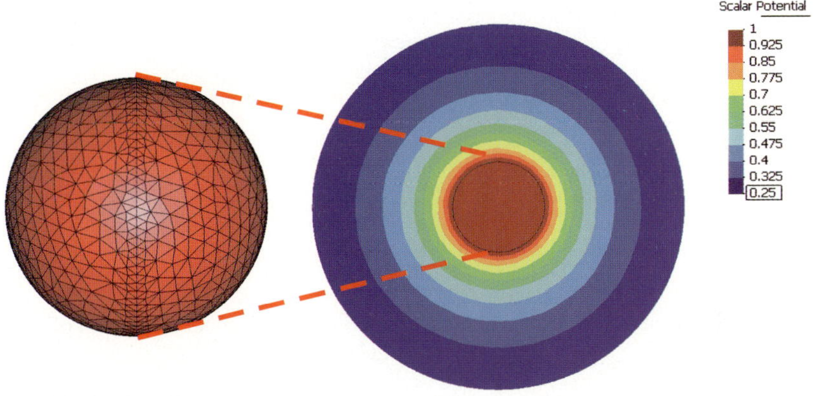

Fig. 2 (p. 285) Geometry (left) and scalar electric potential (right)

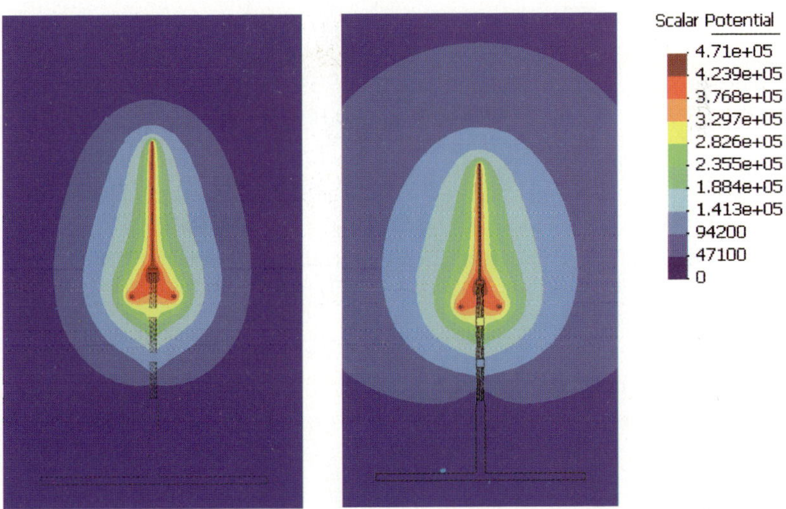

Fig. 2 (p. 285) Geometry, scalar electric potential computed by FEM and by FEM-BEM (from left). While the boundary is set to 0 V in the FEM simulation, the effect of the open boundary in the FEM-BEM simulation is obvious.

Computational Errors in Hysteresis Preisach Modelling

Valentin Ionita, Lucian Petrescu

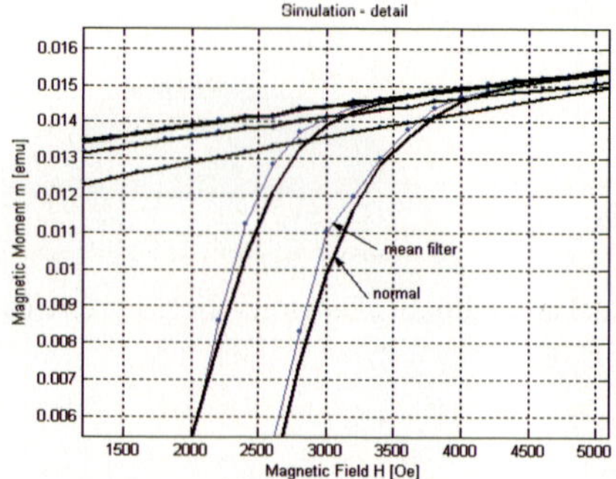

Fig. 1 (p. 319) Effect of the experimental data filtering on numerical simulation of subway magnetic ticket

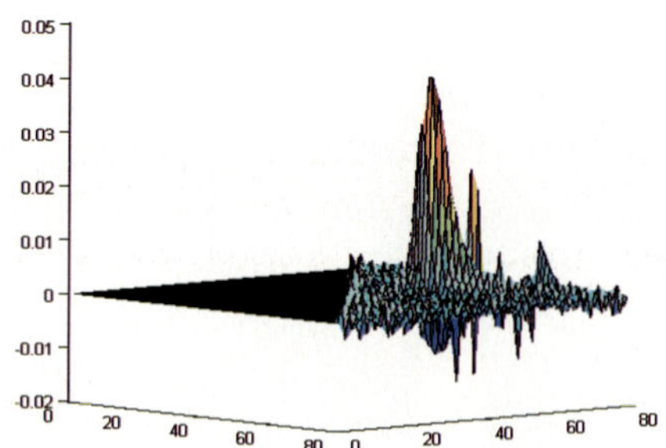

Fig. 2 (p. 319) Preisach function for identification with 80 FORCs

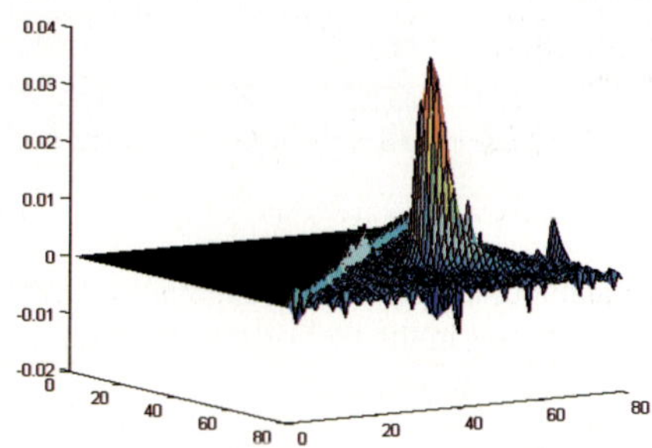

Fig. 3 (p. 320) Preisach function for identification with 80 filtered FORCs

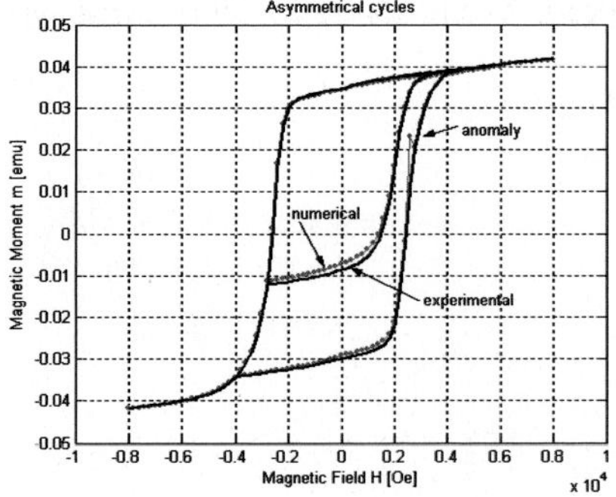

Fig. 4 (p. 321) Experimental and numerical asymmetrical hysteresis cycles for a bank card.

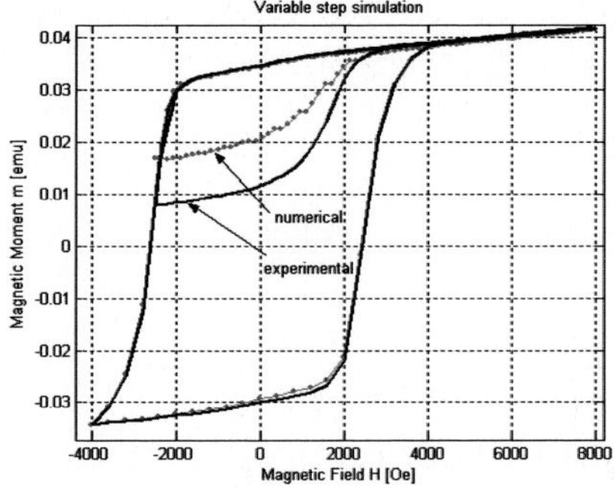

Fig. 5 (p. 321) Experimental and numerical hysteresis curves with variable step field for a magnetic bank card.

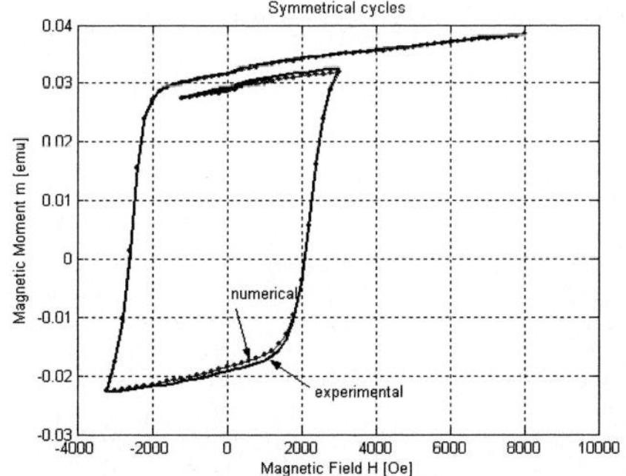

Fig. 6 (p. 321) Experimental and numerical symmetrical cycles for a magnetic bank card.

Manifold Mapping for Multilevel Optimization

Pieter W. Hemker, David Echeverría

The surrogate model: $\mathbf{c}(\mathbf{p}(\mathbf{x})) \approx \mathbf{f}(\mathbf{x})$.

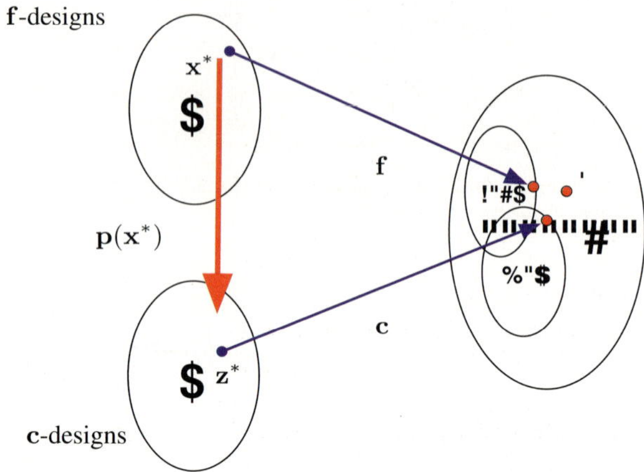

Fig. 1 (p. 327) The space mapping function $\mathbf{p}(\mathbf{x}) = \underset{\mathbf{z}\in Z}{\operatorname{argmin}} \, \| \, \mathbf{c}(\mathbf{z}) - \mathbf{f}(\mathbf{x}) \| \,.$

The surrogate model: $\mathbf{S}_k \circ \mathbf{c} \circ \overline{\mathbf{p}} \approx \mathbf{f}$.

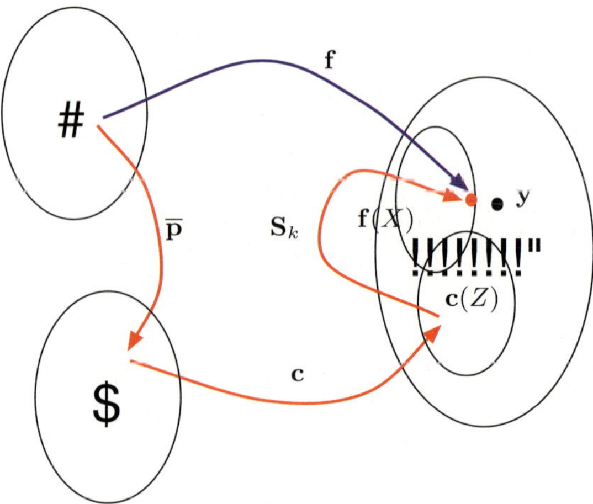

Fig. 2 (p. 330) Manifold Mapping.

Software Package for Multi-Objective Optimal Design of Electromagnetic Devices

Calin Munteanu, Gheorghe Mates, Vasile Topa

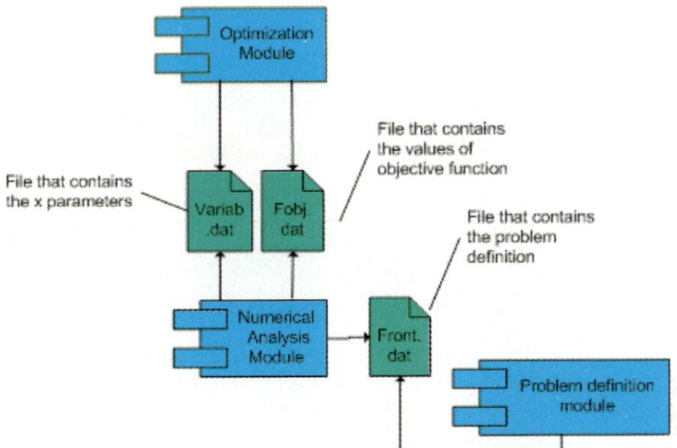

Fig. 1 (**p. 333**) Flowchart of the MOOP Software package

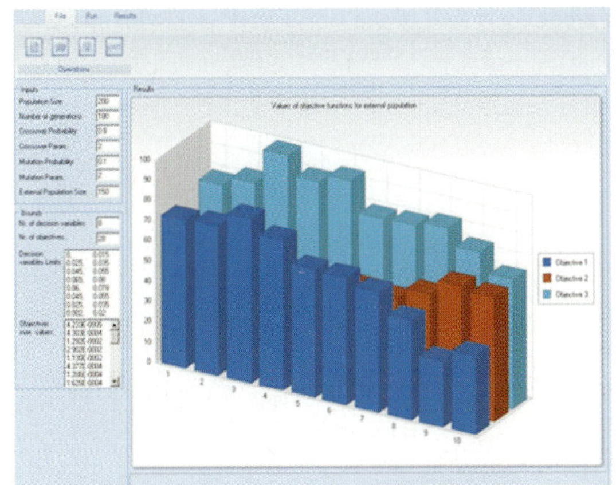

Fig. 2 (**p. 333**) Screenshot of the MOOP software package

Optimal Design of Monolithic ESBT® Device carried out by Multiobjective Optimization

Salvatore Spinella, Vincenzo Enea, Daniele Kroell, Michele Messina, Cesare Ronsisvalle

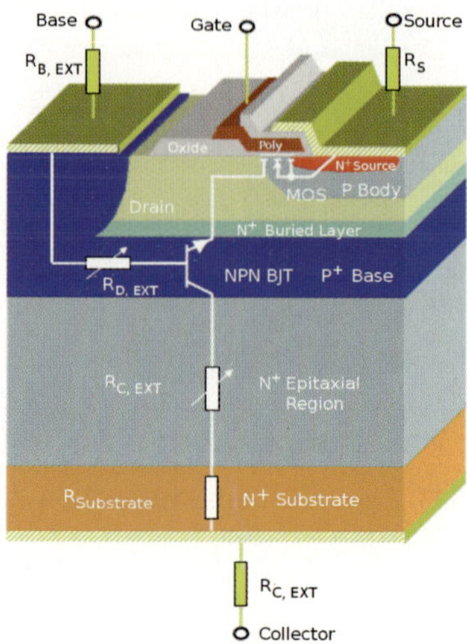

Fig. 2 (p. 340) Half elementary cell of the ESBT® device with superimposed the equivalent electrical circuit.

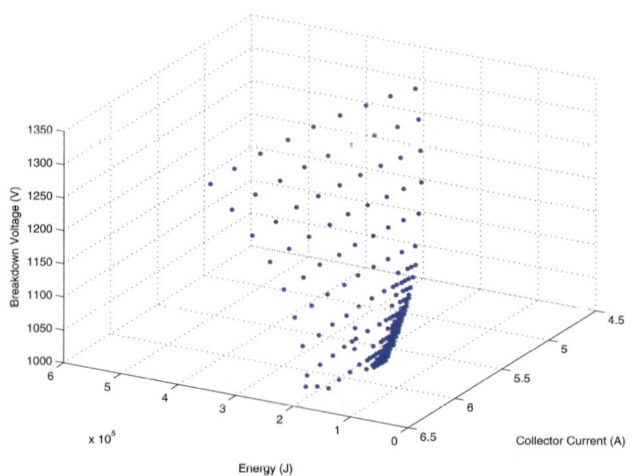

Fig. 5 (p. 345) Sampling in the objective space.

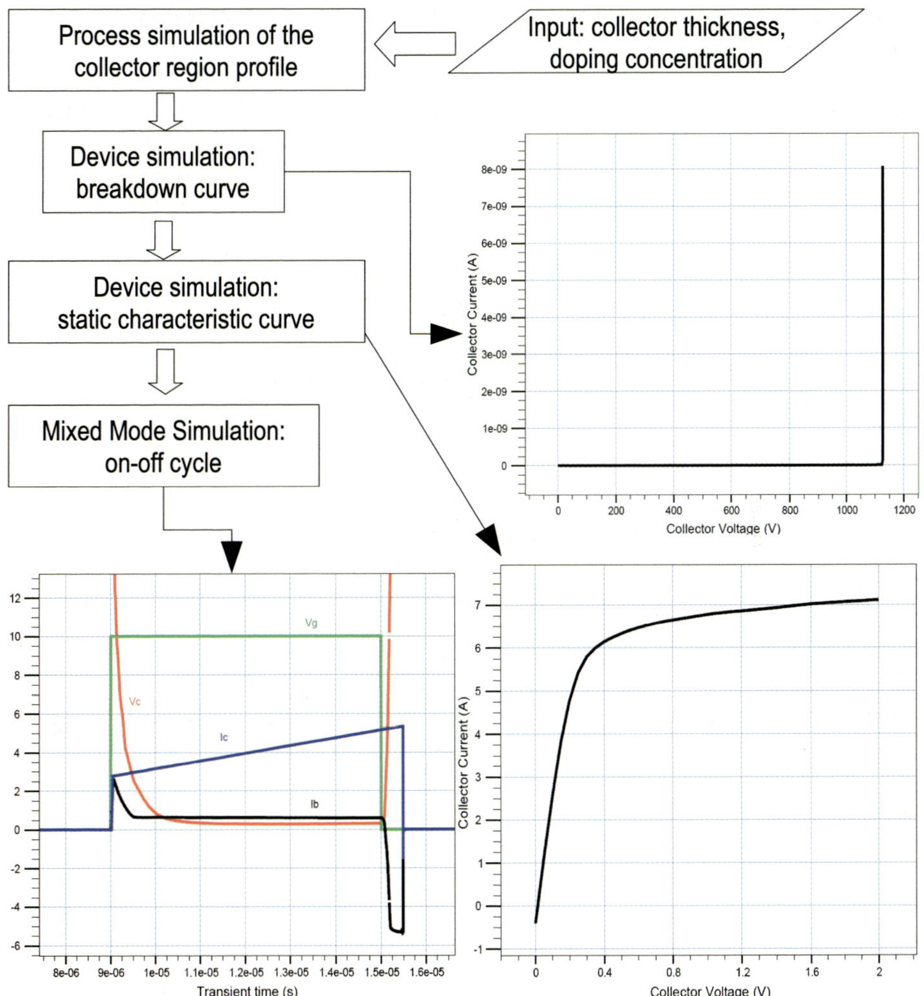

Fig. 3 (p. 343) The simulation flow.

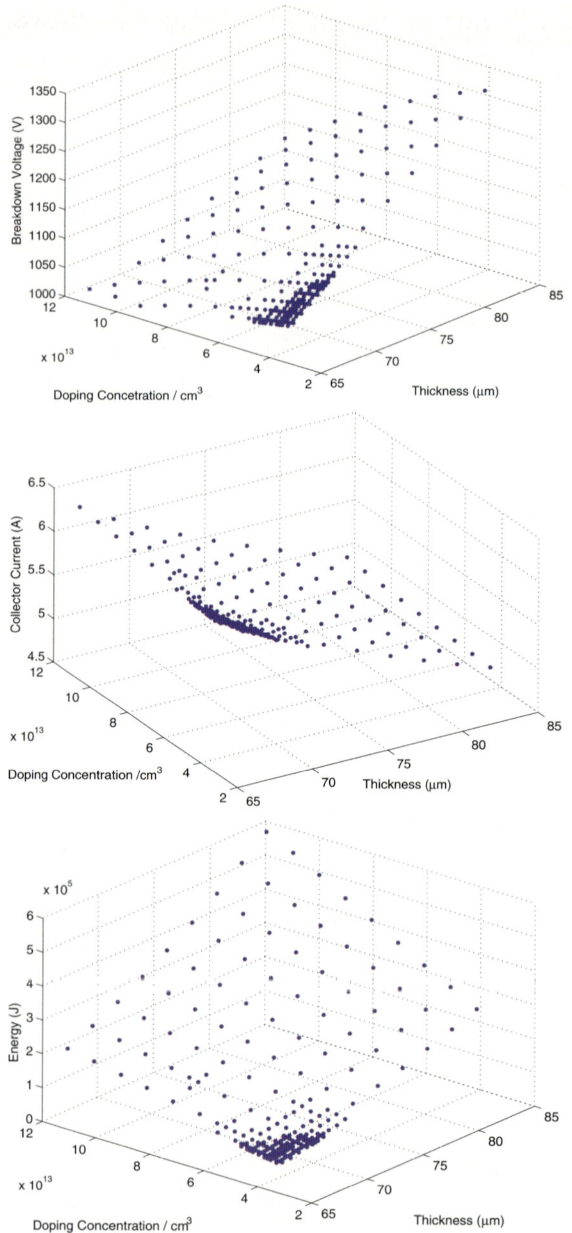

Fig. 4 (**p. 344**) The design variable space against performances.

On Fast Optimal Control for Energy-Transport-based Semiconductor Design

C. R. Drago

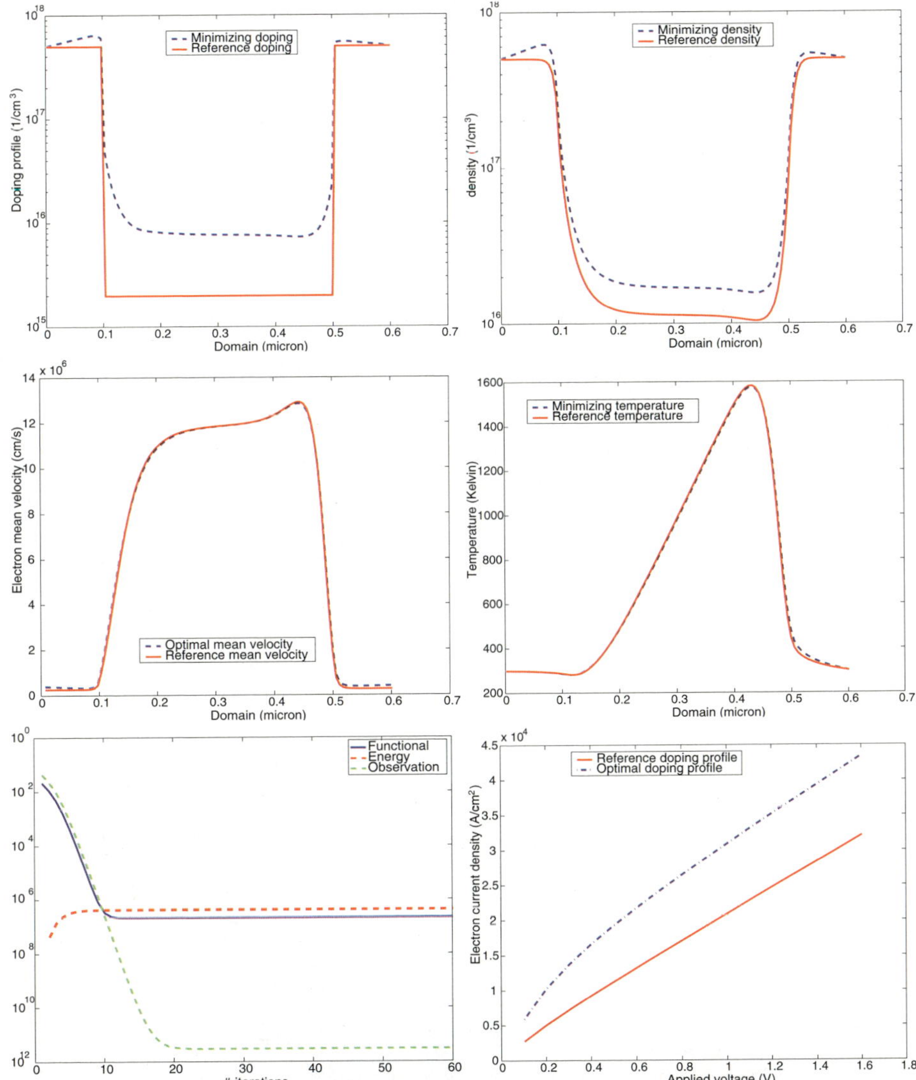

Fig. 1 (p. 354) Optimized doping profile, electron density, electron mean velocity, temperature, evolution of the cost functional for a biasing voltage of 1 V, and the corresponding IVCs

Extended Hydrodynamical Models for Charge Transport in Si

Roberto Beneduci, Giovanni Mascali, Vittorio Romano

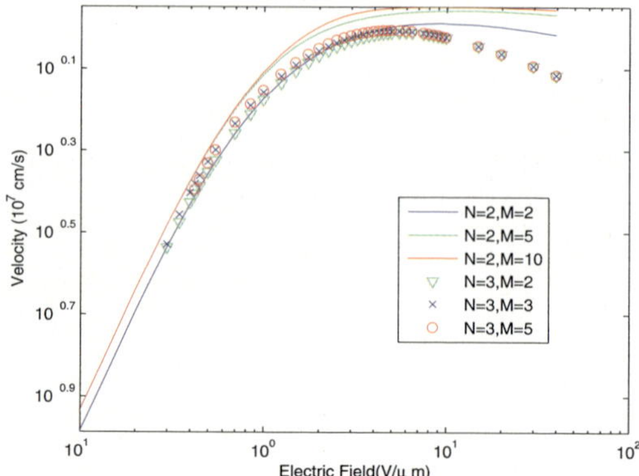

Fig. 1 (p. 363) Drift velocity vs the electric field

On the Implementation of a Delaunay-based 3-dimensional Mesh Generator

K.J. van der Kolk, N.P. van der Meijs

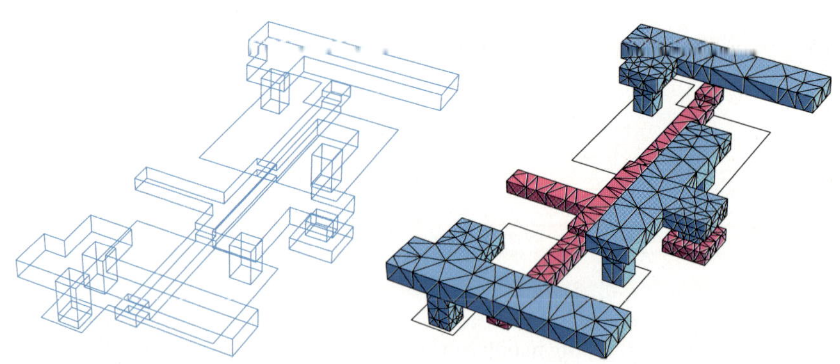

Fig. 1 (p. 366) Example PLC and corresponding mesh. The structure is contained in a bounding box (not shown) and the exterior of the structure is meshed as well in this case.

A Hierarchical Preconditioner within Edge Based BE-FE Coupling in Electromagnetism

K. Straube, I. Ibragimov, V. Rischmüller, S. Rjasanow

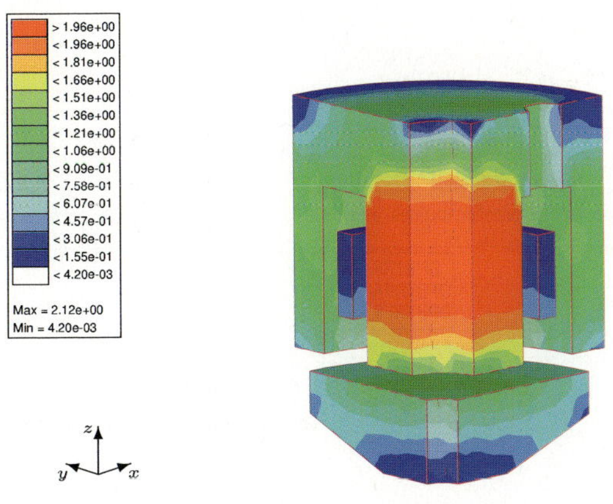

Fig. 3 (p. 384) A quarter of the valve geometry, where only material components and the coil are shown. The colour scale indicates the magnitude of the magnetic induction.

Solution of Band Linear Systems in Model Reduction for VSLI Circuits

Alfredo Remón, Enrique S. Quintana-Ortí, Gregorio Quintana-Ortí

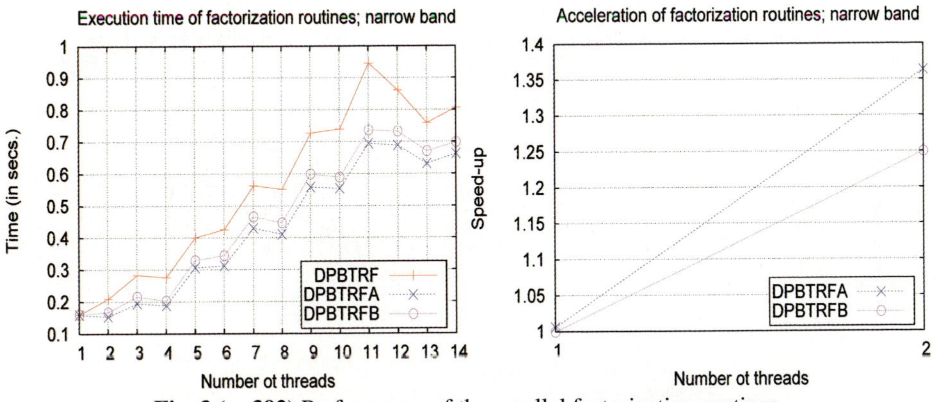

Fig. 3 (p. 392) Performance of the parallel factorization routines.

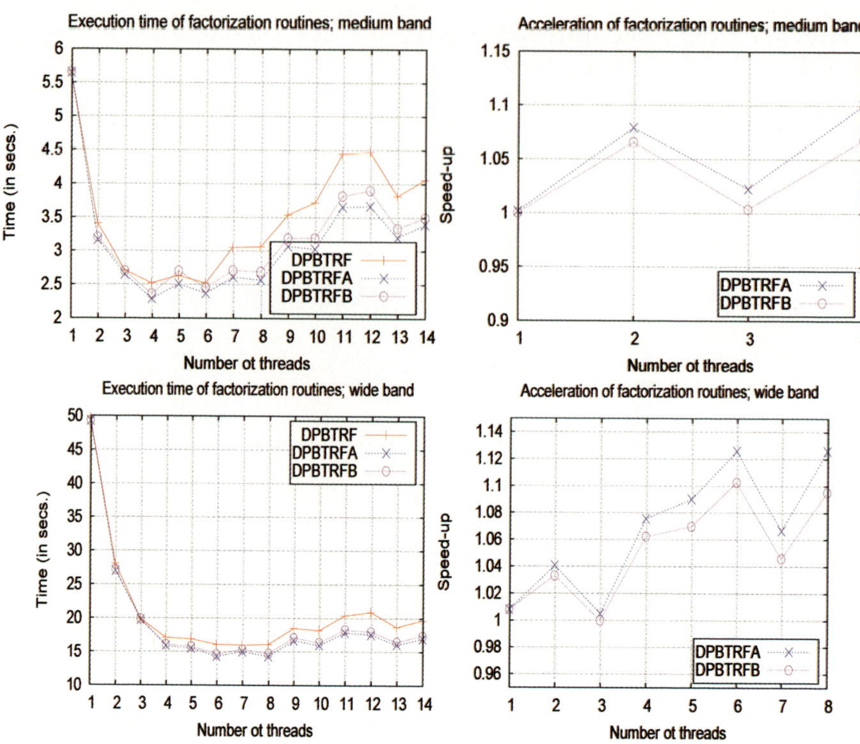

Fig. 3 (**p. 392**) Performance of the parallel factorization routines.

MOESP Algorithm for Converting One-dimensional Maxwell Equation into a Linear System

E. F. Yetkin, H. Dağ, W. H. A. Schilders

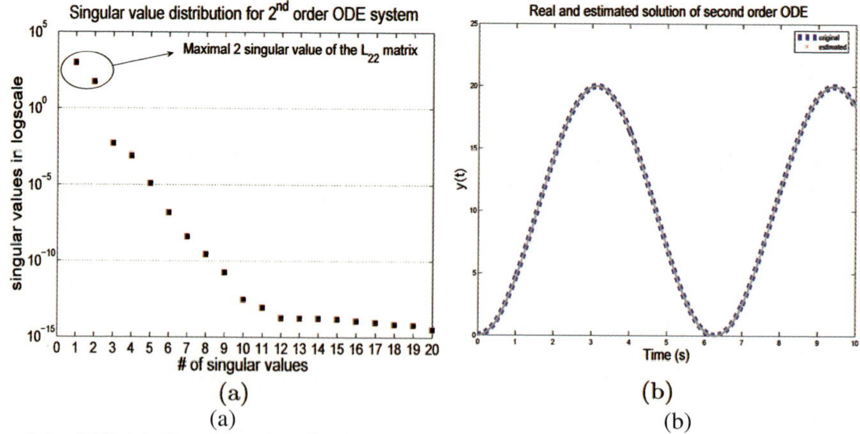

(a) (b)

Fig. 2 (**p. 399**) (a) Singular value distribution of data matrix for u(t)=10 and estimated system order $n = 2$, (b) Original and estimated outputs for a estimated system order $n = 2$

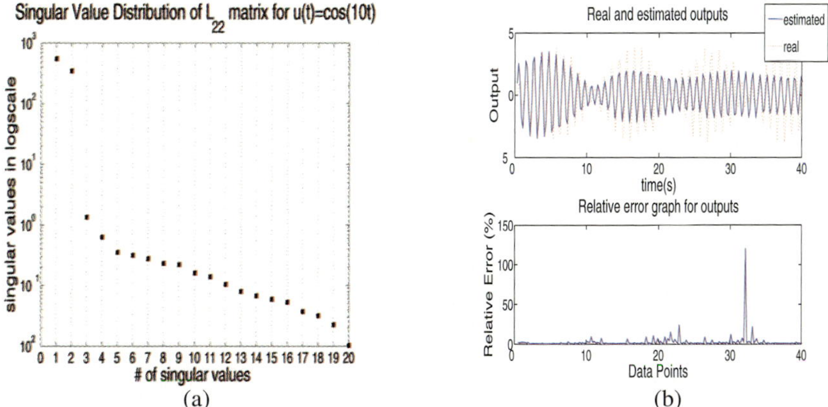

Fig. 3 (p. 400) (a) Singular value distribution of L_{22} matrix for $u(t) = cos(10t)$ (b) Original and estimated outputs and relative error for $u(t) = cos(10t)$ where the estimated system order $n = 2$

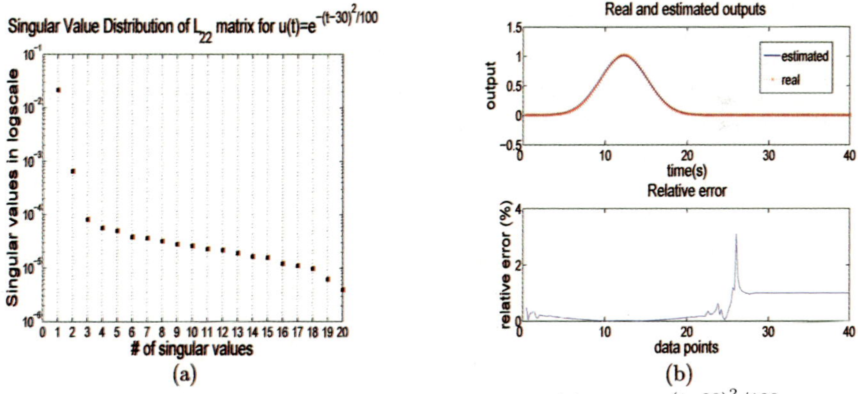

Fig. 4 (p. 400) (a) Singular value distribution of L_{22} for $u(t) = exp^{-(t-30)^2/100}$, (b) Original and estimated outputs and relative error for estimated system order $n = 2$

Adaptive Methods for Transient Noise Analysis

Thorsten Sickenberger, Renate Winkler

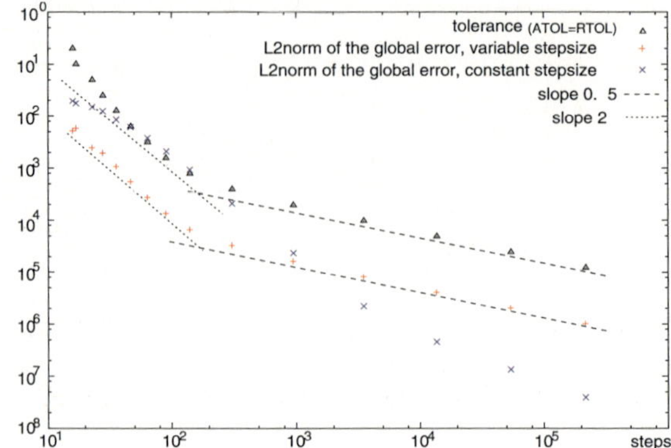

Fig. 1 (**p. 407**) Tolerance and accuracy versus steps for a test-SDE.

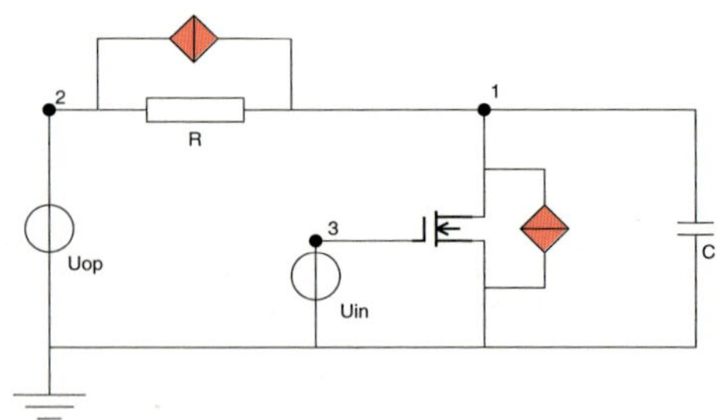

Fig. 2 (**p. 407**) Thermal noise sources in a MOSFET inverter circuit

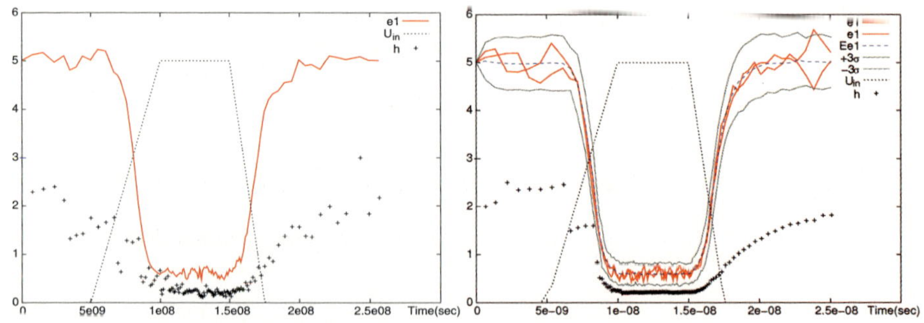

Fig. 3 (**p. 407**) Simulation results for the noisy inverter circuit:
1 path 127(+29 rejected) steps; 100 paths 134(+11 rejected) steps

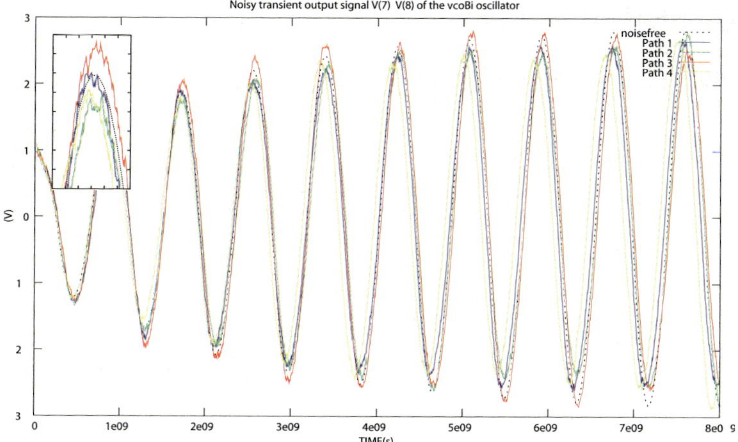

Fig. 4 (**p. 408**) Noisy transient output signal of a VCO.

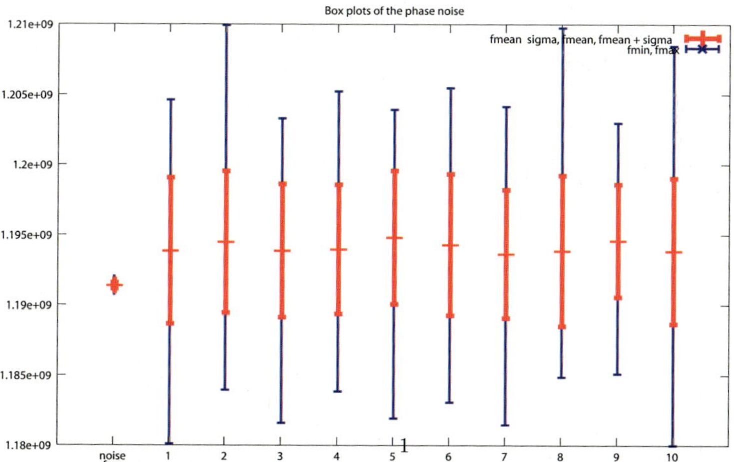

Fig. 5 (**p. 409**) Boxplots of the phase noise, scaled by a factor of 500

Index

Printing: Krips bv, Meppel
Binding: Stürtz, Würzburg